ORGANOMETALLIC REACTIONS

Volume 2

Advisory Board

ORGANOMETALLIC REACTIONS

Volume 2

EDITED BY

Ernest I. Becker

Department of Chemistry
University of Massachusetts
Boston, Massachusetts

Minoru Tsutsui

Department of Chemistry
Texas A & M University
College Station, Texas

Wiley-Interscience

A Division of John Wiley & Sons, Inc.

NEW YORK · LONDON · SYDNEY · TORONTO

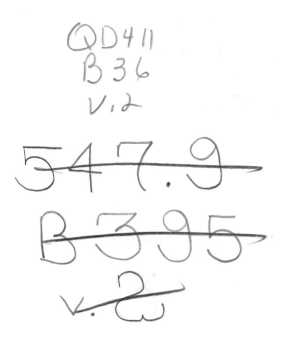
Library of Congress Catalog Card Number: 74-92108

ISBN: 0-471-06130-1

Printed in the United States of America

10 9 8 7 6 5 4 3 2 1

Preface

The primary literature on organometallic chemistry has undergone phenomenal growth. The number of papers published from 1955 to 1970 is about equal to all prior literature. Together with this intense activity there has developed a complexity in the literature. Thus specialized texts and teaching texts, a review journal, an advances series, and a research journal have all appeared during this period. The present series also reflects this growth and recognizes that many categories of organometallic compounds now have numerous representatives in the literature.

The purpose of *Organometallic Reactions* is to provide complete chapters on selected categories of organometallic compounds, describing the methods by which they have been synthesized and the reactions they undergo. The emphasis is on the preparative aspects, although structures of compounds and mechanisms of reactions are briefly discussed and referenced. Tables of all of the compounds prepared in the category under consideration and detailed directions for specific types make these chapters particularly helpful to the preparative chemist. While the specific directions have not been refereed in the same way as are those in *Organic Syntheses* and *Inorganic Syntheses*, the personal experiences of the authors often lend special merit to the procedures and enable the reader to avoid many of the pitfalls frequently encountered in selecting an experimental procedure from the literature.

We acknowledge a debt of gratitude to the contributing authors whose dedication and skill in preparing the manuscripts cannot adequately be rewarded. It has been gratifying to note that virtually all invitations to contribute have been accepted at once. We also owe thanks to the publisher for encouragement and even the "gentle prod" when necessary to see these volumes to their completion.

Ernest I. Becker
Minoru Tsutsui
Editors

September 1970

Contents

Contents for Volume 1

ORGANOMETALLIC REACTIONS

Volume 2

The Redistribution Reaction

KURT MOEDRITZER

Central Research Department, Monsanto Company, St. Louis, Missouri

I. INTRODUCTION

The term "redistribution reaction" generally describes a chemical reaction in which two (or more) kinds of exchangeable substituents change sites with each other on one (or more) kinds of polyfunctional central moieties. The substituents as well as the central moieties may be made up of a number of atoms which, however, under the specified experimental conditions, remain as intact entities. If allowed to proceed a sufficiently long time at a proper reaction temperature, these reactions generally reach an equilibrium state.

As used throughout this review, "redistribution reaction" is meant to encompass such terms as exchange, substituent exchange, scrambling, and ligand exchange reactions or ligand interchange, reorganization, and distribution reactions and also the terms symmetrizations, dissymmetrizations, disproportionations, comproportionations, or metatheses. Some of these latter ones have been used to name special phases of the general redistribution reaction (e.g., if the term comproportionation is used to describe a given reaction, its back reaction is then a disproportionation).

Whereas previous reviews [1-9] on this subject have described special aspects of this class of reactions or have given a phenomenological treatment of the subject matter, it is the objective of this article to present the preparative utility of this reaction and thus give the chemist in the laboratory a reference source for the redistribution reactions observed in organometallic systems.

In line with the definition of what generally is considered as organometallic chemistry, this article is limited to metal–carbon and metal–hydrogen bonded systems. Excluded therefore are halogen-, sulfur-, nitrogen-, and phosphorus-based systems and, by choice of the author, also systems involving transition metals. Only in instances which were considered essential for a better understanding of the general scope of the redistribution reaction, systems were included which do not involve compounds having a direct carbon–metal bond but rather deal with species in which the carbon is bonded via a heteroatom to the metal.

Since the general definition of a redistribution reaction as given above is rather broad and thus embraces a tremendously large number of reactions, the scope of this chapter required certain additional limitations. This review therefore will be limited to redistribution reactions as defined above, with the participating central atoms or moieties of the components involved all based on the same element.

Redistribution reactions originally served as a synthetic method for the preparation of certain classes of compounds. Recently,[5-9] however, a large body of information was collected which was aimed toward the quantitative study of redistribution equilibria by physical tools such as nuclear magnetic resonance, gas chromatography, mass spectrometry, etc. In many of the latter studies, however, these reactions have not been examined in detail with respect to their synthetic utility.

II. MECHANISMS OF THE REACTION

Although redistribution reactions have been observed as early as more than a hundred years ago, it is rather surprising that it was not until about 1940 that the equilibrium nature of this type of reaction was recognized. In a series of now classical papers[1,10–19] entitled "The Redistribution Reaction," Calingaert and his co-workers have applied the laws of probability to a quantitative interpretation of the "random distribution" of exchanging substituents found in certain systems such as the lead alkyls. An outstanding early landmark in developing the logic of this type of reaction as applied to the redistribution of molecular segments in large moecules was made by Flory,[20] also in this period. However, further experimental advances in developing a basic understanding of the redistribution reaction were not made until new and powerful physical tools such as nuclear magnetic resonance, chromatography, computers, and others had become available to the chemist. Therefore it is not surprising that at present relatively few studies have been published which are devoted to kinetic measurements of redistribution reactions and which are aimed at establishing reaction mechanisms.

Redistribution reactions in general may be acid catalyzed, base catalyzed, or thermally induced. In many cases, redistribution reactions proceed quite rapidly at or below room temperature. The rates of redistribution reactions may vary greatly depending on the nature of the central atoms or moieties and of the exchanging substituents. As a rule of thumb, redistributions involving a First-Row element as the central atom generally proceed slower than those of Second-Row elements, and the latter are slower than those of a Third- or Fourth-Row element. Also, exchanges involving carbon–metal bonds are much slower than exchanges involving hydrogen, halogen, sulfur, oxygen, or nitrogen bonds to the metal.

It is generally assumed that the interchange of substituents between central moieties are bimolecular and proceed through a four-center activated complex. For the system antimony trichloride-trimethylantimony,[21] initial rate measurements have shown that the first step reaction

$$(CH_3)_3Sb + SbCl_3 \rightleftharpoons \underset{\underset{CH_3}{\diagup}\ \underset{Cl}{\diagdown}}{Sb} \overset{\overset{CH_3}{\diagdown}\ \overset{Cl}{\diagup}}{} \underset{\underset{Cl}{\diagdown}}{Sb} \overset{\overset{Cl}{\diagup}}{} \rightleftharpoons (CH_3)_2SbCl + CH_3SbCl_2 \quad (1)$$

is first order in each component and that the rate constants are consistent with a mechanism involving the formation of the four-center transition state as shown in Eq. (1). The relatively large negative activation entropy (-25

e.u.) found for this reaction is in agreement with values obtained for other bimolecular reactions believed to involve two sites for attachment in the transition state.[22,23]

Also, catalytic reactions very likely proceed via four-center activated intermediates. Exchanges involving boron–carbon bonds generally require strongly forcing conditions, often temperatures in the range of 200–300°C. However, in the presence of B—H containing substances, such as diborane or tetra-alkyldiboranes, the exchange of B—C with B—X bonds (X = alkyl, halogen, OR, SR) proceeds under very mild conditions probably with the catalyst participating in a bridge mechanism.[24,25]

$$\begin{array}{ccc} \diagdown\diagup^{X}\diagup & \diagdown\diagup^{X}\diagup & \diagdown\diagup^{X}\diagup \\ B \quad B & \rightleftharpoons \quad B \quad B & \rightleftharpoons \quad B \quad B \\ \diagup\quad\diagdown & \diagup\quad\diagdown & \diagup\quad\diagdown \\ H & H & H \end{array} \qquad (2)$$

Another example of a catalytic disproportionation is the redistribution of methyl groups and chlorines in the methylchlorosilanes in the presence of $AlCl_3$. Based on a thorough kinetic study,[26] any mechanism will have to take into account the following facts: (a) The rate is a linear function of the $AlCl_3$ concentration; (b) the reaction is first order in chlorosilanes; (c) the rate equations must reduce to equilibrium constants when the change in concentration with time is zero; (d) substantially all the aluminum chloride is associated in some manner with the chlorosilanes.

A most reasonable mechanism which fits all these requirements is the following. The reaction

$$(CH_3)_2SiCl_2 + AlCl_3 \longrightarrow (CH_3)_2SiCl_2 \cdot AlCl_3 \qquad (3)$$

goes to completion before either one of the two reactions

$$(CH_3)_2SiCl_2 + (CH_3)_2SiCl_2 \cdot AlCl_3 \rightleftharpoons (CH_3)_3SiCl + CH_3SiCl_3 \cdot AlCl_3 \qquad (4)$$

$$CH_3SiCl_3 \cdot AlCl_3 + (CH_3)_2SiCl_2 \rightleftharpoons CH_3SiCl_3 + (CH_3)_2SiCl_2 \cdot AlCl_3 \qquad (5)$$

starts to take place. The latter reaction, which merely involves the transfer of aluminum chloride from one silane to another, must be very much more mobile than the preceding reaction, where rupture and formation of both silicon–carbon and silicon–chlorine bonds are involved. Very probably the aluminum chloride is attached to the chloride of the chlorosilane. This weakens the Si—Cl bond and when a collision occurs with another chlorosilane molecule, the two silicons exchange $AlCl_4^-$ and CH_3^- or $AlCl_4^-$ and Cl^-. Since halogens attached to silicon exchange quite easily, only a relatively few collisions will have the correct orientation to permit rupture of an Si—C bond to give the methyl interchange. This high degree of necessary orientation implies a low entropy of activation.

These few studies make it appear very likely that mechanisms involving four-center transition states will be common in redistribution reactions. However, it is not yet possible to predict a mechanism on an *a priori* basis.

Somewhat more complicated mechanisms are involved when one of the exchanging substituents is difunctional, thus resulting in various kinds of oligomeric and polymeric structures. Although the synthetic utility of these kinds of redistribution reactions is obvious, very little work has been done toward the elucidation of mechanisms in these systems.

III. SCOPE AND LIMITATIONS

A. Equilibrium Constants

Redistribution reactions sooner or later reach an equilibrium state. Therefore, when the reactions are performed in a single-fluid phase, the yields of the desired redistribution products are controlled by equilibrium constants. Considering the simplest case, the scrambling of two kinds of *monofunctional* substituents about a given kind of central atom or nonexchangeable moiety may be exemplified by the system QZ_ν versus QT_ν, where Q is the central atom or moiety, Z and T the monofunctional exchangeable substituents, and ν the number of exchangeable sites on Q. The overall process merely is a rearrangement of covalent bonds, with the total number of σ-bonds of any type remaining constant. However, other effects may also be quite significant —effects such as hybridization changes or nonbonding interactions, including steric factors and dipole interactions and solvent effects.

In the system QZ_ν versus QT_ν, there are $(\nu + 1)$ possible reaction products of the formula $QZ_iT_{\nu-i}$, where $i = 0, 1, 2, 3, \ldots, \nu$. The system thus contains the molecules QT_ν, $QZT_{\nu-1}$, $QZ_2T_{\nu-2}\cdots QZ_{\nu-2}T_2$, $QZ_{\nu-1}T$, and QZ_ν, in which only Q—Z and Q—T bonds redistribute. The resulting equilibria may be treated in terms of the following set of general equations corresponding to $i = 1, 2, 3, \ldots, (\nu - 1)$:

$$2\,QZ_iT_{\nu-i} \rightleftharpoons QZ_{i-1}T_{\nu-i+1} + QZ_{i+1}T_{\nu-i-1} \tag{6}$$

Accordingly, redistribution equilibria of the type described in the above equation are characterized by $(\nu - 1)$ equilibrium constants of the following form:

$$K_i = [QZ_{i-1}T_{\nu-i+1}][QZ_{i+1}T_{\nu-i-1}]/[QZ_iT_{\nu-i}]^2 \tag{7}$$

with i having any integer value between 0 and ν; that is, $1, 2, \ldots, (\nu - 1)$.

Although the values of K_i will depend on the nature of Q, Z, T, and ν, there is, for each ν, a special situation, the "ideal random case," where the sorting of the substituents about the central atom follows the laws of random statistics. For this case, the Z and T substituents become arranged about the Q in a completely random fashion irrespective of other substituents which are attached to Q. The K_i values for the ideal random case may be derived mathematically from the binominal series and are shown in Table I for various values of ν.

TABLE I

Values of Equilibrium Constants for the Ideal Random Case

Number of sites for exchange, ν	Value of K_i						
	K_1	K_2	K_3	K_4	K_5	K_6	K_7
2	0.250						
3	0.333	0.333					
4	0.375	0.444	0.375				
5	0.400	0.500	0.500	0.400			
6	0.417	0.533	0.563	0.533	0.417		
7	0.429	0.556	0.600	0.600	0.556	0.429	
8	0.438	0.571	0.625	0.640	0.625	0.571	0.438

Considering the random case as a reference point, we may term equilibria for which K_i is different from $K_{i(\text{rand.})}$ as "nonrandom" cases, with K_i assuming "smaller than random" or "larger than random" values. If the values of the equilibrium constants K_i are smaller than random, the equilibrium concentrations of the species $QZ_iT_{\nu-i}$ are larger than expected for random distribution. On the other hand, if the values of K_i are larger than the random value, the equilibrium concentrations of the species $QZ_iT_{\nu-i}$ become smaller than at random equilibrium.

B. Thermodynamics

A system which undergoes random redistribution is one in which the scrambling equilibria are entirely entropy controlled and the only contribution to the entropy is that due to the rearrangements of the substituents. This means that the deviations from random behavior are attributable to the reaction enthalpy and/or to entropy contributions due to solution nonideality, etc. In cases where the solvent molecules do not enter coordination spheres of the atoms involved in the exchange process, contribution to the entropy other than the redistribution itself can almost always be ignored so that the deviations from randomness afford a relatively good approximation to the enthalpy of the equilibrium in question. Thus,

$$\Delta F^\circ = -RT \ln K = \Delta H^\circ - T\Delta S^\circ \tag{8}$$

$$\Delta F^\circ_{\text{rand}} = -RT \ln K_{\text{rand}} = -T\Delta S^\circ_{\text{rand}} \tag{9}$$

If

$$\Delta S° \approx \Delta S°_{rand} \tag{10}$$

then

$$\Delta H° \approx \delta \Delta F° \equiv - RT \ln (K/K_{rand}) \tag{11}$$

In redistribution reactions involving the scrambling of two or more kinds of substituents on a single kind of polyfunctional atom or moiety, the number of any kind of chemical bond is conserved; that is, there is no generation or loss of any specified bonds during substituent exchange. Therefore, to the approximation that bond energies are additive, ΔH in Eq. (8) must be equal to zero. This means that deviations from the random distribution are a measure of the effect of changing other bonds in a molecule on the energy of a given bond; that is, the nonadditivity of bond energies.

From the quantitative data on redistribution equilibria obtained during the last five years or so, it is possible to set up several generalities (a–d below) dealing with the value of $\delta \Delta F°$ corresponding to exchange of various pairs of monofunctional substituents on a single kind of central atom: (a) Deviations from randomness for redistributions are primarily a function of the substituent pairs involved, with the term $\delta \Delta F°$ (on a per-mole-of-product basis) referring to the reaction in which the unmixed (or end member) compounds form the desired mixed compound, that is,

$$i \, QZ_v + (v - i) \, QT_v \rightleftharpoons v \, QZ_i T_{v-i} \tag{12}$$

As would be expected from their great similarity, the deviation from randomness is inappreciably small for the exchange[9] of methoxyl groups with ethoxyl groups, for example on Si. However, for the exchange of methoxyl groups with chlorine on the same central moiety, $\delta \Delta F°$ is around 3–4 kcal/mole and it is even greater for the exchange of dimethylamino groups with chlorine (6–9 kcal/mole). (b) Indeed one may set up a series of exchangeable substituent pairs ranging from those for which $\delta \Delta F°$ is nearly zero (e.g., Cl/Br or CH_3O/C_2H_5O) to those for which $\delta \Delta F°$ is large (e.g., Cl/NR_2 or Br/NR_2). (c) It is interesting to note that deviations from random behavior are almost always in the direction of exothermic formation of the mixed compound(s) from the unmixed ones. (d) When going from a central atom which does not have readily occupyable empty orbitals (the elements of main groups IV–VII, not at the very bottom of the periodic table) to one which does (transition metals, group III elements, etc.), there appears to be an increase in the value of $\delta \Delta F°$ for the exchange of a given pair of substituents. Thus the electronic structure of the central atom does have some bearing on the value of the deviation from randomness.

The deviations from randomness in substituent scrambling on a single kind of central atom are due to subtle electronic effects which are presently not well understood but must be about the same as those postulated in reaction-mechanism speculations. These electronic effects also show up in bond-distance changes. It is shown[27] that, for a C—H or Si—H bond, the length is dependent on what other substituents are attached to the carbon or silicon atom and that there is a consistent increase or decrease in the length of the other bond when going across a substitution series of molecules. Even in the case of the methyl-substituted silanes, $SiH_i(CH_3)_{4-i}$, where there should be no π bonding, the Si—C bond distance varies in a regular manner with increase in i.

In order to develop a proper electronic theory for explaining these kinds of effects, it will be necessary to consider both sigma and pi orbitals, as pi character has even been attributed to the C—F bond for the formation of which carbon has no low-lying d orbitals. Although the concept of pi and sigma bonds is very clear-cut in the case of a diatomic molecule, the definition of a pi bond in a polyatomic molecule has some problems. Theoretical work on the deviations from randomness in redistribution reactions is now underway[9] in our laboratory employing this model of a polyatomic molecule in conjunction with a transformation which is used to go back and forth between the chemists' usual localized-orbital treatment and the hard-to-visualize delocalized-orbital representations appropriate for quantitative quantum chemistry. Although it is too early to make any statements concerning the results of this work, it is worth noting that there are certain symmetry restrictions on the nature of the pi bonding—restrictions which change from molecule to molecule across a substitution series.

C. General Considerations

Consistent with the equilibrium approach the relative equilibrium concentration of a species QZ_iT_{v-i} in the system QZ_v versus QT_v is at a maximum at the overall composition of the equilibrating mixture corresponding to the Z/T mole ratio in the desired compound. Thus the compound QZ_iT_{v-i} will be obtained in a maximum yield—but still determined by the value of the equilibrium constant—when the reactants QZ_v and QT_v are combined in the mole ratio $i/(v-i)$. This is very important for synthetic purposes but applies only, of course, when reactants and products interact in a single phase. A redistribution product QZ_iT_{v-i} may be obtained in yields larger than would correspond to the equilibrium concentration only if by some means, either by crystallization or distillation, the desired species may be removed continually from the equilibrating system. Then, as a consequence of the principle of Le Chatelier, the removed species is regenerated continuously by redistribution until one or both of the reactants are used up.

Another important fact to consider when preparing a compound by a redistribution reaction is the rate of equilibration. Generally a redistribution reaction will yield the whole series of compounds QZ_iT_{v-i} in addition to the starting compounds QZ_v and QT_v in amounts determined by the equilibrium constants of the type of Eq. (7) and the overall Z/T ratio. In the ideal case, the rate of equilibration is very small at the temperature where separation of the reaction products, for example, by distillation, is performed. In this case, therefore, disproportionation of the compounds QZ_iT_{v-i} during the separations procedure will not interfere with the isolation process and pure products may be isolated. An equally favorable situation exists when the redistribution may be effected by a catalyst which may be destroyed after reaching equilibrium and before separating the products. This again creates an ideal situation for separating individual species generated by redistribution of substituents.

If, however, the rate of equilibration or reequilibration is of the same order of magnitude as the rate of separation of the reaction products, continuous reorganization of the substituents will occur, so that, for example, in the case of distillation as the separations method, the lowest boiling product will be regenerated as quickly as it is removed from the mixture by distillation. This usually results in pure products or in complete decomposition of the desired mixed compounds to yield the two species having only one kind of substituent attached to the central atoms or moiety.

In the latter types of systems, the redistribution products may be isolated by low-temperature separations methods, such as vacuum distillation, chromatography, low-temperature crystallization, and others. However, if the rate of re-equilibration is significant at room temperature, the products thus isolated will disproportionate upon storage at room temperature, regenerating an equilibrium mixture consisting of more than one species.

Finally, there are redistribution reactions which proceed very rapidly at room temperature and which only recently have been discovered by NMR techniques. Half lives in these systems range from 10^{-4} to several seconds at room temperature. Unless one is interested in the equilibrium mixture consisting of a whole spectrum of molecules, the isolation of individual species from these mixtures generally is not practical.

D. Oligomeric and Polymeric Systems

More complicated redistribution reactions have been observed when one of the exchanging substituents Z or T in the systems QT_v versus QZ_v is a difunctional atom or moiety. Let us again consider Q as the central atom or moiety of the functionality v, T the monofunctional exchangeable substituent, and Z one half of a difunctional (bridging) atom or group such as $O_{1/2}$, $S_{1/2}$, or $(NR)_{1/2}$, or simply "half" of a bond to another Q atom. Then the $(v + 1)$ kinds of species of the general formula QZ_iT_{v-i} resulting from exchange of Z and T on the central moiety are no longer simple molecules (except for $i = 0$) but represent parts of molecules ("building units"). The functionality of each building unit is dependent on i and ranges from 0 for the "neso" molecule, QT_v,—a term taken from silicate chemistry meaning a small "island" structure (from the Greek word for island) which in equilibrium chemistry denotes the smallest member of a family of compounds—and the building units given below to a most branched unit.

$$Z-QT_{v-1} \qquad Z-QT_{v-2}-Z \qquad Z-\overset{\displaystyle Z}{\underset{\displaystyle |}{Q}}T_{v-3}-Z \quad \cdots \qquad Q(Z)_v$$

Z—QT$_{v-1}$	Z—QT$_{v-2}$—Z	Z—QT$_{v-3}$—Z	Q(Z)$_v$
end group, e	middle group, m	threeway branch, b_3	multiway branch, b_v
(monofunctional)	(difunctional)	(trifunctional)	(v functional)

These parts of molecules in the redistribution mixture are rearranged to oligomers and polymers of various size and shape. Again the concentration of a particular molecule in the equilibrated mixture is determined by equilibrium constants and the overall composition of the mixture. These molecules may be linear, branched, cyclic, fused-ring structures, or others. In view of the complexity of the composition of such mixtures, the equilibrium concentration of any such species—except perhaps for the low molecular weight ones—is very small.

E. General Discussion of Tabular Survey

In the following, a discussion of the observed redistribution processes is given subdivided in sections on the basis of the pair of element substituent bonds involved. In addition, the reported redistribution reactions of compounds or systems based on the elements that this article is restricted to are summarized in Tables III–XXIII. Each table is a compilation of literature data on redistributions involving a given pair of element substituent bonds, with the elements ordered in the sequence of their group number in the Periodic Table. In other words, the redistribution reactions involving compounds of elements of Group I will appear first in each table, with the elements within the same group listed in the order of increasing atomic weight.

Wherever studies of complete systems QT_v versus QZ_v were made, the resulting equilibria are described in terms of equilibrium constants K as defined by Eq. (7). The values of constants K_i are listed in parentheses following the formulas of the resulting individual compounds QZ_iT_{v-i}. If the constants are random, this is indicated by (r). These constants represent the most general way of expressing the equilibrium concentration of a compound QZ_iT_{v-i} in a system QZ_v versus QT_v since these constants are independent of the overall composition of the system. They may be used to calculate[28,29] the equilibrium concentration of the given species at any overall composition of the system.

TABLE II

Equilibrium Concentrations in Mole Per Cent in Systems QT_v versus QZ_v as a Function of the Values of the Equilibrium Constants

Functionality of central atom Q	$v = 2$	$v = 3$		$v = 4$		
Compound to be prepared	QZT	QZT_2	QZ_2T	QZT_3	QZ_2T_2	QZT_3
$R = Z/Q$ in starting materials	1	1	2	1	2	3
Equilibrium constants K_i	Equilibrium concentration in mole per cent					
10	13.7	4.8	4.8	1.7	0.5	1.7
5	18.3	8.9	8.9	4.6	1.8	4.6
1	33.3	27.7	27.7	26.1	20.0	26.1
0.5	41.4	38.3	38.3	38.0	34.3	38.0
Random	50.0	44.5	44.5	42.2	37.5	42.2
0.1	61.3	60.9	60.9	60.9	60.5	60.9
0.05	69.1	69.0	69.0	69.0	68.9	69.0
0.01	83.3	83.3	83.3	83.3	83.3	83.3
1×10^{-3}	94.1	94.1	94.1	94.1	94.1	94.1
1×10^{-4}	98.0	98.0	98.0	98.0	98.0	98.0
1×10^{-5}	99.4	99.4	99.4	100.0	100.0	100.0

In Table II, maximally obtainable yields of a desired compound QZ_iT_{v-i} as a function of the equilibrium constant are given for the case of the compound QZ_iT_{v-i} being prepared by equilibrium redistribution from its end members QT_v and QZ_v through exchange of Z and T. The equilibrium concentrations given in Table II for the species QZ_iT_{v-i} pertain to the overall composition where QZ_iT_{v-i} maximizes; that is, when the overall mole ratio $Z/T = i/v - i$. Since the overall composition in a redistributing system generally is defined by a composition parameter of the form $R \equiv [Z]/[Q]$, the R-value at which the equilibrium concentration of QZ_iT_{v-i} maximizes— irrespective of the value of the equilibrium constants—is $R = i$.

From Table II it is seen that highest yields are obtained when the equilibrium constant K_i of the desired species QZ_iT_{v-i} approaches very small values. Since the yield of QZ_iT_{v-i} when prepared from QT_v and QZ_v is thermodynamically controlled, these yields sometimes are lower than when other synthetic methods are used. Another feature of this general method is the fact that if $K_i >$ ca. 10^{-3}, a mixture of products is obtained which sometimes is difficult to separate into pure products.

Most of the results of recent studies of redistribution equilibria are expressed in terms of such equilibrium constants. However, there is a large body of data describing single experiments—corresponding to only one R-value— where the yields of redistribution products are listed in terms of mole % yield based on the stoichiometric reaction. In many other cases such as in the system[30] $Ge(C_6H_5)_4$ versus $GeCl_4$, the reported experiments were directed toward the synthesis of each mixed compound in a maximum yield and therefore the reactants were combined in ratios corresponding to the overall composition of the desired compounds $Ge(C_6H_5)_3Cl$, $Ge(C_6H_5)_2Cl_2$, and $Ge(C_6H_5)Cl_3$. For cases like these, each experiment is listed as a separate entry in Tables III–XXIII since the conditions may vary for the various compositions studied. In both of the above instances, the yield in mole % is given in Tables III–XXIII in parentheses following the formula of the prepared compound.

Many other literature references are limited to a discussion of the "disproportionation" of a compound QZ_iT_{y-i} into its neighbor compounds $QZ_{i+1}T_{y-i-1}$ and $QZ_{i-1}T_{y-i+1}$. In such cases, the compounds QZ_iT_{y-i} are listed in the reactants column in Tables III–XXIII and the resulting disproportionation products in the products column. Brief statements of this kind regarding the instability of a compound with respect to disproportionation are scattered throughout the chemical literature and therefore are difficult to retrieve.

F. Redistribution Reactions Involving Exchange of Element–Hydrogen with Element–Carbon Bonds

The reaction between diborane and trimethylborane[31] leads to the formation of mono-, di-, tri-, and tetramethyl derivatives of diborane. But only one dimethyldiborane, namely the unsymmetrical compound, is obtained in this reaction. The symmetrical dimethyldiborane[32] is obtained by disproportionation of monomethyldiborane in the presence of dimethyl ether.

$$2\,CH_3B_2H_5 + 2\,(CH_3)_2O \rightleftharpoons 2\,BH_3 \cdot (CH_3)_2O + CH_3BH_2BH_2CH_3 \qquad (14)$$

In the earlier work, separation of the equilibration products was achieved by high-vacuum techniques, whereas recent studies of the equilibria used gas chromatography as separations method.[33,34] The kinetics and the equilibria have been studied in detail and a separate study[35] of the disproportionation of tetramethyldiborane is reported also. Exchange of boron–hydrogen with boron–methyl bonds furthermore has been observed in organoboron derivatives such as in mixtures[36] of $BH_2[N(CH_3)_2]$ with $B(CH_3)_2[N(CH_3)_2]$, borazzines[37,38] as shown by Eqs. 15 and 16, and in mixtures of alkaliborohydrides

$$(15)$$

$$
2 \;
\begin{array}{c}
\text{CH}_3 \diagdown \overset{\text{H}}{\underset{|}{\text{B}}} \diagup \text{CH}_3 \\
\text{N} \quad \text{N} \\
| \quad\quad | \\
\text{B} \quad\quad \text{B} \\
\text{CH}_3 \diagup \;\; \overset{|}{\underset{\text{CH}_3}{\text{N}}} \;\; \diagdown \text{CH}_3
\end{array}
\;\rightleftharpoons\;
\begin{array}{c}
\text{CH}_3 \diagdown \overset{\text{H}}{\underset{|}{\text{B}}} \diagup \text{CH}_3 \\
\text{N} \quad \text{N} \\
| \quad\quad | \\
\text{B} \quad\quad \text{B} \\
\text{H} \diagup \;\; \overset{|}{\underset{\text{CH}_3}{\text{N}}} \;\; \diagdown \text{CH}_3
\end{array}
\;+\;
\begin{array}{c}
\text{CH}_3 \diagdown \overset{\text{CH}_3}{\underset{|}{\text{B}}} \diagup \text{CH}_3 \\
\text{N} \quad \text{N} \\
| \quad\quad | \\
\text{B} \quad\quad \text{B} \\
\text{CH}_3 \diagup \;\; \overset{|}{\underset{\text{CH}_3}{\text{N}}} \;\; \diagdown \text{CH}_3
\end{array}
\quad (16)
$$

and trimethylborane.[35,39] The latter system presents a convenient method for preparing methyldiboranes and alkyldiboranes in general when the reaction is performed in the presence of hydrochloric acid.

$$(6\text{-}n)\text{LiBH}_4 + n\,\text{BR}_3 + (6\text{-}n)\text{HCl} \longrightarrow 3\,\text{B}_2\text{H}_{6-n}\text{R}_n + (6\text{-}n)\text{LiCl} + (6\text{-}n)\text{H}_2 \quad (17)$$

$$(6\text{-}n)\text{NaBH}_4 + (n+2)\text{BR}_3 + (6\text{-}n)\text{HCl} \longrightarrow$$
$$4\,\text{B}_2\text{H}_{6\text{-}n}\text{R}_n + (6\text{-}n)\text{NaCl} + (6\text{-}n)\text{CH}_4 \quad (18)$$

where $R = \text{CH}_3$, C_2H_5, $n\text{-C}_3\text{H}_7$, and $n = 1, 2, 3,$ and 4.

Recently the diborane-trimethylborane equilibria were studied by multiple-scan interference spectroscopy[40] extending earlier infrared studies.[41]

Similar exchange behavior of boron–hydrogen with boron–carbon bonds was observed in the system diborane/triethylborane and diborane/tri-n-propylborane yielding alkyldiboranes[42] analogously to the methyldiboranes described above. Also, these compounds resemble in their stability the analog methyldiboranes. For the ethyl compound,[43] the disproportionation of mono-ethyldiborane to form diborane and symmetrical diethyldiborane has been studied in some detail. Whereas the symmetrical diethyldiborane was found to be quite stable with respect to disproportionation at room temperature—in 0.5 hr only 4.5% had disproportionated—the unsymmetrical diethyldiborane equilibrated rapidly at room temperature to give an equilibrium mixture of ethyldiboranes. Reactions similar to Eqs. (17) and (18) have also been reported[44] for ethyl and n-propyldiboranes. Tetra-n-propyldiborane and tetra-n-butyldiborane were prepared[45] in diethyl ether from diborane and tri-n-propyl or tri-n-butylborane, respectively. Exchange of boron–hydrogen bonds with boron–carbon as well as boron–oxygen bonds present in the same molecule has been observed[46–48] when diborane was reacted with a series of alkyl dialkylboronites. Exchange of B—H bonds with B—C bonds concurrently with exchange of B—H bonds with B—Cl bonds has also been observed.[49] Redistributions involving boron compounds bearing higher alkyl groups[50] or aryl[47–49,51–53] groups follow the same general mechanism.

Exchange reactions involving aluminum–hydrogen bonds and aluminum–carbon bonds occur quite readily. Alane as the trimethylamine addition product redistributes readily[54] with trimethylalane or triethylalane, respectively, with the trialkylalanes present in the form of their addition products with trimethylamine. Redistribution has also been observed in the absence of trialkylamine,[55] however disproportionation of the resulting uncoordinated alkylanes seems to occur quite readily—catalyzed by impurities—resulting in insoluble polymer alane which will shift the equilibrium of Eq. (19) to the right.

$$3 \text{ AlHR}_2 \rightleftharpoons \text{AlH}_3 + 2 \text{ AlR}_3 \tag{19}$$

Phenyl groups also undergo redistribution with alane yielding phenylalanes. Instead of alane, lithiumalanate[56,57] may also be used in a similar reaction to prepare alkyl or arylalanes. Sodium alkylalanates[58] of the general formula $\text{NaAlH}_{4-n}\text{R}_n$ have been prepared by the reaction of LiAlH_4 with LiAlR_4 and compounds for $n = 2$ and 3 have been verified by phase-diagram studies, but no evidence was found for the presence of the compound for $n = 1$ in the equilibrium mixture.

In silicon chemistry, exchange of silicon–hydrogen bonds with silicon–carbon bonds was observed to proceed in the absence[59] as well as in the presence of catalysts, such as aluminum halides,[61–63] platinum, or platinum compounds,[59,60] sodium, and/or potassium,[64,65] radical initiators,[66] and other catalysts.[61] Generally elevated temperatures are required and although undoubtedly an equilibrium mixture will result, the reactions almost exclusively were performed in such a manner that upon heating of an alkyl or aryl silane, volatile redistribution products such as silane were allowed to escape thus forcing the reaction towards a complete disproportionation.

An interesting exchange reaction involving exchange of tin–hydrogen bonds and tin–carbon bonds[67] has been observed when trialkylstannanes were reacted with vinyltrialkylstannanes. For the reactants being trimethylstannane and triethylvinylstannane, the resultant products were trimethylvinylstannane and triethylstannane. In another study, redistribution of hydrogen atoms and methyl groups on tin must have preceded the observed thermal decomposition of dimethylstannane[68] according to Eq. (20):

$$3 \text{ (CH}_3)_2\text{SnH}_2 \longrightarrow 2 \text{ (CH}_3)_3\text{SnH} + \text{Sn} + 2 \text{ H}_2 \tag{20}$$

G. Redistribution Reactions Involving Exchange of Element–Hydrogen with Element–Nitrogen Bonds

The ability of B—H bonds to exchange with B—N linkages in the system B_2H_6 versus $\text{B[N(CH}_3)_2]_3$ was demonstrated[69] to occur at room temperature or above room temperature. Also, in aluminum chemistry, exchange of dimethylamino groups in tris(dimethylamino)alane with hydrogen atoms in the trimethylamine adduct of alane was observed[70] and was utilized for the preparation of $\text{AlH}_2[\text{N(CH}_3)_2]$. Exchanges of this type in silicon chemistry take place when disilylamine, $\text{(H}_3\text{Si)}_2\text{NH}$, disproportionates[71] into silane and a polymer of the composition $[\text{SiH}_2\text{NH}]_x$. A similar base-catalyzed interchange,[72,73] however, was observed when trisilylamine was treated with small quantities of bases such as NH_3, CH_3NH_2, $(CH_3)_3N$, and LiH to give $(SiH_2NSiH_3)_3$ and silane. In a kinetic study,[73] it was shown that the elimination of silane from trisilylamine indeed is base catalyzed and probably an intermolecular reaction. For tin[74,75] as the central atom, exchanges involving tin–hydrogen bonds with tin–nitrogen bonds are obscured by condensation reactions of the type shown below.

$$2\ R'_3SnH + R_2Sn[N(C_2H_5)_2]_2 \longrightarrow R'_3Sn\!-\!SnR_2\!-\!SnR'_3 + 2\ HN(C_2H_5)_2 \quad (21)$$

However, preceding the actual condensation, exchange of the substituents H with $N(C_2H_5)_2$ occurs. The exchange reaction is the predominant one for methyl and ethyl derivatives and for butyl compounds it is less significant. Generally, reactions of the type of Eq. (21) ("Hydrostannolysis") proceed much faster than exchange reactions.

H. Redistribution Reactions Involving Exchange of Element–Hydrogen with Element–Phosphorus Bonds

Only very few examples of reactions of the above kind were reported. In contrast to the preceding exchanges involving tin–nitrogen linkages, the analogous reactions with tin–phosphorus linkages proceed considerably faster, which is also true for the condensation reactions. For the reactants $(C_6H_5)_3SnH$ and $(C_4H_9)_3SnP(C_4H_9)_2$, the exchange[74] of H with $P(C_4H_9)_2$ is completed within two hours at $23°$. It appears that these are equilibrium reactions[76] with the equilibrium being independent of the polarity of the solvent. A four-center mechanism has been postulated with the organotin hydride acting as nucleophile.

I. Redistribution Reactions Involving Exchange of Element–Hydrogen with Element–Oxygen Bonds

The reversible interchange of boron–hydrogen bonds with boron–oxygen bonds was studied first in the system[77] $B_2H_6/B(OCH_3)_3$. Dimethoxyborane, $BH(OCH_3)_2$, decomposes reversibly to form diborane and trimethylborate, without indication of forming any other substances. On the other hand, diborane and trimethylborate when reacted at room temperature yield dimethoxyborane, although this reaction is not recommended as a means of preparation. A study of the rate[78] of disproportionation of dimethoxyborane in the gas phase by the manometric method showed that the reaction

$$3\ BH(OCH_3)_3 \rightleftharpoons 2\ B(OCH_3)_3 + \tfrac{1}{2}\ B_2H_6 \quad (22)$$

is heterogeneous and the rate is proportional to the square of the partial pressure of dimethoxyborane. On the basis of these measurements, a most probable mechanism is discussed. Similarly, 1,3,2-dioxaborolane[79] in the liquid state rapidly disproportionates at room temperature to give a non-volatile glassy solid. In the vapor phase, disproportionation occurs reversibly according to the equation shown below:

$$(CH_2O)_2BH \rightleftharpoons \tfrac{1}{6} B_2H_6 + \tfrac{1}{3} (CH_2O)_2BOCH_2CH_2OB(OCH_2)_2 \qquad (23)$$

to give diborane and the completely oxygenated product of Eq. (23). The initial rate of disproportionation seems to depend on the presence of trace impurities. Similar studies were reported for 1,3,2-dioxaborinane,[80] B(OCH$_2$-CH$_2$CH$_2$O)H. In addition to these results, a large number of reactions has been reported involving exchange[46–49,81] of hydrogen in alkyl and arylboranes with alkoxyl or aroxyl groups in alkyl or aryl boronates and boronites. These reactions proceed quite readily, however, in many instances in addition to exchange of B—H with B—O bonds also exchange of B—C with B—O bonds takes place, resulting in a variety of products. From some of these reactions, certain products may be isolated in relatively high yield.

The observation that, upon reacting methanol with lithium alanate in tetrahydrofurane or diglyme, the initially formed precipitate slowly redissolved,[82] suggested that alkoxy-substituted lithium alanate derivatives might be realized through a redistribution reaction of lithium alanate with the corresponding lithium tetraalkoxyaluminate. At room temperature, lithium tetramethoxyaluminate readily dissolved in the solutions of lithium alanate forming lithium trimethoxyalanate. The tetraethoxy and tetraisopropoxy derivatives did not react as readily and the tetra-t-butoxy derivative showed only slight reaction. Consequently it appears that only in the case of the methoxy compounds does this reaction provide a good synthetic path.

The redistribution of silicon–hydrogen with silicon–oxygen bonds is perhaps one of the earliest references[83] of a redistribution reaction in general. Upon placing a piece of metallic sodium into triethoxysilane, volatile silane was evolved and tetraethoxysilane was formed. The overall reaction was correctly interpreted as being sodium catalyzed. The redistribution of disiloxane according to Eq. (24), however,

$$H_3Si—O—SiH_3 \rightleftharpoons SiH_4 + 1/n \, (H_2SiO)_n \qquad (24)$$

was found to be acid catalyzed[71] ("partially hydrolyzed phosphorus pentoxide" as catalyst). Bis(halophenyl)disiloxanes[84] were found to redistribute in the presence of calcium sulfate or—at a much faster rate—potassium carbonate to yield the halophenylsilane and a resin of the overall composition corresponding to a halophenylsilsesquioxane. Other catalysts effecting exchange of Si—H with Si—O bonds are butyllithium[85] and aluminum chloride.[86] It is interesting to note that the mixed compound $(C_6H_5)(CH_3)SiH(OCH_3)$ when prepared by redistribution from phenylmethylsilane and dimethoxyphenylmethylsilane at equilibrium is formed in amounts which are considerably less than expected for a random distribution. Some alkoxysilanes[87] and alkoxyhalosilanes have been found to disproportionate easily, probably catalyzed by trace amounts of acid since in specially treated glass vessels, the disproportionations of some of these could be prevented.

In organotin chemistry, the exchange of tin–hydrogen with tin–oxygen bonds which was first discovered in the system di-n-butyltin dihydride and di-n-butyltin diacetate[88,89] is competing with condensation reactions. This reaction leads to an equilibrium mixture consisting of the two starting compounds and di-n-butylacetoxytin hydride and the latter in a secondary reaction will eliminate molecular hydrogen and condense to a ditin compound. Later it was found that organotin hydrides undergo similar reaction with organotin alkoxides and oxides.[74,75,90,91] Any transition between substituent exchange on one hand and condensation on the other may be realized. In some cases, the rates of these two types of reactions differ so widely that they may be run separately. For example, mixtures of $(C_6H_5)_2SnH_2$ and $[(C_6H_5)_3Sn]_2O$ as well as $(i\text{-}C_4H_9)_2SnH_2$ and $[(i\text{-}C_4H_9)_3Sn]_2O$ exchange substituents quite rapidly at room temperature. The dialkyltin oxide precipitates and thus makes the reaction quantitative. However, upon heating, condensation and elimination of water occurs and mixtures of polystannanes are obtained. The rate of the exchange[76,92] reaction is much less solvent dependent than the condensation reaction suggesting a four-center intermediate for the exchange reaction with the organotin hydride acting as nucleophile.

J. Redistribution Reactions Involving Exchange of Element–Hydrogen with Element–Sulfur Bonds

A white solid of the composition CH_3SBH_2 consisting of a series of polymers disproportionates upon heating—redistributing boron-hydrogen with boron-sulfur bonds[93]—to form $B(SCH_3)$ and B_2H_6 with destruction of the latter by its usual thermal decomposition to hydrogen and B—H polymers. Similarly, the trimethylamine complex, $CH_3SBH_2 \cdot N(CH_3)_3$, undergoes limited disproportionation according to Eq. (25) below:

$$3\ CH_3SBH_2 \cdot N(CH_3)_3 \longrightarrow N(CH_3)_3 + 2\ BH_3 \cdot N(CH_3)_3 + B(SCH_3)_3 \quad (25)$$

Disilylsulfide, $(H_3Si)_2S$, in the presence of NH_3 as well as $N(CH_3)_3$ disproportionates upon standing at room temperature with exchange of silicon–hydrogen with silicon–sulfur bonds[72] yielding silane and a polymer.

$$n\ (H_3Si)_2S \longrightarrow (H_2Si\!-\!S)_n + n\ SiH_4 \quad (26)$$

Whereas tin–hydrogen with tin–oxygen bonds were found to redistribute quite readily at room temperature, the analog reaction with tin–sulfur[75] bonds required 110° and at least four hours for complete hydrogen–sulfur exchange. The exchange process is accompanied by degradation.

K. Redistribution Reactions Involving Exchange of Element–Hydrogen with Element–Halogen Bonds

Hydride chlorides,[94] bromides,[95] and iodides[96] of the alkaline earth metals, calcium, strontium, and barium were obtained by melting together the alkaline earth hydride with the corresponding anhydrous halide. The thermal analysis as well as X-ray spectroscopy indicated that the materials indeed are hydride halides and not merely mixtures of the hydride and halide.

In contrast to the facile hydride exchange reactions observed at room temperature between diborane and boron trichloride or tribromide, the reaction involving boron trifluoride[97,98] occurs in the absence of a catalyst only under relatively vigorous conditions accompanied by pyrolysis of diborane. The reported procedure involves pyrolysis of B_2H_6—BF_3 mixtures at $250°$ for short periods under which condition BHF_2—BF_3 mixtures containing 30–35% BHF_2 were obtained. No other fluorohydride species were found. At room temperature, BHF_2 disproportionates slowly into B_2H_6 and BF_3.

Monochlorodiborane[99] was identified as a product of the interaction of diborane and boron trichloride. Since this reaction[100] proceeds without change in the number of moles of gas present, it may be explained by equilibria such as the one given below:

$$BCl_3 + B_2H_6 \rightleftharpoons B_2H_5Cl + BHCl_2 \tag{27}$$

Monochlorodiborane when isolated disproportionates[101] at $35°$ to $BHCl_2$ which, in turn, disproportionates into B_2H_6 and BCl_3.

In "diglyme" (diethyleneglycol dimethyl ether) as solvent, diborane and boron trichloride[102] yielded $BHCl_2 \cdot R_2O$ and $BH_2Cl \cdot R_2O$; in diethyl ether[103] only $BHCl_2 \cdot R_2O$ was observed as the sole product. Starting with diborane, triethylamine[104] and boron trichloride, the amine adducts $BHCl_2 \cdot NR_3$ and $BH_2Cl \cdot NR_3$ were obtained. Hydrogen–chlorine exchange was also observed in the system borazine/boron trichloride[105] resulting in mono- and dichloroborazine as well as in systems involving organoboron compounds.[49,106,107] Similar reactions have also been observed with boron tribromide.[105,108,109]

Facile exchange of aluminum–hydrogen bonds with aluminum–halogen bonds has been observed in comproportionation reactions of alane with aluminum halides. Depending on the mole ratio of the starting materials, the reaction generally goes to completion forming haloalanes or dihaloalanes which may be isolated as diethyl etherates,[110–113] tetrahydrofuranates,[114] and trimethyl[110] or triethylaminates.[115] Upon distillation, the diethyl etherates and trimethylaminates were found to disproportionate quite readily, with the tendency to disproportionate being the largest for the chloride compounds and the least for the iodine compounds.

Evidence for the exchange of silicon–hydrogen bonds with silicon–fluorine bonds is the observation that trifluorosilane at low temperature dispropor-tionates into silane and silicon tetrafluoride.[116] Difluorosilane as well as monofluorosilane[117] disproportionate at room temperature, the latter more rapidly than the other. Quantitative measurements[85,118] in the system methyl-phenylsilane/methylphenyldifluorosilane in the liquid as well as the gas phase indicate a strong tendency toward accumulation of like substituents in the redistribution of hydrogen and fluorine.

Evidence for hydrogen–chlorine redistributions in silicon chemistry was found in the aluminum chloride-catalyzed comproportionation of silane and dichlorosilane[119] yielding monochlorosilane. On the other hand, trichloro-silane[120] was found to be stable when heated for 48 hr at 175° in the presence of aluminum chloride and only temperatures between 300 and 400° caused disproportionation into silicon tetrachloride and dichlorosilane. In addition to aluminum chloride, tetraalkylammonium salts, nitrogen bases, nitriles, amides, and tertiary phosphines have been used as catalysts for this type of reaction. Generally silicon–carbon bonds remain intact under conditions where exchange of silicon–hydrogen with silicon–chlorine bonds occurs. Numerous reactions[62,121–125] exemplifying hydrogen–chlorine exchange have been reported for alkyl or arylsilane/alkyl or arylchlorosilane systems. Quantitative equilibrium measurements[85,126,127] indicate that in contrast to the hydrogen–fluorine exchange, the hydrogen–chlorine exchange favors the mixed compounds. Furthermore, if several alkyl or aryl moieties of silicon are participating in the equilibria, the chlorine atoms show a preference for the silicon atom having the larger number of silicon–carbon bonds. This means that trialkylsilanes when reacted with alkyl or dialkylhalosilanes have a tendency to transfer silicon-bonded hydrogen to the alkyl or dialkylsilicon moiety in exchange for silicon-bonded chlorine—a generalization which is useful for synthetic purposes.

Hydrogen–bromine exchange[124,125] on silicon follows the same general pattern, although only a few examples were described. Very little is known about hydrogen–iodine exchange[128] on silicon.

Dialkyltindihydrides react with dialkyltin dihalides to give dialkyltin-halide hydrides.[129–133] The reaction is equilibrium controlled, with the equi-librium lying completely on the side of the halogen hydride. The reaction proceeds rapidly at room temperature and in the absence of catalysts. Ex-change of hydrogen with halogen has also been studied in systems of unlike alkyl moieties of tin;[130,132] for example, it was found that tri-*n*-butyltinhydride reacts with di-*n*-butyltindifluoride, di-*n*-butyltindichloride, di-*n*-butyltin-dibromide, or d-*n*-butyltindiiodide in a 1:1 mole ratio to give the respective di-*n*-butyltin halide hydride. Subsequent reactions with excess tri-*n*-butyltin-hydride yields di-*n*-butyltindihydride in each case. Many variations of these general reactions were investigated.

L. Redistribution Reactions Involving Exchange of Element–Carbon with Element–Carbon Bonds

Alkyl or aryl compounds of lithium in solution exist as polymers with alkyl groups being stronger bridge bonding groups than aryl groups. Co-solution of ethyllithium and t-butyllithium in benzene leads to the formation of complex organolithium compositions[134] containing both types of alkyl groups bonded to lithium. These compounds generated by alkyl exchange are electron-deficient polymers of the type $(LiC_2H_5)_n(LiC_4H_9\text{-}t)_{m-n}$, where m is a small number such as 4 or 6 and $n < m$. Extensive studies[134–139] of the ^7Li and ^1H nuclear magnetic spectra as well as X-ray data indicate random distribution of the two kinds of alkyl groups in the tetramer or hexamer. The exchange reactions are quite fast at room temperature in solution, with intra- as well as intermolecular exchange taking place.

Similar studies of the proton nuclear magnetic resonance established rapid exchange in mixtures of dialkylmagnesium compounds with approximate statistical distribution of products. Pre-exchange lifetimes in mixtures of dimethylmagnesium and dicyclopentadienemagnesium[140] are of the order of 10^{-3} sec at ca. 30°C with the equilibrium favoring the mixed species, which is in contrast to the methyl–phenyl exchange of magnesium.

In boron chemistry, the exchange of carbon with carbon-bonded sub-stituents generally does not occur at room temperature or below ca. 100° in the absence of a catalyst. Therefore, mixed trialkyl boranes when distilled at temperatures below 100° do not disproportionate. Rapid exchange of alkyl groups, however, takes place at room temperature when small amounts of aluminum alkyls or tetraalkyl diborane are added as catalysts.[141,142] These catalysts, via intermediate bridged compounds, provide a path for the exchange processes. Mixed trialkylboranes may be isolated from such mixtures by distillation after the catalyst has been destroyed hydrolytically or by other means. Detection of mixed trialkylboranes has been facilitated by gas chromatography.[34,142] Thermal redistribution quite often is accompanied by isomerization of alkyl groups, for example, $B(i\text{-}C_4H_9)_2(t\text{-}C_4H_9)$ is stable below ca. 60° but isomerizes above this temperature to give $B(i\text{-}C_4H_9)_3$.[143]

The kinetics of the thermal isomerization of α-branched trialkylboranes was studied in some detail with first-order kinetics observed in all cases.[144,145] Examples of the stability of mixed trialkylboranes with respect to redistribution as well as isomerization at low temperatures are the compounds $B(t\text{-}C_4H_9)(i\text{-}C_4H_9)(C_6H_5)$, $B(t\text{-}C_4H_9)_2(C_6H_5)$, $B(t\text{-}C_4H_9)_2(n\text{-}C_4H_9)$, and others which could be distilled *in vacuo* without decomposition.[146–149] Therefore, whenever boron–carbon/boron–carbon bond exchange is observed at relatively low temperatures, this perhaps might be due to impurities acting as redistribution catalysts.[150–152] Alkyl–alkyl symmetrizations have also been observed in tetraalkylboronate ions.[153]

Trialkylalanes exist as dimers with two alkyl groups forming bridges between the aluminum atoms. Proton as well as ^{27}Al nuclear magnetic resonance studies at variable temperatures indicate that the alkyl groups exchange rapidly between the terminal and the bridging position.[154–159] Similarly, rapid exchange of alkyl groups takes place in mixtures of aluminum alkyls.[155–157] From nuclear magnetic resonance measurements, it was estimated that the average lifetime [157] for the intermolecular exchange of bridging and terminal methyl groups at $-10°$ is of the order of 10^{-4} sec. From measurements of the heat of dilution, it was estimated that in a one molar solution of triethylalane, ca. 0.08% are dissociated into the monomer.[160]

The mechanism of the alkyl–alkyl exchange proceeding quite rapidly at room temperature is postulated to involve the following equilibria:

$$Al_2R_6 \rightleftharpoons 2\,AlR_3 \tag{28}$$

$$AlR_3 + AlR_3' \rightleftharpoons R_2Al\underset{R'}{\overset{R}{\diagup\diagdown}}AlR_2' \rightleftharpoons R_2AlR' + RAlR_2' \tag{29}$$

Various pairs of alkyl compounds[155–157,161–164] have been studied in detail by nuclear magnetic resonance, for example, $Al(CH_3)_3$–$Al[CH_2CH(CH_3)_2]_3$, $Al(CH_3)_3$–$Al(C_2H_5)_3$, $Al(CH_3)_3$–$Al(C_6H_5)_3$. In the latter case, the exchange is sufficiently slow so that dimethylphenylaluminum [161] may be isolated from the reaction mixture. Coordination with ethers has been found to slow down the exchange process. In the case of dimethylphenylaluminum half lives of less than 10^{-1} seconds were observed for benzene solutions, whereas in diethyl ether or tetrahydrofuran, this value was in the range of 10^{-1} to 10^{+2} sec.

Rapid exchange of ethyl and methyl groups also occurs in mixtures of triethyl and trimethylthallium at room temperature and nuclear magnetic evidence has been found for the existence of the mixed thallium alkyls.[165] Also, the system $Tl(CH_3)_3$—$Tl(CH{=}CH_2)_3$ was investigated [166] and the effect of trimethylamine or dimethyl ether in slowing the exchanges in this system was noted.

The exchange of silicon–carbon with silicon–carbon bonds is catalyzed by Lewis acids, such as $ZnCl_2$, $AlCl_3$, and similar compounds. Probably the first such observation [167] was made when phenyltrichlorosilane was alkylated with diethylzinc and, in addition to phenyltriethylsilane, also tetraethylsilane and diphenyldiethylsilane were obtained. Later studies[10,12,15,168–172] have confirmed the facile redistribution of alkyl or aryl groups on silicon at elevated temperatures under the influence of aluminum chloride. Generally no isomerization of *n*-alkyl to *i*-alkyl groups was observed. The distribution of the mixed silanes follows random statistics. Detailed studies of the catalytic effect of metal halides, of co-catalysts, solvent, and temperature upon the rate and equilibrium of these reactions were published. Recently a large number of systems was studied using gas chromatography[172] as quantitative means of separating the mixed tetraalkylsilane reaction products.

Similarly to silicon-based systems, the exchange of germanium–carbon with germanium–carbon bonds is catalyzed by aluminum chloride and results in a random distribution of alkyl groups.[172,173] The exchange of methyl and ethyl groups in hexaalkyldigermanes[174] according to Eq. (30)

$$(CH_3)_3GeGe(CH_3)_3 + (C_2H_5)_3GeGe(C_2H_5)_3 \rightleftharpoons (CH_3)_3GeGe(C_2H_5)_3, \qquad (30)$$

however, is base catalyzed and results in one product only suggesting a mechanism involving cleavage of germanium–germanium bonds.

Alkyl–alkyl exchange in tin tetra-alkyls is also catalyzed by aluminum chloride and leads to a random distribution of alkyl groups. The alkyl–alkyl exchange in lead tetra-alkyls[10,12,13,17] catalyzed by aluminum chloride and other Lewis acids, is one of the most thoroughly investigated examples of a random redistribution reaction. Fractional distillation as well as gas chromatography was used to separate the mixed tin as well as lead alkyls generated by random exchange.

Considering the elements of Group IV (excluding carbon) it is evident that exchange of element–carbon bonds with element–carbon bonds becomes more facile as one goes from silicon to lead. In all cases, however, catalysts are required.

For the elements of Group V, only one case of exchange involving element–carbon with element–carbon bonds has been reported.[175] Reacting triphenyl bismuth with tri-α-naphthylbismuth resulted in diphenyl-α-naphthyl bismuth.

M. Redistribution Reactions Involving Exchange of Element–Carbon with Element–Nitrogen Bonds

A quantitative study of the exchange of boron–carbon with boron–nitrogen bonds appears to be the only reported example of an exchange reaction of this kind. Phenyl groups exchange smoothly with dimethylamino groups with the compound $B(C_6H_5)[N(CH_3)_2]_2$ at equilibrium appearing in larger than statistical amounts, whereas the amount of $B(C_6H_5)_2[N(CH_3)_2]$ at equilibrium deviates only slightly from the random distribution.[176]

N. Redistribution Reactions Involving Exchange of Element–Carbon with Element–Oxygen Bonds

Lithium ethoxide is quite soluble in diethyl ether in the presence of ethyl-lithium and nuclear magnetic resonance spectra of such solutions have been interpreted in terms of the tetramer model of ethyllithium with lithium ethoxide incorporated into the tetrameric structure[136] giving species of the form $Li_4(C_2H_5)_3(OC_2H_5)$. It is possible that exchange of ethyl for ethoxyl groups has occurred. Another mechanism would involve dissociation of the two kinds of oligomers and subsequent association to mixed species.

A dimeric methoxide, $[CH_3Be(OCH_3)]_2$ is obtained from dimethylberyllium and beryllium methoxide.[177] This mixed compound does not disproportionate although the corresponding phenoxides do.[178] However, upon addition of pyridine to the methylmethoxide, disproportion was observed, yielding insoluble dimethoxyberyllium and soluble dimethylberyllium pyridine adduct.

Redistribution of alkyl groups and alkoxyl groups on boron was observed in mixtures of trialkylboranes and trialkylborates. This reaction may be induced thermally[176,179-183] or by catalysis with boron–hydrogen bond-containing compounds, such as tetraalkyldiboranes.[25,184-186] Upon heating above 100°, compounds of the type $RB(OR')_2$ were found to disproportionate[187] into trialkylborane and trialkoxyborane. Boron compounds having one or more boron–hydrogen bonds as well as boron–carbon bonds in the molecule are self-catalyzed when reacted with boron–oxygen compounds and in addition to exchange of B—C with B—O bonds also exchange of B—H with B—O takes place, leading to a variety of products.[81] A quantitative study[176] of the exchange of phenyl groups with methoxyl groups on boron by proton magnetic resonance indicates that the exchange process at 200° essentially leads to a random mixture of methoxyphenylboranes.

Exchange reactions also occur when boron trioxide is reacted with trialkylboranes.[188,189] The resulting trialkylboroxines, $[RBO]_3$, have been found to undergo disproportionation[190] to a very slight extent above 200° into boric oxide and trialkylboranes, but are completely stable on distillation at lower temperatures. An improved method for the preparation of trialkylboroxine, also based on a redistribution process, involves the reaction of trialkoxyboroxine with trialkylboranes,[191] a reaction which proceeds at considerably lower temperatures.

Facile interchange of alkyl and alkoxyl groups has been observed when trialkylalanes are reacted with trialkoxyalanes.[192,193] Depending upon the mole ratio of the starting materials, dialkylalkoxyalanes or alkyldialkoxyalanes are obtained. The exchange processes[155,194] in mixtures of AlR_3 and $AlR_2(OR')$ have been studied by proton nuclear magnetic resonance and the pre-exchange lifetimes were estimated to be larger than 0.05 sec. The order of reactivity of OR' is $C_2H_5O > i\text{-}C_3H_7O > t\text{-}C_4H_9$. Thallium[195] is the only other element of Group III where exchange of carbon bonded with oxygen-bonded substituents was observed.

In organosilicon chemistry, the interchange of alkoxy groups and aryl groups[196] is catalyzed by sodium. At the temperatures employed in these experiments (200°), it is significant that methyl groups directly bonded to silicon do not take part in the disproportionation. The actual catalyst in these systems is very likely the hydroxide or methoxide ion generated by reaction of sodium with moisture or free methanol which are certainly present in these systems. This general reaction has been extended to a procedure useful for preparing any number of phenyl-substituted alkoxysilanes.

$$CH_3Si(OCH_3)_3 + (C_6H_5)_nSi(OCH_3)_{4-n} \longrightarrow Si(OCH_3)_4 + (C_6H_5)_mCH_3Si(OCH_3)_{3-m}$$
$$n = 1, 2; m = 0, 1, 2, \text{ or } 3 \quad (31)$$

Alkyl[196] groups participate in the exchange reactions only when—in the presence of catalytic amounts of an alkali metal or a strong alkali base—the reaction temperature is in the range of 230–280°. Potassium, sodium, or lithium metal as well as bases such as sodium methoxide, sodium oxide, and alkali metal silanolates all bring about this reaction. Base catalyzed redistribution of phenyl groups has also been observed in rearrangements of phenylsiloxanes at high temperatures, for example, methylphenylpolysiloxane, $[(C_6H_5)(CH_3)SiO]_n$, essentially gave $[(C_6H_5)_2(CH_3)Si]_2O$ and $(C_6H_5)_3SiCH_3$ when heated in the presence of a base.[197] Exchange of methyl and trimethylsiloxy groups[198] surprisingly has been found to occur below 140° in the presence of chloroplatinic acid as catalyst. The base-catalyzed disproportionation reaction involving exchange of carbon with oxygen-bonded substituents on silicon has been used to synthesize polyphenylsilanes.[199]

$$\begin{matrix} CH_3 & CH_3 \\ | & | \\ ROSi-C_6H_4Si-OR \\ | & | \\ CH_3 & CH_3 \end{matrix} \xrightarrow[230°]{KOH} (CH_3)_2Si(OR)_2 + 1/n \begin{bmatrix} CH_3 \\ | \\ -Si-C_6H_4- \\ | \\ CH_3 \end{bmatrix}_n \quad (32)$$

Redistribution of tin–carbon with tin–oxygen bonds[200] was observed when bis(triphenyltin)oxide was heated for 8 hr at 140° yielding equimolar amounts of tetraphenyltin and diphenyltin oxide. Also triphenyltin hydroxide[201] undergoes the above reaction upon thermal treatment and some trialkyltin hydroxides[202,203] methoxides,[204] or esters[205] were found to behave similarly.

In organolead chemistry, redistributions of the above type are manifested by the thermal disproportionation of triethyllead acetate and diethyllead diacetate.[206]

For compounds of Group V elements, exchange of carbon-bonded for oxygen-bonded substituents was observed when phenylarsenous oxide, C_6H_5AsO, is thermally decomposed to yield triphenylarsine and arsenic trioxide.[207] Arylantimonous oxides[208–210] at moderately elevated temperature (ca. 200°) yield bis(diarylantimony)oxides and antimony trioxide. The reactions are bimolecular autocatalytic with the rate of disproportionation greatly influenced by the character of the substituent on the aryl group. The order of stability for some para substituents is $NO_2 > CH_3CO > Br > Cl > H > CH_3$. At temperatures of the order of 200°, the disproportionation products are triarylstibine and antimony trioxide.

O. Redistribution Reactions Involving Exchange of Element–Carbon with Element–Sulfur Bonds

Methylthio groups were found to exchange with methyl groups[24] when mixtures of trimethylborane and trimethylthioborane were heated in the presence of diborane as catalyst. The reaction was found to be an equilibrium reaction since the reaction products of the above reaction, $B(CH_3)(SCH_3)_2$ and $B(CH_3)_2(SCH_3)$, upon heating in the presence of diborane disproportionate quite readily to an extent of 20% into the original starting materials. Redistribution of methyl groups with SH groups[211] in the system trimethylborane and thiometaboric acid, $[B(SH)S]_3$, results in trimethylborthiin, $[B(CH_3)S]_3$, which upon standing converts to a tetramer. On the other hand, upon heating, these compounds disproportionate into trimethylborane and boron trisulfide. The corresponding phenyl derivative,[212] $[B(C_6H_5)S]_3$, which cannot be prepared analogously is more stable with respect to thermal disproportionation into triphenylborane and boron trisulfide.

P. Redistribution Reactions Involving Exchange of Element–Carbon with Element–Halogen Bonds

Redistribution of element–carbon with element–halogen bonds is perhaps one of the most intensively studied exchange reactions, the reason being that many of the resulting products are important products of commerce or valuable intermediates.

Equilibria[213] of the type

$$BeR_2 + BeX_2 \rightleftharpoons 2 BeRX \tag{33}$$

where R = alkyl and X = halogen are established in mixtures of dimethylberyllium and beryllium dibromide in diethyl ether as evidenced by low-temperature nuclear magnetic resonance spectra, ebullioscopic molecular-weight data, and selective precipitation experiments with 1,4-dioxane. Upon heating alkylberyllium halides, dialkylberyllium[214] was obtained, indicating facile disproportionation.

The problem of establishing the precise nature of the Grignard reagent in ethers has received considerable attention in the past years. Although this reagent generally is prepared from metallic magnesium and alkyl halide, it is of interest to discuss the equilibria existing among the various species present in mixtures of MgR_2 and $MgHal_2$ and in the Grignard reagent. The literature on this subject was reviewed earlier.[215] Recent data show[216–218] that dilute ether solutions of MgR_2 and $MgHal_2$, where $R = C_2H_5$ or C_6H_5 and Hal = Br or I, react almost instantaneously in a $1:1$ mole ratio upon mixing, with evolution of heat (2.0–4.9 kcal/mole of either reactant converted). For this equimolar mixture, the conversion is quite high, of the order of 82–98%, indicating nonrandom behavior. In dilute solution, the product is largely monomeric and therefore can only be MgRHal. Equilibrium constants for the reaction

$$2\ MgRHal \rightleftharpoons MgR_2 + MgHal_2 \qquad\qquad (34)$$

have been calculated. The Grignard reagent and solutions of equimolar mixtures of MgR_2 and $MgHal_2$ essentially are equivalent. Similar conclusions were obtained for tetrahydrofuran as solvent.

The redistribution of carbon-bonded substituents on boron with halogens bonded to boron may be achieved thermally,[219-235] requiring in most instances temperatures well above 100°. Recently it was found that this reaction is catalyzed by compounds containing B—H bonds,[25,236,237] such as diborane tetraalkyldiborane, and others, thus allowing lower reaction temperatures. In the latter case, the catalyst has to be destroyed before the isolation of the products. The resultant alkyl or arylhaloboranes generally are quite stable below ca. 100°. At higher temperatures, they disproportionate, establishing an equilibrium between the mixed compounds and the trialkyborane and boron trihalide. The rate of the thermal disproportionation depends, among others, on steric effects, for example, bromine derivatives react slower than chlorine derivatives. A mechanism may be postulated assuming a dimeric intermediate formed by overlap of σ orbitals with empty p orbitals of boron. From the published yield data, it appears that at equilibrium, the mixed compounds are favored with respect to the species having all like substituents.

Heterocyclic organohaloboranes[106,232,236,237] are readily obtained by cleavage of boron–carbon bonds in suitable compounds with boron trihalide, for example, B-alkylborolanes react with boron trihalides to give B-haloborolanes and upon further addition of boron trihalide acyclic structures are formed. The exchange of chlorine and methyl groups occurring on the borazine[38] ring structure is related to the above reactions.

Quite early it was established that organohaloalanes[192] of the type $R_4Al_2X_2$ and $R_2Al_2X_4$, where R = alkyl or aryl and X = halogen are formed upon combining two appropriate aluminum compounds. The bromides and chlorides so prepared did not disproportionate readily and, in many cases, could be distilled at temperatures below 100°. Methyl groups in methylchloroalanes were found by nuclear magnetic resonance techniques to exchange rapidly[238] in cyclohexane but much slower in coordinating solvents such as ethers.[239] Alkylfluoralanes[240] were prepared from trialkylalanes and amorphous aluminum fluoride and the exchange of fluorine and alkyl groups[241] in compounds of the type $NaAlR_2F_2$ at elevated temperatures was utilized to synthesize trialkylalanes according to the equation given below:

$$3 NaAlR_2F_2 \longrightarrow Na_3AlF_6 + 2 AlR_3 \tag{35}$$

A related redistribution was extended to chlorine, bromine, and iodine compounds and is based on the disproportionation of aluminum sesquihalides[242, 243] under the influence of sodium halides according to Eq. (36):

$$2 R_3Al_2X_3 + n NaX \longrightarrow 3 R_2AlX + AlX_3 \cdot n NaX \tag{36}$$

Diethylchloroalane containing ^{14}C in the molecules was prepared by comproportionation of $Al(C_2H_5)_3$—obtained from ^{14}C–ethylene and $LiAlH_4$—with aluminum chloride.[244] Phenylchloroalanes were prepared from triphenylalane and aluminum chloride and appear quite stable thermally.[245] Proton nuclear magnetic resonance shows that the compounds known as "aluminum alkyl sesquihalides" at room temperature exist as association complexes[246] of $RAlX_2$ and R_2AlX with AlX_2Al bridges, which latter compounds are in statistical equilibrium with $RAlX_2$ and R_2AlX.

Also gallium triaklyls react with gallium trihalides to give alkylgallium halides.[231] Diphenylgallium bromide and phenylgallium dibromide were prepared by this method and both were found to be dimeric in benzene solution with halogen bridges being assumed.[247]

It has been claimed[248,249] that alkylthallium halides could be prepared by comproportionation of dialkylthallium halides and thallium trihalides. However, later investigators[250,251] failed to confirm the earlier work. Only in the case of diphenylthallium chloride and thallium trichloride the resulting component was identified as the phenylthallium dichloride. Vinylthallium dihalides are claimed to have been prepared by the same method.[252]

In organosilicon chemistry, redistribution of silicon–carbon bonds with silicon–halogen bonds,[253] for example, in the system of methylchlorosilanes,[120] was shown to proceed at 250–400° in the presence of aluminum chloride as catalyst. At equilibrium,[26] the mixed species are favored, with the equilibrium constants being smaller than for the random case. Furthermore, rate measurements were reported, various other variables were studied,[61] and a mechanism was proposed which has been discussed earlier in this article. Amines[254] were found to be effective catalysts for the disproportionation of certain trichloromethyl derivatives of halosilanes, also involving exchange of carbon with halogen bonds on silicon.

$$4 CCl_3SiCl_3 \longrightarrow Si(CCl_3)_4 + 3 SiCl_4 \tag{37}$$

Also phenyl groups[255] were found to undergo redistribution reactions with halogens bonded to silicon in the presence of silica-alumina and other catalysts.

For germanium[256-260] as the central moiety, exchange of alkyl or aryl groups with halogens required catalysts such as aluminum chloride, organo-aluminum halides, germanium dihalides, or gallium trihalides. These reactions were applied to the preparation of alkylhalogermanes. The equilibria[260] in these systems generally favor the mixed compounds, with the equilibrium constants being considerably smaller than for the random case. The kinetics of the comproportionation of tetramethylgermane with germanium tetrachloride was studied by nuclear magnetic resonance. The first exchange, according to Eq. (38),

$$Ge(CH_3)_4 + GeCl_4 \longrightarrow (CH_3)_3GeCl + CH_3GeCl_3 \tag{38}$$

is much faster than the succeeding ones. The next most rapid reaction is

$$(CH_3)_3GeCl + CH_3GeCl_3 \longrightarrow 2\,(CH_3)_2GeCl_2 \tag{39}$$

and this reaction predominates when equimolar mixtures of reactants were used. The exchange of phenyl[105,257,261-264] groups with halogens on germanium as central atom was also described in some detail and was utilized for the preparation of phenylhalogermanes.

The redistribution of tin tetraalkyls or aryls with tin tetrahalides is perhaps the most thoroughly studied reaction of this type. As early as 1859 and 1862, reports[265,266] have appeared describing the reaction of tetrabutyltin and tin tetrachloride. Subsequently many papers[267-298] were published reporting the preparation of alkyl or arylchlorostannanes by comproportionation of the tetraalkyl or aryl compounds with the tetrahalides. These reactions generally proceed with remarkable ease, sometimes even at room temperature. Kinetic studies[272,299,300] of the system $Sn(CH_3)_4$ versus $SnCl_4$, $Sn(CH_3)_4$ versus $SnBr_4$, and $Sn(CH_3)_4$ versus SnI_4 by proton nuclear magnetic resonance and of the system $Sn(C_2H_5)_4$ versus $SnCl_4$ by gas chromatography show that the predominant reaction is described by Eq. (40):

$$SnR_4 + SnHal_4 \longrightarrow SnR_3Hal + SnRHal_3 \tag{40}$$

At all concentrations and temperatures studied, the above reaction is completed before any further reactions take place. For the 1:1 mole ratio of starting materials, the subsequently occurring reaction is shown below:

$$SnR_3Hal + SnRHal_3 \rightleftharpoons 2\,SnR_2Hal_2 \tag{41}$$

If there is an excess of SnR_4 over that of the equimolar amounts of starting materials, the mixture resulting after completion of reaction 40 and consisting of SnR_4, SnR_3Hal, and $SnRHal_3$ undergoes further reaction to give an increase in the amount of SnR_3Hal and some SnR_2Hal_2. This is expressed in the following reactions:

$$SnR_4 + SnRHal_3 \longrightarrow SnR_3Hal + SnR_2Hal_2 \tag{42}$$

$$SnR_4 + SnR_2Hal_2 \longrightarrow 2\,SnR_3Hal \tag{43}$$

For excess of tin tetrahalide in the mixture of starting materials, after completion of reaction 40, the following reactions become important:

$$SnR_3Hal + SnHal_4 \longrightarrow SnR_2Hal_2 + SnRHal_3 \qquad (44)$$

$$SnR_2Hal_2 + SnHal_4 \longrightarrow 2\,SnRHal_3 \qquad (45)$$

The twelve interactions corresponding to the forward and backward reactions of Eqs. (40–45) represent the total number of chemically nontrivial exchange processes. The rates of the redistribution decreases in the sequence Cl > Br > I. Complexing solvents[293] exhibit a retarding effect on the rate of redistribution and also affect the thermodynamics in some instances, for example, the high yield of $Sn(CH_3)Cl_3$ as dimethylsulfoxide complex from $Sn(CH_3)_2Cl_2$ and $SnCl_4$ is due to the insolubility of the complex in the solvent system. The equilibrium constants determined in the system $R = CH_3$ and $Hal = Cl$ indicate that the mixed compounds are greatly preferred at equilibrium with equilibrium constants smaller than the random case. The redistribution reactions involving concurrent exchange of tin–carbon with tin–halogen as well as tin–carbon bonds[301–304] were utilized for the preparation of mixed alkyl or aryl tin halides, for example,[304]

$$Sn(CH{=}CH_2)_4 + Sn(CH_3)(C_4H_9)Cl_2 \rightleftharpoons$$
$$Sn(CH_3)(CH{=}CH_2)(C_4H_9)Cl + Sn(CH{=}CH_2)_3Cl \quad (46)$$

Although these reactions are equilibrium controlled, the reactions may be shifted in the desired direction by removal of volatile equilibrium components by distillation.

For lead as central atom, only few examples of exchange of carbon-bonded substituents with halogens were reported. Refluxing of tetraphenyllead[293,305] with diphenyllead dichloride and of tetramethyllead[305] with dimethyllead dichloride yielded the respective triaryl or trialkyl halogen derivatives.

The thermal redistribution of arsenic–carbon bonded substituents with arsenic–halogen bonds has been known for a considerable time.[306,307] Later, exchange reactions involving a series of carbon-bonded groups[308–313] were investigated and their preparative utility was reported. The system $As(C_6H_5)_3$ versus $AsCl_3$ was studied in detail[314,315] in terms of the kinetics of the comproportionation of $As(C_6H_5)_3$ with $AsCl_3$ and the disproportionation of $As(C_6H_5)_2Cl$ and $As(C_6H_5)Cl_2$ as well as in terms of the resulting equilibria. Various metals catalyze these reactions, such as iron, tin, cobalt, nickel, chromium, copper, glass, with the catalysts listed in decreasing order of activity. The reactions are bimolecular and proceed at reasonable rates above ca. 250°. The equilibrium constants calculated for this system indicate preference of the mixed species in the equilibrium mixture of products.

For antimony, exchange reactions involving redistribution of aryl groups[316–322] with halogens have found an early application in the synthesis of arylhalostibines. The reactions generally proceed smoothly at ca. 250° and aryldihalostibines as well as diarylhalostibines were prepared. Also alkyl-dihalostibines and dialkylhalostibines were obtained by this method.[21,312] The kinetic study[21] in the system $Sb(CH_3)_3$ versus $SbCl_3$ was discussed earlier.

In contrast to the analog reactions of arsenic and antimony redistribution of carbon-bonded substituents with halogens on bismuth[175,323-328] as central atom proceed at room temperature or slightly elevated temperature, with the reaction products generally being insoluble in the solvent. Alkyl as well as aryl derivatives of the type BiR_2Hal and $BiRHal_2$ were prepared in this manner from trialkyl or triarylbismuth and bismuth trihalides.

Q. Redistribution Reactions Involving Exchange of Element–Nitrogen with Element–Oxygen Bonds

Exchange reactions of the above kind have been reported for some organo-boron compounds[329] and for the system[188,330] $B[N(CH_3)_2]_3$ versus B_2O_3. As part of a study of the latter system, cyclic tris(dimethylamino)boroxine was obtained in quantitative yield by heating $B[N(CH_3)_2]_3$ with B_2O_3. On the other hand, the attempt to make $(R_2N)_2B$—O—$B(NR_2)_2$, $R = CH_3$ or $n\text{-}C_3H_7$, resulted in the disproportionation products of the expected compound.

$$(R_2N)_2B\text{—O—}B(NR_2)_2 \longrightarrow B(NR_2)_3 + \tfrac{1}{3}[B(NR_2)O]_3 \qquad (47)$$

In silicon chemistry, exchange of nitrogen-bonded substituents with oxy-gen-bonded substituents[331,332] follows random statistics and requires the presence of aluminum chloride as catalyst. Without a catalyst considerable decomposition occurred at the required higher temperature with the reaction being quite slow even under these conditions. A study of the equilibria in the ternary system[333] $Si[N(CH_3)_2]_4$ versus $Si(OCH_3)_4$ versus $SiCl_4$ showed that, in this case, equilibria were established under relatively mild conditions, with $SiCl_4$ providing a path for the exchange of $N(CH_3)_2$ with OCH_3 groups. If the exchanging oxygen atom is part of an oligomeric system, such as octa-methylcyclotetrasiloxane,[332] polymeric species are obtained with dialkylamino groups as chain terminators and oxygens as bridging atoms.

In contrast to the results obtained for silicon, exchange of nitrogen–bonded with oxygen–bonded groups on germanium[260] as well as arsenic[334,335] proceed quite rapidly at room temperature, with the reaction at room tem-perature being completed within ca. 1 min for the germanium case and ca. 1 hr for the arsenic case. Equilibrium mixtures of poly(dimethylaminoarsen-ites)[335] are obtained from $As[N(CH_3)_2]_3$ and As_2O_3 as a consequence of redistribution and temperatures studies of the proton nuclear magnetic resonance spectra of such mixtures indicate rapid exchanges processes taking place with half lives at room temperature ranging from 1 to 300 sec.

R. Redistribution Reactions Involving Exchange of Element–Nitrogen with Element–Sulfur Bonds

Exchange of SH groups bonded to boron in borthiin six-membered ring structures with dimethylamino groups is easily accomplished by $B[N(CH_3)_2]_3$ and therefore this reaction is used as a method for the preparation of tris-(dimethylamino)borthiin.[336,337] Another case[338,339] reported thus far deals with the exchange of bridging methylimino groups in nonamethylcyclotrisilazane with bridging sulfur atoms in hexamethylcyclotrisilthiane or tetramethylcyclodisilthiane. This reaction results in cyclic silicon compounds having bridging methylimino groups as well as sulfur atoms in the same molecule.

S. Redistribution Reactions Involving Exchange of Element–Nitrogen with Element–Halogen Bonds

The interaction of boron trihalides or substances containing boron–halogen bonds with boron–nitrogen bonded compounds results in facile substituent exchange.[230,337,340-345] If the nitrogen-bearing group is a dialkylamino group, bis- and monodialkylaminohaloboranes are obtained in good yield. If the group is a monoalkylamino group,[345] the primary products are monalkylamino haloboranes, which for the alkyl group being $i\text{-}C_3H_7$ or $t\text{-}C_4H_9$ are stable but which for other alkyl groups undergo the secondary reaction shown below:

$$2 \text{ RHNBHal}_2 \longrightarrow \text{RNH}_2 \cdot \text{BHal}_3 + \tfrac{1}{3} (\text{RNBHal})_3 \qquad (48)$$

A similarly facile exchange of dimethylamino groups and halogen atoms was observed in the case of aluminum yielding bis(dimethylamino)haloalane and dimethylaminodihaloalane.[70]

Quantitative equilibrium measurements of the exchange of dimethylamino groups with halogens on silicon or methyl and dimethylsilicon moieties[331,346-349] indicate that this process is nonrandom with a great preference for formation of the mixed species. The equilibrium constants are of the order of 10^{-4} and the rates of equilibration at room temperature are quite rapid, with exchanges on the dimethylsilicon moiety occurring faster than in methylsilicon derivatives. The kinetics of some of these reactions has been studied in a semiquantitative manner, indicating the occurrence of reaction sequences of different rates. If the nitrogen–bonded substituent is a difunctional group, such as an imino group, NH, or alkylimino group, NR, redistribution with halogens results in oligomeric and polymeric halogen terminated polydialkylsilazanes.[350-355] Depending on the reaction conditions and the ratio of reactants, species of varying molecular size may be prepared. Another nitrogen–bonded group,[356] the azide group, has been found to exchange readily with halogens on silicon, and these reactions have been utilized for the synthesis of certain alkylsilicon azides.

A similar situation with respect to exchange of nitrogen–bonded groups, such as dimethylamino groups with halogens exists for germanium. The equilibrium constants indicate highly nonrandom exchange[260] with preference at equilibrium of the mixed species. The rates of equilibration at room temperature are quite fast. Oligomeric and polymeric N-methylated perchloropolygermazanes[357] are obtained when halogens on germanium are exchanged with methylimino groups. In this system, the isolation of the hexatomic ring species $[GeCl_2(NCH_3)_3]_3$ is noteworthy.

Analogous redistribution reactions on arsenic[334] are highly nonrandom also, with the mixed species being preferred at equilibrium. From nuclear magnetic resonance linewidth studies, it was concluded that the exchange processes proceed at half lives of around 0.1 sec at room temperature in 0.1 molar solutions. Families of oligomeric and polymeric haloarsenouspolymethylimides[358,359] were obtained from arsenic trihalide and tetraarsenic hexamethylimide, the latter is a structure consisting of methylimino-bridged arsenic atoms.

T. Redistribution Reactions Involving Exchange of Element–Oxygen with Element–Oxygen Bonds

There are several classes of redistribution reactions involving exchange of boron–oxygen with boron–oxygen bonds. First, there are the simple ester–ester interchanges, where OR groups are exchanged with OR′ groups. This exchange is quite rapid[360,361] at room temperature, as was shown for mixtures of $B(OR)_3$ and $B(OR')_3$ as well as $R_2B(OR')$ and $RB(OR'')_2$ and $B(OR)_3$ and $R_2'B(OR'')$. The minimum lifetime of a species in such mixtures is of the order of 0.7–3.8 sec. The rapid exchange is taken advantage of in transesterification reactions, where the ester–ester equilibrium is shifted in the desired direction by removal of the lowest boiling equilibrium component by distillation. Several examples of this general principle are reported in the literature.[329]

The second type involves exchanges of OR groups with the bridging oxygen[188,191,362–364] atoms in B_2O_3, $(BRO)_3$ or R_2BOBR_2. Refluxing of a trialkyl borate with boron trioxide—a more rapid reaction is observed when "active B_2O_3" is used—results in alkyl metaborates. In a nuclear magnetic resonance study of the $B(OCH_3)_3$—B_2O_3 system, the experimental data are interpreted in terms of equilibria between trimethyl borate and middle, end, and branch groups. It was found that at equilibrium, there are hardly any end groups present, with the middle groups predominating, the total amount of the latter being much larger than expected for a random process. From linewidth studies, it was estimated that the half life for exchange of methoxy groups in this system at room temperature is ca. 0.04 sec. These studies are in agreement with the observation that attempted distillation of metaborates, [B(OR)O]₃, results in disproportionation into trialkyl borate and B_2O_3.

A similar situation is found for analog exchanges on silicon. Ester–ester interchanges[331,346,365–368] are random processes initiated by heat with or without a catalyst. The exchange process is sufficiently slow as to allow separation of the mixed esters by fractional distillation or gas chromatography. Then there are the oligomers and polymers obtained by redistribution of alkoxyl groups with bridging oxygen atoms.[332,369,370] In these systems at equilibrium, a near-random distribution of building units occurs and the properties of the thus obtained polysilicates depend on the overall ratio of alkoxy groups versus silicon atoms. Also these systems were studied by nuclear magnetic resonance. Certain species could be isolated from the equilibrium mixture by fractional distillation since the rates of redistribution are quite slow.

The third group of oxygen–oxygen exchange reactions involves the exchange of bridging oxygen atoms with other bridging atoms in oligomeric and polymeric systems of silicon.[371–381] The equilibria resulting from hexamethyldisiloxane with octamethylcyclotetrasiloxane were studied quite extensively in view of the commercial importance of the "silicones." Equilibria between linear and cyclic molecules are established under the influence of catalysts such as acids, bases, acid clay, or simply by heating, and detailed studies of these equilibria were reported. Many linear and cyclic molecules on the basis of "silicones" were prepared and characterized. Low polymer organosilsesquioxanes,[382–387] $(RSiO_{1.5})_n$, may be prepared by alkali-catalyzed siloxane rearrangement of organotrichlorosilane hydrolyzates. By rearrangement of Si—O—Si bonds, crystalline octamers ($n = 8$), hexagonal dodecamers ($n = 12$) and several unidentified crystalline species were obtained. Similar phenylsilsesquioxanes were also studied.

In germanium chemistry, equilibria in solutions of dimethylgermanium oxide[388] are established quite rapidly. Nuclear magnetic resonance studies were reported and the constant for the equilibrium between the cyclic trimer and tetramer was determined. A polymer form of dimethylgermanium oxide upon heating converts into a mixture of the trimer and tetramer. All interconversions involve exchange of germanium–oxygen with germanium–oxygen bonds. Further evidence for the facile exchange of such bonds may be derived from the instability upon distillation of mixed tetraalkoxygermanes.[389]

Interchange of tin–oxygen with tin–oxygen bonds[390–393] occurs quite easily in mixtures of dialkyltin dialkoxides with other dialkyltin dialkoxides or dialkyl tin salts of organic acids, such as acetic acid. Recrystallization of the reaction product generally leads to substantial yields of the mixed product. Also the bridging oxygen[390,394] in dialkyltin oxides may undergo exchange with other tin–oxygen bonds yielding distannoxanes, as described by the equation below:

$$R_2SnO + R_2Sn(OR')_2 \longrightarrow (R'O)R_2Sn\!-\!O\!-\!SnR_2(OR') \qquad (49)$$

Exchange reactions involving arsenic–oxygen bonds proceed quite rapidly as exemplified by the study of the system As_2O_3 versus $As(OCH_3)_3$, where polymethylarsenites[335,395] are the result of the redistribution of bridging oxygen with methoxyl groups. In this family of compounds, exchange is quite rapid at room temperature as evidenced by line broadening seen in nuclear magnetic resonance spectra at this temperature. These spectra were interpreted in terms of exchanging structure building units, such as end groups, $-\!O\!-\!As(OCH_3)_2$, middle groups, $-\!O\!-\!As(OCH_3)_3\!-\!O\!-$, and branch groups, $As(O\!-\!)_3$. For the homolog element antimony, it was claimed that redistribution of $Sb(OR)_3$ with $Sb(OR')_3$ results[396] in mixed esters which were isolated by distillation of the reaction product.

U. Redistribution Reactions Involving Exchange of Element–Oxygen with Element–Sulfur Bonds

Reactions involving exchange of boron–oxygen with boron–sulfur bonds are reversible and the equilibrium can be displaced by the removal of one component from the reaction mixture, generally by distillation of the most volatile compound. Within several hours at ca. $100°$, the equilibrium may be shifted almost completely to the side of the most volatile component and thus yield the expected pair of compounds[329] almost quantitatively. The SH groups in trimeric metathioboric acid, $[BS(SH)]_3$, may be replaced analogously by OCH_3 groups by simply reacting metathioboric acid with trimethylborate.[336] The resulting trimethoxyborthiin has a tendency to disproportionate into B_2S_3 and $B(OCH_3)_3$.

In the case of silicon, the exchange of oxygen–bonded with sulfur–bonded substituents requires the presence of aluminum chloride and for $Si(CH_3)_2$ or $SiCH_3$ as the central moiety, it was found that the redistribution equilibrium involving exchange of OCH_3 with SCH_3 groups is nonrandom, with the mixed species at equilibrium being present in amounts which are smaller than calculated for the random case.[397] Oligomers and polymers based on Si—S—Si units were obtained when dimethoxydimethylsilane was scrambled with hexamethylcyclotrisilthiane, at equilibrium cyclic structures are favored.[398]

In general, like exchange reactions on germanium as central atom or moiety proceed faster than on silicon but the equilibrium constants generally are of the same order of magnitude.[397] Exchange of bridging oxygen atoms in the trimeric cyclic dimethylgermanium oxide with bridging sulfur in cyclic trimeric dimethylgermanium sulfide resulted in sulfur as well as oxygen containing cyclic germanium compounds.[399] Also in organotin chemistry,[400] exchange reactions of this type were reported and were found to proceed smoothly at room temperature.

V. Redistribution Reactions Involving Exchange of Element–Oxygen with Element–Halogen Bonds

Redistribution of boron–oxygen bonded substituents, for example, alkoxy groups, with boron–halogen bonds proceeds readily at or below room temperature. Depending on the mole ratio of the reactants, alkoxyhaloboranes or dialkoxyhaloboranes are formed in almost quantitative yield.[98,401-409] The reaction was reported for the halogen being fluorine and chlorine. There are two causes for the instability of certain chloro compounds, (*a*) decomposition according to reactions 50 and 51:

$$3 \ B(OR)Cl_2 \longrightarrow 3 \ RCl + B_2O_3 + BCl_3 \tag{50}$$

$$3 \ B(OR)_2Cl \longrightarrow 3 \ RCl + B_2O_3 + B(OR)_3 \tag{51}$$

or (*b*) disproportionation according to reactions 52 and 53:

$$3 \ B(OR)Cl_2 \longrightarrow 2 \ BCl_3 + B(OR)_3 \tag{52}$$

$$3 \ B(OR)_2Cl \longrightarrow BCl_3 + 2 \ B(OR)_3 \tag{53}$$

Also, phenoxyl groups were found to redistribute with boron–halogen bonds and quantitative equilibrium measurements[176,408] have shown that the equilibrium constants for phenoxydichloroborane and diphenoxychloroborane are both of the order of 10^{-4} indicating nonrandom exchange with the mixed compounds being preferred at equilibrium.

Exchange of bridging oxygen[224,410-416] in alkyl or arylboroxines, $(BRO)_3$, with halogens in the reaction with boron halides (fluorides, chlorides, and bromides) is an excellent method for the preparation of alkyl or aryldihaloboranes. While the alkylchloro and fluoro compounds are obtained in good yields, the yield of arylfluoroboranes is relatively low due to partial cleavage of triarylboroxine by boron trifluoride to form hydrocarbons. Therefore, the aryl difluoroboranes are best prepared by the action of fluorinating agents, such as SbF_3, on the aryldichloroborane which in turn is readily obtained from triphenylboroxine.

Qualitative and quantitative equilibrium measurements of the exchange of silicon–alkoxyl groups with silicon–halogen bonds (chlorine as well as bromine) have shown nonrandom distribution, with the mixed species being greatly preferred.[331,346,370,417-425] For the halogen being bromine, the equilibration process in some cases, depending on the reaction conditions, is accompanied by a side reaction involving formation of alkyl bromide and Si—O—Si linkages. In contrast to the equilibrium constants for the alkoxyl/halogen exchange, the analogous constants for the aroxyl/halogen exchange, are generally larger by a factor of ten and also the reaction rates are considerably smaller. Rapid disproportionation involving interchange of ethoxyl groups with fluorine atoms [426] was observed for the compounds $Si(OC_2H_5)F_3$ and $Si(OC_2H_5)_2F_2$, while $Si(OC_2H_5)_3F$ appeared to be stable at room temperature.

Redistribution involving bridging oxygen atoms with silicon–halogen bonds results in oligomeric and polymeric structures with siloxane linkages and halogens as end groups.[332,370,427-435] These reactions generally proceed slower than exchanges involving alkoxyl groups and, in many instances, catalysts such as $AlCl_3$, $AlBr_3$, H_3PO_4, H_2SO_4, or relatively high temperatures are required to effect complete redistribution.

The exchange of germanium–oxygen bonds with germanium–halogen bonds is quite rapid at room temperature.[260,436,437] Mixed methoxyhalogermanes cannot be isolated by conventional ambient temperature methods since continuous redistribution will take place. A similar statement may be made for the oligomeric and polymeric products obtained from scrambling bridging oxygen in dimethylgermanium oxides with halogens in dimethyldihalogermanes. From nuclear magnetic resonance linewidth measurements, it was concluded that the half life of the halogen-terminated polydimethylgermanium oxides is of the order of less than one second at room temperature.

Dialkyltindialkoxides, diacetates, or dioxinates when reacted with dialkyltin dihalides readily give the mixed compounds in essentially quantitative yield.[282,390,438-440] Addition of bipyridyl to the methoxyl derivatives reverses the comproportionation giving the dialkyltin dihalide bipyridine adduct and presumably the dialkyltin dimethoxide. When the bridging oxygen in dialkyltin oxides is exchanged with halogen atoms bonded to tin, functionally substituted distannoxanes[393,394,441] are obtained.

$$R_2SnO + SnR_2Hal \longrightarrow HalR_2Sn—O—SnR_2Hal \qquad (54)$$

Proton nuclear magnetic resonance spectra showed[334] that exchange of methoxyl groups in trimethylarsenite with halogens in arsenic trihalides proceeds readily at room temperature with the lifetime for exchange of a methoxyl group in the neat-liquid mixture being of the order of 10^{-2} sec. The distribution of products at equilibrium is nonrandom with the mixed species preferred, as was found generally for these types of exchange processes. The exchange of bridging oxygen with halogens in the system $AsO(C_6H_5)$ versus $AsCl_2(C_6H_5)$ appeared to be somewhat slower. As indicated by the reaction products, $AsCl(C_6H_5)_2$ and As_2O_3 however, this appears to be due to the additional exchange of arsenic–halogen as well as oxygen bonds with arsenic—carbon bonds.[442-444] The participation of arsenic–carbon bonds accounts for the rather severe reaction conditions. On the other hand, in the systems[395,445] As_2O_3 versus AsF_3 exchange again is quite rapid at room temperature and the slowest reaction—when arsenic trioxide is dissolved in arsenic trifluoride —appears to be the dissolution of the arsenic trioxide.

W. Redistribution Reactions Involving Exchange of Element–Sulfur with Element–Sulfur Bonds

Polymers obtained by the reaction of alkanethiols and diborane yield polymeric alkylthioboranes, $[B(SR)H_2]_n$, with B—S—B rather than B—H—B bonds serving as the associating units.[446,447] These polymers are metastable and depolymerize and rearrange at room temperature to mixtures of $B(SR)_3$ and $[B(SR)H_2]_3$, exchanging boron–sulfur with boron–sulfur as well as boron–hydrogen bonds. Exchange of thioalkyl with thioalkyl groups[329] was utilized to synthesize a series of alkylthioboranes by transthiolation—that is, distillation of the lowest boiling product from an equilibrium mixture as represented by the equation below.

$$3\ BR(SR')_2 + 2\ B(SR'')_3 \rightleftharpoons 3\ BR(SR'')_2 + 2\ B(SR')_3 \qquad (55)$$

The exchange of SH groups with SR groups[448] is exemplified in the preparation of trimethylthioborthiin from metathioboric and tri(methylthio)borane. Bridging sulfur in hexamethylcyclotrisilthiane at 200° redistributes with the alkylthio groups in dimethyldi(methylthio)silane yielding an equilibrium mixture of methylthio substituted polydimethylsilthianes[398] and large amounts of cyclic silthianes.

X. Redistribution Reactions Involving Exchange of Element–Sulfur with Element–Halogen Bonds

Exchange of boron–sulfur bonded substituents with boron–halogen bonds was postulated[449] to account for the instability of $B(SCH_3)_2Br$ with respect to disproportionation into $B(SCH_3)Br_2$ and $B(SCH_3)_3$. The compound $B(SCH_3)Br_2$, however, was found to be quite stable. From quantitative measurements of the equilibria resulting from exchange of methylthio groups and halogens on methyl- and dimethylsilicon moieties[423–425] it was found that the distribution is nonrandom, with the mixed species being preferred. The rates of equilibration in the above reactions generally are slower than those of the corresponding methoxyl derivatives. The redistributions of methylthio groups with pseudohalogens[424,425] however, appear to follow random statistics. Polymeric and oligomeric Si—S—Si bonded species were obtained when the bridging sulfur in hexamethylcyclotrisilthiane was scrambled with the halogen atoms in dimethyldihalosilanes.[398,450]

Also in germanium chemistry the exchange of methylthio groups with halogens is a nonrandom process, with the mixed species being preferred at equilibrium.[436,451] Polymeric species result when the bridging sulfur in cyclic hexamethyltrigermanium trisulfide is scrambled with the halogens in dimethyl-dihalogermanes.[451] The resulting equilibria were studied by proton nuclear magnetic resonance and interpreted in terms of equilibria between linear and cyclic molecules based on Ge—S—Ge structures.

In analogy to the equilibria in the above silicon and germanium–sulfur systems, nuclear magnetic resonance spectra of mixtures of $SnS(CH_3)_2$ and $SnHal_2(CH_3)_2$ were also interpreted in terms of equilibria between rapidly exchanging halogen-terminated polydimethyltin sulfides.[452] The simplest species in this system, generated according to the equation below

$$SnSR_2 + SnHal_2R_2' \rightleftharpoons HalR_2Sn\text{—}S\text{—}SnR_2'Hal \qquad (56)$$

was obtained in several cases as a crystalline product from the reaction mixture under given conditions.[390,400]

Y. Redistribution Reactions Involving Exchange of Element–Halogen with Element–Halogen Bonds

Mixtures of simple boron trihalides are known to exist only in equilibrium with the appropriate mixed halides.[453,454] The same is true for mixtures of alkyl or aryl dihaloboranes[223,455–458] where from the lack of broadening of the single [11]B nuclear magnetic resonance line seen in spectra of such mixtures, an average lifetime of [11]B in these environments of 10^{-2} sec is inferred. Similarly, a mass spectrometric study of the exchange reactions between boron trihalides and organohaloboranes, $BHal_2R$ and $BHalR_2$, provided evidence for the presence of mixed species which exist only as part of an equilibrium mixture.

Halogen–halogen exchanges on silicon[425,459–468] as central atom proceed slower and may be effected either thermally or by catalysts such as aluminum halides. Under proper precautions, mixed compounds may be isolated from the equilibrium mixture. In many instances, the equilibrium constants were determined and near-random distribution of products generally was observed. Pseudohalogens behave analogously giving random mixtures upon redistribution with halogens as well as with other pseudohalogens. Silicon–carbon or silicon–hydrogen bonds generally remain intact under the conditions where halogen–halogen exchange was observed. In some instances, almost quantitative "transhalogenations" may be performed by removing the lowest boiling product from the equilibrium mixture by distillation.

Organohalogermanes rapidly exchange halogens at room temperature, chlorine with bromine, bromine with iodine, as well as chlorine with iodine.[466,467] Except for the latter case where the equilibrium constants have been found about 2–3 times larger than the random value, the resulting equilibria generally follow quite closely random statistics.

Halogen–halogen exchange reactions have been studied between $SnCl_4$,

$SnBr_4$, and SnI_4 by ^{119}Sn nuclear magnetic resonance[469] as well as by Raman spectroscopy.[470] As a result of these studies it was concluded that complete randomization of the halogen atoms takes place. An analog random distribution has been found to occur in mixtures of alkyltrihalostannanes and also in di- and trialkyhalostannanes, respectively.[282,471,472] From these proton nuclear magnetic resonance studies, it is further concluded that the rate of halogen exchange between bromides and chlorides is always very high. It decreases in mixtures of bromides and iodides and decreases still further for the systems of chlorides and iodides, regardless of the number of the methyl groups bonded to tin.

Solutions of two different trialkyl or triaryl arsenic dihalides, AsR_3Hal_2 and $AsR_3Hal'_2$, generated a mixed halide,[473] $AsR_3HalHal'$. The equilibrium constants in these systems approach statistical values as the electronegativity between Hal and Hal' decreases and as the temperature increases. Furthermore, the exchange processes at room temperature are quite rapid causing line broadening of the proton nuclear magnetic resonance lines for $R = CH_3$. Similar redistribution reactions have been observed for analog antimony compounds.[474] Also in these cases, line broadening is observed, indicating rapid exchange in solution. Attempts to isolate pure samples of the mixed compounds were unsuccessful, even at low temperatures.

Z. Redistribution Reactions Involving Exchange of Element–Element Bonds

Diboron compounds, such as $B_2(OR)_4$ or B_2Cl_4 upon standing disproportionate[401,475] under exchange of boron–boron bonds to give reactive elemental boron and $B(OR)_3$ or BCl_3, respectively. The decomposition of the esters is catalyzed by moisture. In other instances, boron polymers may be formed.[476]

Disilanes and alkyldisilanes[477–480] in the presence of potassium or lithium hydride or lithium chloride initially disproportionate into silane and a polymer of the composition $(SiH_2)_n$. These reactions involve redistribution of silicon–silicon bonds with silicon–hydrogen or silicon–carbon bonds respectively. Similar reactions with silicon–oxygen bonds[481–483] were observed in the base-catalyzed redistribution of compounds of the series $(CH_3O)[Si(CH_3)_2]_n(OCH_3)$ where $n = 2$ to 5. The lower members, $n = 2$ or 3, undergo facile base-catalyzed redistribution as exemplified by the equation given below:

$$2\,(CH_3O)[Si(CH_3)_2]_2(OCH_3) \longrightarrow$$
$$Si(CH_3)_2(OCH_3)_2 + (CH_3O)[Si(CH_3)_2]_3(OCH_3) \quad (57)$$

Similarly, sym-dimethyltetramethoxydisilane rearranges thermally via an analog mechanism.

$$(CH_3O)_2(CH_3)SiSi(CH_3)(OCH_3)_2 \longrightarrow$$
$$Si(CH_3)(OCH_3)_3 + Si(CH_3)[Si(CH_3)(OCH_3)_2]_3 \quad (58)$$

Hexamethoxydisilane is even less stable thermally. Upon heating below 200°, it disproportionates according to the following reaction:

$$(CH_3O)_3SiSi(OCH_3)_3 \longrightarrow Si(OCH_3)_4 + Si[Si(OCH_3)_3]_4 \qquad (59)$$

Kinetic, mass spectral, and chemical evidence[483] support the intermediate formation of a divalent silicon (silylene) species in the redistribution of these methoxypolysilane derivatives.

Other disilane and polysilane derivatives undergo similar catalytic redistributions resulting in monosilane derivatives and oligomers having two or more silicon–silicon bonds. Examples are hexachlorodisilane,[484,485] various alkylchlorodisilanes[484,486,487] and alkylisocyanodisilane.[486] Methylchlorodisilanes[488,489] readily undergo aluminum chloride catalyzed Si/Si redistribution at room temperature while the disproportionation of alkylchlorosilanes catalyzed by aluminum chloride (involving Si—C and Si—Hal bonds) generally proceeds quite slowly at much higher temperatures. This indicates that the mechanism of Eq. (60) involves

$$(CH_3)_3SiSi(CH_3)_3 + Cl(CH_3)_2SiSi(CH_3)_2Cl \rightleftharpoons 2 (CH_3)_3SiSi(CH_3)_2Cl \qquad (60)$$

interchange of Si—Si bonds rather than Si—C and Si—Cl bonds.

Related exchange reactions involving element–element bonds were observed also for the other elements of Groups IV, for example, digermane[479] disproportionates in the presence of basic catalysts to GeH_4 and a polymer of the composition $(GeH_2)_n$. Hexachloroditin[490] yields tin tetrachloride and tin dichloride, hexaalkyl or aryldilead compounds[491,492] give tetraalkyl or aryllead and elemental lead. Similarly, tetraalkyl or tetraaryldiarsines[493] disproportionate into the trialkyl or triarylarsine and metallic arsenic. Rapidly exchanging linear and cyclic permethylpolysarsines[494] were obtained from pentamethylpentaarsine and tetramethyldiarsine as a result of As—As interchanges.

IV. EXPERIMENTAL PROCEDURES

1. Diphenylalane,[57] $(C_6H_5)_2AlH$

When a mixture of two moles of triphenylalane and one mole of alane etherate is heated for 1.5 hr in refluxing benzene, a clear solution results. Concentration *in vacuo*, followed by treatment with light petroleum, affords diphenylalane in 78% yield. After recrystallization from benzene–light petroleum, the product melts at 156–157°.

2. Dimethylaminoalane,[70] $(CH_3)_2NAlH_2$

A sample of 0.8 g (5 mmole) of tris-(dimethylamino)alane and 0.9 g (10 mmole) of trimethylamine alane are heated together under dry nitrogen at 80°

for two hr. During this procedure, 9.9 mmoles of trimethylamine are evolved. The product sublimes at 40° and melts at 89°.

3. n-Butylsilane,[124] n-C$_4$H$_9$SiH$_3$

Upon distilling a mixture of 26.6 g (0.155 mole) of n-butyltrichlorosilane, 48.5 g (0.410 mole) of triethylsilane and 1.5 g of aluminum chloride on a fractionating column, the following products are obtained. Fraction I: b.p. 52–56°, n-butylsilane in a quantity of 10.8 g (0.12 mole) corresponding to a 98% yield based on the butyltrichlorosilane reacted; Fraction II: b.p. 56–143°; Fraction III: b.p. 143–147°, consisting of a mixture of 95.2% triethyl-chlorosilane and 4.8% butyltrichlorosilane in a quantity of 55.7 g (93.9% yield of triethylchlorosilane based on the butyltrichlorosilane reacted); Fraction IV: b.p.: above 147%, 6.1 g.

4. Mixed Methyl–Ethyl Lead Alkyls,[13] Pb(CH$_3$)$_n$(C$_2$H$_5$)$_{4-n}$

Depending on the desired compound, a mixture of tetramethyl lead and tetraethyl lead in the molar ratios 3:1 for Pb(CH$_3$)$_3$(C$_2$H$_5$), 1:1 for Pb(CH$_3$)$_2$-(C$_2$H$_5$)$_2$ or 1:3 for Pb(CH$_3$)(C$_2$H$_5$)$_3$ is placed in a three-necked distilling flask fitted with a reflux condenser, thermometer, and a nitrogen inlet tube while maintaining an atmosphere of nitrogen within the system and cooling the flask in ice water. Prior to the addition of the lead alkyls, 0.5 g of anhydrous aluminum chloride per 0.3 mole of tetraalkyl lead compound is added as catalyst. The mixture is kept at 80° for two hours. After cooling, an equal volume of water is added to destroy the catalyst and after shaking, the alkyl lead layer is separated, filtered, and fractionated. Redistribution generally is complete, yielding the methyl ethyl lead alkyls in amounts corresponding to random statistics, as exemplified in Table II.

5. Triisobutylboroxine,[191] (i-C$_4$H$_9$BO)$_3$

A quantity of 78 g (0.42 mole) of trimethoxyboroxine, (CH$_3$OBO)$_3$, is heated with 76 g (0.42) mole of triisobutylborane at atmospheric pressure and the resulting trimethyoxyborane is removed as rapidly as it is formed by distillation through a 28-theoretical plate spinning band column. Complete removal of the trimethoxyborane (0.39 mole) requires 2.5 hr during which time the temperature of the reaction mixture rises from 131 to 207°. Distillation of the remaining reaction mixture gives a few grams of a forerun containing isobutyldimethoxyborane and unchanged triisobutylborane and 77 g of crude triisobutylboroxine. Unchanged boric oxide (7 g) remained as residue. Redistillation of the crude material gave 68 g (64% yield) of pure triisobutylboroxine, b.p. 90° at 3.6 mm.

Other trialkylboroxines containing primary alkyl groups, for example, tri-*n*-propylboroxine, tri-*n*-butylboroxine, and tri-*n*-amylboroxine may be prepared by similar procedures in yields above 60%.

6. Diethylchloroborane,[495] $(C_2H_5)_2BCl$

A quantity of 900 g (7.6 moles) of boron trichloride is passed into a mixture of 1470 g (15 moles) of triethylborane and 5 ml of tetraethyldiborane in a 4 liter two-necked flask fitted with a gas inlet tube, a 30-cm distilling column with condenser and receiving flasks. The reaction is exothermic and therefore the introduction of the gas should be adjusted to avoid boiling of the flask contents. After the reaction is completed, 15 ml of 1-octene are added to destroy the catalyst and the mixture is distilled. After a small amount of forerun, 2300 g of diethylchloroborane distills at 81° in 98% yield.

7. Ethyldichloroborane,[495] $C_2H_5BCl_2$

A quantity of 240 g (2.05 mole) of boron trichloride is passed into a mixture of 98 g (1 mole) of triethylborane and 2 ml of tetraethyldiborane. After the addition of 6 ml of 1-octene, the mixture is distilled to yield a quantitative yield of ethyldichloroborane boiling at 50–51°.

Alkylchloroboranes having higher alkyl groups may be prepared similarly, also alkylfluoroboranes, using $BF_3 \cdot O(C_2H_5)_2$ instead of BCl_3, at a reaction temperature of ca. 100°.

8. Diphenylchloroborane,[495] $(C_6H_5)_2BCl$

From $B(C_6H_5)_3$ and BCl_3: A quantity of 55.3 g (0.22 mole) of triphenylborane is dissolved in 200 ml of xylene and after addition of 0.5 ml of bis-borolane, the mixture is heated at 120°. Under these conditions, the bis-borolane dissociates into borolane. Then 13 g (0.11 mole) of boron trichloride are passed into the solution which after the addition of 1 ml of 1-octene is distilled resulting in a small forerun consisting of phenyldichloroborane and as main fraction 46.6 g (0.23 mole) of diphenylchloroborane, b.p. 101–102° at 0.2 mm, corresponding to a 70% yield. Upon cooling, the product crystallizes in plates with a melting point of 32°. The residue is triphenylborane which may be combined with the forerun and subjected to the procedure given above resulting in additional 17 g of diphenylchloroborane giving a total yield of 53.6 g (93%).

From $B(C_6H_5)_3$ and $C_6H_5BCl_2$: A variation of the above procedure consists in the use of phenyldichloroborane which is easily available from the reaction of benzene with boron trichloride. A quantity of 107 g (0.44 mole) of triphenylborane, 72 g (0.45 mole) of phenyldichloroborane and 1 ml of bis-borolane are heated in a simple distilling apparatus for two hours at 140°. Instead of bis-borolane tetraethyldiborane may be used at a reaction temperature of 40°. During the heating, the reaction mixture should change color only slightly. Distillation at 4 mm pressure results in a small forerun (phenyldichloroborane and bis-borolane) and, as main fraction, diphenyl-chloroborane distilling at 100–105°. The above procedure is repeated with the combined forerun and residue resulting in additional product giving a total yield of 166 g (94%). Upon cooling, the compound crystallizes as large crystals having a melting point of 31°.

9. Phenyltrichlorogermane,[30] $C_6H_5GeCl_3$

A quantity of 11.4 g (0.03 mole) of tetraphenylgermane and 19.3 g (0.09 mole) of germanium tetrachloride in the presence of 0.8 g (0.006 mole) of freshly sublimed aluminum chloride is heated at 120° for 1.5 hr. The resulting clear solution does not yield a precipitate of tetraphenylgermane upon cooling, indicating that at least partial reaction has occurred. Additional heating for 5 hr and subsequent distillation yields 25.3 g (0.10 mole) of phenyltri-chlorogermane (b.p. 103° at 12 mm) corresponding to 82% yield.

10. Triethylchlorostannane,[272] $(C_2H_5)_3SnCl$

A quantity of 23.5 g (0.1 mole) of tetraethylstannane is added dropwise to 12.7 g (0.05 mole) of ethyltrichlorostannane which is kept at 200°. Distillation of the reaction mixture after one hour yields 35.8 g (0.14 mole) of triethyl-chlorostannane (corresponding to a yield of 94%) having a boiling point of 97° at 12 mm.

11. Tris(dimethylamino)boroxine,[188] $[(CH_3)_2NBO]_3$

A tube is charged with 2.02 g (0.029 mole) of boron trioxide and 4.3 g (0.030 mole) of tris(dimethylamino)borane. The tube is then evacuated and sealed and subsequently heated at 280° for two hours. A well-crystallized material is formed in almost quantitative yield. Sublimation at high vacuum yields the boroxine in high purity having a melting point at 64° and a boiling poing at 221°/752 mm. The compound is sensitive to moisture and should be handled in a dry atmosphere.

12. Dimethylaminodifluoroborane,[344] $(CH_3)_2NBF_2$

A solution of 4.1 g (0.03 mole) of tris(dimethylamino)borane in 200 ml of dry pentane is placed in a flask equipped with a gas bubbler, magnetic stirrer, and outlet tube which terminates in a mercury seal. A mixture of boron tri-fluoride and nitrogen is bubbled into the solution; a white powder is immediately precipitated and the flow of gas is continued until precipitation ceases. The solvent is removed by distillation and 6.6 g (0.07 mole) of the product, corresponding to a yield of 82%, is obtained. The latter is purified either by vacuum sublimation at 30° (96% recovery) or by recrystallization from benzene (90% recovery). The compound crystallizes as well-formed, elongated needles.

13. Dimethylaminodichloroborane,[344] $(CH_3)_2NBCl_2$

Boron trichloride replaces boron trifluoride in the preceding preparation. After removal of the solvent, the residue is a colorless liquid which deposits crystals upon standing at room temperature for five days. The crystals correspond to a yield of 68% and are obtained pure after filtering and washing with dry pentane.

14. Dimethylaminodibromoborane,[344] $(CH_3)_2NBBr_2$

A quantity of 14.4 g (0.06 mole) of boron tribromide is added slowly to a pentane solution of 4.1 g (0.03 mole) of tris(dimethylamino)borane. The solvent is removed under a vacuum, leaving a colorless liquid. This is sealed under a vacuum in an ampoule and heated at 65° for seven days. The large crystals which are formed are then filtered off and dried *in vacuo*. Very little more solid is obtained when the filtrate is heated at 65° for another 14 days. The overall yield of products is 63%.

15. Dimethylaminodiiodoborane,[344] $(CH_3)_2NBI_2$

A solution of 12.6 g (0.03 mole) of boron triiodide in 250 ml of dry pentane in a flask is immersed in an ice–salt bath at −10° and 4.8 g (0.03 mole) of tris(dimethylamino)borane is added slowly from a syringe while the flask is being shaken. Immediate precipitation occurs. The solvent is decanted in a glove box and the yellow oily residue is purified by washing with pentane to give 14.8 g (85% yield) of product in the form of a yellow powder which is stable indefinitely under vacuum at 0°. At higher temperature or with different molar proportions, the reaction gives dark brown residues.

16. Mixed Ethyl n-Butyl Orthosilicates,[366] $Si(C_2H_5O)_n(OC_4H_9\text{-}n)_{4-n}$

Mixtures of ethyl orthosilicate and *n*-butyl orthosilicate in the presence of 0.03 mole of magnesium aluminum ethoxide per mole of silicon compound give essentially a statistical distribution of alkoxy groups within a refluxing period of 4.5 hr. The resulting products may be separated by fractional distillation after the catalyst is destroyed by washing the cold reaction mixture successively with dilute hydrochloric acid, water and sodium carbonate solution, and the solution is dried over calcium chloride. The boiling points of the products which are obtained in quantities corresponding to random distribution (Table II) are as followed: $Si(OC_2H_5)_4$, 77–77.5° at 32 mm; $Si(OC_2H_5)_3$-$(OC_4H_9\text{-}n)$, 102–102.5° at 32 mm; $Si(OC_2H_5)_2(OC_4H_9\text{-}n)_2$, 128–128.5° at 32 mm; $Si(OC_2H_5)(OC_4H_9\text{-}n)_3$, 150–150.5° at 32 mm and $Si(OC_4H_9\text{-}n)_4$, 141–142° at 3 mm.

17. Di-n-butyltin Acetate Chloride,[390] $(n\text{-}C_4H_9)_2SnCl(OOCCH_3)$

A quantity of 1.47 g (0.0048 mole) of di-n-butyltin dichloride and 1.70 g (0.0048 mole) of di-n-butyltin diacetate are mixed in carbon tetrachloride. An exothermic reaction occurs giving di-n-butyltin acetate chloride as white crystals which are recrystallized from carbon tetrachloride, m.p. 56.5–57.5°.

18. n-Butyl Di-n-propylthioborinate,[329] $(n\text{-}C_3H_7)_2B(SC_4H_9\text{-}n)$

A mixture of 23.5 g (0.183 mole) of $(n\text{-}C_3H_7)_2BOCH_3$ and 17.3 g (0.062 mole) of $(n\text{-}C_4H_9S)_3B$ is heated at ca. 130–110° in a distilling appratus for several hours. During this time, 13.0 g (0.056 mole) of tri-n-butyl borate distills over corresponding to a 91% yield. The residue consists of $(n\text{-}C_3H_7)_2$-$BSC_4H_9\text{-}n$ which may be purified by distillation, b.p. 57–59° at 2 mm, to yield 23.0 g (0.16 mole) of product corresponding to a yield of 87.5%.

19. 1,3-Dimethoxyhexamethyltrisilane,[482] $(CH_3O)[Si(CH_3)_2]_3(OCH_3)$

To a 100 ml flask, equipped with a septum cap inlet and attached to an 18-in. Nester Faust spinning-band column, is added 3 ml of dimethyldimethoxy-silane and 4.4 ml of n-butyllithium (1%). Following the resultant exothermic reaction, 44.6 g (0.24 mole) of 1,2-dimethoxytetramethyldisilane is added. The flask is heated using a 100° constant-temperature bath. After 10–11 hr, 15 g of dimethyldimethoxysilane is collected. The reaction residue is cooled, quenched with 2 ml of allyl bromide and distilled to give 18.1 g (61%) of 1,3-dimethoxyhexamethyltrisilane. The residue (5 g) consists of 90% of 1,4-dimethoxyoctamethyltetrasilane.

V. TABULAR SURVEY

TABLE III

Redistribution Reactions Involving Exchange of Element–Hydrogen with Element–Carbon Bonds

Reactants	Products	Remarks	References
$B_2H_6/B(CH_3)_3$	$B_2H_5CH_3$, $B_2H_4(CH_3)_2$, $B_2H_3(CH_3)_3$, $B_2H_2(CH_3)_4$	At r.t.	31,32
$BH_2[N(CH_3)_2]/B(CH_3)_2[N(CH_3)_2]$	$BHCH_3[N(CH_3)_3]$, (35%)	3 hr at 100°	36
$[B_3N_3H_6]/[B_3H_3N_3(CH_3)_3]$	$[B_3H_2(CH_3)N_3(CH_3)_3]$, $[B_3H(CH_3)_2N_3(CH_3)_3]$	At 175°	38
$[BH_4]Li/B(CH_3)_3$	$B_2H_2(CH_3)_4$, $B_2H_3(CH_3)_3$, $B_2H_4(CH_3)_2$, $B_2H_5(CH_3)$	In the presence of HCl at 100–140°	35,39
$[BH_4]Na/B(CH_3)_3$	$B_2H_2(CH_3)_4$, $B_2H_3(CH_3)_3$, $B_2H_4(CH_3)_2$, $B_2H_5(CH_3)$	In the presence of HCl at 150–175°	35,39
$B_2H_6/B(C_2H_5)_3$	$B_2H_5(C_2H_5)$, $B_2H_4(C_2H_5)_2$, $B_2H_3(C_2H_5)_3$, $B_2H_2(C_2H_5)_4$	At r.t.	42
$B_2H_5(C_2H_5)$	B_2H_6; sym-$B_2H_4(C_2H_5)_2$, ($K = 0.04$, gas phase; $K = 0.17$, liquid phase)	1–5 min at r.t.	43
$B_2H_4(C_2H_5)_2$-$unsym.$	$B_2H_n(C_2H_5)_{6-n}$; ($n = 1$–4)	At r.t.	43
$[BH_4]Na/B(C_2H_5)_3$	$B_2H_n(C_2H_5)_{6-n}$; ($n = 1$–4)	In the presence of HCl at 150–175°	39,44
$B_2H_6/B(n$-$C_3H_7)_3$	$B_2H_5(n$-$C_3H_7)$, $B_2H_4(n$-$C_3H_7)_2$		42
$B_2H_6/B(n$-$C_3H_7)_3$	$B_2H_2(n$-$C_3H_7)_4$, (70%); $unsym.$-$B_2H_4(n$-$C_3H_7)_2$, (16%)	In diethyl ether	45
$B_2H_6/B(n$-$C_3H_7)_2(OC_6H_{13}$-$n)$	$B(n$-$C_3H_7)(OC_6H_{13}$-$n)_2$, (60%); $BH(n$-$C_3H_7)_2$	Also B—C/B—O exchange	46
$B_2H_6/B(n$-$C_3H_7)_2(OC_7H_{15}$-$n)$	$B(n$-$C_3H_7)(OC_7H_{15}$-$n)_2$, (78%); $[BH(n$-$C_3H_7)_2]_2$, (43%)	Also B—C/B—O exchange	46

$B_2H_n(n\text{-}C_3H_7)_{6-n}(n = 1\text{-}4)$	At 150–175°	39,44
$B_2H_6/B(n\text{-}C_4H_9)_4$, (87%)	In diethyl ether	45
$BH(n\text{-}C_6H_{13})_2$	In the presence of $N(C_2H_5)_3$	50
$[BH_2(C_6H_5)]_2$, (74%)	At 40–100°, 2–4 atm	49
$[BH_2(C_6H_5)]_2$		51
B_2H_6, $B(C_6H_5)_3$		52
$[BH_2(C_6H_5)_2]$, (60%)	Also B—H/B—Cl exchange	49
$B(C_6H_5)_2[N(C_2H_5)_2]$, $BH_2[N(C_2H_5)_2]$	At 90–150°	53
$[BH_2(C_6H_5)_2]$, (71%)	Also B—O/B—H exchange	48
$[BH_2C_6H_4CH_3\text{-}o]_2$, (51%)	Also B—O/B—H exchange	47
$[BH_2C_6H_4Cl\text{-}p]_2$, (79%)	Also B—O/B—H exchange	48
$[BH_2(C_{10}H_7\text{-}\alpha)]_2$, (81%)	Also B—O/B—H exchange	48
$AlH_2(CH_3)$	As $N(CH_3)_3$ adduct	54
$AlH(CH_3)_2$	As $N(CH_3)_3$ adduct	54
$AlH(CH_3)_2$		56
$AlH_2(C_2H_5)$		55
$AlH(C_2H_5)_2$	As $N(CH_3)_3$ adduct	54
$AlH(C_2H_5)_2$	As $N(CH_3)_3$ adduct	54
$NaAlH(C_2H_5)_3$, $NaAlH_2(C_2H_5)_2$	Maxima in phase diagram	58
$AlH(C_6H_5)_2$, (78%); $AlH_2(C_6H_5)$, (23%)	In diethylether	57
SiH_4, $SiH_3(CH_3)$, $SiH_2(CH_3)_2$, $Si(CH_3)_4$	At 40–60° $AlBr_3$ as catalyst	61
$SiH(CH_3)_2(C_2H_5)$, (81%); $SiH_2(C_2H_5)_2$, (66%), $Si(C_2H_5)_4$	$AlCl_3$ as catalyst	63
$SiH(CH_3)_2$, $SiH_3(CH_3)$, $Si(C_6H_5)_4$	$AlCl_3$ as catalyst	63
$Si(CH_3)(C_6H_5)_3$ (trace); $SiH(CH_3)(C_6H_5)_2$, (35%); $SiH_3(CH_3)$, (21%)	Catalyzed by di-t-butyl peroxide	66

$[BH_4]Li/B(n\text{-}C_3H_7)_3$	
$B_2H_6/B(n\text{-}C_4H_9)_3$	
$B_2H_6/B(n\text{-}C_6H_{13})_3$	
$B_2H_6/B(C_6H_5)_3$	
$B_2H_6/B(C_6H_5)_3$	
$BH_2(C_6H_5)$	
$B_2H_6/B(C_6H_5)_2Cl$	
$BH(C_6H_5)[N(C_2H_5)_2]$	
$B_2H_6/B(C_6H_5)_2(OC_4H_9\text{-}n)$	
$B_2H_6/B(C_6H_4CH_3\text{-}o)_2(OC_4H_9\text{-}i)$	
$B_2H_6/B(C_6H_4Cl\text{-}p)_2(OC_4H_9\text{-}i)$	
$B_2H_6/B(\alpha\text{-}C_{10}H_7)(OC_4H_9\text{-}n)_2$	
$AlH_3/Al(CH_3)_3$	
$AlH_3/Al(CH_3)_3$	
$[AlH_4]Li/Al(CH_3)_3$	
$AlH_3/Al(C_2H_5)_3$	
$AlH_3/Al(C_2H_5)_3$	
$AlH_3/Al(C_2H_5)_3$	
$[AlH_4]Na/[Al(C_2H_5)_4]Na$	
$AlH_3/Al(C_6H_5)_3$	
$SiH(CH_3)_3$	
$SiH(CH_3)(C_2H_5)_2$	
$SiH_2(CH_3)(C_6H_5)$	
$SiH_2(CH_3)(C_6H_5)$	

49

(continued)

TABLE III (*Continued*)

Reactants	Products	Remarks	References
SiH(CH$_3$)(C$_6$H$_5$)$_2$	Si(CH$_3$)(C$_6$H$_5$)$_3$, (25%); SiH$_2$(CH$_3$)(C$_6$H$_5$), (18%)	Catalyzed by di-t-butyl peroxide	66
SiH(CH$_3$)$_2$(C$_6$H$_5$)	Si(CH$_3$)$_2$(C$_6$H$_5$)$_2$, (19%) SiH$_2$(CH$_3$)$_2$, (9%)	Catalyzed by di-t-butyl peroxide	66
SiH(CH$_3$)$_2$(C$_6$H$_4$CH$_3$-m)	Si(CH$_3$)$_2$(C$_6$H$_4$CH$_3$-m)$_2$, (16%); SiH$_2$(CH$_3$)$_2$, (8%)	Catalyzed by 2,5-dimethyl-2,5-di-t-butyl peroxyhexane	66
SiH(C$_2$H$_5$)$_3$	SiH$_2$(C$_2$H$_5$)$_2$, (58%); Si(C$_2$H$_5$)$_4$, (98%)	AlCl$_3$ as catalyst	62
SiH$_3$(C$_6$H$_5$)	Si(C$_6$H$_5$)$_4$, SiH$_4$	AlCl$_3$ as catalyst	63
SiH$_3$(C$_6$H$_5$)	SiH$_2$(C$_6$H$_5$)$_2$, SiH(C$_6$H$_5$)$_3$, Si(C$_6$H$_5$)$_4$	Na or K as catalyst	64
SiH$_3$(C$_6$H$_5$)	Si(C$_6$H$_5$)$_4$, SiH(C$_6$H$_5$)$_3$	14 hr at r.t., Na/K alloy as catalyst	65
SiH$_2$(C$_6$H$_5$)Cl	SiH$_4$, SiH$_3$Cl, Si(C$_6$H$_5$)$_2$Cl$_2$, Si(C$_6$H$_5$)$_3$Cl		63
SiH$_2$(C$_6$H$_5$)$_2$	Si(C$_6$H$_5$)$_4$, SiH(C$_6$H$_5$)$_3$	8 hr at r.t., Na/K alloy as catalyst	65
SiH$_2$(C$_6$H$_5$)$_2$	SiH(C$_6$H$_5$)$_3$, Si(C$_6$H$_5$)$_4$	Na or K as catalyst	64
SiH$_2$(C$_6$H$_5$)$_2$	SiH(C$_6$H$_5$)$_3$, SiH$_3$(C$_6$H$_5$), SiH$_4$	Catalyzed by radical initiators at 130°	66
SiH$_2$(C$_6$H$_5$)$_2$	SiH$_3$(C$_6$H$_5$), SiH$_2$(C$_6$H$_5$)$_2$, SiH(C$_6$H$_5$)$_3$, Si(C$_6$H$_5$)$_4$	24 hr at 100–160°; Pt compounds as catalyst	59
SiH$_2$(C$_6$H$_5$)$_2$	SiH$_4$, SiH$_3$(C$_6$H$_5$), SiH$_2$(C$_6$H$_5$)$_2$, SiH(C$_6$H$_5$)$_3$, Si(C$_6$H$_5$)$_4$	At 230° to 300°	59
SiH(C$_6$H$_4$CH$_3$-p)$_3$	SiH$_2$(C$_6$H$_4$CH$_3$-p)$_2$, (7%), [Si(C$_6$H$_4$CH$_3$-p)]$_4$, (35%)	12 hr at 240°, H$_2$PtCl$_6$ as catalyst	60
SnH(CH$_3$)$_3$/Sn(CH=CH$_2$)(C$_2$H$_5$)$_3$	Sn(CH$_3$)$_3$(CH=CH$_2$), SnH(C$_2$H$_5$)$_3$		67
SnH(n-C$_4$H$_9$)$_3$/Sn(CH=CHCH$_3$)(C$_2$H$_5$)$_2$	Sn(C$_4$H$_9$)$_3$(CH=CHCH$_3$), SnH$_2$(C$_2$H$_5$)$_2$		67

TABLE IV

Redistribution Reactions Involving Exchange of Element–Hydrogen with Element–Nitrogen Bonds

Reactants	Products	Remarks	References
$B_2H_6/B[N(CH_3)_2]_3$	$B_2H_5[N(CH_3)_2]$, $BH_2[N(CH_3)_2]$		69
$B_2H_6/BH[N(CH_3)_2]_2$	$B_2H_5[N(CH_3)_2]$, $BH_2[N(CH_3)_2]$		69
$BH_2[N(CH_3)_2]$	$BH[N(CH_3)_2]_2$; $B_2H_5[N(CH_3)_2]$, $(K = 0.102\ [atm^{-1}]$ at $107°$	At $80°$, also in the presence of $N(CH_3)_3$	69
$BH[N(CH_3)_2]_2$	$B[N(CH_3)_2]_3$, B_2H_6	At $140°$, in the presence of $N(CH_3)_3$	69
$AlH_3/Al[N(CH_3)_2]_3$	$AlH_2[N(CH_3)_2]$	2 hr at $80°$	70
$(SiH_3)_2NH$	SiH_4, $[SiH_2(NH)]_n$		71
$(SiH_3)_3N$	$[SiH_2NSiH_3]_3$, SiH_4	Base catalyzed	72,73
$Si_2H_5[N(CH_3)_2]$	$Si_2H_4[N(CH_3)_2]_2$, Si_2H_6	Base catalyzed	72
$SnH(CH_3)_3/Sn[N(C_2H_5)_2](C_2H_5)_3$	$Sn[N(C_2H_5)_2](CH_3)_3$, $SnH(C_2H_5)_3$	At $40°$	76
$SnH(CH_3)_3/Sn[N(C_6H_5)_2](C_2H_5)_3$	$Sn[N(C_6H_5)_2](CH_3)_3$, $SnH(C_2H_5)_3$	At $40°$	76
$SnH(CH_3)_3/Sn[N(C_6H_5)C(O)H](C_2H_5)_3$	$Sn[N(C_6H_5)C(O)H](CH_3)_3$, $SnH(C_2H_5)_3$	At $40°$	76
$SnH(C_2H_5)_3/Sn[N(C_2H_5)_2]_4$	$Sn[N(C_2H_5)_2](C_2H_5)_3$, Sn, $2\ H_2$	At $65°$	75
$SnH(C_6H_5)_3/Sn[N(C_6H_5)C(O)H](C_2H_5)_3$	$Sn[N(C_6H_5)C(O)H](C_6H_5)_3$, $SnH(C_2H_5)_3$	At $40°$	76

TABLE V

Redistribution Reactions Involving Exchange of Element–Hydrogen with Element–Phosphorus Bonds

Reactants	Products	Remarks	References
$SnH(CH_3)_3/Sn[P(C_6H_5)_2](C_2H_5)_3$	$Sn[P(C_6H_5)_2](CH_3)_3$, $SnH(C_2H_5)_3$	At $40°$	76
$SnH(C_6H_5)_3/Sn[P(C_6H_5)_2](C_2H_5)_3$	$Sn[P(C_6H_5)_2](C_6H_5)_3$, $SnH(C_2H_5)_3$	At $40°$	76
$SnH(C_6H_5)_3/Sn[P(n\text{-}C_4H_9)_2](n\text{-}C_4H_9)_3$	$Sn[P(n\text{-}C_4H_9)_2](C_6H_5)_3$, $SnH(n\text{-}C_4H_9)_3$	2 hr at $23°$	74

TABLE VI

Redistribution Reactions Involving Exchange of Element–Hydrogen with Element–Oxygen Bonds

Reactants	Products	Remarks	References
$B_2H_6/B(OCH_3)_3$	$BH(OCH_3)_2$	r.t.	77,78
$B_2H_6/[(CH_2O)_2BOCH_2]_2$	$BH(OCH_2)_2$ $(K = 0.4)$		79,80
$B_2H_6/$ [cyclic structure: $BO(CH_2)_3OB$]	$BH(OCH_2CH_2CH_2O)$ $(K = 11)$		80
$[BH(n\text{-}C_3H_7)_2]_2/B(OCH_3)_3$	$B(n\text{-}C_3H_7)(OCH_3)_2$, (89%)	2 hr reflux. Also B—C/ B—O exchange	81
$B_2H_6/B(OCH_3)(n\text{-}C_4H_9)_2$	$[BH(n\text{-}C_3H_7)_2]_2$, $B(OCH_3)_2(n\text{-}C_4H_9)$, (68%)	Also B—H/B—C exchange	46
$[BH(n\text{-}C_4H_9)_2]_2/B(OCH_3)_3$	$B(n\text{-}C_4H_9)(OCH_3)_2$, (73%)	2 hr at 70–112°. Also B—C/B—O exchange	46
$B_2H_6/B(OC_4H_9\text{-}n)_2(C_6H_5)$	$[BH_2(C_6H_5)]_2$, (40%)	In diethyl ether	48
$B_2H_6/B(OC_4H_9\text{-}n)(C_6H_5)_2$	$[BH_2C_6H_5)]_2$, (71%)	Also B—C/B—H exchange	48
$B_2H_6/B(OC_4H_9\text{-}n)(\alpha\text{-}C_{10}H_7)$	$[BH_2(\alpha\text{-}C_{10}H_7)]_2$, (81%)	Also B—C/B—H exchange	48
$[BH(n\text{-}C_3H_7)_2]_2/B(OC_4H_9\text{-}n)_3$	$B(n\text{-}C_3H_7)(OC_4H_9\text{-}n)_2$, (89%)	25 hr at 100°. Also B—C/B—O exchange	81
$(BH_2C_6H_5)_2/B(OC_4H_9\text{-}n)_3$	$BC_6H_5(OC_4H_9\text{-}n)_2$, (44%); $B(C_6H_5)_2$-$(C_4H_9O\text{-}n)$, $B(C_6H_5)_3$, B_2H_6		49
$B_2H_6/B(OC_4H_9\text{-}i)(C_6H_4Cl\text{-}p)$	$[BH_2(C_6H_4Cl\text{-}p)]_2$, (79%)	Also B—C/B—H exchange	48
$B_2H_6/B(OC_4H_9\text{-}i)_2(C_6H_4CH_3\text{-}o)$	$[BH_2(C_6H_4CH_3\text{-}o)]_2$, (51%)	Also B—C/B—H exchange	47
$B_2H_6/B(OC_4H_9\text{-}i)_2(C_6H_4Cl\text{-}p)$	$[BH_2(C_6H_4Cl\text{-}p)]_2$, (50%)	In diethyl ether	48
$B_2H_6/B(OC_4H_9\text{-}i)_2(\alpha\text{-}C_{10}H_7)$	$[BH_2(\alpha\text{-}C_{10}H_7)]_2$, (55%)	In diethyl ether	48
$B_2H_6/B(n\text{-}C_3H_7)_2(OC_6H_{13}\text{-}n)$	$[BH(n\text{-}C_3H_7)_2]$, (43%); $B(n\text{-}C_3H_7)(OC_6H_{13}\text{-}n)_2$, (60%)	Also B—H/B—C exchange	46
$B_2H_6/B(n\text{-}C_3H_7)_2(OC_7H_{15}\text{-}n)$	$[BH(n\text{-}C_3H_7)_2]_2$, $B(n\text{-}C_3H_7)(OC_7H_{15}\text{-}n)_2$, (78%)	Also B—H/B—C exchange	46

52

[BH(n-C$_3$H$_7$)$_2$]$_2$/B(OC$_7$H$_{15}$-n)$_3$	B(n-C$_3$H$_7$)(OC$_7$H$_{15}$)$_2$, (80%)	2.5 hr at 100°, also B—C/B—O exchange	81
[AlH$_4$]Li/[Al(OCH$_3$)$_4$]Li	LiAlH(OCH$_3$)$_3$	At r.t.	82
[AlH$_4$]Li/[Al(OC$_2$H$_5$)$_4$]Li	LiAlH$_n$(OC$_2$H$_5$)$_{4-n}$	At r.t.	82
[AlH$_4$]Li/[Al(OC$_3$H$_7$-i)$_4$]Li	LiAlH$_n$(OC$_3$H$_{7-n}$)$_{4-n}$	At r.t.	82
(SiH$_3$)$_2$O	SiH$_4$, [SiH$_2$O]$_n$	In the presence of P$_2$O$_5$	71
SiH$_2$(C$_6$H$_5$)(CH$_3$)/Si(OCH$_3$)$_2$(C$_6$H$_5$)(CH$_3$)	SiH$_2$(OCH$_3$)(C$_6$H$_5$)(CH$_3$), ($K = 27.0$)	C$_4$H$_9$Li as catalyst	85
SiH$_3$(OC$_2$H$_5$)	SiH$_4$, SiH$_2$(OC$_2$H$_5$)$_2$	At r.t.	87
SiH$_2$(OCH$_3$)$_2$	SiH$_3$(OCH$_3$), SiH(OCH$_3$)$_3$	At r.t.	87
SiH$_2$(OC$_2$H$_5$)(CH$_3$)	SiH$_3$(CH$_3$), SiH(OC$_2$H$_5$)$_2$(CH$_3$)	At r.t.	87
SiH(OC$_2$H$_5$)$_3$	Si(OC$_2$H$_5$)$_4$, SiH$_4$	Refluxing with Na	83
SiH(OC$_3$H$_7$-n)$_3$	Si(OC$_3$H$_{7-n}$)$_4$, (20%)	20 hr reflux, AlCl$_3$ as catalyst	80
[SiH$_2$(C$_6$H$_4$Br-m)]$_2$O	SiH$_3$(m-C$_6$H$_4$Br), [Si(m-C$_6$H$_4$Br)O$_{1.5}$]	Base catalyzed	84
[SiH$_2$(C$_6$H$_4$Cl-p)]$_2$O	SiH$_3$(p-C$_6$H$_4$Cl) [Si(p-C$_6$H$_4$Cl)O$_{1.5}$]	Base catalyzed	84
SnH$_2$(C$_2$H$_5$)$_2$/[Sn(C$_2$H$_5$)$_3$]$_2$O	Sn(C$_2$H$_5$)$_2$O, SnH(C$_2$H$_5$)$_3$	1 hr at 20°	75
SnH$_2$(C$_2$H$_5$)$_2$/[Sn(n-C$_4$H$_9$)$_3$]$_2$O	Sn(C$_2$H$_5$)$_2$O, SnH(n-C$_4$H$_9$)$_3$	At 25°	90
SnH$_2$(C$_2$H$_5$)$_2$/[Sn(i-C$_4$H$_9$)$_3$]$_2$O	Sn(C$_2$H$_5$)$_2$O, SnH(i-C$_4$H$_9$)$_3$	At 25°	74,90
SnH$_2$(n-C$_4$H$_9$)$_2$/Sn(OOCCH$_3$)$_2$(n-C$_4$H$_9$)$_2$	SnH(OOCCH$_3$)(n-C$_4$H$_9$)$_2$		88,89
SnH$_2$(n-C$_4$H$_9$)$_2$/[Sn(n-C$_4$H$_9$)$_3$]$_2$O	Sn(n-C$_4$H$_9$)$_2$O, SnH(n-C$_4$H$_9$)$_3$	At r.t.	91
SnH$_2$(i-C$_4$H$_9$)$_2$/Sn(n-C$_4$H$_9$)$_3$]$_2$O	SnH$_2$(i-C$_4$H$_9$)$_2$, SnH(n-C$_4$H$_9$)$_3$	At 25°	90
SnH$_2$(C$_6$H$_5$)$_2$/[Sn(C$_6$H$_5$)$_3$]$_2$O	Sn(C$_6$H$_5$)$_2$O, SnH(C$_6$H$_5$)$_3$	4 hr at 20°	75
SnH(CH$_3$)$_3$/Sn(OCH$_3$)(C$_2$H$_5$)$_3$	Sn(OCH$_3$)(CH$_3$)$_3$, SnH(C$_2$H$_5$)$_3$	At 40°	74
SnH(CH$_3$)$_3$/Sn(OC$_6$H$_5$)(C$_2$H$_5$)$_3$	Sn(OC$_6$H$_5$)(CH$_3$)$_3$, SnH(C$_2$H$_5$)$_3$	At 40°	74
SnH(i-C$_4$H$_9$)$_3$/[Sn(C$_2$H$_5$)$_3$]$_2$O	[Sn(i-C$_4$H$_9$)$_3$]$_2$O, SnH(C$_2$H$_5$)$_3$	At 60°	90
SnH(C$_6$H$_5$)$_3$/Sn(OC$_6$H$_5$)(C$_2$H$_5$)$_3$	Sn(OC$_6$H$_5$)(C$_6$H$_5$)$_3$, SnH(C$_2$H$_5$)$_3$	At 40°	74
SnH(C$_6$H$_5$)$_3$/Sn(OR)(C$_2$H$_5$)$_3$	Sn(OR)(C$_6$H$_5$)$_3$, SnH(C$_2$H$_5$)$_3$	At 40°	92

TABLE VII
Redistribution Reactions Involving Exchange of Element–Hydrogen with Element–Sulfur Bonds

Reactants	Products	Remarks	References
[BH$_2$(SCH$_3$)]$_n$	B$_2$H$_6$, B(SCH$_3$)$_3$	At 140°	93
(SiH$_3$)$_2$S	H$_3$SiS[SiH$_2$S]$_n$SiH$_3$, SiH$_4$	Catalyzed by NH$_3$ or N(CH$_3$)$_3$	72
SnH$_2$(i-C$_4$H$_9$)$_2$/[Sn(n-C$_4$H$_9$)$_3$]$_2$S	[Sn(i-C$_4$H$_9$)$_2$S]$_3$, SnH(n-C$_4$H$_9$)$_3$	4 hr at 100°	75

TABLE VIII
Redistribution Reactions Involving Exchange of Element–Hydrogen with Element–Halogen Bonds

Reactants	Products	Remarks	References
CaH$_2$/CaCl$_2$	CaHCl	At 900° in H$_2$ atm.	94
CaH$_2$/CaBr$_2$	CaHBr	At 900° in H$_2$ atm.	95
CaH$_2$/CaI$_2$	CaHI	At 900° in H$_2$ atm.	96
SrH$_2$/SrCl$_2$	SrHCl	At 900° in H$_2$ atm.	94
SrH$_2$/SrBr$_2$	SrHBr	At 900° in H$_2$ atm.	95
SrH$_2$/SrI$_2$	SrHI	At 900° in H$_2$ atm.	96
BaH$_2$/BaCl$_2$	BaHCl	At 900° in H$_2$ atm.	94
BaH$_2$/BaBr$_2$	BaHBr	At 900° in H$_2$ atm.	95
BaH$_2$/BaI$_2$	BaHI	At 900° in H$_2$ atm.	96
B$_2$H$_6$/BF$_3$	BHF$_2$, (30–35%)	0.5 hr at 250°	98
B$_2$H$_6$/BCl$_3$	B$_2$H$_5$Cl	At r.t.	99
B$_2$H$_6$/BCl$_3$	BHCl$_2$·R$_2$O	R = C$_2$H$_5$	103
B$_2$H$_6$/BCl$_3$	BHCl$_2$·OR$_2$, BH$_2$Cl·OR$_2$	In diglyme	102
[BH$_3$·N(C$_2$H$_5$)$_3$]/BCl$_3$	BH$_2$Cl·N(C$_2$H$_5$)$_3$, BHCl$_2$·N(C$_2$H$_5$)$_3$	At 0°	104
B$_3$N$_3$H$_6$]/BCl$_3$	[B$_3$N$_3$H$_5$Cl], [B$_3$N$_3$H$_4$Cl$_2$]	At r.t.	105

54

B_2H_6/$B(C_6H_5)_2Cl$		Also B—C/B—H exchange	49
[$B(C_6H_5)_3H$]Na/	[$B(C_6H_5)_2H_2$]$_2$, (60%)		106
B_2H_6/BBr_3	B_2H_5Br	At r.t.	108
$BH_3 \cdot N(CH_3)_3$/BBr_3	$BH_2Br \cdot N(CH_3)_3$, (80%); $BHBr_2 \cdot N(CH_3)_3$, (85%)		109
[$B_3N_3H_6$]/BBr_3	[$B_3N_3H_5Br$], [$B_3N_3H_4Br_2$]	At r.t.	105
AlH_3/$AlCl_3$	AlH_2Cl, $AlHCl_2$	As tetrahydrofurane or diethyl ether adducts	113,114
AlH_3/$AlCl_3$	AlH_2Cl, $AlHCl_2$	As mono- and bis-trimethylaminates	110
[AlH_4]Li/$AlCl(CH_3)_2$	$AlH(CH_3)_2$, $AlH_2(CH_3)$	As $N(CH_3)_3$ adducts	54
[AlH_4]Li/$AlCl(C_2H_5)_2$	$AlH(C_2H_5)_2$, $AlH_2(C_2H_5)$	As $N(CH_3)_3$ adducts	54
AlH_3/$AlBr_3$	AlH_2Br, $AlHBr_2$	As tetrahydrofurane or diethyl ether adducts	112,114
AlH_3/$AlBr_3$	AlH_2Br, $AlHBr_2$	As mono- and bis-trimethylaminates	110
AlH_3/AlI_3	AlH_2I, $AlHI_2$	As tetrahydrofurane or diethyl ether adducts	111,114
AlH_3/AlI_3	AlH_2I, $AlHI_2$	As mono- and bis-trimethylaminates	110
$SiH_2(C_6H_5)(CH_3)$/$SiF_2(C_6H_5)(CH_3)$	$SiHF(C_6H_5)(CH_3)$, ($K = 150$)	At 25°, $N(C_4H_9)_4F$ as catalyst	118

(continued)

TABLE VIII (*Continued*)

Reactants	Products	Remarks	Reference
SiH₂(CH₃)(C₆H₅)/SiF₂(CH₃)(C₆H₅)	SiHF(CH₃)(C₆H₅), (K = 43.5)	At 100°. (C₄H₉)₄NF as catalyst	85
SiH₄/SiH₂Cl₂	SiH₃Cl	7 d at 100°, AlCl₃ as catalyst	119
SiH₃(CH₃)/SiCl₃(CH₃)	SiClH₂(CH₃), (K = 0.48); SiCl₂H(CH₃), (K = 0.041)	At 100°, N(C₄H₉-n)₄Cl as catalyst	127
SiH₃(C₆H₅)/SiCl₃(C₆H₅)	SiClH₂(C₆H₅), (17%); SiCl₂H(C₆H₅), (26%)	18 hr at 100°, (C₄H₉)₄NCl as catalyst	126
SiH₂(CH₃)₂/SiCl₂(CH₃)₂	SiClH(CH₃)₂, (K = 0.098)	< 23 hr at 100°, N(C₄H₉-n)₄Cl as catalyst	127
SiH₂(CH₃)(C₆H₅)/SiCl₂(CH₃)(C₆H₅)	SiClH(CH₃)(C₆H₅), (K = 3.3)	(C₄H₉)₄NCl as catalyst	85
SiHCl₃	SiCl₂H₂, (11–14%); SiCl₄, (11–14%)	1 hr at 25°, (C₄H₉)₄NCl as catalyst	126
SiHCl₂(CH₃)	SiClH₂(CH₃), (11–18%); SiCl₃(CH₃), (11–18%)	1 hr at 100°, (C₄H₉)₄NCl as catalyst	126
SiHCl₂(CH₃)/SiClH(CH₃)₂	SiCl₂(CH₃)₂, (32%); SiCl₃(CH₃), (3%); SiH₃(CH₃), (7%); SiClH₂(CH₃), (22%)	20 hr at 100°, (C₄H₉)₄NCl as catalyst	126
SiHCl(CH₃)₂	SiH₂(CH₃)₂, (17–20%); SiCl₂(CH₃)₂, (17–20%)	6 hr at 100°, (C₄H₉)₄NCl as catalyst	126
SiHCl(CH₃)₂/SiCl₃(C₂H₄CF₃)	SiCl₂H(C₂H₄CF₃), (33%); SiCl₂(CH₃)₂, (> 36%)	24 hr reflux, (C₄H₉)₄NCl as catalyst	126
SiHCl(CH₃)₂/SiCl₃(C₆H₅)	SiCl₂H(C₆H₅), (22%); SiCl₂(CH₃)₂, (> 30%)	24 hr reflux, (C₄H₉)₄NCl as catalyst	126
SiH(CH₃)(C₂H₅)₂/SiCl₃(n-C₄H₉)	SiH₃(n-C₄H₉), SiH(CH₃)₂(C₂H₅), SiCl(CH₃)₂(C₂H₅), SiCl(CH₃)(C₂H₅)₂, SiCl(C₂H₅)₃	At 75–44°, AlCl₃ as catalyst	62

SiH(CH3)2(C6H5)/SiCl3(CH3)	SiCl(CH3)2(C6H5), (>46%); SiCl2H(CH3), (38%); SiClH2(CH3), (8%)	24 hr at 100°, (C4H9)4NCl as catalyst	126
SiHCl(CH3)(n-C3H7)	SiH2(CH3)(n-C3H7), (57%); SiCl2(CH3)(n-C3H7), (79%)	AlCl3 as catalyst	123
SiHCl(CH3)(i-C3H7)	SiH2(CH3)(i-C3H7), (48%); SiCl2(CH3)(i-C3H7)	AlCl3 as catalyst	123
SiHCl2(C2H5)	SiClH2(C2H5), SiCl3(C2H5)	AlCl3 as catalyst	122
SiHCl(C2H5)2	SiH2(C2H5)2, SiCl2(C2H5)2	AlCl3 as catalyst	122,123
SiH(C2H5)3/SiCl2(C2H5)2	SiH2(C2H5)2, (26%); SiCl(C2H5)3, (98%)	AlCl3 as catalyst	123,125
SiH(C2H5)3/SiCl2(C2H5)2	SiH2(C2H5)2, (43%); SiClH(C2H5)2, (44%)	AlCl3 as catalyst	121
SiH(C2H5)3/SiCl2(C2H5)2	SiH2(C2H5)2, (53%)	AlCl3 as catalyst	124
SiH(C2H5)3/SiCl2(CH3)(n-C3H7)	SiH2(CH3)(n-C3H7), (62%)	AlCl3 as catalyst	124
SiH(C2H5)3/SiCl2(CH3)(n-C3H7)	SiH2(CH3)(n-C3H7), (62%); SiCl(C2H5)3, (98%)	AlCl3 as catalyst	124,125
SiH(C2H5)3/SiCl3(n-C4H9)	SiH3(n-C4H9), (98%); SiCl(C2H5)3, (94%)	AlCl3 as catalyst	124,125
SiH(C2H5)3/SiCl3(i-C4H9)	SiH3(i-C4H9), (98%); SiCl(C2H5)3, (97%)	AlCl3 as catalyst	124,125
SiH(C2H5)3/SiCl3(s-C4H9)	SiH3(s-C4H9), (98%); SiCl(C2H5)3, (94%)	AlCl3 as catalyst	124,125
SiHCl2(n-C4H9)	SiClH2(n-C4H9), SiCl3(n-C4H9)	AlCl3 as catalyst	122
SiH(C2H5)3/SiBr2(C2H5)2	SiH2(C2H5)2, (59.2%); SiBr(C2H5)3, (89%)	AlCl3 as catalyst	124,125
SnH2(n-C4H9)2/SnF2(n-C4H9)2	SnHF(n-C4H9)2, (70%)	At r.t.	131
SnH2(CH3)2/SnCl2(CH3)2	SnClH(CH3)2	At r.t.	133
SnH2(C2H5)2/SnCl2(C2H5)2	SnClH(C2H5)2	At r.t.	133
SnH2(n-C3H7)2/SnCl2(n-C3H7)2	SnClH(n-C3H7)2	At r.t.	133
SnH2(n-C4H9)2/SnCl2(n-C4H9)2	SnClH(n-C4H9)2, (100%)	At r.t.	129,130, 131,133

(continued)

57

TABLE VIII (Continued)

Reactants	Products	Remarks	References
SnH₂(i-C₄H₉)₂/SnCl₂(i-C₄H₉)₂	SnClH(i-C₄H₉)₂	At r.t.	133
SnH₂(CH₃)₂/SnBr₂(CH₃)₂	SnBrH(CH₃)₂	At r.t.	133
SnH₂(C₂H₅)₂/SnB_2(C₂H₅)₂	SnBrH(C₂H₅)₂	At r.t.	133
SnH₂(n-C₃H₇)₂/SnBr₂(n-C₃H₇)₂	SnBrH(n-C₃H₇)₂	At r.t.	133
SnH₂(n-C₄H₉)₂/SnBr₂(n-C₄H₉)₂	SnBrH(n-C₄H₉)₂, (100%)	At r.t.	131,133
SnH₂(i-C₄H₉)₂/SnBr₂(i-C₄H₉)₂	SnBrH(i-C₄H₉)₂	At r.t.	133
SnH₂(CH₃)₂/SnI₂(CH₃)₂	SnIH(CH₃)₂	At r.t.	133
SnH₂(C₂H₅)₂/SnI₂(C₂H₅)₂	SnIH(C₂H₅)₂	At r.t.	133
SnH₂(n-C₃H₇)₂/SnI₂(n-C₃H₇)₂	SnIH(n-C₃H₇)₂	At r.t.	133
SnH₂(n-C₄H₉)₂/SnI₂(n-C₄H₉)₂	SnIH(n-C₄H₉)₂	At r.t.	131,133
SnH₂(i-C₄H₉)₂/SnI₂(i-C₄H₉)₂	SnIH(i-C₄H₉)₂	At r.t.	133
SnH(n-C₄H₉)₃/SnF₂(n-C₄H₉)₂	SnF(n-C₄H₉)₃, SnFH(n-C₄H₉)₂	At r.t.	132
SnH(n-C₄H₉)₃/SnF₂(n-C₄H₉)₂	SnF(n-C₄H₉)₃, SnH₂(n-C₄H₉)₂	At r.t.	132
SnH(n-C₄H₉)₃/SnCl₃(n-C₄H₉)	SnCl₂H(n-C₄H₉), SnClH₂(n-C₄H₉), SnH₃(n-C₄H₉), SnCl(n-C₄H₉)₃	At r.t.	132
SnH(n-C₄H₉)₃/SnCl₂(n-C₄H₉)₂	SnCl(n-C₄H₉)₃, SnClH(n-C₄H₉)₂	At r.t.	132
SnH(n-C₄H₉)₃/SnClH(n-C₄H₉)₂	SnCl(n-C₄H₉)₃, SnH₂(n-C₄H₉)₂	At r.t.	132
SnH(n-C₄H₉)₃/SnBr₂(n-C₄H₉)₂	SnBr(n-C₄H₉)₃, SnBrH(n-C₄H₉)₂	At r.t.	132
SnH(n-C₄H₉)₃/SnBrH(n-C₄H₉)₂	SnBr(n-C₄H₉)₃, SnH₂(n-C₄H₉)₂	At r.t.	132
SnH(n-C₄H₉)₃/SnI₂(n-C₄H₉)₂	SnI(n-C₄H₉)₃, SnIH(n-C₄H₉)₂	At r.t.	132
SnH(n-C₄H₉)₃/SnIH(n-C₄H₉)₂	SnI(n-C₄H₉)₃, SnH₂(n-C₄H₉)₂	At r.t.	132
SnH(i-C₄H₉)₃/SnCN(C₂H₅)₃	SnCN(i-C₄H₉)₃, SnH(C₂H₅)₃	10 hr at 150°	74
SnH(C₆H₅)₃/SnCN(C₂H₅)₃	SnCN(C₆H₅)₃, SnH(C₂H₅)₃	At 20°	74

58

TABLE IX

Redistribution Reactions Involving Exchange of Element–Carbon with Element–Carbon Bonds

Reactants	Products	Remarks	References
$LiCH_3/LiC_2H_5$	$(LiCH_3)_n(LiC_2H_5)_m$, (r)	X-ray, ^1H, ^7Li NMR	135,136
$LiCH_3/LiC_6H_5$	$[Li_2(CH_3)(C_6H_5)]_n$, $Li_4(CH_3)_3(C_6H_5)$	^1H, ^7Li NMR	139
$LiC_2H_5/LiC(CH_3)_3$	$(LiC_2H_5)_n[LiC(CH_3)_3]_m$	^7Li NMR	134
$LiC(CH_3)_3/LiCH_2Si(CH_3)_3$	$[LiC(CH_3)_3]_n[LiCH_2Si(CH_3)_3]_m$, (r)	^1H, ^7Li NMR	138
$Mg(CH_3)_2/Mg(C_5H_5)_2$	$Mg(CH_3)(C_5H_5)$	Rapid exchange	140
$Mg(CH_3)_2/Mg(C_6H_5)_2$	$Mg(CH_3)(C_6H_5)$	Rapid exchange	140
$Mg(C_5H_5)_2/Mg[CH_2C(CH_3)_3]_2$	$Mg(C_5H_5)[CH_2C(CH_3)_3]$	Rapid exchange	140
$Mg(C_5H_5)_2/Mg(C_6H_5)_2$	$Mg(C_5H_5)(C_6H_5)$	Rapid exchange	140
$B(CH_3)_2(CHCH_2)$	$B(CH_3)_3$, $B(CHCH_2)_3$	Slow at room temp.	151
$B(CH_3)(CHCH_2)_2$	$B(CH_3)_3$, $B(CHCH_2)_3$	Slow at room temp.	151
$B(C_2H_5)_3/B(n-C_3H_7)_3$	$B(C_2H_5)(n-C_3H_7)_2$, $B(C_2H_5)_2(n-C_3H_7)$	$Al(C_2H_5)_3$ as catalyst	141
$B(n-C_3H_7)(n-C_4H_9)_2$	$B(n-C_3H_7)_3$, $B(n-C_4H_9)_3$	150–165°	152
$[B(CH_3)(C_6H_5)_3]Li$	$[B(C_6H_5)_4]Li$	In H_2O	153
$[B(C_2H_5)(C_{10}H_7)_3]Li$	$[B(C_{10}H_7)_4]Li$	In H_2O	153
$Al(CH_3)_3/Al(C_6H_5)_3$	$Al(CH_3)_2(C_6H_5)$, (73%)		161
$Al(C_2H_5)_3/Al(C_6H_5)_3$	$Al(C_2H_5)_2(C_6H_5)$		161
$Al(i-C_4H_9)_3/Al(C_6H_5)_3$	$Al(i-C_4H_9)_2(C_6H_5)$, (80%)		161
$Al(i-C_4H_9)_3/Al(C_6H_5)_3$	$Al(i-C_4H_9)(C_6H_5)_2$, (58%)		161
$Si(CH_3)_3(C_2H_5)$	$Si(CH_3)_4$, $Si(CH_3)_2(C_2H_5)_2$, $Si(CH_3)(C_2H_5)_3$, $Si(C_2H_5)_4$	At 80–100°, $AlCl_3$ as catalyst	10,168, 172
$Si(CH_3)_3(C_2H_5)$	$Si(CH_3)_4$, $Si(CH_3)_3(C_2H_5)$, $Si(CH_3)_2(C_2H_5)_2$	30 hr at 280°	169
$Si(CH_3)_3(C_6H_5)$	$Si(CH_3)_4$, $Si(CH_3)_2(C_6H_5)_2$	48 hr at 300°	169
$Si(C_2H_5)_4/Si(n-C_3H_7)_4$	$Si(C_2H_5)(n-C_3H_7)_3$, (r); $Si(C_2H_5)_2(n-C_3H_7)_2$, (r); $Si(C_2H_5)_3(n-C_3H_7)$, (r)	$AlCl_3$ as catalyst	12,15,170

(continued)

59

TABLE IX (Continued)

Reactants	Products	Remarks	References
$Si(C_2H_5)_3(n\text{-}C_3H_7)$	$Si(C_2H_5)_4$, $Si(C_2H_5)_2(n\text{-}C_3H_7)_2$	48 hr at 300°	169
$Si(C_2H_5)_3(i\text{-}C_4H_9)$	$Si(C_2H_5)_4$, $Si(C_2H_5)_2(i\text{-}C_4H_9)_2$	48 hr at 300°	169
$Si(C_2H_5)_3(i\text{-}C_5H_{11})$	$Si(C_2H_5)_4$, $Si(C_2H_5)_2(i\text{-}C_5H_{11})_2$	48 hr at 300°	169
$Si(C_2H_5)(C_6H_5)_3$	$Si(C_6H_5)_4$, $Si(C_2H_5)_2(C_6H_5)_2$	25 hr at 300°	169
$Si(C_2H_5)_3(C_6H_5)$	$Si(C_6H_5)_2(C_2H_5)_2$, $Si(C_2H_5)_4$	At 175° in the presence of $ZnCl_2$	167
$Ge(CH_3)_4/Ge(C_2H_5)_4$	$Ge(CH_3)_3(C_2H_5)$, (r); $Ge(CH_3)_2(C_2H_5)_2$, (r); $Ge(CH_3)(C_2H_5)_3$, (r)	2 hr at 120°; $AlCl_3$	172
$Ge_2(CH_3)_6/Ge_2(C_2H_5)_6$	$(CH_3)_3GeGe(C_2H_5)_3$ (K = 0.12)	At r.t.; base catalyzed	174
$Ge(CH_3)_4/Ge(n\text{-}C_4H_9)_4$	$Ge(CH_3)_3(n\text{-}C_4H_9)$, (r); $Ge(CH_3)_2(n\text{-}C_4H_9)_2$, (r); $Ge(CH_3)(n\text{-}C_4H_9)_3$, (r)	6 hr at 120°; $AlCl_3$ as catalyst	172
$Ge(C_2H_5)_4/Ge(n\text{-}C_3H_7)_4$	$Ge(C_2H_5)_3(n\text{-}C_3H_7)$, (r); $Ge(C_2H_5)_2(n\text{-}C_3H_7)_2$, (r); $Ge(C_2H_5)(n\text{-}C_3H_7)_3$, (r)	6 hr at 120°; $AlCl_3$ as catalyst	172
$Ge(C_2H_5)_4/Ge(n\text{-}C_3H_7)_4$	$Ge(C_2H_5)_3(n\text{-}C_3H_7)$, (r); $Ge(C_2H_5)_2(n\text{-}C_3H_7)_2$, (r); $Ge(C_2H_5)(n\text{-}C_3H_7)_3$, (r)	$AlCl_3$ as catalyst	173
$Ge(C_2H_5)_4/Ge(n\text{-}C_4H_9)_4$	$Ge(C_2H_5)_3(n\text{-}C_4H_9)$, (r); $Ge(C_2H_5)_2(n\text{-}C_4H_9)_2$, (r); $Ge(C_2H_5)(n\text{-}C_4H_9)_3$, (r)	6 hr at 120°; $AlCl_3$ as catalyst	172
$Ge(C_2H_5)_4/Ge(n\text{-}C_6H_{13})_4$	$Ge(C_2H_5)_3(n\text{-}C_6H_{13})$, (r); $Ge(C_2H_5)_2(n\text{-}C_6H_{13})_2$, (r); $Ge(C_2H_5)(n\text{-}C_6H_{13})_3$, (r)	3 min at 170°; $AlCl_3$ as catalyst	172
$Ge(n\text{-}C_3H_7)_4/Ge(n\text{-}C_5H_{11})_4$	$Ge(n\text{-}C_3H_7)_3(n\text{-}C_5H_{11})$, (r); $Ge(n\text{-}C_3H_7)_2(n\text{-}C_5H_{11})_2$, (r); $Ge(n\text{-}C_3H_7)(n\text{-}C_5H_{11})_3$, (r)	2 min at 170°; $AlCl_3$ as catalyst	172
$Ge(n\text{-}C_3H_7)_4/Ge(n\text{-}C_6H_{13})_4$	$Ge(n\text{-}C_3H_7)_3(n\text{-}C_6H_{13})$, (r); $Ge(n\text{-}C_3H_7)_2(n\text{-}C_6H_{13})_2$, (r); $Ge(n\text{-}C_3H_7)(n\text{-}C_6H_{13})_3$, (r)	3 min at 170°; $AlCl_3$ as catalyst	172
$Ge(n\text{-}C_4H_9)_4/Ge(n\text{-}C_5H_{11})_4$	$Ge(n\text{-}C_4H_9)_3(n\text{-}C_5H_{11})$, (r); $Ge(n\text{-}C_4H_9)_2(n\text{-}C_5H_{11})_2$, (r); $Ge(n\text{-}C_4H_9)(n\text{-}C_5H_{11})_3$, (r)	2 min at 170°; $AlCl_3$ as catalyst	172

Ge(n-C$_4$H$_9$)$_4$/Ge(n-C$_6$H$_{13}$)$_4$	Ge(n-C$_4$H$_9$)$_3$(n-C$_6$H$_{13}$), (r); Ge(n-C$_4$H$_9$)$_2$(n-C$_6$H$_{13}$)$_2$, (r); Ge(n-C$_4$H$_9$)(n-C$_6$H$_{13}$)$_3$, (r)	2 min at 170°; AlCl$_3$ as catalyst	172
Ge(n-C$_5$H$_{11}$)$_4$/Ge(n-C$_6$H$_{13}$)$_4$	Ge(n-C$_5$H$_{11}$)$_3$(n-C$_6$H$_{13}$), (r); Ge(n-C$_5$H$_{11}$)$_2$(n-C$_6$H$_{13}$)$_2$, (r); Ge(n-C$_5$H$_{11}$)(n-C$_6$H$_{13}$)$_3$, (r)	2 min at 170°; AlCl$_3$ as catalyst	172
Sn(CH$_3$)$_4$/Sn(C$_2$H$_5$)$_4$	Sn(CH$_3$)$_3$(C$_2$H$_5$), (r); Sn(CH$_3$)$_2$(C$_2$H$_5$)$_2$, (r); Sn(CH$_3$)(C$_2$H$_5$)$_3$, (r)	12 hr at 120°; AlCl$_3$ as catalyst; 5–7 hr at 60– 80°; AlCl$_3$	172 10,13,17
Sn(CH$_3$)$_4$/Sn(n-C$_3$H$_7$)$_4$	Sn(CH$_3$)$_3$(n-C$_3$H$_7$), (r); Sn(CH$_3$)$_2$(n-C$_3$H$_7$)$_2$, (r); Sn(CH$_3$)(n-C$_3$H$_7$)$_3$, (r)	2 hr at 160°; AlCl$_3$ as catalyst	172
Sn(CH$_3$)$_4$/Sn(n-C$_4$H$_9$)$_4$	Sn(CH$_3$)$_3$(n-C$_4$H$_9$), (r); Sn(CH$_3$)$_2$(n-C$_4$H$_9$)$_2$, (r); Sn(CH$_3$)(n-C$_4$H$_9$)$_3$, (r)	1 hr at 160°; AlCl$_3$ as catalyst	172
Sn(n-C$_3$H$_7$)$_4$/Sn(n-C$_4$H$_9$)$_4$	Sn(n-C$_3$H$_7$)$_3$(n-C$_4$H$_9$), (r); Sn(n-C$_3$H$_7$)$_2$(n-C$_4$H$_9$)$_2$, (r); Sn(n-C$_3$H$_7$)(n-C$_4$H$_9$)$_3$, (r)	2 min at 170°; AlCl$_3$ as catalyst	172
Pb(CH$_3$)$_4$/Pb(C$_2$H$_5$)$_4$	Pb(CH$_3$)$_3$(C$_2$H$_5$), (r); Pb(CH$_3$)$_2$(C$_2$H$_5$)$_2$, (r); Pb(CH$_3$)(C$_2$H$_5$)$_3$, (r)	5–7 hr at 60–80°; AlCl$_3$ as catalyst	10,12, 13
Pb(CH$_3$)$_3$Cl/Pb(C$_2$H$_5$)$_3$Cl	Pb(CH$_3$)$_2$(C$_2$H$_5$)Cl, (r); Pb(CH$_3$)(C$_2$H$_5$)$_2$Cl, (r)	5–7 hr at 60–80°; AlCl$_3$ as catalyst	10,17
Pb(CH$_3$)$_4$/Pb(n-C$_3$H$_7$)$_4$	Pb(CH$_3$)$_3$(n-C$_3$H$_7$); (r); Pb(CH$_3$)$_2$(n-C$_3$H$_7$)$_2$, (r); Pb(CH$_3$)(n-C$_3$H$_7$)$_3$, (r)	5 hr at 80°; AlCl$_3$ as catalyst	13
Pb(CH$_3$)$_4$/Pb(C$_6$H$_5$)$_4$	Pb(CH$_3$)$_3$(C$_6$H$_5$), (r); Pb(CH$_3$)$_2$(C$_6$H$_5$)$_2$, (r); Pb(CH$_3$)(C$_6$H$_5$)$_3$, (r)	5–7 hr at 60–80°; AlCl$_3$ as catalyst	10,13
Pb(C$_2$H$_5$)$_4$/Pb(n-C$_3$H$_7$)$_4$	Pb(C$_2$H$_5$)$_3$(n-C$_3$H$_7$), (r); Pb(C$_2$H$_5$)$_2$(n-C$_3$H$_7$)$_2$, (r); Pb(C$_2$H$_5$)(n-C$_3$H$_7$)$_3$, (r)	5 hr at 80°; AlCl$_3$ as catalyst	13
Pb(C$_6$H$_5$)$_4$/Pb(C$_6$H$_4$CH$_3$-p)$_4$	Pb(C$_6$H$_5$)$_3$(C$_6$H$_4$CH$_3$-p), (r); Pb(C$_6$H$_5$)$_2$(C$_6$H$_4$CH$_3$-p)$_2$, (r); Pb(C$_6$H$_5$)(C$_6$H$_4$CH$_3$-p)$_3$, (r)	5–7 hr at 60–80°; AlCl$_3$ as catalyst	10,13
Bi(C$_6$H$_5$)$_3$/Bi(α-C$_{10}$H$_7$)$_3$	Bi(C$_6$H$_5$)$_2$(C$_{10}$H$_7$-α)	2 hr at 190°	175

61

TABLE X

Redistribution Reactions Involving Exchange of Element–Carbon with Element–Nitrogen Bonds

Reactants	Products	Remarks	References
$B(C_6H_5)_3/B[N(CH_3)_2]_3$	$BC_6H_5[N(CH_3)_2]_2]_2$, $[K = (3 \pm 8) \times 10^{-3}]$; $B(C_6H_5)_2[N(CH_3)_2]$, $(K = (0.2 \pm 0.2)$	5 hr at 120°	176

TABLE XI

Redistribution Reactions Involving Exchange of Element–Carbon with Element–Oxygen Bonds

Reactants	Products	Remarks	References
$LiC_2H_5/LiOC_2H_5$	$[(LiC_2H_5)_n(LiOC_2H_5)_m]$, $(n + m = 4)$		136
$Be(CH_3)_2/Be(OC_2H_5)_2$	$[BeCH_3(OCH_3)]_2$		177
$B(CH_3)_3/B_2O_3$	$[B(CH_3)O]_3$, (95%)	6 hr at 300–330°	188
$B(CH_2)_4(i\text{-}C_4H_9)/B(OCH_3)_3$	$B(CH_2)_4(OCH_3)$, $B(i\text{-}C_4H_9)(OCH_3)_2$	At 200°	180
$B(C_2H_5)_3/B_2O_3$	$[B(C_2H_5)O]_3$	24 hr reflux	190
$B(C_3H_5)_3/B(OCH_3)_3$	$B(C_3H_5)(OCH_3)_2$, (79%)	10 hr reflux	186
$B(C_6H_5)_3/B(OC_4H_9\text{-}n)_3$	$B(C_3H_5)(OC_4H_9\text{-}n)_2$, (80%)	16 hr at 160–170°	181
$[B(n\text{-}C_3H_7)_2H]_2/B(OCH_3)_3$	$B(n\text{-}C_3H_7)(OCH_3)_2$, (89%)	Also B–H/B–O exchange	81
$[B(n\text{-}C_3H_7)_2H]_2/B(OC_4H_9\text{-}n)_3$	$B(n\text{-}C_3H_7)(OC_4H_9\text{-}n)_2$, (89%)	Also B–H/B–O exchange	81
$[B(n\text{-}C_3H_7)_2H]_2/B(OC_7H_{15}\text{-}n)_3$	$B(n\text{-}C_3H_7)(OC_7H_{15}\text{-}n)_2$, (80%)	Also B–H/B–O exchange	81
$B(n\text{-}C_3H_7)_3/B(OCH_3)_3$	$B(n\text{-}C_3H_7)(OCH_3)_2$, (70%)	$[B(C_3H_7)_2H]_2$ as catalyst	185
$B(n\text{-}C_3H_7)_3/B(OC_4H_9\text{-}n)_3$	$B(n\text{-}C_3H_7)(OC_4H_9\text{-}n)_2$, (89%)	$[B(C_3H_7)_2H]_2$ as catalyst	185

Reactants	Product (%)	Conditions	Ref.
$B(n\text{-}C_3H_7)_3/B(OC_7H_{15})_3$	$B(n\text{-}C_3H_7)(OC_7H_{15})_2$, (74%)	$[B(C_3H_7)_2H]_2$ as catalyst, 2 hr at 156–204°	185
$B(n\text{-}C_3H_7)_3/B_2O_3$	$[B(n\text{-}C_3H_7)O]_3$, (52%)		190
$B(n\text{-}C_3H_7)_3/[B(OCH_3)O]_3$	$[B(n\text{-}C_3H_7)O]_3$, (>60%)		191
$B(n\text{-}C_3H_7)_3/B_2(OCH_2CH_2O)_3$	$B(n\text{-}C_3H_7)(OCH_2CH_2O)_3$	>150°	179
$[B(n\text{-}C_4H_9)_2H]_2/B(OCH_3)_3$	$B(n\text{-}C_4H_9)(OCH_3)_2$, (73%)	Also B—H/B—O exchange	81
$B(n\text{-}C_4H_9)_3/B(OCH_3)_3$	$B(n\text{-}C_4H_9)(OCH_3)_2$, (91%)	$[B(C_4H_9)_2H]_2$ as catalyst	185
$B(n\text{-}C_4H_9)_3/B(OC_7H_{15})_3$	$B(n\text{-}C_4H_9)(OC_7H_{15})_2$, (70%)	$[B(C_4H_9)_2H]_2$ as catalyst	185
$B(n\text{-}C_4H_9)_3/[B(OCH_3)O]_3$	$[B(n\text{-}C_4H_9)O]_3$, (>60%)		191
$B(n\text{-}C_4H_9)_3/B_2O_3$	$[B(n\text{-}C_4H_9)O]_3$, (69%)	40 hr reflux	190
$B(i\text{-}C_4H_9)_3/[B(OCH_3)O]_3$	$[B(i\text{-}C_4H_9)O]_3$, (64%)		191
$B(i\text{-}C_4H_9)_3/B_2O_3$	$[B(i\text{-}C_4H_9)O]_3$, (83%)	3 d reflux	190
$B(s\text{-}C_4H_9)_3/[B(OCH_3)O]_3$	$[B(s\text{-}C_4H_9)O]_3$, (35%)		191
$B(i\text{-}C_5H_{11})_3/B(OCH_3)_3$	$B(i\text{-}C_5H_{11})(OCH_3)_2$, (75%)	$[B(C_5H_{11})_2H]_2$ as catalyst	185
$B(n\text{-}C_5H_{11})_3/[B(OCH_3)O]_3$	$[B(n\text{-}C_5H_{11})O]_3$, (>60%)		191
$B(C_6H_5)_2(OC_4H_9\text{-}n)/B(OC_4H_9\text{-}n)_3$	$B(C_6H_5)(OC_4H_9\text{-}n)_2$	100 hr at 200°	183
$B(C_6H_5)_3/B(OCH_3)_3$	$BC_6H_5(OCH_3)_2$, ($K = 0.19$; $B(C_6H_5)_2(OCH_3)$, ($K = 0.5$)	2 d at 200°	176
$B(C_6H_5)_3/B_2O_3$	$[B(C_6H_5)O]_3$, (85%)	48 hr reflux	190
$B(C_6H_{11})_3/B_2O_3$	$[B(C_6H_{11})O]_3$, (85%)	48 hr reflux	190
$B(n\text{-}C_6H_{13})_3/B(OCH_3)_3$	$B(n\text{-}C_6H_{13})(OCH_3)_2$, (72%)	$[B(C_3H_7)_2H]_2$ as catalyst	185
$B(C_{12}H_{23})_3/B(OCH_3)_3$	$B(C_{12}H_{23})(OCH_3)_2$, (81.5%)	3 hr at 250°	182

63

(continued)

TABLE XI (*Continued*)

Reactants	Products	Remarks	References
[⬠ BCH$_2$CH$_2$]$_2$ /B(OCH$_3$)$_3$	⬠ BOCH$_3$, (81%)	Catalyzed by B—H compounds	184
[⬠ BCH$_2$CH$_2$]$_2$ /B(OCH$_3$)$_3$	[B(OCH$_3$)$_2$CH$_2$CH$_2$]$_2$, (75%)	Catalyzed by B—H compounds	184
[⬡ BCH$_2$CH$_2$CH$_2$]$_2$ /B(OCH$_3$)$_3$	⬡ B(OCH$_3$), (81%)	Catalyzed by B—H compounds	184
[⬡ BCH$_2$CH$_2$CH$_2$]$_2$ /B(OCH$_3$)$_3$	[B(OCH$_3$)$_2$CH$_2$CH$_2$CH$_2$]$_2$, (63%)	Catalyzed by B—H compounds	184
BR$_3$/B(OR')$_3$	BR(OR')$_2$, BR$_2$(OR')		179
Al(CH$_3$)$_3$/Al(OCH$_3$)$_3$	Al(CH$_3$)$_2$(OCH$_3$)	At 200°	192
Al(CH$_3$)$_3$/Al(OC$_3$H$_7$-i)	Al(CH$_3$)$_2$(OC$_3$H$_7$-i)	20 min at 135°	193
Al(C$_2$H$_5$)$_3$/Al(OC$_2$H$_5$)$_3$	Al(C$_2$H$_5$)$_2$(OC$_2$H$_5$)	At 170°	192
Al(C$_2$H$_5$)$_3$/Al(OC$_2$H$_5$)$_3$	Al(C$_2$H$_5$)$_2$(OC$_2$H$_5$)	At 170°	192
Tl(C$_6$H$_5$)$_2$[OOCH(CH$_3$)$_2$]/Tl(OOCH(CH$_3$)$_2$)$_3$	Tl(C$_6$H$_5$)[OOCH(CH$_3$)]$_2$		195
Si(CH$_3$)(OCH$_3$)$_3$	Si(CH$_3$)$_3$(OCH$_3$), Si(CH$_3$)$_2$(OCH$_3$)$_2$, Si(OCH$_3$)$_4$	Various catalysts: Li, Na, K, NaOCH$_3$, Na$_2$O, KOSi(CH$_3$)$_3$	196

Si(CH$_3$)(OC$_2$H$_5$)$_3$	Si(CH$_3$)$_3$(OC$_2$H$_5$), Si(CH$_3$)$_2$(OC$_2$H$_5$)$_2$, Si(OC$_2$H$_5$)$_4$	16.5 hr at 250°; sodium as catalyst	196
Si(CH$_3$)(OCH$_3$)$_3$/Si(C$_6$H$_5$)(OCH$_3$)$_3$	Si(CH$_3$)(OCH$_3$)$_3$, Si(OCH$_3$)$_4$, Si(C$_6$H$_5$)(CH$_3$)(OCH$_3$)$_2$, Si(C$_6$H$_5$)$_2$(OCH$_3$)$_2$	19 hr at 200°; sodium as catalyst	196
Si(CH$_3$)(OCH$_3$)$_3$/Si(C$_6$H$_5$)$_2$(OCH$_3$)$_2$	Si(C$_6$H$_5$)(CH$_3$)(OCH$_3$)$_2$, Si(C$_6$H$_5$)(OCH$_3$)$_3$, Si(OCH$_3$)$_4$	60 hr at 190°; sodium as catalyst	196
Si(CH$_3$)H[OSi((CH$_3$)$_2$]$_2$	Si(CH$_3$)$_2$H[OSi(CH$_3$)$_3$], SiH[OSi((CH$_3$)$_3$]$_3$	At <138°; H$_2$PtCl$_6$ as catalyst	198
Si(CH$_3$)$_2$(OCH$_3$)$_2$	Si(CH$_3$)$_3$(OCH$_3$), Si(CH$_3$)(OCH$_3$)$_3$	19 hr at 240°; sodium as catalyst	196
[Si(CH$_3$)$_2$(OC$_2$H$_5$)]$_2$C$_6$H$_4$	[Si(CH$_3$)$_2$(C$_6$H$_4$)]$_n$, Si(CH$_3$)$_2$(OC$_2$H$_5$)$_2$	At 230° with KOH as catalyst	199
Si(CH$_3$)(n-C$_3$H$_7$)(OCH$_3$)$_2$	Si(CH$_3$)$_2$(n-C$_3$H$_7$)(OCH$_3$), Si(n-C$_3$H$_7$)$_2$(OCH$_3$)$_2$	38 hr at 270°; sodium as catalyst	196
Si(CH$_3$)(C$_6$H$_5$)(OCH$_3$)$_2$	Si(CH$_3$)(OCH$_3$)$_3$, Si(C$_6$H$_5$)$_2$(CH$_3$)(OCH$_3$), Si(C$_6$H$_5$)$_3$(CH$_3$)	200 hr at 200°; sodium as catalyst	196
[Si((CH$_3$)(C$_6$H$_5$)O]$_n$	[Si((CH$_3$)(C$_6$H$_5$)$_2$]$_2$O, Si(CH$_3$)(C$_6$H$_5$)$_3$	At 300° with LiOH as catalyst	197
Si(n-C$_3$H$_7$)$_2$(OCH$_3$)$_3$	Si(n-C$_3$H$_7$)$_2$(OCH$_3$)$_3$	20 hr at 278°; sodium as catalyst	196
Si(C$_6$H$_5$)(OCH$_3$)$_3$	Si(OCH$_3$)$_4$, Si(C$_6$H$_5$)(OCH$_3$)$_3$, Si(C$_6$H$_5$)$_2$(OCH$_3$)$_3$	18 hr at 173°; sodium as catalyst	196
[Si(C$_6$H$_5$)O$_{1.5}$]$_n$	Si(C$_6$H$_5$)$_4$, [Si(C$_6$H$_5$)$_3$]$_2$O	At 500°, base catalyzed	197
Si(CH$_2$C$_6$H$_5$)(OC$_2$H$_5$)$_3$	Si(OC$_2$H$_5$)$_4$, Si(CH$_2$C$_6$H$_5$)$_2$(OC$_2$H$_5$)$_2$	10 hr at 190–240, sodium as catalyst	196
[Sn(C$_6$H$_5$)$_3$]$_2$O	Sn(C$_6$H$_5$)$_4$, [Sn(C$_6$H$_5$)$_2$O]	8 hr at 140°	200,201
Pb(C$_2$H$_5$)$_3$(OOCCH$_3$)	Pb(C$_2$H$_5$)$_2$(OOCCH$_3$)$_2$, Pb(C$_2$H$_5$)$_4$	4 hr at 130°	206

(continued)

65

TABLE XI (*Continued*)

Reactants	Products	Remarks	References
Pb(C$_2$H$_5$)$_2$(OOCCH$_3$)$_2$	Pb(C$_2$H$_5$)$_3$(OOCCH$_3$), Pb(C$_2$H$_5$)(OOCCH$_3$)$_3$	1 hr at 130°	206
As(C$_6$H$_5$)O	As(C$_6$H$_5$)$_3$, As$_2$O$_3$		207
Sb(C$_6$H$_5$)O	[Sb(C$_6$H$_5$)$_2$]$_2$O, (100%); Sb$_2$O$_3$	At 100°	208,209, 210
Sb(C$_6$H$_4$CH$_3$-p)O	[Sb(C$_6$H$_4$CH$_3$-p)$_2$]$_2$O, Sb$_2$O$_3$	At 100°	210
Sb(C$_6$H$_4$Br-p)O	[Sb(C$_6$H$_4$Br-p)$_2$]$_2$O, Sb$_2$O$_3$	At 100°	210
Sb(C$_6$H$_4$Cl-p)O	[Sb(C$_6$H$_4$Cl-p)$_2$]$_2$O, Sb$_2$O$_3$	At 100°	210
Sb(C$_6$H$_4$Cl-m)O	[Sb(C$_6$H$_4$Cl-m)$_2$]$_2$O, Sb$_2$O$_3$	At 100°	210
Sb(C$_6$H$_4$OCCH$_3$-p)O	[Sb(C$_6$H$_4$OCCH$_3$-p)$_2$]$_2$O, Sb$_2$O$_3$	At 100°	210
Sb(C$_6$H$_4$NO$_2$-p)O	[Sb(C$_6$H$_4$NO$_2$-p)$_2$]$_2$O, Sb$_2$O$_3$	At 100°	210

TABLE XII
Redistribution Reactions Involving Exchange of Element–Carbon with Element–Sulfur Bonds

Reactants	Products	Remarks	References
B(CH$_3$)$_3$/B(SCH$_3$)$_3$	B(CH$_3$)$_2$(SCH$_3$)	Catalyzed by B$_2$H$_6$	24
B(CH$_3$)$_3$/B(SCH$_3$)$_3$	B(CH$_3$)(SCH$_3$)$_2$	Catalyzed by B$_2$H$_6$	24
B(CH$_3$)$_3$/(BSSH)$_3$	[BS(CH$_3$)]$_{3,4}$	At r.t.	211

TABLE XIII

Redistribution Reactions Involving Exchange of Element–Carbon with Element–Halogen Bonds

Reactants	Products	Remarks	References
$Be(CH_3)_2/BeBr_2$	$Be(CH_3)Br$		213
$BeRHal$	BeR_2, $BeHal_2$		214
$Mg(C_2H_5)_2/MgCl_2$	$Mg(C_2H_5)Cl$, $(K = 0.18)$	At 25° in THF	218
$Mg(C_2H_5)_2/MgBr_2$	$Mg(C_2H_5)Br$, $(K = 0.2)$	At 25° in THF	216,218
$Mg(C_2H_5)_2/MgBr_2$	$Mg(C_2H_5)Br$, $(K = 2 \times 10^{-3})$	At 25° in diethyl ether	217
$Mg(C_2H_5)_2/MgI_2$	$Mg(C_2H_5)I$, $(K = 4 \times 10^{-4})$	At 25° in diethyl ether	217
$Mg(C_6H_5)_2/MgCl_2$	$Mg(C_6H_5)Cl$, $(K = 0.6)$	At 25° in THF	218
$Mg(C_6H_5)_2/MgBr_2$	$Mg(C_6H_5)Br$, $(K = 0.3)$	At 25° in THF	218
$Mg(C_6H_5)_2/MgBr_2$	$Mg(C_6H_5)Br$, $(K = 2 \times 10^{-2})$	At 25° in diethyl ether	217
$Mg(C_6H_5)_2/MgI_2$	$Mg(C_6H_5)I$, $(K = 6 \times 10^{-3})$	At 25° in diethyl ether	217
$MgR_2/MgHal_2$	$MgRHal$		215,217
$BCl_2(CHCHCH)BCl_2$	BCl_3, $[ClBCHCH]_n$	At 110° or in the presence of amines	233
$B(CH_3)_2[N(CH_3)_2]/BCl_2[N(CH_3)_2]$	$BCl(CH_3)[N(CH_3)_2]$, $(60-100\%)$		228
$B(CH_3)_3/BF_3$	$B(CH_3)F_2$, $B(CH_3)_2F$	36 hr at 160–170°	220
$B(CH_3)_3/BCl_3$	$B(CH_3)Cl_2$, $B(CH_3)_2Cl$	8 hr at 300°	219
$B(CH_3)_3/BBr_3$	$B(CH_3)Br_2$, $B(CH_3)_2Br$, $(15-20\%)$	At 390° in gas phase	222
$B(CH_3)(C_2H_3)_2/BCl_3$	$B(CH_3)(C_2H_3)Cl$, (5%)	At r.t.	227
$B(C_2H_5)_3/BCl_3$	$B(C_2H_5)_2Cl$, (61%); $B(C_2H_5)Cl_2$, (30%)	4 hr at 200°	223
$B(n\text{-}C_3H_7)_3/BBr_3$	$B(n\text{-}C_3H_7)_3Br_2$, (55%)	At 128–156°	234

67

(continued)

TABLE XIII (Continued)

Reactants	Products	Remarks	References
B(n-C₃H₇)₃/BBr₃	B(n-C₃H₇)₂Br, (75%)	At 128–156°	234
B(i-C₃H₇)₃/BBr₃	B(i-C₃H₇)₂Br	At 101–145°, isomerization of iso to n-propyl groups	234
B(n-C₄H₉)₃/BF₃	B(n-C₄H₉)F₂, (82%)	At 200°	234
B(n-C₄H₉)₃/BF₃	B(n-C₄H₉)₂F	At 200°	235
B(n-C₄H₉)₂F/BF₃	B(n-C₄H₉)F₂	At 150°	235
B(n-C₄H₉)₃/BF₃	B(n-C₄H₉)₂F, (31%)	20 hr at 200°	223
B(n-C₄H₉)₃/BF₃	B(n-C₄H₉)F₂, (91%)	20 hr at 200°	223
B(n-C₄H₉)₃/BCl₃	B(n-C₄H₉)₂Cl, (69%)	20 hr at 200°	223
B(n-C₄H₉)₃/BCl₃	B(n-C₄H₉)Cl₂, (74%)	20 hr at 200°	223
B(n-C₄H₉)₃/BCl₃	B(n-C₄H₉)₂Cl, (91%)	At 160°	221
B(n-C₄H₉)₃/B(n-C₄H₉)Cl₂	B(n-C₄H₉)₂Cl, (80%)	48 hr reflux	221
B(i-C₄H₉)₃/BCl₃	B(i-C₄H₉)₂Cl	At 160°	221
B(s-C₄H₉)₃/BCl₃	B(s-C₄H₉)₂Cl, (82%)		221
[B(CH₂)₄C₂H₄]₂/BCl₃	BCl₂(CH₂)₄BCl₂, (88%)		229
B(n-C₄H₉)₃/BBr₃	B(n-C₄H₉)₂Br, (56%)	At 129–141°	234
B(i-C₄H₉)₃/BBr₃	B(i-C₄H₉)Br₂, (55–73%); B(i-C₄H₉)₂Br, (55–73%)	From 1:1 mixture at 146–175°	234
B(s-C₄H₉)₃/BBr₃	B(C₄H₉)₂Br, (85% n- and 15% s-butyl groups)	Partial isomerization at 170°	234
B(n-C₅H₁₁)₃/BF₃	B(n-C₅H₁₁)F₂	At 200°	235
B(n-C₅H₁₁)₃/BCl₃	B(n-C₅H₁₁)₂Cl, (87%)		226
B(n-C₅H₁₁)₃/BBr₃	B(n-C₅H₁₁)Br₂, (85%)	At 175–192°	234
B(n-C₅H₁₁)₃/BBr₃	B(n-C₅H₁₁)₂Br, (28%)	At 175–192°	234
B(i-C₅H₁₁)₃/BF₃·O(C₂H₅)₂	B(i-C₅H₁₁)F₂, (77%)	At 200–210°	225
B(C₆H₅)₂Cl/BCl₃	B(C₆H₅)Cl₂, (90%)	75 hr at 200°	224
B(C₆H₅)₂Br/BBr₃	B(C₆H₅)Br₂, (69%)	10 hr at 200°	224

Reactants	Products	Conditions	Ref.
$B(C_6H_5)_3/B(C_6H_5)Cl_2$	$B(C_6H_5)_2Cl$, (77%)	3 hr reflux	230
$B(n\text{-}C_6H_{13})_3/BF_3$	$B(n\text{-}C_6H_{13})F_2$	At 200°	235
$B(n\text{-}C_6H_{13})_3/BF_3 \cdot O(C_2H_5)_2$	$B(n\text{-}C_6H_{13})F_2$, (85%)	At 200–215°	225
[fluorene] $\overset{B}{\underset{R}{}}$ /BCl_3	[fluorene] $\overset{B}{\underset{Cl}{}}$, $RBCl_2$ $R = C_2H_5, C_3H_7, i\text{-}C_4H_9, C_6H_5$	In the presence of BH compounds	106
[cyclopentane] $B(CH_2)_3B$ /BCl_3	$ClB[(CH_2)_3]_2BCl$, (82%)	With $B_2(C_2H_5)_4H_2$ as catalyst	237
[cyclopentane] $B(CH_2)_3B$ /BCl_3	$BCl_2(CH_2)_3BCl_2$, (89%)	With $B_2(C_2H_5)_4H_2$ as catalyst	237
[cyclopentane] $\left[BCH_2CH_2\right]_2$ /BCl_3	[cyclopentane] BCl, (77%)	At 140° with $B_2(C_2H_5)_4H_2$) as catalyst	237
[cyclopentane] $\left[BCH_2CH_2\right]_2$ /BCl_3	$[BCl_2(CH_2CH_2)]_2$, (90%)	With $B_2(C_2H_5)_4H_2$ as catalyst	237
[cyclohexane] $\left[B\text{-}CH_2\text{-}CH_2\right]_2$ /BCl_3	[cyclohexane] BCl, (90%)	Catalyzed by B—H compounds	236

(continued)

TABLE XIII (*Continued*)

Reactants	Products	Remarks	References
$\left[\text{cyclohexane-CH}_3, \text{B—CH}_2\text{CH}_2\right]_2 / \text{BCl}_3$	H₃C-cyclohexane-BCl	Catalyzed by B—H compounds	236,231
$\left[\text{cyclohexane-CH}_3, \text{B—CH}_2\text{—CH}_2\right]_2 / \text{BCl}_3$	$BCl_2CH_2CH_2CH_2CH(CH_3)CH_2BCl_2$	Catalyzed by B—H compounds	236,231
$B(CH_2)_5B$ (cyclohexane) / BCl_3	cyclohexane-BCl, (85%)	With $B_2(C_2H_5)_4H_2$ as catalyst	232,237
$B(CH_2)_5B$ (cyclohexane) / BCl_3	$BCl_2(CH_2)_5BCl_2$, (90%)	With $B_2(C_2H_5)_4H_2$ as catalyst	232,237
$\left[\text{BCH}_2\text{CH}_2\text{CH}_2 \text{(cycloheptane)}\right]_2 / \text{BCl}_3$	cycloheptane-BCl, (86%)	With $B_2(C_2H_5)_4H_2$ as catalyst	232,237
$\left[\text{BCH}_2\text{CH}_2\text{CH}_2 \text{(cycloheptane)}\right]_2 / \text{BCl}_3$	$[BCl_2(CH_2CH_2CH_2)]_2$, (88%)	With $B_2(C_2H_5)_4H_2$ as catalyst	232,237
$[B_3Cl_3N_3(CH_3)_3]/[B_3N_3(CH_3)_6]$	$[B_3N_3Cl(CH_3)_5]$, $[B_3N_3Cl_2(CH_3)_4]$	At 250–350°	38

$[Al(CH_3)_2F_2]Na$	$Al(CH_3)_3$, (88%); $[AlF_6]Na_3$	At 250°	241
$Al_2(CH_3)_3Cl_3$	$Al(CH_3)_2Cl$, (84%); $AlCl_3$	2 hr at 220°, NaCl as catalyst	242
$Al_2(CH_3)_3I_3$	$Al(CH_3)_3$, (80%); AlI_3	At 180°, NaI as catalyst	242
$Al(CH_3)_3/AlI_3$	$Al(CH_3)I_2$		192
$Al_2(CH_3)_3I_3$	$Al(CH_3)_3$, (85%)	In the presence of Mg	243
$[Al(C_2H_5)_2F_2]Na$	$Al(C_2H_5)_3$, (93%); $[AlF_6]Na_3$	At 130–135°	241
$Al(C_2H_5)_3/AlF_3$	$Al(C_2H_5)F_2$		241
$Al(C_2H_5)_3/AlCl_3$	$Al(C_2H_5)_2Cl$, (77%); $Al(C_2H_5)Cl_2$	From the sesquichloride	192
$Al_2(C_2H_5)_3Cl_3$	$Al(C_2H_5)_2Cl$, (85%); $AlCl_3$	2 hr at 220°; NaCl as catalyst	242
$Al(C_2^{14}H_5)_3/AlCl_3$	$Al(C_2^{14}H_5)_2Cl$, (97%)		244
$Al(C_2H_5)_3/AlBr_3$	$Al(C_2H_5)Br_2$	$\frac{1}{2}$ hr at 80–100°	192
$Al_2(C_2H_5)_3Br_3$	$Al(C_2H_5)_2Br$, (92%); $Al(C_2H_5)Br_2$	2 hr at 220°, NaBr as catalyst	242
$Al(C_2H_5)_3/AlI_3$	$Al(C_2H_5)_2I$, $Al(C_2H_5)I_2$		192
$Al_2(C_2H_5)_3I_3$	$Al(C_2H_5)_2I$, (98%); AlI_3	2 hr at 220°, NaI as catalyst	242
$[Al(n\text{-}C_3H_7)_2F_2]Na$	$Al(n\text{-}C_3H_7)_3$, (91%); $[AlF_6]Na_3$	At 160–200°	241
$Al_2(n\text{-}C_3H_7)_3I_3$	$Al(n\text{-}C_3H_7)_2I$, (75%); AlI_3	2 hr at 220°, NaI as catalyst	242
$Al(C_6H_5)_3/AlCl_3$	$Al(C_6H_5)Cl_2$, $Al(C_6H_5)_2Cl$	$\frac{1}{2}$ hr at 180°	245
$Al(C_6H_5)_3/AlCl_3$	$Al(C_6H_5)Cl_2$	20 min at 200°	192
$Al(C_6H_5)_3/AlBr_3$	$Al(C_6H_5)Br_2$	20 min at 200°	192
$Al(C_6H_5)_3/AlI_3$	$Al(C_6H_5)I_2$	From the sesquiiodide	192

71

(continued)

TABLE XIII (*Continued*)

Reactants	Products	Remarks	References
Al(C₆H₄CH₃-p)₃/AlI₃	Al(C₆H₄CH₃-p)I₂	From the sesquiiodide	192
AlR₃/AlF₃	AlR₂F		240
Ga(C₂H₅)₃/GaCl₃	Ga(C₂H₅)₂Cl, (98%)	1 hr at 100°	231
Ga(C₆H₅)₃/GaBr₃	Ga(C₆H₅)Br₂, Ga(C₆H₅)₂Br	3–4 hr at 170°	247
Tl(CH₃)₂Br/TlBr₃	Tl(CH₃)Br₂		248,250,251
Tl(C₂H₅)₂Cl/TlCl₃	Tl(C₂H₅)Cl₂		248,250,251
Tl(CHCH₂)₂Br/TlBr₃	Tl(CHCH₂)Br₂		252
Tl(CHCH₂)₂Cl/TlCl₃	Tl(CHCH₂)Cl₂		252
Tl(C₂H₅)₂Br/TlBr₃	Tl(C₂H₅)Br₂		248,250,251
Tl(i-C₅H₁₁)₂Cl/TlCl₃	Tl(i-C₅H₁₁)Cl₂		248,250,251
Tl(i-C₅H₁₁)₂Br/TlBr₃	Tl(i-C₅H₁₁)Br₂		248,250,251
Tl(C₆H₅)₂Cl/TlCl₃	Tl(C₆H₅)Cl₂		248,250,251
Tl(C₆H₅)₂Br/TlBr₃	Tl(C₆H₅)Br₂		248,250,251
Tl(C₆H₄Cl-p)₂Cl/TlCl₃	Tl(C₆H₄Cl-p)Cl₂		248,250,251
Tl(C₆H₄Br-p)₂Cl/TlCl₃	Tl(C₆H₄Br-p)Cl₂		248,250,251
Tl(C₆H₄NO₂)₂Cl/TlCl₃	Tl(C₆H₄NO₂)Cl₂		249
Tl(C₆H₄CH₃-p)₂Cl/TlCl₃	Tl(C₆H₄CH₃-p)Cl₂		248,250,251
Tl(C₆H₄C₆H₅)₂Br/TlBr₃	Tl(C₆H₄C₆H₅)Br₂		249
Si(CH₃)Cl₂H	SiClH₃, SiCl₂H₂, SiCl₃H, Si(CH₃)Cl₂H, Si(CH₃)₂Cl₂	At 325°, AlCl₃ as catalyst (also Si—H exchange)	120
Si(CCl₃)Cl₃	Si(CCl₃)₄, SiCl₄	At 130–150°; tri-alkylamines as catalysts	254
Si(CH₃)(CCl₃)Cl₂	Si(CH₃)(CCl₃)₃, Si(CH₃)Cl₃	At 130–150°; tri-alkylamines as catalysts	254

$Si(CH_3)_3Br$		At 80°; $AlBr_3$ as catalyst	61
$Si(CH_3)_4/SiCl_4$	$Si(CH_3)_3Cl$, ($K = 0.013$); $Si(CH_3)_2Cl_2$, ($K = 0.025$); $Si(CH_3)Cl_3$	At 350°; $AlCl_3$ as catalyst	26,120,253
$Si(C_6H_5)_2Cl_2$	$Si(C_6H_5)Cl_3$, $Si(C_6H_5)_3Cl$, $Si(C_6H_5)_2Cl(C_6H_4)Si(C_6H_5)Cl_2$	At 480–540°; with SiO_2/Al_2O_3 as catalyst	255
$Ge(CH_3)_2Cl_2/GeCl_4$	$Ge(CH_3)_2Cl_2$, (92%)	16 hr at 235°; $GaCl_3$ as catalyst	259
$Ge(CH_3)_3Cl/Ge(CH_3)Cl_3$	$Ge(CH_3)_2Cl_2$, (100%)	16 hr at 235°; $GaCl_3$ as catalyst	259
$Ge(CH_3)_3Cl/GeCl_4$	$Ge(CH_3)_2Cl_2$, (100%)	16 hr at 235°; $GaCl_3$ as catalyst	259
$Ge(CH_3)_4/GeCl_4$	$Ge(CH_3)_3Cl$, (96%)	16 hr at 150°; $GaCl_3$ as catalyst	259
$Ge(CH_3)_4/GeCl_4$	$Ge(CH_3)_2Cl_2$, (100%)	64 hr at 235°,; $GaCl_3$ as catalyst	259
$Ge(CH_3)_4/GeCl_4$	$Ge(CH_3)_3Cl$, ($K = 0.02$); $Ge(CH_3)_2Cl_2$, ($K = 0.005$); $Ge(CH_3)Cl_3$, ($K = 10^{-3}$)	At 300°; $AlCl_3$ as catalyst	260
$Ge(CH_3)_4/Ge(CH_3)Cl_3$	$Ge(CH_3)_3Cl$, (93%)	5 hr at 240°; $GaCl_3$ as catalyst	259
$Ge(CH_3)_4/Ge(CH_3)Cl_3$	$Ge(CH_3)_2Cl_2$, (94%)	24 hr at 225°; $GaCl_3$ as catalyst	259
$Ge(CH_3)_4/Ge(CH_3)_2Cl_2$	$Ge(CH_3)_3Cl$, (95%)	3–6 hr at 50–97°; $GaCl_3$ as catalyst	259
$Ge(CH_3)_4/GeBr_4$	$Ge(CH_3)_2Br_2$, (83%)	17 hr at 165°; $GaBr_3$ as catalyst	259
$Ge(C_2H_5)_4/GeCl_4$	$Ge(C_2H_5)_2Cl_2$, (65%)	92 hr at 175°; $GaCl_3$ as catalyst	259

(continued)

73

TABLE XIII (*Continued*)

Reactants	Products	Remarks	References
$Ge(C_2H_5)_4/GeCl_4$	$Ge(C_2H_5)_3Cl$	At 200–280°; $AlCl_3$ as catalyst	257
$Ge(C_2H_5)_4/GeCl_4$	$Ge(C_2H_5)_2Cl_2$	At 200–280°; $AlCl_3$ as catalyst	257
$Ge(C_2H_5)_4/GeCl_4$	$Ge(C_2H_5)Cl_3$	At 200–280°; $AlCl_3$ as catalyst	257
$Ge(C_2H_5)_4/GeBr_4$	$Ge(C_2H_5)_2Br_2$, (85%)	16 hr reflux, $AlBr_3$ as catalyst	263
$Ge(C_2H_5)_4/GeBr_4$	$Ge(C_2H_5)_3Br$, (70%)	16 hr reflux, $AlBr_3$ as catalyst	263
$Ge(n\text{-}C_3H_7)_4/GeCl_4$	$Ge(n\text{-}C_3H_7)_3Cl$	At 200°; $AlCl_3$ as catalyst	257
$Ge(n\text{-}C_3H_7)_4/GeCl_4$	$Ge(n\text{-}C_3H_7)_2Cl_2$	At 200°; $AlCl_3$ as catalyst	257
$Ge(n\text{-}C_3H_7)_4/GeCl_4$	$Ge(n\text{-}C_3H_7)Cl_3$	At 200°; $AlCl_3$ as catalyst	257
$Ge(n\text{-}C_4H_9)_4/GeCl_4$	$Ge(n\text{-}C_4H_9)_3Cl$; $Ge(n\text{-}C_4H_9)_2Cl_2$; $Ge(n\text{-}C_4H_9)Cl_3$	At 100–200°; $AlCl_3$ as catalyst	256, 257
$Ge_2(n\text{-}C_4H_9)_6/GeCl_4$	$Ge_2(n\text{-}C_4H_9)_5Cl$	6.5 hr at 200° $AlBr_3$ as catalyst	258
$Ge(n\text{-}C_4H_9)_4/GeBr_4$	$Ge(n\text{-}C_4H_9)_2Br_2$, (73%)	At 205°; $AlCl_3$ as catalyst	263
$Ge(n\text{-}C_5H_{11})_4/GeCl_4$	$Ge(n\text{-}C_5H_{11})_3Cl$	At 205°; $AlCl_3$ as catalyst	257
$Ge(C_6H_5)_4/GeCl_4$	$Ge(C_6H_5)Cl_3$, (82%)	1.5 hr at 120°; $AlCl_3$ as catalyst	30
$Ge(C_6H_5)_4/GeCl_4$	$Ge(C_6H_5)_2Cl_2$, (74%)	15 hr at 140°; $AlCl_3$ as catalyst	30

Ge(C$_6$H$_5$)$_4$/GeCl$_4$	Ge(C$_6$H$_5$)$_3$Cl, (84%)	15 min at 140°; AlCl$_3$ as catalst	30
Ge(C$_6$H$_5$)$_4$/GeCl$_4$	Ge(C$_6$H$_5$)$_3$Cl	At 60–120°; AlCl$_3$ as catalyst	257
Ge(C$_6$H$_5$)$_4$/GeCl$_4$	Ge(C$_6$H$_5$)$_2$Cl$_2$	At 60–120°; AlCl$_3$ as catalyst	257
Ge(C$_6$H$_5$)$_4$/GeCl$_4$	Ge(C$_6$H$_5$)Cl$_3$	At 60–120°; AlCl$_3$ as catalyst	257
Ge(C$_6$H$_5$)$_4$/GeCl$_4$	Ge(C$_6$H$_5$)Cl$_3$, (75%)	36 hr at 350°	262,261
Ge(C$_6$H$_5$)$_4$/GeBr$_4$	Ge(C$_6$H$_5$)$_2$Br$_2$, (65%)	AlBr$_3$ as catalyst	264
Ge(n-C$_6$H$_{13}$)$_4$/GeCl$_4$	Ge(n-C$_6$H$_{13}$)$_3$Cl	At 200°; AlCl$_3$ as catalyst	257
Ge(n-C$_7$H$_{15}$)$_4$/GeCl$_4$	Ge(n-C$_7$H$_{15}$)$_3$Cl	At 200–210°; AlCl$_3$ as catalyst	257
Sn(CH$_3$)$_2$Cl$_2$/SnCl$_4$	Sn(CH$_3$)Cl$_3$, (90%)	In dimethylsulfoxide	293
Sn(CH$_3$)$_4$/SnCl$_4$	Sn(CH$_3$)$_3$Cl, ($K = 3 \times 10^{-3}$); Sn(CH$_3$)$_2$Cl$_2$, ($K = 10^{-4}$); Sn(CH$_3$)Cl$_3$, ($K = 7 \times 10^{-2}$)	At 175°	299
Sn(CH$_3$)$_3$Br/Sn(CH$_3$)Br$_3$	Sn(CH$_3$)$_2$Br$_2$, (~100%)	2 hr at 220°	267
Sn(CH$_3$)$_4$/SnBr$_4$	Sn(CH$_3$)$_2$Br$_2$, (~100%)	1 hr at 185°	267
Sn(CH$_3$)$_4$/Sn(CH$_3$)Br$_3$	Sn(CH$_3$)$_2$Br$_2$	2 hr at 200°	267
Sn(CH$_3$)$_4$/SnI$_4$	Sn(CH$_3$)I$_3$	At 100°	268
Sn(CH=CH$_2$)$_4$/SnCl$_4$	Sn(CH=CH$_2$)Cl$_3$, (86%)	2 hr at 70°	269–271
Sn(CH=CH$_2$)$_4$/SnCl$_4$	Sn(CH=CH$_2$)$_2$Cl$_2$, (98%)	2 hr at 30°	269–271
Sn(CH=CH$_2$)$_4$/SnCl$_4$	Sn(CH=CH$_2$)$_3$Cl, (96%)	1 hr at 30°	269–271
Sn(C$_2$H$_5$)$_4$/SnCl$_4$	Sn(C$_2$H$_5$)Cl$_3$, (49%); Sn(C$_2$H$_5$)$_2$Cl$_2$, (49%)	1 hr at 0°	272
Sn(C$_2$H$_5$)$_3$Cl/SnCl$_4$	Sn(C$_2$H$_5$)$_2$Cl$_2$, (~100%)	2 hr at 195°	267,273
Sn(C$_2$H$_5$)$_4$/SnCl$_4$	Sn(C$_2$H$_5$)$_3$Cl, (86%)	2 hr at 210°	265–267, 274,275

(continued)

75

TABLE XIII (*Continued*)

Reactants	Products	Remarks	References
Sn(C₂H₅)₄/SnCl₄	$Sn(C_2H_5)_2Cl_2$, (~100%)	0.5 hr at 210°	266,267, 274,275
Sn(C₂H₅)₄/SnCl₄	$Sn(C_2H_5)Cl_3$, (48%)	At 0°	272,275
Sn(C₂H₅)₄/SnCl₄	$Sn(C_2H_5)Cl_3$, (94%)	48 hr at 130°	272
Sn(C₂H₅)₄/Sn(C₂H₅)Cl₃	$Sn(C_2H_5)_3Cl$, (99%)	1 hr at 200°	272
Sn(C₂H₅)₄/Sn(C₂H₅)₂Cl₂	$Sn(C_2H_5)_3Cl$, (75%)	2 hr at 215°	267,274
Sn(C₂H₅)₃Br/SnBr₄	$Sn(C_2H_5)_2Br_2$, (~100%)	2 hr at 220°	267
Sn(C₂H₅)₄/SnBr₄	$Sn(C_2H_5)Br_3$, (67%); $Sn(C_2H_5)_2Br_2$, (25%)	12 hr at 100°	272
Sn(C₂H₅)₄/SnBr₄	$Sn(C_2H_5)_3Br$, (91%)	2 hr at 225°	267,274
Sn(C₂H₅)₄/SnBr₄	$Sn(C_2H_5)_2Br_2$, (~100%)	1 hr at 185°	267,274
Sn(C₂H₅)₄/Sn(C₂H₅)₂Br₂	$Sn(C_2H_5)_3Br$, (70%)	2 hr at 215°	267,274
Sn(CH₂CH=CH₂)₄/SnBr₄	$Sn(CH_2CH=CH_2)_2Br_2$	18 hr at 50°	276
Sn(n-C₃H₇)₄/SnCl₄	$Sn(n\text{-}C_3H_7)_2Cl_2$, (85%)	1.5 hr at 200°	267
Sn(n-C₃H₇)₄/SnCl₄	$Sn(n\text{-}C_3H_7)_3Cl$, (88%)	2.5 hr at 200°	277
Sn(i-C₃H₇)₄/SnBr₄	$Sn(i\text{-}C_3H_7)_3Br$, (~90%)	125–165°	278
Sn(n-C₄H₉)₃Cl/SnCl₄	$Sn(n\text{-}C_4H_9)Cl_3$, (48%); $Sn(n\text{-}C_4H_9)_2Cl_2$, (48%)	3 hr at 100°	272,279
Sn(n-C₄H₉)₄/SnCl₄	$Sn(n\text{-}C_4H_9)_2Cl_2$, (95%)	3 hr at 250°	280,281
Sn(n-C₄H₉)₄/SnCl₄	$Sn(n\text{-}C_4H_9)Cl_3$, (63%); $Sn(n\text{-}C_4H_9)Cl_2$, (33%)	At 100°	272
Sn(n-C₄H₉)₄/SnBr₄	$Sn(n\text{-}C_4H_9)_2Br_2$	3 hr at 220°	282
Sn(i-C₄H₉)₄/SnCl₄	$Sn(i\text{-}C_4H_9)Cl_3$, (43%); $Sn(i\text{-}C_4H_9)_3Cl$, (51%)	At 0°	272
Sn(i-C₄H₉)₄/SnCl₄	$Sn(i\text{-}C_4H_9)Cl_3$, (61%); $Sn(i\text{-}C_4H_9)_2Cl_2$, (33%)	At 150°	272
Sn(α-C₄H₃S)₄/SnCl₄	$Sn(\alpha\text{-}C_4H_3S)_2Cl_2$		283
Sn(C₆H₅)₂Cl₂/SnCl₄	$Sn(C_6H_5)Cl_3$, (85–90%)	1 hr at 220°	284
Sn(C₆H₅)₄/SnCl₄	$Sn(C_6H_5)Cl_3$, (80%)	2 hr at 220°	284–287
Sn(C₆H₅)₄/SnCl₄	$Sn(C_6H_5)_2Cl_2$, (~100%)	1.5 hr at 220°	265,266, 284–286

Reactant	Product (yield)	Conditions	Ref.
$Sn(C_6H_5)_4/SnCl_4$	$Sn(C_6H_5)_3Cl$, (76%)	20 hr at 240°	273,279, 287,291
$Sn(C_6H_5)_4/SnCl_4$	$Sn(C_6H_5)_3Cl$, (70%)	3 hr at 210°	288
$Sn(C_6H_5)_4/SnCl_4$	$Sn(C_6H_5)_2Cl_2$, (~100%)	UV irradiation	289
$Sn(C_6H_4Cl\text{-}p)_2Cl_2/SnCl_4$	$Sn(C_6H_4Cl\text{-}p)Cl_3$, (~100%)	2 hr at 150°	290
$Sn(C_6H_4Br\text{-}p)_2Cl_2/SnCl_4$	$Sn(C_6H_4Br\text{-}p)Cl_3$	1.5 hr at 150°	290
$Sn(C_6H_4Br\text{-}p)_2Br_2/SnBr_4$	$Sn(C_6H_4Br\text{-}p)Br_3$, (91%)		292
$Sn(C_6H_4I\text{-}p)_2Cl_2/SnCl_4$	$Sn(C_6H_4I\text{-}p)Cl_3$	35 min at 165°	290
$Sn(C_6H_4I\text{-}p)_2Br_2/SnBr_4$	$Sn(C_6H_4I\text{-}p)Br_3$	2 hr at 150°	290
$Sn(C_6H_{11})_4/SnCl_4$	$Sn(C_6H_{11})_3Cl$, $Sn(C_6H_{11})_2Cl_2$	1–20 hr at 25–190°	293
$Sn(n\text{-}C_6H_{13})_4/SnCl_4$	$Sn(n\text{-}C_6H_{13})_2Cl_2$, (88%)	3 hr at 225°	294
$Sn(C_6H_4CH_3\text{-}m)_4/SnCl_4$	$Sn(C_6H_4CH_3\text{-}m)_2Cl_2$	2 hr at 205°	291
$Sn(C_6H_4CH_3\text{-}o)_4/SnCl_4$	$Sn(C_6H_4CH_3\text{-}o)_3Cl$, (72%)	3 hr at 210°	288,296
$Sn(C_6H_4CH_3\text{-}o)_4/SnCl_4$	$Sn(C_6H_4CH_3\text{-}o)_2Cl_2$	2 hr at 205°	297
$Sn(C_6H_4CH_3\text{-}o)_4SnCl_4$	$Sn(C_6H_4CH_3\text{-}o)Cl_3$, (65%)	2 hr at 215°	297
$Sn(C_6H_4CH_3\text{-}p)_2Cl_2/SnCl_4$	$Sn(C_6H_4CH_3\text{-}m)Cl_3$, (75%)	3 hr at 210°	288,296
$Sn(C_6H_4CH_3\text{-}p)_2Cl_2/SnCl_4$	$Sn(C_6H_4CH_3\text{-}p)_2Cl_2$, (45%)	2 hr at 215°	297
$Sn(C_6H_4CH_3\text{-}p)/SnCl_4$	$Sn(C_6H_4CH_3\text{-}p)_3Cl$, (81%)	3 hr at 210°	267,296
$Sn(C_6H_4CH_3\text{-}p)_4/SnCl_4$	$Sn(C_6H_4CH_3\text{-}p)Cl_3$, (40%)	2 hr at 215°	297
$Sn(C_6H_4CH_3\text{-}p)_4/SnCl_4$	$Sn(C_6H_4CH_3\text{-}p)_2Cl_2$	2 hr at 205°	297
$Sn(anisyl\text{-}p)_4/SnCl_4$	$Sn(anisyl\text{-}p)_2Cl_2$, (87%)	2 hr at 185°	298
$Sn(anisyl\text{-}p)_4/SnBr_4$	$Sn(anisyl\text{-}p)_2Br_2$, (74%)	2 hr at 185°	298
$Sn(phenetyl\text{-}p)_4/SnCl_4$	$Sn(phenetyl\text{-}p)_2Cl_2$, (66%)	2 hr at 185°	298
$Sn(n\text{-}C_8H_{17})_4/SnCl_4$	$Sn(n\text{-}C_8H_{17})_2Cl_2$, (84%)	2 hr at 240°	294
$Sn(CH_2CH(C_2H_5)C_4H_9)_4/SnCl_4$	$Sn(CH_2CH(C_2H_5)C_4H_9)_2Cl_2$, (65%)	3 hr at 230°	294
$Sn(n\text{-}hexyl\text{-}3,5,5\text{-}trimethyl)_4/SnCl_4$	$Sn(n\text{-}hexyl\text{-}3,5,5\text{-}trimethyl)_2Cl_2$, (75%)	3 hr at 230°	294
$Sn(biphenyl\text{-}p)_4/SnCl_4$	$Sn(biphenyl\text{-}p)_2Cl_2$, (80%)	2 hr at 185°	298
$Sn(biphenyl\text{-}p)_4/SnBr_4$	$Sn(biphenyl\text{-}p)_2Br_2$, (71%)	2 hr at 185°	298

(continued)

TABLE XIII (*Continued*)

Reactants	Products	Remarks	References
$Sn(C_{14}H_{12})_2/SnCl_4$[a]	$Sn(C_{14}H_{12})Cl_2$	3 hr at 150°	295
$Pb(CH_3)_4/PbCl_2(CH_3)_2$	$Pb(CH_3)_3Cl$, (39%)	0.5 hr at 80°	305
$Pb(C_6H_5)_4/PbCl_2(C_6H_5)_2$	$Pb(C_6H_5)_3Cl$, (86%)	18 hr at 117°	305
$As(CF_3)_3/AsCl_3$	$As(CF_3)_2Cl$, $As(CF_3)Cl_2$	20 hr at 210°	309
$As(CF_3)_3/AsI_3$	$As(CF_3)_2I$, (16%); $As(CF_3)I_2$, (1%)	48 hr at 240°	310,311
$As(CH=CH_2)_3/AsCl_3$	$As(CH=CH_2)_2Cl$, (23%); $As(CH=CH_2)Cl_2$, (53%)	5 hr at 100°	312
$As(CH=CH_2)_3/AsBr_3$	$As(CH=CH_2)_2Br$, (34%); $As(CH=CH_2)Br_2$, (38%)	3 hr at 130°	312
$As(C_2H_5)_3/AsBr_3$	$As(C_2H_5)Br_2$, (51%)	At r.t.	312
$As(C_4H_3O-\alpha)_3/AsCl_3$	$As(C_4H_3O-\alpha)_2Cl$, (36); $As(C_4H_3O-\alpha)Cl_2$		313
$As(C_6H_5)_3/AsCl_3$	$As(C_6H_5)_2Cl$, ($K = 0.08$); $As(C_6H_5)Cl_2$, ($K = 0.07$)	30 hr at 250°	307,308, 314,315
$As(CH_2C_6H_5)_3/AsCl_3$	$As(CH_2C_6H_5)Cl_2$	10–12 hr at 160–180°	306
$Sb(CH_3)_3/SbCl_3$	$Sb(CH_3)_2Cl$, $Sb(CH_3)Cl_2$	At 72°	21
$Sb(CH=CH_2)_3/SbCl_3$	$Sb(CH=CH_2)Cl_2$	At r.t.	312
$Sb(CH=CH_2)_3/SbBr_3$	$Sb(CH=CH_2)_2Br$, $Sb(CH=CH_2)Br_2$	0.5 hr at 65°	312
$Sb(C_6H_5)_3/SbCl_3$	$Sb(C_6H_5)_2Cl$, $Sb(C_6H_5)Cl_2$	48 hr at 240°	316–322
$Sb(C_6H_4CH_3-p)_3/SbCl_3$	$Sb(C_6H_4CH_3-p)Cl_2$	48 hr at 245°	317
$Bi(CH_3)_3/BiCl_3$	$Bi(CH_3)Cl_2$	At r.t.	323
$Bi(CH_3)_3/BiBr_3$	$Bi(CH_3)Br_2$	At r.t.	323
$Bi(C_2H_5)_3/BiCl_3$	$Bi(C_2H_5)Cl_2$	At r.t.	323
$Bi(i-C_4H_9)_3/BiBr_3$	$Bi(i-C_4H_9)Br_2$	At r.t.	324
$Bi(i-C_5H_{11})_3/BiBr_3$	$Bi(i-C_5H_{11})Br_2$	At r.t.	324
$Bi(C_6H_5)_3/BiCl_3$	$Bi(C_6H_5)_2Cl$	At r.t.	322,325, 327

78

$Bi(C_6H_5)_3/BiBr_3$	$Bi(C_6H_5)_2Br$	At r.t.	325,326
$Bi(C_6H_4Cl\text{-}p)_3/BiBr_3$	$Bi(C_6H_4Cl\text{-}p)_2Br$, $Bi(C_6H_4Cl\text{-}p)Br_2$	At r.t.	175
$Bi(C_6H_4CH_3\text{-}p)_3/BiCl_3$	$Bi(C_6H_4CH_3\text{-}p)_2Cl$, $Bi(C_6H_4CH_3\text{-}p)Cl_2$	At r.t.	326
$Bi(\alpha\text{-}C_{10}H_7)_3/BiCl_3$	$Bi(\alpha\text{-}C_{10}H_7)_2Cl$	At r.t.	326
$Bi(\alpha\text{-}C_{10}H_7)_3/BiBr_3$	$Bi(\alpha\text{-}C_{10}H_7)_2Br$, $Bi(\alpha\text{-}C_{10}H_7)Br_2$	At r.t.	326

[a] $Sn(C_{14}H_{12})_2$ =

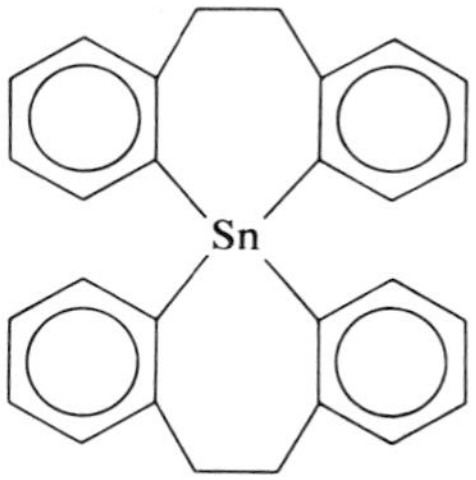

TABLE XIV

Redistribution Reactions Involving Exchange of Element–Nitrogen with Element–Oxygen Bonds

Reactants	Products	Remarks	References
$B[N(CH_3)_2]_3/B_2O_3$	$[B[N(CH_3)_2]O]_3$, (100%)	3 hr at 260–300°	188
$B(NHC_6H_5)(n\text{-}C_3H_7)_2/B(OCH_3)_2(n\text{-}C_6H_{13})$	$B(OCH_3)(n\text{-}C_3H_7)_2$, (95%); $B(NHC_6H_5)_2(C_6H_5)$	2 hr at 85–120°	329
$\{B[N(CH_3)_2]_2\}_2O$	$B[N(CH_3)_2]_3$, (86%); $[B[N(CH_3)_2]O]_3$, (90%)	3 hr at 80°	330
$\{B[N(n\text{-}C_3H_7)_2]_2\}_2O$	$B[N(n\text{-}C_3H_7)_2]_3$, (87%); $[B[N(n\text{-}C_3H_7)_2]O]_3$, (90%)	Reflux	330
$Si[N(CH_3)_2]_3(CH_3)/Si(OCH_3)_3(CH_3)$	$Si[N(CH_3)_2](OCH_3)_2(CH_3)$, (r); $Si[N(CH_3)_2]_2(OCH_3)(CH_3)$, (r)	At 120°; $AlCl_3$ as catalyst	331
$Si[N(CH_3)_2]_2(CH_3)_2/Si(OCH_3)_2(CH_3)_2$	$Si[N(CH_3)_2](OCH_3)(CH_3)_2$, (r)	At 120°; $AlCl_3$ as catalyst	331
$Si[N(CH_3)_2]_2(CH_3)_2/[SiO(CH_3)_2]$	$[(CH_3)_2N][Si(CH_3)_2O]_nSi(CH_3)_2[N(CH_3)_2]$	1 day at 200°; $AlCl_3$ as catalyst	332
$Ge[N(CH_3)_2]_4/Ge(OCH_3)_4$	$Ge[N(CH_3)_2]_3(OCH_3)$, ($K = 0.4$); $Ge[N(CH_3)_2]_2(OCH_3)_2$, ($K = 0.08$); $Ge[N(CH_3)_2](OCH_3)_3$, ($K = 0.019$)	At r.t.	260
$As[N(CH_3)_2]_3/As(OCH_3)_3$	$As[N(CH_3)_2](OCH_3)_2$, ($K = 0.31$); $As[N(CH_3)_2]_2(OCH_3)$, ($K = 0.1$)	At r.t.	334
$As[N(CH_3)_2]_3/As_2O_3$	Poly(dimethylamino arsenites)	8 hr at 150°	335

80

TABLE XV

Redistribution Reactions Involving Exchange of Element–Nitrogen with Element–Sulfur Bonds

Reactants	Products	Remarks	References
$B[N(CH_3)_2]_3/[BS(SH)]_3$	$\{BS[N(CH_3)_2]\}_3$, " $B(SH)_3$ "	At 80–140°	336,337
$[SiNCH_3(CH_3)_2]_3/[SiS(CH_3)_2]_3$	(structure) $K = 0.98$ (structure) $K = 0.32$	<360 hr at 120°	338,339

TABLE XVI

Redistribution Reactions Involving Exchange of Element–Nitrogen with Element–Halogen Bonds

Reactants	Products	Remarks	References
B[N(CH₃)₂]₃/BF₃	B[N(CH₃)₂]F₂, (90–96%)		344
BF₂[N(CH₃)₂]	BF₃, B[N(CH₃)₂]F	In the presence of N(CH₃)₃	340
B[N(CH₃)₂](CH₃)₂/BF₃	B(CH₃)₂F, B[N(CH₃)₂]F₂	At 0°	340
B[N(CH₃)₂](CH₃)₂/BF₃	B(CH₃)₂F, B[N(CH₃)₂]F₂	At r.t.	341
B(NHCH₃)₃/BF₃	B(NHCH₃)F₂, (47%)	At −30°	345
B(NHC₂H₅)₃/BF₃	B(NHC₂H₅)F₂, (small yield)		345
B(NHC₃H₇-n)/BF₃	B(NHC₃H₇-n)F₂, (25%)		345
B(NHC₃H₇-i)₃/BF₃	B(NHC₃H₇-i)F₂, (80%)	At −20°	345
B(NHC₄H₉-i)₃/BF₃	B(NHC₄H₉-i)F₂, (27%)		345
B(NHC₄H₉-t)₃/BF₃	B(NHC₄H₉-t)F₂, (94%)		345
B(NH₂)(CH₃)₂/BF₃	B(CH₃)₂F, B(NH₂)F₂	At r.t.	341
B[N(CH₃)₂]₃/BCl₃	B[N(CH₃)₂]₂Cl, (68%)		344,345a
B[N(CH₃)₂]₃/BCl₃/B(OC₄H₉-n)₃	B[N(CH₃)₂]Cl(OC₄H₉-n), (68%)	At −78°	342
B[N(CH₃)₂](CH₃)₂/BCl₃	B(CH₃)₂Cl, BCl₂[N(CH₃)₂]	At r.t.	341
B[N(C₂H₅)₂]₃/BCl₃	B[N(C₂H₅)₂]Cl₂	At r.t.	345a
B[N(C₆H₅)(CH₃)]₃/BCl₃	B[N(C₆H₅)(CH₃)]Cl₂, (46%)	At −78°	342
B[N(C₂H₅)₂]₂(C₆H₅)/BCl₃	B(C₆H₅)Cl₂, (~100%)	At r.t.	230
B[N(C₂H₅)₂]₂(C₆H₅)/BCl₃	B[N(C₂H₅)₂]Cl₂, (~100%)	At r.t.	230
B[N(CH₃)₂]₃/BBr₃	B[N(CH₃)₂]Br₂, (63%)		344
B[N(CH₃)₂]₃/[BSBr]₃	{BS[N(CH₃)₂]}₃, BBr₃	2 hr at 5°	337
B[N(CH₃)₂]₃/BI₃	B[N(CH₃)₂]I₂, (85%)		344
Al[N(CH₃)₂]₃/AlCl₃	Al[N(CH₃)₂]Cl₂	1 hr at 90°	70
Al[N(CH₃)₂]₃/AlCl₃	Al[N(CH₃)₂]₂Cl	1 hr at 90°	70

Si[N(CH$_3$)$_2$]$_4$/SiCl$_4$	Si[N(CH$_3$)$_2$]$_3$Cl, ($K = 6 \times 10^{-6}$); Si[N(CH$_3$)$_2$]$_2$Cl$_2$, ($K = 6 \times 10^{-4}$); Si[N(CH$_3$)$_2$]Cl$_3$, ($K = 6 \times 10^{-4}$)	At r.t.	346
Si[N(CH$_3$)$_2$]$_3$(CH$_3$)/SiCl$_3$(CH$_3$)	Si[N(CH$_3$)$_2$]$_2$Cl$_2$(CH$_3$), ($K \sim 10^{-3}$); Si[N(CH$_3$)$_2$]$_2$Cl(CH$_3$), ($K \sim 10^{-3}$);	At r.t.	331
Si[N(CH$_3$)$_2$]$_3$(CH$_3$)/SiBr$_3$(CH$_3$)	Si[N(CH$_3$)$_2$]Br$_2$(CH$_3$), ($K \sim 10^{-3}$); Si[N(CH$_3$)$_2$]$_2$Br(CH$_3$), ($K \sim 10^{-4}$)	<5 hr at 120°	347
Si[N(CH$_3$)$_2$]$_2$(CH$_3$)$_2$/SiCl$_2$(CH$_3$)$_2$	Si[N(CH$_3$)$_2$]Cl(CH$_3$)$_2$, ($K \sim 10^{-4}$)	At r.t.	331
Si[N(CH$_3$)$_2$]$_2$(CH$_3$)$_2$/SiBr$_2$(CH$_3$)$_2$	Si[N(CH$_3$)$_2$]Br(CH$_3$)$_2$, ($K \sim 10^{-4}$)	<15 hr at 120°	348
[SiNH(CH$_3$)$_2$]/SiCl$_2$(CH$_3$)$_2$	Si(CH$_3$)$_2$ClNHSi(CH$_3$)$_2$Cl and higher polymers	At reflux temperature	351–355
[SiNCH$_3$(CH$_3$)$_2$]$_3$/SiCl$_2$(CH$_3$)$_2$	Cl[Si(CH$_3$)$_2$NCH$_3$]$_n$Si(CH$_3$)$_2$Cl	<80 hr at 120°	350
[SiNCH$_3$(CH$_3$)$_2$]/SiCl$_2$(CH$_3$)$_2$	Cl[Si(CH$_3$)$_2$NCH$_3$]$_2$Si(CH$_3$)$_2$Cl, (32%)	96 hr at reflux	354
Si(NHC$_6$H$_5$)$_2$(CH$_3$)$_2$/Si(NCS)$_3$(C$_6$H$_5$)	Si(NCS)$_2$(CH$_3$)$_2$, Si(NHC$_6$H$_5$)$_3$(C$_6$H$_5$)		349
Si(NHC$_6$H$_5$)$_2$(CH$_3$)$_2$/Si(NCO)$_2$(C$_6$H$_5$)$_2$	Si(NCO)$_2$(CH$_3$)$_2$, Si(NHC$_6$H$_5$)$_2$(C$_6$H$_5$)$_2$		349
SiN$_3$(CH$_3$)$_3$/SiCl$_2$(CH$_3$)$_2$	Si(N$_3$)(CH$_3$)$_3$, (~100%)	AlCl$_3$ as catalyst	356
SiN$_3$(C$_6$H$_5$)$_3$/SiCl(C$_6$H$_5$)$_3$	Si(N$_3$)(C$_6$H$_5$)$_3$, (~100%)	AlCl$_3$ as catalyst	356
SiN$_3$(CH$_3$)$_3$/SiCl$_2$(C$_6$H$_5$)$_2$	Si(N$_3$)$_2$(C$_6$H$_5$)$_2$		356
SiN$_3$(CH$_3$)$_3$/SiCl$_3$(C$_6$H$_5$)	Si(N$_3$)$_3$(C$_6$H$_5$)		356
Ge[N(CH$_3$)$_2$]$_4$/GeCl$_4$	Ge[N(CH$_3$)$_2$]$_3$Cl, ($K = 10^{-4}$); Ge[N(CH$_3$)$_2$]$_2$Cl$_2$, ($K = 10^{-4}$); Ge[N(CH$_3$)$_2$]Cl$_3$, ($K = 10^{-3}$)	At r.t.	260
[Ge(NCH$_3$)Cl$_2$]/GeCl$_4$	N-methyl perchlorogermazanes		357
As[N(CH$_3$)$_2$]$_3$/AsF$_3$	As[N(CH$_3$)$_2$]F$_2$, ($K = 10^{-8}$); As[N(CH$_3$)$_2$]$_2$F, ($K = 1.6 \times 10^{-2}$)	At r.t.	334
As[N(CH$_3$)$_2$]$_3$/AsCl$_3$	As[N(CH$_3$)$_2$]Cl$_2$, ($K = 10^{-5}$); As[N(CH$_3$)$_2$]$_2$Cl, ($K = 10^{-5}$)	At r.t.	334
As[N(CH$_3$)$_2$]$_3$/AsBr$_3$	As[N(CH$_3$)$_2$]Br$_2$, ($K = 10^{-6}$); As[N(CH$_3$)$_2$]$_2$Br, ($K = 10^{-7}$)	At r.t.	334
As$_4$[N(CH$_3$)]$_6$/AsF$_3$	Poly(fluoroarsenous methylimides)	8 hr at 130°	358
As$_4$[N(CH$_3$)]$_6$/AsCl$_3$	Poly(chloroarsenous methylimides)	9 hr at 130°	359

TABLE XVII

Redistribution Reactions Involving Exchange of Element–Oxygen with Element–Oxygen Bonds

Reactants	Products	Remarks	References
B(OCH$_3$)$_3$/B$_2$O$_3$	[BO(OCH$_3$)]$_3$	Equilibrium mixture	188,364
B(OCH$_3$)$_3$/B$_2$O$_3$	[BO(OCH$_3$)]$_3$, (98%)	6 hr at 180°	363
B(OCH$_3$)$_3$/[B(n-C$_4$H$_9$)O]$_3$	B(n-C$_4$H$_9$)(OCH$_3$)$_2$, (6%)	Refluxing	191
B(OCH$_3$)$_3$/B(OC$_7$H$_{15}$-n)$_2$(n-C$_4$H$_9$)	B(OC$_7$H$_{15}$-n)$_3$, (83%)		329
B(OCH$_3$)$_2$(n-C$_6$H$_{13}$)/B(OC$_7$H$_{15}$-n)$_2$(n-C$_3$H$_7$)	B(OCH$_3$)$_2$(n-C$_3$H$_7$), (94%); B(OC$_7$H$_{15}$-n)$_2$(n-C$_6$H$_{13}$), (92%)	3 hr at 110–180°	329
B(OCH$_3$)$_2$(n-C$_4$H$_9$)/B(OC$_7$H$_{15}$-n)$_3$	B(OCH$_3$)$_3$, (80%); B(OC$_7$H$_{15}$-n)$_2$(n-C$_4$H$_9$), (77%)	2.5 hr at 160°	329
B(OCH$_3$)(n-C$_4$H$_9$)$_2$/B(OC$_7$H$_{15}$-n)$_3$	B(OCH$_3$)$_3$, (84%); B(OC$_7$H$_{15}$-n)(n-C$_4$H$_9$)$_2$, (72%)	4 hr at 125–190°	329
B(OCH$_3$)(n-C$_4$H$_9$)$_2$/B(OC$_7$H$_{15}$-n)$_2$(n-C$_3$H$_7$)$_2$	B(OCH$_3$)(n-C$_3$H$_7$)$_2$, (98%); B(OC$_7$H$_{15}$-n)(n-C$_4$H$_9$)$_2$, (98%)	At 75–150°	329
B(OC$_2$H$_5$)$_3$/B$_2$O$_3$	[BO(OC$_2$H$_5$)]$_3$, (98%)	10 hr at 250°	363
B(OC$_3$H$_7$-n)$_3$/B$_2$O$_3$	[BO(OC$_3$H$_7$-n)]$_3$, (93%)	3 hr refluxing	363
B(OC$_3$H$_7$-i)$_3$/B$_2$O$_3$	[BO(OC$_3$H$_7$-i)]$_3$, (93%)	3 hr refluxing	363
B(OC$_4$H$_9$-n)$_3$/B$_2$O$_3$	[BO(OC$_4$H$_9$-n)]$_3$, (96%)	8 hr at 240°	362,363
[B(OC$_4$H$_9$-n)(C$_6$H$_5$)]$_2$O	[BO(OC$_4$H$_9$-n)]$_3$, (96%); B(OC$_4$H$_9$-n)$_2$(C$_6$H$_5$), (96%)	At r.t.	330
B(OC$_4$H$_9$-i)$_3$/B$_2$O$_3$	[BO(OC$_4$H$_9$-i)]$_3$, (95%)	8 hr refluxing	363
B(OC$_4$H$_9$-s)$_3$/B$_2$O$_3$	[BO(OC$_4$H$_9$-s)]$_3$, (96%)	8 hr refluxing	363
B(OC$_6$H$_5$)$_3$/B$_2$O$_3$	[BO(OC$_6$H$_5$)]$_3$, (88%)	14 hr at 400°	363
B(OC$_7$H$_{15}$)$_3$(n-C$_3$H$_7$)$_2$/B(OCH$_3$)$_2$(n-C$_6$H$_{13}$)	B(OCH$_3$)(n-C$_3$H$_7$)$_2$, (100%); B(OC$_7$H$_{15}$-n)$_2$(n-C$_6$H$_{13}$), (94%)	3 hr at 75–180°	329
Si(OCH$_3$)$_4$/Si(OC$_2$H$_5$)$_4$	Si(OCH$_3$)$_3$(OC$_2$H$_5$), (r); Si(OCH$_3$)$_2$(OC$_2$H$_5$)$_2$, (r); Si(OCH$_3$)(OC$_2$H$_5$)$_3$, (r)	4 days at 150°	346
Si(OCH$_3$)$_4$/SiO$_2$	Polymeric methylsilicates	Several months at 110–135°	369
Si(OCH$_3$)$_3$(CH$_3$)/Si(OC$_2$H$_5$)$_3$(CH$_3$)	Si(OCH$_3$)$_2$(OC$_2$H$_5$)$_2$(CH$_3$), (r); Si(OCH$_3$)$_2$(OC$_2$H$_5$)(CH$_3$), (r)	7 days at 150°	331
Si(OCH$_3$)$_3$(CH$_3$)/[SiO$_{1.5}$(CH$_3$)]	Polymeric methoxypolymethylsiloxanes	At 150°	370
Si(OCH$_3$)$_3$(CH$_2$CH$_2$Cl)/Si(OC$_3$H$_7$-n)$_3$(CH$_2$CH$_2$Cl)	Si(OCH$_3$)$_2$(OC$_3$H$_7$-n)(CH$_2$CH$_2$Cl), Si(OCH$_3$)(OC$_3$H$_7$-n)$_2$(CH$_2$CH$_2$Cl),	16 hr at 150°	367
Si(OCH$_3$)$_3$(CH$_2$CH$_2$Cl)/Si(OC$_4$H$_9$-n)$_3$(CH$_2$CH$_2$Cl)	Si(OCH$_3$)$_2$(OC$_4$H$_9$-n)(CH$_2$CH$_2$Cl), Si(OCH$_3$)(OC$_4$H$_9$-n)$_2$(CH$_2$CH$_2$Cl)	16 hr at 150°	367

Si(OCH$_3$)$_3$(CH$_2$CH$_2$Cl)/Si(OC$_5$H$_{11}$-n)$_3$(CH$_2$CH$_2$Cl)	Si(OCH$_3$)$_3$(OC$_5$H$_{11}$-n)(CH$_2$CH$_2$Cl), Si(OCH$_3$)(OC$_5$H$_{11}$-n)$_2$(CH$_2$CH$_2$Cl)	16 hr at 150°	367
Si(OCH$_3$)$_2$(CH$_3$)$_2$/Si(OC$_2$H$_5$)$_2$(CH$_3$)$_2$	Si(OCH$_3$)(OC$_2$H$_5$)(CH$_3$)$_2$, (r)	7 days at 150°	331
Si(OCH$_3$)$_2$(CH$_3$)$_2$/SiO(CH$_3$)$_2$	CH$_3$O[Si(CH$_3$)$_2$O]$_n$Si(CH$_3$)$_2$(OCH$_3$)	2 days at 200°; AlCl$_3$ as catalyst	332
Si(OC$_2$H$_5$)$_4$/Si(OC$_4$H$_9$-n)$_4$	Si(OC$_2$H$_5$)$_3$(OC$_4$H$_9$-n), (r); Si(OC$_2$H$_5$)$_2$(OC$_4$H$_9$-n)$_2$, (r); Si(OC$_2$H$_5$)(OC$_4$H$_9$-n)$_3$, (r)	4 hr refluxing, MgAl(OC$_2$H$_5$)$_5$ as catalyst	366
Si(OC$_2$H$_5$)$_4$/Si(OC$_5$H$_{11}$-n)$_4$	Si(OC$_2$H$_5$)$_3$(OC$_5$H$_{11}$-n), (r); Si(OC$_2$H$_5$)$_2$(OC$_5$H$_{11}$-n)$_2$, (r); Si(OC$_2$H$_5$)(OC$_5$H$_{11}$-n)$_3$, (r)	4 hr refluxing, MgAl(OC$_2$H$_5$)$_5$ as catalyst	366
Si(OC$_2$H$_5$)$_4$/SiO$_2$	Polymeric ethyl silicates	Several months at 110–135°	369
Si(OC$_2$H$_5$)$_3$(CH$_2$CH$_2$Cl)/Si(OC$_4$H$_9$-n)$_3$(CH$_2$CH$_2$Cl)	Si(OCH$_3$)$_3$(OC$_4$H$_9$-n)(CH$_2$CH$_2$Cl), Si(OCH$_3$)(OC$_4$H$_9$-n)$_2$(CH$_2$CH$_2$Cl)	16 hr at 150°	367
Si(OC$_3$H$_7$-n)$_4$/SiO$_2$	Polymeric n-propyl silicates	Several months at 110–135°	369
Si(OC$_4$H$_9$-n)$_4$/SiO$_2$	Polymeric n-butyl silicates	Several months at 110–135°	369
[SiO(CH$_3$)$_2$]$_4$	Cyclics (ca. 15%); linears (ca. 85%)	At 110–178°; K$^+$ and other bases as catalyst	374,377, 379
[SiO(CH$_3$)$_2$]$_4$	Long-chain methylpolysiloxanes	H$_2$SO$_4$ as catalyst, at r.t.	372
[SiO(CH$_3$)$_2$]$_4$/[Si(CH$_3$)$_3$]$_2$O	Linear and cyclic methylpolysiloxanes	Acid clay catalysis	380,381
[SiO(CH$_3$)$_2$]$_4$/[Si(CH$_3$)$_3$]$_2$O	Linear and cyclic methylpolysiloxanes	At r.t., concentrated H$_2$SO$_4$ as catalyst	374
[SiO(CH$_3$)$_2$]$_4$/[Si(CH$_3$)$_3$]$_2$O	Long-chain methylpolysiloxanes	H$_2$SO$_4$ as catalyst, at r.t.	371
[SiO(CH$_3$)$_2$]$_4$/[Si(CH$_3$)$_2$(CH$_2$Br)]$_2$O	Linear and cyclic methylpolysiloxanes	5 hr at 150°; acid catalysts	376
Si(CH$_3$)$_3$O[Si(CH$_3$)$_2$O]$_2$Si(CH$_3$)$_3$	Cyclic and linear methylpolysiloxanes	HCl and solvent, at r.t.	375
[SiO(CH$_3$)$_2$]$_n$, (n = large)	[SiO(CH$_3$)$_2$]$_n$, (n = 4–8), (94%)	At 350–400°, NaOH as catalyst	371,372, 373

85

(continued)

TABLE XVII (*Continued*)

Reactants	Products	Remarks	References
[SiO(CH$_3$)(CH$_2$CH$_2$CF$_3$)]$_3$	Cyclics (87%), linears (13%)	49 hr at 60–150°, KOH as catalyst	378
[SiO$_{1.5}$(CH$_3$)]$_n$	Cage silsesquioxanes	300–380°, KOH as catalyst	382
[SiO$_{1.5}$(C$_2$H$_5$)]$_n$	Cage silsesquioxanes	300–380°, KOH as catalyst	382
[SiO$_{1.5}$(n-C$_3$H$_7$)]$_n$	Cage silsesquioxanes	300–380°, KOH as catalyst	382
[SiO$_{1.5}$(n-C$_4$H$_9$)]$_n$	Cage silsesquioxanes	300–380°, KOH as catalyst	382
[SiO$_{1.5}$(C$_6$H$_5$)]$_8$	Double chain polymers	At 250–275°, base catalyzed	383
[SiO$_{1.5}$(C$_6$H$_5$)]$_n$	Oligomeric cage molecules and double chain polymers	At high temperature, base catalysis	382,384, 385
[SiO$_{1.5}$(C$_6$H$_{11}$)]$_n$	Cage silsesquioxanes	300–380°, KOH as catalyst	382
[GeO(CH$_3$)$_2$]$_n$	[GeO(CH$_3$)$_2$]$_3$; [GeO(CH$_3$)$_2$]$_4$		388
Sn(OCH$_3$)$_2$(n-C$_4$H$_9$)$_2$/Sn(OOCCH$_3$)$_2$(n-C$_4$H$_9$)$_2$	Sn(OCH$_3$)(OOCCH$_3$)(n-C$_4$H$_9$)$_2$	At r.t.	390
Sn(OCH$_3$)$_2$(n-C$_4$H$_9$)$_2$/Sn(laurate)$_2$(n-C$_4$H$_9$)$_2$	Sn(OCH$_3$)(laurate)(n-C$_4$H$_9$)$_2$	At r.t.	390
Sn(OCH$_3$)$_2$(n-C$_4$H$_9$)$_2$/Sn(camphor-sulfonate)$_2$(n-C$_4$H$_9$)$_2$	Sn(OCH$_3$)(camphorsulfonate)(n-C$_4$H$_9$)$_2$,	At 40°	390
Sn(OC$_2$H$_5$)(n-C$_4$H$_9$)$_3$/Sn(OC$_4$H$_9$-n)(C$_2$H$_5$)$_3$	Sn(OC$_4$H$_9$-n)(n-C$_4$H$_9$)$_3$, Sn(OC$_2$H$_5$)(C$_2$H$_5$)$_3$		391
Sn(OC$_2$H$_5$)(n-C$_4$H$_9$)$_3$/Sn(OOCCH$_3$)(C$_2$H$_5$)$_3$	Sn(OOCCH$_3$)(n-C$_4$H$_9$)$_3$, Sn(OOCCH$_3$)(C$_2$H$_5$)$_3$		391
Sn(Ox)$_2$(CH$_3$)$_2$/Sn(tropolon)$_2$(CH$_3$)$_2$[a]	Sn(CH$_3$)$_2$(Ox)(tropolon)	Rapid exchange	392
[SnO(C$_2$H$_5$)$_2$]/Sn(OCS)$_2$(C$_2$H$_5$)$_2$[b]	(SCO)Sn(C$_2$H$_5$)$_2$OSn(C$_2$H$_5$)$_2$(OCS)	At ~ 80°	393
[SnO(n-C$_4$H$_9$)$_2$]/Sn(OOCCH$_3$)$_3$(n-C$_4$H$_9$)	(CH$_3$COO)(n-C$_4$H$_9$)$_2$SnOSn(n-C$_4$H$_9$)(OOCCH$_3$)$_2$	At r.t.	394
[SnO(n-C$_4$H$_9$)$_2$]/Sn(OOCCH$_3$)$_2$(n-C$_4$H$_9$)$_2$	(CH$_3$COO)Sn(n-C$_4$H$_9$)$_2$OSn(n-C$_4$H$_9$)$_2$(OOCCH$_3$)	At ~ 80°	393
[SnO(n-C$_4$H$_9$)$_2$]/Sn(OOCC$_7$H$_{15}$)$_2$(n-C$_4$H$_9$)$_2$	(C$_7$H$_{15}$COO)Sn(n-C$_4$H$_9$)$_2$OSn(n-C$_4$H$_9$)$_2$(OOCC$_7$H$_{15}$)	At ~ 80°	393
As(OCH$_3$)$_3$/As$_2$O$_3$	Polymethylarsenites	At r.t.	335,395
Sb(OC$_2$H$_5$)$_3$/Sb(OC$_6$H$_{13}$)$_3$	Sb(OC$_2$H$_5$)$_2$(OC$_6$H$_{13}$), Sb(OC$_2$H$_5$)(OC$_6$H$_{13}$)$_2$	18 hr at 95°	396

[a] Ox = 8-hydroxyxyquinolinate group; tropolon = tropolonate group.

[b] OCS = Camphorsulfonate.

TABLE XVIII
Redistribution Reactions Involving Exchange of Element–Oxygen with Element–Sulfur Bonds

Reactants	Products	Remarks	References
$B(OCH_3)_3/[BS(SH)]_3$	$[BS(OCH_3)]_3$	At 40°	336
$B(OCH_3)_3/B(SC_4H_9-n)_2(n-C_4H_9)$	$B(SC_4H_9-n)_3$, (81%); $B(OCH_3)_2(n-C_4H_9)$, (93%)		329
$B(OCH_3)_2(n-C_4H_9)/B(SC_4H_9-n)_3$	$B(OCH_3)_3$, (88%); $B(SC_4H_9-n)_2(n-C_4H_9)$, (91%)	25 hr at 75–125°	329
$B(OCH_3)_2(n-C_6H_{13})/B(SC_2H_5)_3$	$B(OCH_3)_3$, (95%); $B(n-C_6H_{13})(SC_2H_5)_2$, (93%)	At 110–150°	329
$B(OCH_3)_2(n-C_6H_{13})/B(SC_2H_5)(n-C_3H_7)_2$	$B(OCH_3)(n-C_3H_7)_2$, (97%); $B(SC_2H_5)_2(n-C_6H_{13})$, (91%)	2 hr at 80–125°	329
$B(OCH_3)_2(n-C_6H_{13})/B(SC_2H_5)_2(n-C_3H_7)$	$B(OCH_3)_2(n-C_3H_7)$, (94%); $B(SC_2H_5)_2(n-C_6H_{13})$, (93%)	At 80–150°	329
$B(OCH_3)(n-C_3H_7)_2/B(SC_2H_5)_3$	$B(SC_2H_5)(n-C_4H_9)_2$, (88%); $(BOCH_3)_3$, (84%)	At 75–150°	329
$B(OCH_3)(n-C_3H_7)_2/B(SC_4H_9-n)_3$	$B(OCH_3)_3$, (91%); $B(SC_4H_9-n)(n-C_3H_7)_2$, (88%)	At 75–150°	329
$B(OCH_3)(n-C_4H_9)_2/B(SC_2H_5)(n-C_3H_7)_2$	$B(OCH_3)(n-C_3H_7)_2$, (97%); $B(SC_2H_5)_2(n-C_6H_{13})$, (91%)	2 hr at 80–125°	329
$B(OCH_3)(n-C_4H_9)_2/B(SC_4H_9-n)(n-C_3H_7)_2$	$B(OCH_3)(n-C_3H_7)_2$, (98%); $B(SC_4H_9-n)(n-C_4H_9)_2$, (93%)	3 hr at 75–175°	329
$B(OC_7H_{15}-n)_2(n-C_3H_7)/B(SC_2H_5)_3$	$B(OC_7H_{15}-n)_3$, $B(SC_2H_5)_2(n-C_3H_7)$, (94%)	2.5 hr at 100–170°	329
$Si(OCH_3)_2(CH_3)_2/Si(SCH_3)_2(CH_3)_2$	$Si(OCH_3)(SCH_3)(CH_3)_2$, $(K = 2.9)$	<24 hr at 120°; $AlCl_3$ as catalyst	397
$Si(OCH_3)_2(CH_3)_2/[Si:S(CH_3)_2]_3$	$CH_3O[Si(CH_3)_2S]_nSi(CH_3)_2(OCH_3)$	20 days at 200°	398
$Ge(OCH_3)_2(CH_3)_2/Ge(SCH_3)_2(CH_3)_2$	$Ge(OCH_3)(SCH_3)(CH_3)_2$, $(K = 2.6)$	<46 hr at 120°	397
$[GeO(CH_3)_2]_{3,4}/[GeS(CH_3)_2]_3$	$Ge(CH_3)_2OGe(CH_3)_2OGe(CH_3)_2S$, $(K = 0.6)$; $Ge(CH_3)_2OGe(CH_3)_2SGe(CH_3)_2S$, $(K = 0.9)$	At 120°	399
$Sn(OCH_3)_2(n-C_4H_9)_2/[SnS(n-C_4H_9)_2]_3$	$(CH_3O)Sn(n-C_4H_9)_2SSn(n-C_4H_9)_2(OCH_3)$	At r.t.	400
$Sn(OOCC_7H_{15})_2(n-C_4H_9)_2/[SnS(n-C_4H_9)_2]_3$	$(C_7H_{15}COO)Sn(n-C_4H_9)_2SSn(n-C_4H_9)_2(OOCC_7H_{15})$	At r.t.	400

TABLE XIX

Redistribution Reactions Involving Exchange of Element–Oxygen with Element–Halogen Bonds

Reactants	Products	Remarks	References
$B(OCH_3)_2H/BF_3$	BHF_2		409
$B(OCH_3)_2H/BF_3$	BHF_2		98
$B(OCH_3)_2(C_6H_5)/BCl_2(C_6H_5)$	$B(OCH_3)Cl(C_6H_5)$, (92%)	At r.t.	408
$B(OC_2H_5)_3/BCl_3$	$B(OC_2H_5)Cl_2$, (92%)	1–3 hr at r.t.	406
$B(OC_2H_5)_3/BCl_3$	$B(OC_2H_5)_2Cl$, (100%)	At $-80°$	406
$B(OC_2H_5)_3/BCl_3$	$B(OC_2H_5)Cl_2$, $B(OC_2H_5)_2Cl$	At $-80°$	401
$B(OC_2H_5)_2(C_6H_5)/BCl_3$	$BCl_2(C_6H_5)$, (80%)	1 hr at r.t.	403
$B(OC_2H_5)_2(C_6H_5)/BCl_2(C_6H_5)$	$B(OC_2H_5)Cl(C_6H_5)$, (93%)	At 20°; FeCl$_3$ as catalyst	408
$B(OCH_2CH_2Cl)_3/BCl_3$	$B(OCH_2CH_2Cl)_2Cl$, $B(OCH_2CH_2Cl)Cl_2$	1–3 hr at r.t.	402
$B(OCH_2CH_2Cl)_2(C_6H_5)/BCl_2(C_6H_5)$	$B(OCH_2CH_2Cl)Cl(C_6H_5)$, (81%)	1–3 hr at r.t.	408
$B(OC_3H_7\text{-}i)_2(C_6H_5)/BCl_2(C_6H_5)$	$B(OC_3H_7\text{-}i)Cl(C_6H_5)$, (96%)	1–3 hr at r.t.	408
$B(OC_4H_9\text{-}n)_2(n\text{-}C_4H_9)/BF_3$	$BF_2(n\text{-}C_4H_9)$, (81%); $B(OC_4H_9\text{-}n)F_2$, (85%)	At 0°	403
$B(OC_4H_9\text{-}n)_2(n\text{-}C_4H_9)/BF_3$	$BF_2(n\text{-}C_4H_9)$, (88%); $B(OC_4H_9\text{-}n)_3$, (98%)	At 0°	403
$B(OC_4H_9\text{-}n)_2(C_6H_5)/BF_3$	$B(OC_4H_9\text{-}n)F(C_6H_5)$, (72%); $B(OC_4H_9\text{-}n)F_2$	At 0°	403
$B(OC_4H_9\text{-}n)_2(n\text{-}C_4H_9)/B(OC_4H_9\text{-}n)F_2$	$B(OC_4H_9\text{-}n)F(n\text{-}C_4H_9)$, (93%); $B(OC_4H_9\text{-}n)_3$, (97%)	At 23°	403
$B(OC_4H_9\text{-}n)F(n\text{-}C_4H_9)$	$BF_2(n\text{-}C_4H_9)$, (72%); $B(OC_4H_9\text{-}n)_2(n\text{-}C_4H_9)$, (98%)	In the presence of pyridine	404
$B(OC_4H_9\text{-}n)_2(n\text{-}C_4H_9)/BCl_3$	$B(OC_4H_9\text{-}n)Cl_2$, (93%); $B(n\text{-}C_4H_9)(OC_4H_9\text{-}n)Cl$, (64%)	At 70°	403
$B(OC_4H_9\text{-}n)_2(C_6H_5)/BCl_3$	$B(OC_4H_9\text{-}n)Cl_2$, (70%); $B(OC_4H_9\text{-}n)Cl(C_6H_5)$, (48%)	At 15°	403
$B(OC_4H_9\text{-}n)_2(C_6H_5)/BCl_3$	$BCl_2(C_6H_5)$, (81%)	At 20°; FeCl$_3$ as catalyst	403

B(OC$_4$H$_9$-n)$_2$(n-C$_4$H$_9$)/B(OC$_4$H$_9$-n)Cl$_2$	B(OC$_4$H$_9$-n)$_2$Cl(n-C$_4$H$_9$), (76%);	13 hr at 20°	403
B(OC$_4$H$_9$-n)$_2$(C$_6$H$_5$)/BCl$_2$(C$_6$H$_5$)	B(OC$_4$H$_9$-n)$_3$, B(OC$_4$H$_9$-n)Cl(C$_6$H$_5$), (97%);	At 10°	403
B(OC$_4$H$_9$-n)Cl(n-C$_4$H$_9$)	BCl$_2$(n-C$_4$H$_9$), (96%); B(OC$_4$H$_9$-n)$_2$(C$_4$H$_9$), (93%)	In the presence of pyridine	404
B(OC$_4$H$_9$-i)$_2$(C$_6$H$_5$)/BClBCl$_2$(C$_6$H$_5$)	B(OC$_4$H$_9$-i)Cl(C$_6$H$_5$), (93%)	1–3 hr at r.t.	408
B(OC$_4$H$_9$-s)$_2$(C$_6$H$_5$)/BCl$_3$	BCl$_2$(C$_6$H$_5$), (76%)	At 20°	403
B(OC$_4$H$_9$-s)$_2$(C$_6$H$_5$)/BCl$_2$(C$_6$H$_5$)	B(OC$_4$H$_9$-s)Cl(C$_6$H$_5$), (94%)	1–3 hr at r.t.	408
B(OC$_4$H$_9$-t)$_2$(C$_6$H$_5$)/BCl$_2$(C$_6$H$_5$)	B(OC$_4$H$_9$-t)Cl(C$_6$H$_5$), (89%)	1–3 hr at r.t.	408
B(OC$_5$H$_{11}$-n)$_3$/BCl$_3$	B(OC$_5$H$_{11}$-n)Cl$_2$, (85%)	At −80°	406
B(OC$_5$H$_{11}$-n)$_3$/BCl$_3$	B(OC$_5$H$_{11}$-n)$_2$Cl, (100%)	At −80°	406
B(OC$_6$H$_2$Cl$_3$-2,4,6)$_3$/BCl$_3$	B(OC$_6$H$_2$Cl$_3$-2,4,6)$_2$Cl, (100%)	At −70°	405
B(OC$_6$H$_4$NO$_2$-o)$_3$/BCl$_3$	B(OC$_6$H$_4$NO$_2$-o)Cl$_2$, (98%)	At −70°	405
B(OC$_6$H$_5$)$_3$/BCl$_3$	B(OC$_6$H$_5$)$_2$Cl$_2$, ($K = 1 \times 10^{-4}$); B(OC$_6$H$_5$)$_2$Cl, ($K = 1 \times 10^{-4}$)	At r.t.	176
B(OC$_6$H$_5$)$_2$(C$_6$H$_5$)/BCl$_2$(C$_6$H$_5$)	B(OC$_6$H$_5$)Cl(C$_6$H$_5$), (62%)	1–3 hr at r.t.	408
B(OC$_6$H$_5$)Cl(C$_6$H$_5$)	B(OC$_6$H$_5$)$_2$(C$_6$H$_5$), (94%); BCl$_2$(C$_6$H$_5$), (56%)	At 150°/0.05 mm	408
B(OC$_7$H$_{15}$-n)$_2$(C$_6$H$_5$)/BCl$_2$(C$_6$H$_5$)	B(OC$_7$H$_{15}$-n)Cl(C$_6$H$_5$), (80%)	1–3 hr at r.t.	408
B(OC$_6$H$_4$CH$_3$-o)$_2$(C$_6$H$_5$)/BCl$_2$(C$_6$H$_5$)	B(OC$_6$H$_4$CH$_3$-o)Cl(C$_6$H$_5$), (95%)	1–3 hr at r.t.	408
B(OCH$_2$C$_6$H$_5$)(C$_6$H$_5$)/BCl$_2$(C$_6$H$_5$)	B(OCH$_2$C$_6$H$_5$)Cl(C$_6$H$_5$), (81%)	1–3 hr at r.t.	408
B[OCH(CH$_3$)(C$_6$H$_{13}$-n)]$_2$(C$_6$H$_5$)/BCl$_2$(C$_6$H$_5$)	B[OCH(CH$_3$)(C$_6$H$_{13}$-n)]Cl(C$_6$H$_5$), (77%)	1–3 hr at r.t.	408
B(OCH$_2$)$_2$(OC$_4$H$_9$-n)/BCl$_3$	B(OCH$_2$)$_2$Cl, (94%); B(OC$_4$H$_9$-n)Cl$_2$	At −80°	407
B[O(CH$_2$)$_3$Cl]$_3$/BCl$_3$	B[O(CH$_2$)$_3$Cl]Cl$_2$, (56%)	At −80°	406
B[O(CH$_2$)$_3$Cl]$_3$/BCl$_3$	B[O(CH$_2$)$_3$Cl]$_2$Cl, (100%)	At −80°	406
B[O(CH$_2$)$_4$Cl]$_3$/BCl$_3$	B[O(CH$_2$)$_4$Cl]$_2$Cl, B[O(CH$_2$)$_4$Cl]Cl$_2$		402
B[O(CH$_2$)$_4$Cl]$_3$/BCl$_3$	B[O(CH$_2$)$_4$Cl]Cl$_2$, (100%)	At −80°	406
B[O(CH$_2$)$_5$Cl]$_3$/BCl$_3$	B[O(CH$_2$)$_5$Cl]$_2$Cl$_2$, (100%)	At −80°	406
B[O(CH$_2$)$_5$Cl]$_3$/BCl$_3$	B[O(CH$_2$)$_5$Cl]$_2$Cl, (100%)	At −80°	406

(continued)

89

TABLE XIX (*Continued*)

Reactants	Products	Remarks	References
[BO(CH₃)]₃/BF₃	BF₂(CH₃), (~90%)	10–40 hr at 25–100°	410
[B(CH₃)₂]₂O/BF₃	BF(CH₃)₂, (60–80%)	1–27 hr at 25°	410
[BO(C₂H₅)]₃/BCl₃	BCl₂(C₂H₅), (60%)	At reflux temp.	413
[BO(C₂H₅)]₃/BBr₃	BBr₂(C₂H₅), (58%)		415
[BO(n-C₃H₇)]₃/BCl₃	BCl₂(n-C₃H₇), (50–60%)		413
[BO(n-C₃H₇)]₃/BBr₃	BBr₂(n-C₃H₇), (58%)		415
[BO(i-C₃H₇)]₃/BCl₃	BCl₂(i-C₃H₇), (50–60%)		413
[BO(n-C₄H₉)]₃/BF₃	BF₂(n-C₄H₉)		411
[BO(n-C₄H₉)]₃BBr₃	BBr₂(n-C₄H₉), (72%)		415
[BO(i-C₄H₉)]₃/BCl₃	BCl₂(i-C₄H₉), (50–60%)		413
[BO(i-C₄H₉)]₃/BBr₃	BBr₂(i-C₄H₉), (42%)		416
[BO(s-C₄H₉)]₃/BCl₃	BCl₂(s-C₄H₉), (50–60%)		413
[BO(t-C₄H₉)]₃/BCl₃	BCl₂(t-C₄H₉), (50–60%)		413
[BO(n-C₅H₁₁)]₃/BF₃	BF₂(n-C₅H₁₁)		411
[BO(n-C₅H₁₁)]₃/BCl₃	BCl₂(n-C₅H₁₁), (50–60%)		413
[BO(i-C₅H₁₁)]₃/BBr₃	BBr₂(i-C₅H₁₁), (77.4%)		415
[BO(2-sec-C₅H₁₁)]₃/BF₃	BF₂(2-sec-C₅H₁₁)		411
[BO(t-C₅H₁₁)]₃/BF₃	BF₂(t-C₅H₁₁)		411
[BO(C₆H₅)]₃/BF₃	BF₂(C₆H₅), (16%)		414
[BO(C₆H₅)]₃/BCl₃	BCl₂(C₆H₅), (55%)	At r.t.	414
[BO(C₆H₅)]₃/BCl₃	BCl₂(C₆H₅), (90%); B₂O₃	At −80°	412
O[B(C₆H₅)₂]₂/BCl₃	BCl(C₆H₅)₂, (76%); B₂O₃	At −80°	412
[BO(C₆H₅)]₃/BBr₃	BBr₂(C₆H₅), (54%)		224
[BO(cyclo-C₆H₁₁)]₃/BF₃	BF₂(cyclo-C₆H₁₁)		411
[BO(cyclo-C₆H₁₁)]/BCl₃	BCl₂(cyclo-C₆H₁₁), (50–60%)		413
[BO(n-C₆H₁₃)]₃/BF₃	BF₂(n-C₆H₁₃)		411

Reactants	Products	Conditions	Ref.
$[BO(n\text{-}C_6H_{13})]_3/BCl_3$	$BCl_2(n\text{-}C_6H_{13})$, (50%)		413
$[BO(p\text{-}CH_3C_6H_4)]_3/BF_3$	$BF_2(p\text{-}CH_3C_6H_4)$, (10–15%)		414
$[BO(p\text{-}CH_3C_6H_4)]_3/BCl_3$	$BCl_2(p\text{-}CH_3C_6H_4)$, (36–58%)	At r.t.	414
$Si(OCH_3)_4/SiCl_4$	$Si(OCH_3)_3Cl$, ($K = 0.009$); $Si(OCH_3)_2Cl_2$, ($K = 0.042$); $Si(OCH_3)Cl_3$, ($K = 0.094$)	200 hr at 120°	346
$Si(OCH_3)_4/Si(NCO)_4$	Reaction too slow	Through hot tube at 450°	420
$Si(OCH_3)_3(CH_3)/SiCl_3(CH_3)$	$Si(OCH_3)_2Cl_2(CH_3)$, ($K = 0.04$); $Si(OCH_3)_2Cl(CH_3)$, ($K = 0.02$)	At 120°	331
$Si(OCH_3)_3(CH_3)/SiBr_3(CH_3)$	$Si(OCH_3)_2Br(CH_3)$, ($K = 0.013$); $Si(OCH_3)_2Br_2(CH_3)$, ($K = 0.058$)		370
$Si(OCH_3)_2(CH_3)_2/SiCl_2(CH_3)_2$	$Si(OCH_3)Cl(CH_3)_2$, ($K = 0.01$)	At 120°	331
$Si(OCH_3)_2(CH_3)_2/SiBr_2(CH_3)_2$	$Si(OCH_3)Br(CH_3)_2$, ($K = 6.4 \times 10^{-3}$)	<0.5 hr at 120°	423
$Si(OCH_3)_2(CH_3)_2/Si(NCO)_2(CH_3)_2$	$Si(OCH_3)(NCO)(CH_3)_2$, ($K = 0.058$)	<20 hr at 120°	424
$Si(OC_2H_5)_4/SiCl_4$	$Si(OC_2H_5)_3Cl$, $Si(OC_2H_5)_2Cl_2$, $Si(OC_2H_5)Cl_3$	At 160°	417–419
$Si(OC_2H_5)_3(CH_3)/SiCl_3(CH_3)$	$Si(OC_2H_5)_2Cl(CH_3)$, (32%); $Si(OC_2H_5)_2Cl_2(CH_3)$, (41%)	30 hr at 180°	422
$Si(OC_2H_5)_3(C_2H_5)/SiCl_3(C_2H_5)$	$Si(OC_2H_5)_2Cl(C_2H_5)$, (42%); $Si(OC_2H_5)Cl_2(C_2H_5)$, (43%)	40 hr at 200°	422
$Si(OC_2H_5)_2(CH_3)_2/SiCl_2(CH_3)_2$	$Si(OC_2H_5)Cl(CH_3)_2$, (76%)	35 hr at 180°	422
$Si(OC_2H_5)_2(C_2H_5)_2/SiCl_2(C_2H_5)_2$	$Si(OC_2H_5)Cl(C_2H_5)_2$, (82%)	40 hr at 100°	422
$Si(OC_4H_9\text{-}n)_4/SiCl_4$	$Si(OC_4H_9\text{-}n)_3Cl$, $Si(OC_4H_9\text{-}n)_2Cl_2$, $Si(OC_4H_9\text{-}n)Cl_3$	3 hr at 200°	421
$Si(OC_6H_5)_3(CH_3)/SiCl_3(CH_3)$	$Si(OC_6H_5)_2Cl(CH_3)$, ($K = 0.18$); $Si(OC_6H_5)Cl_2(CH_3)$, ($K = 0.22$)	<163 days at 150°	425
$Si(OC_6H_5)_3(CH_3)/SiBr_3(CH_3)$	$Si(OC_6H_5)_2Br(CH_3)$, ($K = 0.28$); $Si(OC_6H_5)Br_2(CH_3)$, ($K = 0.38$)	<27 days at 150°	425
$Si(OC_6H_5)_2(CH_3)_2/SiCl_2(CH_3)_2$	$Si(OC_6H_5)Cl(CH_3)_2$, ($K = 0.16$)	<17 days at 150°	348

91

(continued)

TABLE XIX (Continued)

Reactants	Products	Remarks	References
$Si(OC_6H_5)_2(CH_3)_2/SiBr_2(CH_3)_2$	$Si(OC_6H_5)Br(CH_3)_2$, $(K = 0.14)$	<190 hr at 150°	348
$[SiO_{1.5}(CH_3)]/SiBr_3(CH_3)$	Bromopolymethylsiloxanes	At 150; H_3PO_4 as catalyst	370
$[SiO(CH_3)_2]_3/SiCl_2(CH_3)_2$	$Cl[Si(CH_3)_2O]_nSi(CH_3)_2Cl$	At 200°; $AlCl_3$ as catalyst	332,428, 429,434
$[SiO(CH_3)_2]_3/SiBr_2(CH_3)_2$	$Br[Si(CH_3)_2O]_nSi(CH_3)_2Br$	At 200°	435
$[SiO(CH_3)_2]_3/SiCl_3(CH_2CH_2Cl)$	$Cl[Si(CH_3)_2O]_nSi(CH_2CH_2Cl)Cl_2$	5 hr at 250°	433
$[SiO(CH_3)_2]_3/SiCl_3(CH_2CH_2CN)$	$Cl[Si(CH_3)_2O]_nSi(CH_2CH_2CN)Cl_2$	5 hr at 250°	433
$[SiO(CH_3)_2]_3/SiCl_3(CH_2CH_2CH_2Cl)$	$Cl[Si(CH_3)_2O]_nSi(CH_2CH_2CH_2Cl)Cl_2$	5 hr at 250°	433
$[SiO(CH_3)_2]_3/SiCl_2(CH_3)(CH_2Cl)$	$Cl[Si(CH_3)_2O]_n[Si(CH_3)(CH_2Cl)O]_mSi(CH_3)_2Cl$	5 hr at 250°	433
$[SiO(CH_3)_2]_3/SiCl_2(CH_3)(C_2H_3)$	$Cl[Si(CH_3)_2O]_n[Si(CH_3)(C_2H_3)O]_mSi(CH_3)_2Cl$	5 hr at 250°	432
$[SiO(CH_3)_2]_3/SiCl_2(CH_3)(C_2H_3)$	$Cl[Si(CH_3)(C_2H_3)O]_n[Si(CH_3)_2O]_mSi(CH_3)_2Cl$	5 hr at 250°	434
$[SiO(CH_3)_2]_3/SiCl_2(CH_3)(C_6H_5)$	$Cl[Si(CH_3)_2O]_n[Si(CH_3)(C_6H_5)O]_mSi(CH_3)_2Cl$	5 hr at 250°	432
$[SiO(CH_3)_2]_3/SiCl_2H(CH_3)$	$Cl[Si(CH_3)_2O]_n[Si(CH_3)HO]_mSi(CH_3)_2Cl$	5 hr at 250°	432
$[SiO(CH_3)_2]_3/SiCl_2H(CH_3)$	$Cl[Si(CH_3)HO]_n[Si(CH_3)_2O]_nSi(CH_3)_2Cl$	5 hr at 250°	434
$[SiO(CH_3)_2]_3/SiCl_2(CH_3)(CH_2Cl)$	$Cl[Si(CH_3)_2O]_n[Si(CH_3)(CH_2Cl)O]_mSi(CH_3)_2Cl$	5 hr at 250°	433
$[SiO(CH_3)_2]_4/SiCl_3(CH_3)$	$Cl[Si(CH_3)_2O]_nSiCH_3Cl_2$	5 hr at 250°	431
$[SiO(CH_3)_2]_4/SiCl_3(C_2H_5)$	$Cl[Si(CH_3)_2O]_nSi(C_2H_5)Cl_2$	5 hr at 250°	431
$[SiO(CH_3)_2]_4/SiCl_3(CHCHCH_2)$	$Cl[Si(CH_3)_2O]_nSi(CHCHCH_2)Cl_2$	5 hr at 250°	431
$[SiO(C_2H_5)_2]_3/SiCl_2(CH_3)_2$	$Cl[Si(CH_3)_2O]_m[Si(C_2H_5)_2O]_nSi(CH_3)_2Cl$	5 hr at 250°	430
$[SiO(CH_3)(C_6H_5)]_3/SiCl_2(CH_3)_2$	$Cl[Si(CH_3)_2O]_m[Si(CH_3)(C_6H_5)]_mSi(CH_3)_2Cl$	5 hr at 250°	430,434
$Si_2O(CH_3)_6/SiCl(C_2H_5)_3$	$Si_2O(CH_3)_3(C_2H_5)_3$, $(60-70\%)$	Catalyzed by metal halides	427
$Si_2O[(CH_3)(C_2H_5)_2]_2/SiCl(C_2H_5)_3$	$Si_2O[(CH_3)(C_2H_5)_2](CH_3)_3$	Catalyzed by metal halides	427
$Si_2O(CH_3)_6/SiBr_2(CH_3)_2$	$Si(CH_3)_2[OSi(CH_3)_3]_2$, (75%); $SiBr(CH_3)_3$	Catalyzed by metal halides	427
$Si_2O(CH_3)_6/SiCl_2(C_2H_5)_2$	$Si(C_2H_5)_2[OSi(CH_3)_3]_2$, (54%); $SiCl(CH_3)_3$	Catalyzed by metal salts	427

$Si_2O(CH_3)_6/SiCl_2H(C_2H_5)$	$Si(C_2H_5)H[OSi(CH_3)_3]_2$, (35%); $SiCl(CH_3)_3$	Catalyzed by metal salts	427
$Si_2O(CH_3)_6/SiCl(C_2H_5)_3$	$Si_2O(C_2H_5)_6$, (54%); $SiCl(CH_3)_3$, (75%)	Catalyzed by metal halides	427
$Si_2O(CH_3)_6/SiBr(n\text{-}C_3H_7)_3$	$Si_2O(n\text{-}C_3H_7)_6$, (60%); $SiBr(CH_3)_3$, (80%)	Catalyzed by metal halides	427
$Ge(OCH_3)_4/GeCl_4$	$Ge(OCH_3)_3Cl$, ($K = 0.01$); $Ge(OCH_3)_2Cl_2$, ($K = 0.03$); $Ge(OCH_3)Cl_3$, ($K = 0.05$)	At r.t.	260
$Ge(OCH_3)_3(CH_3)/GeCl_3(CH_3)$	$Ge(OCH_3)_2Cl(CH_3)$, ($K = 0.021$); $Ge(OCH_3)Cl_2(CH_3)$, ($K = 0.013$)	At r.t.	436
$Ge(OCH_3)_3(CH_3)/GeBr_3(CH_3)$	$Ge(OCH_3)_2Br(CH_3)$, ($K = 0.040$); $Ge(OCH_3)Br_2(CH_3)$, ($K = 0.021$)	At r.t.	436
$Ge(OCH_3)_3(CH_3)/GeI_3(CH_3)$	$Ge(OCH_3)_2I(CH_3)$, ($K = 0.15$); $Ge(OCH_3)I_2(CH_3)$, ($K = 0.11$)	At r.t.	436
$[GeO(CH_3)_2]_4/GeCl_2(CH_3)_2$	$Cl[Ge(CH_3)_2O]_nGe(CH_3)_2Cl$	At 120°	437
$[GeO(CH_3)_2]_4/GeBr_2(CH_3)_2$	$Br[Ge(CH_3)_2O]_nGe(CH_3)_2Br$	At 120°	437
$[GeO(CH_3)_2]_4/GeI_2(CH_3)_2$	$I[Ge(CH_3)_2O]_nGe(CH_3)_2I$	At 120°	437
$Sn(OCH_3)_2(n\text{-}C_3H_7)_2/SnF_2(n\text{-}C_3H_7)_2$	$Sn(OCH_3)F(n\text{-}C_3H_7)_2$	At 40°	390
$Sn(OCH_3)_2(n\text{-}C_4H_9)_2/SnF_2(n\text{-}C_4H_9)_2$	$Sn(OCH_3)F(n\text{-}C_4H_9)_2$	At 40°	390
$Sn(OCH_3)_2(CH_3)_2/SnCl_2(CH_3)_2$	$Sn(OCH_3)Cl(CH_3)_2$	At r.t.	390
$Sn(OCH_3)_2(C_2H_5)_2/SnCl_2(C_2H_5)_2$	$Sn(OCH_3)Cl(C_2H_5)_2$	At r.t.	390
$Sn(OCH_3)_2(n\text{-}C_4H_9)_2/SnCl_2(n\text{-}C_4H_9)_2$	$Sn(OCH_3)Cl(n\text{-}C_4H_9)_2$	At r.t.	390
$Sn(OCH_3)_2(C_2H_5)_2/SnBr_2(C_2H_5)_2$	$Sn(OCH_3)Br(C_2H_5)_2$	At r.t.	390
$Sn(OCH_3)_2(n\text{-}C_4H_9)_2/SnBr_2(n\text{-}C_4H_9)_2$	$Sn(OCH_3)Br(n\text{-}C_4H_9)_2$	At r.t.	390
$Sn(OCH_3)_2(CH_3)_2/SnI_2(CH_3)_2$	$Sn(OCH_3)I(CH_3)_2$	At r.t.	390
$Sn(OCH_3)_2(n\text{-}C_4H_9)_2/SnI_2(n\text{-}C_4H_9)_2$	$Sn(OCH_3)I(n\text{-}C_4H_9)_2$	At r.t.	390
$Sn(OCH_3)_2(n\text{-}C_4H_9)_2/Sn(SCN)_2(n\text{-}C_4H_9)_2$	$Sn(OCH_3)(SCN)(n\text{-}C_4H_9)_2$	At r.t.	390
$Sn(OOCCH_3)_2(n\text{-}C_4H_9)_2/SnCl_2(n\text{-}C_4H_9)_2$	$Sn(OOCCH_3)Cl(n\text{-}C_4H_9)_2$	At r.t.	390
$Sn(OOCCH_3)_2(n\text{-}C_4H_9)_2/SnBr_2(n\text{-}C_4H_9)_2$	$Sn(OOCCH_3)Br(n\text{-}C_4H_9)_2$	At r.t.	390
$Sn(OOCCH_3)_2(n\text{-}C_4H_9)_2/SnI_2(n\text{-}C_4H_9)_2$	$Sn(OOCCH_3)I(n\text{-}C_4H_9)_2$	At r.t.	282
$Sn(Ox)_2(CH_3)_2/SnCl_2(CH_3)_2$[a]	$Sn(Ox)Cl(CH_3)_2$	At r.t.	439

(continued)

TABLE XIX (*Continued*)

Reactants	Products	Remarks	References
Sn(Ox)$_2$(C$_6$H$_5$)$_2$/SnCl$_2$(C$_6$H$_5$)$_2$[a]	Sn(Ox)Cl(C$_6$H$_5$)$_2$, (85%)	At 80°	439,440
[SnO(CH$_3$)$_2$]/SnCl$_3$(C$_2$H$_5$)	Cl(CH$_3$)$_2$SnOSn(C$_2$H$_5$)Cl$_2$	At r.t.	394
[SnO(CH$_3$)$_2$]/SnCl$_2$(CH$_3$)$_2$	Cl(CH$_3$)$_2$SnOSn(CH$_3$)$_2$Cl	At r.t.	393,441
[SnO(C$_2$H$_5$)$_2$]/SnCl$_2$(C$_2$H$_5$)$_2$	Cl(C$_2$H$_5$)$_2$SnOSn(C$_2$H$_5$)$_2$Cl	At ~ 70°	393,441
SnO(C$_2$H$_5$)$_2$]/SnBr$_2$(C$_2$H$_5$)$_2$	Br(C$_2$H$_5$)$_2$SnOSn(C$_2$H$_5$)$_2$Br	At ~ 80°	441
[SnO(n-C$_3$H$_7$)$_2$/SnCl$_2$(n-C$_3$H$_7$)$_2$	Cl(n-C$_3$H$_7$)$_2$SnOSn(n-C$_3$H$_7$)$_2$Cl	At ~ 70°	441
[SnO(n-C$_3$H$_7$)$_2$]/SnBr$_2$(n-C$_3$H$_7$)$_2$	Br(n-C$_3$H$_7$)$_2$SnOSn(n-C$_3$H$_7$)$_2$Br	At ~ 80°	441
[SnO(n-C$_4$H$_9$)$_2$]/SnF$_2$(n-C$_4$H$_9$)$_2$	F(n-C$_4$H$_9$)$_2$SnOSn(n-C$_4$H$_9$)$_2$F	At ~ 80°	393
[SnO(n-C$_4$H$_9$)$_2$]/SnCl$_3$(n-C$_4$H$_9$)	Cl(n-C$_4$H$_9$)$_2$SnOSn(n-C$_4$H$_9$)Cl$_2$	At r.t.	394
[SnO(n-C$_4$H$_9$)$_2$]/SnCl$_2$(n-C$_4$H$_9$)$_2$	Cl(n-C$_4$H$_9$)$_2$SnOSn(n-C$_4$H$_9$)$_2$Cl	At ~ 70°	393,441
[SnO(n-C$_4$H$_9$)$_2$]/SnBr$_2$(n-C$_4$H$_9$)$_2$	Br(n-C$_4$H$_9$)$_2$SnOSn(n-C$_4$H$_9$)$_2$Br	At ~ 80°	393,441
[SnO(n-C$_4$H$_9$)$_2$/Sn(SCN)$_2$(n-C$_4$H$_9$)$_2$	(SCN)(n-C$_4$H$_9$)$_2$SnOSn(n-C$_4$H$_9$)$_2$(SCN)	At ~ 80°	393
[SnO(n-C$_4$H$_9$)$_2$]/SnCl(C$_2$H$_5$)$_3$	Cl(n-C$_4$H$_9$)$_2$SnOSn(C$_2$H$_5$)$_3$	At r.t.	394
[SnO(n-C$_4$H$_9$)$_2$]/SnCl(n-C$_4$H$_9$)$_3$	Cl(n-C$_4$H$_9$)$_2$SnOSn(n-C$_4$H$_9$)$_3$	At r.t.	394
[SnO(n-C$_8$H$_{17}$)$_2$]/SnCl$_3$(n-C$_4$H$_9$)	Cl(n-C$_8$H$_{17}$)$_2$SnOSn(n-C$_4$H$_9$)Cl$_2$	At r.t.	394
[SnR$_2$]/SnCl$_4$	SnR$_2$Cl$_2$, SnOCl$_2$	At r.t.	394
[SnR$_2$]/SnBr$_4$	SnR$_2$Br$_2$/SnOBr$_2$	At r.t.	394
As(OCH$_3$)$_3$/AsF$_3$	As(OCH$_3$)$_2$F, (K = 0.07); As(OCH$_3$)F$_2$, (K = 0.02)	At r.t.	334
As(OCH$_3$)$_3$/AsCl$_3$	As(OCH$_3$)$_2$Cl, (K = 0.09); As(OCH$_3$)Cl$_2$, (K = 10^{-3})	At r.t.	334
As(OCH$_3$)$_3$/AsBr$_3$	As(OCH$_3$)$_2$Br, (K = 0.01); As(OCH$_3$)Br$_2$, (K = 0.02)	At r.t.	334
AsO(C$_6$H$_5$)/AsCl$_2$(C$_6$H$_5$)	AsCl(C$_6$H$_5$)$_2$, As$_2$O$_3$	2–4 hr at 200–240°; metal salts as catalyst	442–444

[a] Ox = 8-hydroxyquinolinate group

94

TABLE XX

Redistribution Reactions Involving Exchange of Element–Sulfur with Element–Sulfur Bonds

Reactants	Products	Remarks	References
$[B(SC_2H_5)H_2]_n$	$[B(SC_2H_5)H_2]_3$	At r.t.	446
$B(SC_2H_5)_3/[BS(SH)]_3$	$[BS(SC_2H_5)]_3$, $[BS(SC_2H_5)]_2$	Refluxing	448
$B(SC_2H_5)_3/B(SC_4H_9\text{-}n)_2(n\text{-}C_3H_7)$	$B(SC_2H_5)_2(n\text{-}C_3H_7)$, (87%); $B(SC_4H_9\text{-}n)_3$, (92%)	At 75–130°	329
$B(SC_2H_5)_3/B(SC_4H_9\text{-}n)(n\text{-}C_4H_9)_2$	$B(SC_2H_5)(n\text{-}C_4H_9)_2$, (97%); $B(SC_4H_9\text{-}n)_3$, (92%)	3 hr at 80–145°	329
$B(SC_2H_5)(n\text{-}C_4H_9)_2/B(SC_4H_9\text{-}n)(n\text{-}C_3H_7)_2$	$B(SC_2H_5)(n\text{-}C_3H_7)_2$, (98%); $B(SC_4H_9\text{-}n)(n\text{-}C_4H_9)_2$, (97%)	3.5 hr at 85–125°	329
$B(SC_4H_9)_2(n\text{-}C_4H_9)/B(SC_2H_5)_2(n\text{-}C_6H_{13})$	$B(SC_2H_5)_2(n\text{-}C_4H_9)$, (95%); $B(SC_4H_9\text{-}n)_2(n\text{-}C_6H_{13})$, (95%)	8 hr at 105–145°	329
$B(SC_4H_9\text{-}n)(n\text{-}C_4H_9)_2/B(SC_2H_5)_2(n\text{-}C_6H_{13})$	$B(SC_2H_5)(n\text{-}C_4H_9)_2$, (96%); $B(SC_4H_9\text{-}n)_2(n\text{-}C_6H_{13})$, (97%)	4 hr at 80–145°	329
$B[(SC_4H_9\text{-}n)H_2]_n$	$[B(SC_4H_9\text{-}n)H_2]_3$	At r.t.	446
$[SiS(CH_3)_2]_3/Si(SCH_3)_2(CH_3)_2$	$CH_3S[Si(CH_3)_2S]_nSi(CH_3)_2(SCH_3)$	20 days at 200°	398

TABLE XXI

Redistribution Reactions Involving Exchange of Element–Sulfur with Element–Halogen Bonds

Reactants	Products	Remarks	References
$B(SCH_3)_2Br$	$B(SCH_3)Br_2$, $B(SCH_3)_3$	At 220°	449
$[SiSCl_2]_2$	SiS_2, $SiCl_4$	At 120°	450
$Si(SCH_3)_3(CH_3)/SiCl_3(CH_3)$	$Si(SCH_3)_2Cl(CH_3)$, ($K = 0.12$); $Si(SCH_3)Cl_2(CH_3)$, ($K = 0.063$)		425
$Si(SCH_3)_3(CH_3)/SiBr_3(CH_3)$	$Si(SCH_3)_2Br(CH_3)$, ($K = 0.041$); $Si(SCH_3)Br_2(CH_3)$, ($K = 0.086$)	At 120°	425
$Si(SCH_3)_3(CH_3)/Si(NCO)_3(CH_3)$	$Si(SCH_3)_2(NCO)(CH_3)$, ($K = 0.77$); $Si(SCH_3)(NCO)_2(CH_3)$, ($K = 1.19$)	<26 days at 120°	425
$Si(SCH_3)_3(CH_3)/Si(NCS)_3(CH_3)$	$Si(SCH_3)_2(NCS)(CH_3)$, ($K = 0.24$); $Si(SCH_3)(NCS)_2(CH_3)$, ($K = 0.40$)	<22 hr at 120°	425
$Si(SCH_3)_2(CH_3)_2/SiCl_2(CH_3)_2$	$Si(SCH_3)Cl(CH_3)_2$, ($K = 0.18$)	<200 hr at 120°	423
$Si(SCH_3)_2(CH_3)_2/SiBr_2(CH_3)_2$	$Si(SCH_3)Br(CH_3)_2$, ($K = 0.049$)	<46 hr at 120°	423
$Si(SCH_3)_2(CH_3)_2/Si(NCO)_2(CH_3)_2$	$Si(SCH_3)(NCO)(CH_3)_2$, ($K = 0.014$)	<8 days at 25°	424
$Si(SCH_3)_2(CH_3)_2/Si(NCS)_2(CH_3)_2$	$Si(SCH_3)(NCS)(CH_3)_2$, (r)	<90 hr at 25°	424
$[SiS(CH_3)_2]_3/SiCl_2(CH_3)_2$	$Cl[Si(CH_3)_2S]_nSi(CH_3)_2Cl$	20 days at 200°	398
$Ge(SCH_3)_3(CH_3)/GeCl_3(CH_3)$	$Ge(SCH_3)_2Cl(CH_3)$, ($K = 0.040$); $Ge(SCH_3)Cl_2(CH_3)$, ($K = 0.048$)	<195 hr at 120°	436
$Ge(SCH_3)_3(CH_3)/GeBr_3(CH_3)$	$Ge(SCH_3)_2Br(CH_3)$, ($K = 0.032$); $Ge(SCH_3)Br_2(CH_3)$, ($K = 0.031$)	<42 hr at 120°	436
$Ge(SCH_3)_3(CH_3)/GeI_3(CH_3)$	$Ge(SCH_3)_2I(CH_3)$, ($K = 0.034$); $Ge(SCH_3)I_2(CH_3)$, ($K = 0.043$)	<18 hr at 120°	436
$Ge(SCH_3)_2(CH_3)_2/GeCl_2(CH_3)_2$	$Ge(SCH_3)Cl(CH_3)_2$, ($K = 0.055$)	At 120°	451
$Ge(SCH_3)_2(CH_3)_2/GeBr_2(CH_3)_2$	$Ge(SCH_3)Br(CH_3)_2$, ($K = 0.025$)	At 120°	451
$Ge(SCH_3)_2(CH_3)_2/GeI_2(CH_3)_2$	$Ge(SCH_3)I(CH_3)_2$, ($K = 0.012$)	At 120°	451
$[GeS(CH_3)_2]_3/GeCl_2(CH_3)_2$	$Cl[Ge(CH_3)_2S]_nGe(CH_3)_2Cl$	3 hr at 120°	451

$[GeS(CH_3)_2]_3/GeBr_2(CH_3)_2$	$Br[Ge(CH_3)_2S]_nGe(CH_3)_2Br$	72 hr at 120°	451
$[GeS(CH_3)_2]_3/GeI_2(CH_3)_2$	$I[Ge(CH_3)_2S]_nGe(CH_3)_2I$	130 hr at 120°	451
$[SnS(CH_3)_2]_3/SnCl_2(CH_3)_2$	$Cl[Sn(CH_3)_2S]_n(CH_3)_2Cl$	At r.t.	452
$[SnS(CH_3)_2]_3/SnCl_2(CH_3)_2$	$ClSn(CH_3)_2SSn(CH_3)_2Cl$	At r.t.	400
$[SnS(CH_3)_2]_3/SnBr_2(CH_3)_2$	$Br[SnS(CH_3)_2]_nSn(CH_3)_2Br$	At r.t.	452
$[SnS(CH_3)_2]_3/SnI_2(CH_3)_2$	$I[SnS(CH_3)_2]_nSn(CH_3)_2I$	At r.t.	452
$Sn(SC_4H_9-n)_2(n-C_4H_9)_2/SnCl_2(n-C_4H_9)_2$	$Sn(SC_4H_9-n)Cl(n-C_4H_9)_2$, (65%)	2 hr at 110°	390
$[SnS(n-C_4H_9)_2]_3/SnCl_4$	$SnCl_2(n-C_4H_9)_2$, $SnSCl_2$	At r.t.	400
$[SnS(n-C_4H_9)_2]_3/SnBr_4$	$SnBr_2(n-C_4H_9)_2$, $SnSBr_2$	At r.t.	400
$[SnS(n-C_4H_9)_2]_3/SnCl_3(n-C_4H_9)$	$Cl(n-C_4H_9)_2SnSSn(n-C_4H_9)Cl_2$	At r.t.	400
$[SnS(n-C_4H_9)_2]_3/SnBr_3(n-C_4H_9)$	$Br(n-C_4H_9)_2SnSSn(n-C_4H_9)Br_2$	At r.t.	400
$[SnS(n-C_4H_9)_2]_3/SnCl_3(C_6H_5)$	$Cl(n-C_4H_9)_2SnSSn(C_6H_5)Cl_2$	At r.t.	400
$[SnS(n-C_4H_9)_2]_3/SnF_2(n-C_4H_9)_2$	$F(n-C_4H_9)_2SnSSn(n-C_4H_9)_2F$	At r.t.	400
$[SnS(n-C_4H_9)_2]_3/SnCl_2(n-C_4H_9)_2$	$Cl(n-C_4H_9)_2SnSSn(n-C_4H_9)_2Cl$	At r.t.	400
$[SnS(n-C_4H_9)_2]_3/Sn(SCN)_2(n-C_4H_9)_2$	$(SCN)(n-C_4H_9)_2SnSSn(n-C_4H_9)_2(NCS)$	At r.t.	400

TABLE XXII

Redistribution Reactions Involving Exchange of Element–Halogen with Element–Halogen Bonds

Reactants	Products	Remarks	References
$BF_2(n\text{-}C_4H_9)/BCl_2(n\text{-}C_4H_9)$	$BClF(n\text{-}C_4H_9)$	Rapid halogen interchange	223
$BF_2(C_6H_5)/BCl_2(C_6H_5)$	$BClF(C_6H_5)$	IR evidence	455
$BCl_2(C_6H_5)/BBr_2(C_6H_5)$	$BBrCl(C_6H_5)$	IR evidence	455
$BCl_2(C_6H_5)/BBr_2(C_6H_5)$	$BBrCl(C_6H_5)$	Rapid halogen interchange	456
$BCl_2[N(CH_3)_2]/BBr_2[N(CH_3)_2]$	$BBrCl[N(CH_3)_2]$, (70%)	At 100°	457
$SiF_3(C_6H_5)/SiCl_4$	$SiCl_3(C_6H_5)$, (80–95%)	At 20–70°; NR_3 as catalyst	465
$SiCl(CH_3)_3/SiF_3(C_6H_5)$	$SiCl_3(C_6H_5)$	At 50–100°; NR_3 as catalyst	465
$SiCl_3(CH_3)/SiBr_3(CH_3)$	$SiBrCl_2(CH_3)$, (r); $SiBr_2Cl(CH_3)$, (r)	120 hr at 200°; $AlCl_3$ as catalyst	462,467
$SiClBrH_2$	$SiCl_2H_2$, $SiBr_2H_2$		464
$SiClBrH(C_6H_5)$	$SiCl_2H(C_6H_5)$, $SiBr_2H(C_6H_5)$		464
$SiCl_3(C_2H_5)/SiBr_3(C_2H_5)$	$SiCl_2Br(C_2H_5)$, (r); $SiClBr_2(C_2H_5)$, (r)	50 hr at 200°; $AlCl_3$ as catalyst	462
$SiCl_2(CH_3)_2/SiBr_2(CH_3)_2$	$SiClBr(CH_3)_2$, (r)	80 hr at 200°; $AlCl_3$ as catalyst	462
$SiCl_2(CH_3)_2/SiBr_2(CH_3)_2$	$SiClBr(CH_3)_2$, (r)	< 85 hr at 120°	466
$SiCl_3(CH_3)/SiI_3(CH_3)$	$SiCl_2I(CH_3)$, (r); $SiClI_2(CH_3)$, (r)	20 hr at 300°	461
$SiClIH_2$	$SiCl_2H_2$, SiI_2H_2		464
$SiCl_2(CH_3)_2/SiI_2(CH_3)_2$	$SiClI(CH_3)_2$, (r)	5 hr at 450°	461
$SiCl_2(C_2H_5)_2/SiI_2(C_2H_5)_2$	$SiClI(C_2H_5)_2$, (r)	10 hr at 400°	461
$SiClIH(C_6H_5)$	$SiCl_2H(C_6H_5)$, $SiHI_2(C_6H_5)$		464

Reactants	Products	Conditions	Ref.
$SiCl_2(CH_3)_2/Si(NC)_2(CH_3)_2$	$SiCl(NC)(CH_3)_2$, (r)	<3 hr at 25°	424
$SiCl_4/Si(NCO)_4$	$SiCl(NCO)_3$, (r); $SiCl_2(NCO)_2$, (r); $SiCl_3(NCO)$, (r)	At 600° or 1 week at 135°	459–461
$SiCl_3(CH_3)/Si(NCO)_3(CH_3)$	$SiCl_2(NCO)(CH_3)$, ($K = 0.51$); $SiCl(NCO)_2(CH_3)$, ($K = 0.34$)	20–75 hr at 120°	425
$SiCl_2(CH_3)_2/Si(NCO)_2(CH_3)_2$	$SiCl(NCO)(CH_3)_2$, (r)	7 hr at 370°	424,461
$SiCl_3(CH_3)/Si(NCS)_3(CH_3)$	$SiCl_2(NCS)(CH_3)$, ($K = 0.40$); $SiCl(NCS)_2(CH_3)$, ($K = 0.46$)	<21 hr at r.t.	425
$SiBrIH_2$	$SiBr_2H_2$, SiI_2H_2		464
$SiBr_2(CH_3)_2/Si(NC)_2(CH_3)_2$	$SiBr(NC)(CH_3)_2$, (r)	<3 hr at 25°	424
$SiBr_3(CH_3)/Si(NCO)_3(CH_3)$	$SiBr_2(NCO)(CH_3)$, ($K = 0.64$); $SiBr(NCO)_2(CH_3)$, ($K = 0.57$)	50–100 hr at 120°	425
$SiBr_2(CH_3)_2/Si(NCO)_2(CH_3)_2$	$SiBr(NCO)(CH_3)_2$, (r)	<20 days at 120°	424
$Si(NCS)(CH_3)_3/SiBr_4$	$SiBr(CH_3)_3$, $Si(NCS)_4$		463
$Si(NCS)(CH_3)_3/SiI_3(C_{12}H_{25})$	$SiI(CH_3)_3$, $Si(NCS)_3(C_{12}H_{25})$		463
$Si(NCS)_2(CH_3)_2/Si(NCO)_2(CH_3)_2$	$Si(NCS)(NCO)(CH_3)_2$, (r)	<160 hr at 72°	424
$Si(NCO)_3(CH_3)/Si(NCS)_3(CH_3)$	$Si(NCO)_2(NCS)(CH_3)$, ($K = 0.36$); $Si(NCO)(NCS)_2(CH_3)$, ($K = 0.39$)	<7 days at r.t.	425
$Si(NCS)_3(CH_3)/Si(NCO)_3(CH_2C_6H_5)$	$Si(NCO)_3(CH_3)$, $Si(NCS)_3(CH_2C_6H_5)$		463
$Si(NCO)_2(CH_3)_2/Si(NCS)_2(CH_3)_2$	$Si(NCO)(NCS)(CH_3)_2$, (r)	<100 hr at 25°	424
$GeCl_3(CH_3)/GeBr_3(CH_3)$	$GeBrCl_2(CH_3)$, (r); $GeBr_2Cl(CH_3)$, (r)	At r.t.	467
$GeCl_2(CH_3)_2/GeBr_2(CH_3)_2$	$GeBrCl(CH_3)_2$, (r)	At r.t.	466
$GeCl_3(CH_3)/GeI_3(CH_3)$	$GeClI_2(CH_3)$, ($K = 0.72$); $GeCl_2I(CH_3)$, ($K = 0.79$)	At r.t.	467
$GeCl_2(CH_3)_2/GeI_2(CH_3)_2$	$GeClI(CH_3)_2$, ($K = 0.67$)	At r.t.	466
$GeBr_3(CH_3)/GeI_3(CH_3)$	$GeBrI_2(CH_3)$, (r); $GeBr_2I(CH_3)$, (r)	At r.t.	467
$GeBr_2(CH_3)_2/GeI_2(CH_3)_2$	$GeBrI(CH_3)_2$, (r)	At r.t.	466
$SnCl_3(CH_3)/SnBr_3(CH_3)$	$SnCl_2Br(CH_3)$, $SnClBr_2(CH_3)$	Rapid exchange	472

(continued)

TABLE XXII (*Continued*)

Reactants	Products	Remarks	References
$SnCl_2(CH_3)_2/SnBr_2(CH_3)_2$	$SnBrCl(CH_3)_2$	Rapid exchange	472
$SnCl_2(n\text{-}C_4H_9)_2/SnBr_2(n\text{-}C_4H_9)_2$	$SnBrCl(n\text{-}C_4H_9)_2$	At r.t.	282,471
$SnCl_3(CH_3)/SnI_3(CH_3)$	$SnCl_2I(CH_3), SnClI_2(CH_3)$	Slow exchange	472
$SnCl_2(CH_3)_2/SnI_2(CH_3)_2$	$SnClI(CH_3)_2$	Rapid exchange	472
$SnBr_3(CH_3)/SnI_3(CH_3)$	$SnBr_2I(CH_3), SnBrI_2(CH_3)$	Slow exchange	472
$SnBr_2(CH_3)_2/SnI_2(CH_3)_2$	$SnBrI(CH_3)_2$	Rapid exchange	472
$AsF_2(CH_3)_3/AsCl_2(CH_3)_3$	$AsClF(CH_3)_3$, (r)	At $-53°$	473,474
$AsF_2(CH_2C_6H_5)_3/AsCl_2(CH_2C_6H_5)_3$	$AsClF(CH_2C_6H_5)_3$, ($K = 0.1$)	At $-47°$	473
$SbF_2(C_6H_5)_3/SbCl_2(C_6H_5)_3$	$SbClF(C_6H_5)_3$, ($K = 0.1$)	At $35°$	473
$SbF_2(CH_2C_6H_5)_3/SbCl_2(CH_2C_6H_5)_3$	$SbClF(CH_2C_6H_5)_3$, ($K = 0.2$)	At $-37°$	473
$SbF_2(CH_3)_3/SbBr_2(CH_3)_3$	$SbBrF(CH_3)_3$, (r)	At $35°$	473,474
$SbF_2(C_6H_5)_3/SbBr_2(C_6H_5)_3$	$SbFBr(C_6H_5)_3$, (r)	At $35°$	473
$SbF_2(CH_3)_3/SbI_2(CH_3)_3$	$SbFI(CH_3)_3$, ($K = 0.7$)	At $35°$	473,474
$SbF_2(C_6H_5)_3/SbI_2(C_6H_5)_3$	$SnFI(C_6H_5)_3$, ($K = 1.0$)	At $35°$	473
$SbCl_2(CH_3)_3/SbBr_2(CH_3)_3$	$SbBrCl(CH_3)_3$, (r)	At $35°$	473,474
$SbCl_2(CH_3)_3/SbI_2(CH_3)_3$	$SbClI(CH_3)_3$, (r)	At $35°$	473,474
$SbBr_2(CH_3)_3/SbI_2(CH_3)_3$	$SbBrI(CH_3)_3$, (r)	At $35°$	473,474

TABLE XXIII

Redistribution Reactions Involving Exchange of Element–Element Bonds

Reactants	Products	Remarks	References
$B_2(OCH_3)_4$	$B(OCH_3)_3$, B	At r.t., catalyzed by moisture	401
$B_2(OC_2H_5)_4$	$B(OC_2H_5)_3$, B	At r.t.	401
$B_2(n\text{-}C_4H_9)_2[N(CH_3)_2]_2$	$B(n\text{-}C_4H_9)_2[N(CH_3)_2]$, $B[N(CH_3)_2]_n$		476
Si_2H_6	SiH_4, $(SiH_2)_n$	Base catalyzed	478,479
$Si_2H_5[N(CH_3)_2]$	Si_2H_6, $Si_2H_4[N(CH_3)_2]_2$		477
$Si_2H_5(C_2H_5)$	$Si_3H_7(C_2H_5)$, $SiH_3(C_2H_5)$	Lewis bases as catalysts	480
$Si_2(OCH_3)_6$	$Si(OCH_3)_4$, $Si[Si(OCH_3)_3]_4$	5 hr at 200°	483
$Si_2(CH_3)_2(OCH_3)_4$	$[Si(CH_3)(OCH_3)]_n$, $Si(CH_3)(OCH_3)_3$	At 130°, base catalyzed	481
$Si_2(CH_3)_2(OCH_3)_4$	$Si(CH_3)[Si(CH_3)(OCH_3)_2]_3$, $Si(CH_3)(OCH_3)_3$, higher polysilanes	40 hr at 185°	481,483
$Si_2(CH_3)_4(OCH_3)_2$	$(CH_3O)[Si(CH_3)_2]_n(OCH_3)$, $Si(CH_3)_2(OCH_3)_2$	At 225°; base catalyzed and thermal	481–483
Si_2Cl_6	$SiCl_4$, Si_6Cl_{14}	r.t., $N(CH_3)_3$ as catalyst	484,485
Si_6Cl_{14}	$SiCl_4$, $(SiCl_2)_n$	Elevated temperature, $N(CH_3)_3$ as catalyst	485
$Si_2Cl_5(C_2H_5)$	$Si(C_2H_5)Cl_3$, $(SiCl_2)_n$	NH_4Cl as catalyst	484
$Si_2Cl_4(C_2H_5)_2$	$SiCl_2(C_2H_5)_2$, $(SiCl_2)_n$	NH_4Cl as catalyst	484
$Si_2Cl_4(CH_3)_2$	$CH_3Cl_2Si[Si(CH_3)_2]_nCl$, $SiCl_3(CH_3)$	Reflux, AgCN as catalyst	486
$Si_2Cl_2(CH_3)_4/Si_2Cl_3(CH_3)_3$	Higher silanes with >4 catenated Si atoms	7 hr at 260°; $N(CH_3)_4Cl$ as catalyst	487

(continued)

TABLE XIII (*Continued*)

Reactants	Products	Remarks	References
$Si_2Cl(CH_3)_5$	$(CH_3)_3Si[Si(CH_3)_2]_nCl$, $SiCl(CH_3)_3$	At 200°; AgCN as catalyst	486
$Si_2(CN)(CH_3)_5$	$(CH_3)_3Si[Si(CH_3)_2]_nCN$, $Si(CH_3)_3CN$	7 hr at 175°	486
$Si_2(CH_3)_6/Si_2(CH_3)_4Cl_2$	$Si_2(CH_3)_5Cl$, (68%)	1.5 hr at room temperature, $AlCl_3$ as catalyst	488
Ge_2H_6	GeH_4, $(GeH_2)_x$	At r.t., base catalyzed	479
Sn_2Cl_6	$SnCl_4$, $SnCl_2$		490
$Pb_2(CH_3)_6/Pb_2(C_2H_5)_6$	$Pb(CH_3)_n(C_2H_5)_{4-n}$, Pb	5 hr at 100°	491
$Pb_2(C_6H_5)_6$	$Pb(C_6H_5)_4$, Pb		492
$As_2(CF_3)_4$	$As(CF_3)_3$, As	At 210°	493
$As_2(C_6H_5)_4$	$As(C_6H_5)_3$, As		207
$(AsCH_3)_5/As_2(CH_3)_4$	Linear and cyclic permethylpolyarsines	At r.t.	494

REFERENCES

1. G. Calingaert and H. A. Beatty, *Organic Chemistry. An Advanced Treaty*, Vol. 2, H. Gilman, Ed., Wiley, New York, 1950, p. 1806.
2. J. R. Van Wazer, *Am. Sci.*, **50**, 450 (1962).
3. R. E. Dessy, T. Psarras, and S. Green, *Ann. N. Y. Acad. Sci.*, **125**, Art 1, 43 (1965).
4. J. C. Lockhart, *Chem. Rev.*, **65**, 131 (1965).
5. K. Moedritzer, *Organometal. Chem. Rev.*, **1**, 179 (1966).
6. J. R. Van Wazer and L. C. D. Groenweghe, *NMR in Chemistry*, Academic Press, New York, 1965, pp. 283–298.
7. J. R. Van Wazer and K. Moedritzer, *Angew. Chem.*, **78**, 401 (1966); *Angew. Chem. Intern. Ed. Engl.*, **5**, 341 (1966).
8. K. Moedritzer, *Advan. Organometal. Chem.*, **6**, 171 (1968).
9. "Redistribution Reactions in Chemistry,"—A Conference Monograph—*Ann. N. Y. Acad. Sci.*, **159**, Art 1, pp 1–334 (1969).
10. G. Calingaert and H. A. Beatty, *J. Am. Chem. Soc.*, **61**, 2748 (1939).
11. G. Calingaert, H. A. Beatty, and L. Hess, *J. Am. Chem. Soc.*, **61**, 3300 (1939).
12. G. Calingaert, H. A. Beatty, and H. R. Neal, *J. Am. Chem. Soc.*, **61**, 2755 (1939).
13. G. Calingaert, H. A. Beatty, and H. Soroos, *J. Am. Chem. Soc.*, **62**, 1099 (1940).
14. G. Calingaert and H. Soroos, *J. Am. Chem. Soc.*, **61**, 2758 (1939).
15. G. Calingaert, H. Soroos, and V. Hnizda, *J. Am. Chem. Soc.*, **62**, 1107 (1940).
16. G. Calingaert, H. Soroos, V. Hnizda, and H. Shapiro, *J. Am. Chem. Soc.*, **62**, 1545 (1940).
17. G. Calingaert, H. Soroos, and H. Shapiro, *J. Am. Chem. Soc.*, **62**, 1104 (1940).
18. G. Calingaert, H. Soroos, and H. Shapiro, *J. Am. Chem. Soc.*, **63**, 947 (1941).
19. G. Calingaert, H. Soroos, and G. W. Thomson, *J. Am. Chem. Soc.*, **62**, 1542 (1940).
20. P. J. Flory, *J. Am. Chem. Soc.*, **64**, 2205 (1942).
21. H. Weingarten and J. R. Van Wazer, *J. Am. Chem. Soc.*, **88**, 2700 (1966).
22. A. Wassermann, *Monatsh.*, **83**, 543 (1952).
23. A. F. Reid and C. J. Wilkins, *J. Chem. Soc.*, **1960**, 3857.
24. A. B. Burg and F. M. Graber, *J. Am. Chem. Soc.*, **78**, 1523 (1956).
25. R. Köster, *Angew. Chem.*, **73**, 66 (1961).
26. P. D. Zemany and F. P. Price, *J. Am. Chem. Soc.*, **70**, 4222 (1948).
27. L. E. Sutton et al., "Tables of Interatomic Distances and Configuration in Molecules and Ions," in *Chem. Soc. (London)*, Spec. Publ. **11**, 1958; **18**, 1965.
28. L. C. D. Groenweghe, J. R. Van Wazer, and A. W. Dickinson, *Anal. Chem.*, **36**, 303 (1964).
29. A. R. Conrad and A. G. Lee, *Anal. Chem.*, **40**, 569 (1968).
30. K. Kühlein and W. P. Neumann, *Ann.*, **702**, 17 (1967).
31. H. I. Schlesinger and A. O. Walker, *J. Am. Chem. Soc.*, **57**, 621 (1935).
32. H. I. Schlesinger, W. Flodin, and A. B. Burg, *J. Am. Chem. Soc.*, **61**, 1078 (1939).
33. L. Van Alten, G. R. Seely, J. Oliver, and D. M. Ritter, *Advan. Chem.*, No. 32, 107, (1961).
34. G. Schomburg, *Gas Chromatography*, van Swaay, Ed., Butterworths, 1962.
35. L. H. Long and H. G. H. Wallbridge, *J. Chem. Soc.*, **1965**, 3513.
36. A. B. Burg and J. L. Boone, *J. Am. Chem. Soc.*, **78**, 1521 (1956).
37. R. I. Wagner and J. L. Bradford, *Inorg. Chem.*, **1**, 99 (1962).
38. H. C. Newson, W. G. Woods, and A. L. McCloskey, *Inorg. Chem.*, **2**, 36 (1962).
39. L. H. Long and M. G. H. Wallbridge, *Chem. Ind. London*, 295 (1959).
40. M. J. D. Low, R. Epstein, and A. C. Bond, *Chem. Commun.*, **1967**, 226.

41. W. J. Lehmann, C. O. Wilson, J. F. Ditter, and I. Shapiro, *Advan. Chem.*, No. 32, 139 (1961).
42. H. I. Schlesinger, L. Horvitz, and A. B. Burg, *J. Am. Chem. Soc.*, **58**, 407 (1936).
43. I. J. Solomon, M. J. Klein, and K. Hattori, *J. Am. Chem. Soc.*, **80**, 4520 (1958).
44. L. H. Long and M. G. H. Wallbridge, *J. Chem. Soc.*, **1963**, 2181.
45. B. M. Mikhailov, A. A. Akhnazaryan, and L. S. Vasilev, *Proc. Acad. Sci. USSR, English Transl.*, **36**, 139 (1961).
46. B. M. Mikhailov and L. S. Vasilev, *Bull. Acad. Sci. USSR, Div. Chem. Sci. English Transl.*, 580 (1962).
47. B. M. Mikhailov and V. A. Dorokhov, *Proc. Acad. Sci. USSR, English Transl.*, **133**, 743 (1960).
48. B. M. Mikhailov and V. A. Dorokhov, *Proc. Acad. Sci. USSR, English Transl.*, **130**, 137 (1959).
49. B. M. Mikhailov and V. A. Dorokhov, *J. Gen. USSR, English Transl.*, **31**, 3751 (1961).
50. R. Köster and G. Rotermund, *Angew. Chem.*, **72**, 138 (1960).
51. E. Wiberg, J. E. F. Evans, and H. Nöth, *Z. Naturforsch.*, **13b**, 263 (1958).
52. D. R. Nielsen, W. E. McEwen, and C. A. Vanderwerf, *Chem. Ind. London*, 295 (1959).
53. B. M. Mikhailov and V. A. Dorokhov, *Proc. Acad. Sci. USSR, English Transl.*, **136**, 51 (1961).
54. F. M. Peters, B. Bartocha, and A. J. Bilbo, *Can. J. Chem.*, **41**, 1051 (1963).
55. K. Ziegler, H. G. Gellert, H. Martin, K. Nagel, and J. Schneider, *Ann.*, **589**, 91 (1954).
56. T. Wartik and H. I. Schlesinger, *J. Am. Chem. Soc.*, **75**, 835 (1953).
57. J. R. Surtees, *Chem. Ind. London*, 1260 (1964).
58. P. Kobetz, W. E. Becker, R. C. Pinkerton and J. B. Honeycutt, Jr., *Inorg. Chem.*, **2**, 859 (1963).
59. H. Gilman and D. H. Miles, *J. Org. Chem.*, **23**, 326 (1958).
60. C. G. Pitt and K. R. Skillern, *J. Organometal. Chem.*, **7**, 525 (1967).
61. G. A. Russell, *J. Am. Chem. Soc.*, **81**, 4825 (1959).
62. S. N. Borisov, M. G. Voronkov, and B. N. Dolgov, *J. Gen. Chem. USSR, English Transl.*, **27**, 1416 (1957).
63. J. L. Speier and R. E. Zimmerman, *J. Am. Chem. Soc.*, **77**, 6395 (1955).
64. R. A. Benkeser and D. J. Foster, *J. Am. Chem. Soc.*, **74**, 5314 (1952).
65. R. A. Benkeser and D. J. Foster, *J. Am. Chem. Soc.*, **74**, 4200 (1952).
66. L. E. Nelson, N. C. Angelotti, and D. R. Weyenberg, *J. Am. Chem. Soc.*, **85**, 2662 (1963).
67. W. P. Neumann, *Ann. N. Y. Acad. Sci.*, **159**, Art 1, 56 (1969).
68. H. C. Clark, S. H. Furnival, and J. T. Kwon, *Can. J. Chem.*, **41**, 2889 (1963).
69. A. B. Burg and C. L. Randolph, Jr., *J. Am. Chem. Soc.*, **73**, 953 (1951).
70. J. K. Ruff, *J. Am. Chem. Soc.*, **83**, 2835 (1961).
71. A. Stock, *Ber.*, **54**, 142 (1921).
72. A. G. MacDiarmid, *J. Inorg. Nucl. Chem.*, **25**, 1534 (1963).
73. R. L. Wells and R. Schaeffer, *J. Am. Chem. Soc.*, **88**, 37 (1966).
74. W. P. Neumann, B. Schneider, and R. Sommer, *Ann.*, **692**, 1 (1966).
75. R. Sommer, B. Schneider, and W. P. Neumann, *Ann.*, **692**, 12 (1966).
76. H. M. J. C. Creemers, F. Verbeek, and J. G. Noltes, *J. Organometal. Chem.*, **8**, 469 (1967).

77. A. B. Burg and H. I. Schlesinger, *J. Am. Chem. Soc.*, **55**, 4020 (1933).
78. H. S. Uchida, H. B. Kreider, A. Murchison, and J. F. Masi, *J. Phys. Chem.*, **63**, 1414 (1959).
79. S. H. Rose and S. G. Shore, *Inorg. Chem.*, **1**, 744 (1962).
80. G. E. McAchran and S. G. Shore, *Inorg. Chem.*, **5**, 2044 (1966).
81. B. M. Mikhailov and L. S. Vasilev, *Bull. Acad. Sci. USSR, Div. Chem. Sci. English Transl.*, **1962**, 771.
82. H. C. Brown and C. J. Shoaf, *J. Am. Chem. Soc.*, **86**, 1079 (1964).
83. C. Friedel and A. Ladenburg, *Ann.*, **143**, 124 (1867).
84. W. C. Hammann and C. F. Hobbs, *J. Org. Chem.*, **33**, 1277 (1968).
85. D. R. Weyenberg, A. E. Bey, H. F. Stewart, and W. H. Atwell, *J. Organometal. Chem.*, **6**, 583 (1966).
86. A. F. Reilly and H. W. Post, *J. Org. Chem.*, **16**, 387 (1951).
87. O. J. Klejnot, *Inorg. Chem.*, **2**, 825 (1963).
88. A. K. Sawyer and H. G. Kuivila, *J. Am. Chem. Soc.*, **82**, 5958 (1960).
89. A. K. Sawyer and H. G. Kuivila, *J. Org. Chem.*, **27**, 837 (1962).
90. W. P. Neumann and B. Schneider, *Angew. Chem.*, **76**, 891 (1964).
91. A. K. Sawyer, *J. Am. Chem. Soc.*, **87**, 537 (1965).
92. H. M. J. C. Creemers and J. G. Noltes, *Recueil*, **84**, 1589 (1965).
93. A. B. Burg and R. I. Wagner, *J. Am. Chem. Soc.*, **76**, 3307 (1954).
94. P. Ehrlich, B. Alt, and L. Gentsch, *Z. Anorg. Allgem. Chem.*, **283**, 58 (1956).
95. P. Ehrlich and H. Görz, *Z. Anorg. Allgem. Chem.*, **288**, 148 (1956).
96. P. Ehrlich and H. Kulke, *Z. Anorg. Allgem. Chem.*, **288**, 156 (1956).
97. T. D. Coyle, J. J. Ritter, and T. C. Farrar, *Proc. Chem. Soc.*, 25 (1964).
98. T. D. Coyle, J. Cooper, and J. J. Ritter, *Inorg. Chem.*, **7**, 1014 (1968).
99. A. Stock, *Hydrides of Boron and Silicon*, Cornell University Press, Ithaca, N. Y., 1933.
100. J. V. Kerrigan, *Inorg. Chem.*, **3**, 908 (1964).
101. H. W. Myers and R. F. Putnam, *Inorg. Chem.*, **2**, 655 (1963).
102. H. C. Brown and P. A. Tierney, *J. Am. Chem. Soc.*, **80**, 1552 (1958).
103. T. Onak, H. Landesmann, and I. Shapiro, *J. Phys. Chem.*, **62**, 1605 (1958).
104. S. Ratajczak, *Bull. Soc. Chim. France.*, **1960**, 487.
105. G. W. Schaeffer, R. Schaeffer, and H. I. Schlesinger, *J. Am. Chem. Soc.*, **73**, 1612 (1951).
106. R. Köster and G. Benedikt, *Angew. Chem.*, **75**, 419 (1963).
107. R. Köster, *Ann. N. Y. Sci.*, **159**, Art 1, 73 (1969).
108. H. I. Schlesinger and A. B. Burg, *J. Am. Chem. Soc.*, **53**, 4321 (1931).
109. H. Nöth and H. Beyer, *Chem. Ber.*, **93**, 2251 (1960).
110. E. Wiberg, K. Moedritzer, and R. Uson, *Rev. Acad. Cienc. Exact. Fis. Quim. Nat. Zaragoza*, [1], **9**, 91 (1954); *Chem. Abstr.*, **52**, 3584 (1958).
111. E. Wiberg and M. Schmidt, *Z. Naturforsch.*, **6b**, 458 (1951).
112. E. Wiberg and M. Schmidt, *Z. Naturforsch.*, **6b**, 459 (1951).
113. E. Wiberg and M. Schmidt, *Z. Naturforsch.*, **6b**, 460 (1951).
114. D. L. Schmidt and E. E. Fagg, *Inorg. Chem.*, **6**, 1262 (1967).
115. E. C. Ashby and J. Prather, *J. Am. Chem. Soc.*, **88**, 729 (1966).
116. H. S. Booth and W. D. Stillwell, *J. Am. Chem. Soc.*, **56**, 1531 (1934).
117. H. J. Emeleus and A. G. Maddock, *J. Chem. Soc.*, **1944**, 293.
118. L. G. Mahone and D. R. Weyenberg, *J. Organometal. Chem.*, **12**, 231 (1968).
119. A. Stock and C. Somieski, *Ber.*, **52**, 695 (1919).
120. R. O. Sauer and E. M. Hadsell, *J. Am. Chem. Soc.*, **70**, 3590 (1948).

121. F. C. Whitmore, E. W. Pietrusza, and L. H. Sommer, *J. Am. Chem. Soc.* **69**, 2108 (1947).
122. B. N. Dolgov, M. G. Voronkov, and S. N. Borisov, *J. Gen. Chem. USSR, English Transl.*, **27**, 781 (1957).
123. B. N. Dolgov, S. N. Borisov, and M. G. Voronkov, *J. Gen. Chem. USSR, English Transl.*, **27**, 2117 (1957).
124. B. N. Dolgov, S. N. Borisov, and M. G. Voronkov, *J. Gen. Chem. USSR, English Transl.*, **27**, 2733 (1959).
125. S. N. Borisov, M. G. Voronkov, and B. N. Dolgov, *Proc. Acad. Sci. USSR, Chem. Sec. English Transl.*, **114**, 433 (1957).
126. D. R. Weyenberg, A. E. Bey, and P. J. Ellison, *J. Organometal. Chem.*, **3**, 489 (1965).
127. K. Moedritzer and J. R. Van Wazer, *J. Organometal. Chem.*, **12**, 69 (1968).
128. H. J. Emeleus, A. G. Maddock, and C. Reid, *J. Chem. Soc.*, **1941**, 353.
129. A. K. Sawyer and H. G. Kuivila, *Chem. Ind. London*, **1961**, 260.
130. W. P. Neumann and J. Pedain, *Tetrahedron Letters*, **1964**, 2461.
131. A. K. Sawyer, J. E. Brown, and E. L. Hanson, *J. Organometal. Chem.*, **3**, 464 (1965).
132. A. K. Sawyer and J. E. Brown, *J. Organometal. Chem.*, **5**, 438 (1966).
133. K. Kawakami, T. Saito, and R. Okawara, *J. Organometal. Chem.*, **8**, 377 (1967).
134. M. A. Weiner and R. West, *J. Am. Chem. Soc.*, **85**, 485 (1963).
135. E. Weiss, *Chem. Ber.*, **97**, 3241 (1964).
136. L. M. Seitz and T. L. Brown, *J. Am. Chem. Soc.*, **88**, 2174 (1966).
137. L. M. Seitz and T. L. Brown, *J. Am. Chem. Soc.*, **88**, 4140 (1966).
138. G. E. Hartwell and T. L. Brown, *J. Am. Chem. Soc.*, **88**, 4625 (1966).
139. L. M. Seitz and T. L. Brown, *J. Am. Chem. Soc.*, **89**, 1607 (1967).
140. H. O. House, R. A. Latham, and G. M. Whiteside, *J. Org. Chem.*, **32**, 2481 (1967).
141. R. Köster and G. Bruno, *Ann.*, **629**, 89 (1960).
142. G. Schomburg, R. Köster, and D. Henneberg, *Z. Anal. Chem.*, **170**, 285 (1959).
143. G. F. Hennion, P. A. McCusker, and A. J. Rutkowski, *J. Am. Chem. Soc.*, **80**, 617 (1958).
144. F. M. Rossi, P. A. McCusker, and G. F. Hennion, *J. Org. Chem.*, **32**, 450 (1967),
145. P. A. McCusker, F. M. Rossi, J. H. Bright, and G. F. Hennion, *J. Org. Chem.*, **28**, 2889 (1963).
146. G. F. Hennion, P. A. McCusker, and J. V. Marra, *J. Am. Chem. Soc.*, **80**, 3481 (1958).
147. G. F. Hennion, P. A. McCusker, and J. V. Marra, *J. Am. Chem. Soc.*, **81**, 1768 (1959).
148. G. F. Hennion, P. A. McCusker, E. C. Ashby, and A. J. Rutkowski, *J. Am. Chem. Soc.*, **79**, 5190 (1957).
149. P. A. McCusker, J. V. Marra, and G. F. Hennion, *J. Am. Chem. Soc.*, **83**, 1924 (1961).
150. E. Krause and P. Nobbe, *Ber.*, **64**, 2112 (1931).
151. T. D. Parsons and D. M. Ritter, *J. Am. Chem. Soc.*, **76**, 1710 (1954).
152. B. M. Mikhailov and T. A. Shchegoleva, *Proc. Acad. Sci. USSR, Chem. Sec., English Transl.*, **108**, 481 (1956).
153. G. A. Razuvaev and T. G. Brikina, *J. Gen. Chem. USSR, Eng. Transl.*, **24**, 1397 (1954).
154. N. Muller and D. E. Pritchard, *J. Am. Chem. Soc.*, **82**, 248 (1960).
155. E. G. Hoffmann, *Trans. Faraday Soc.*, **58**, 642 (1962).
156. E. G. Hoffmann, *Bull. Soc. Chim. France*, **1963**, 1467.
157. K. C. Ramey, J. F. O'Brien, I. Hasegawa, and A. E. Borchert, *J. Phys. Chem.*, **69**, 3418 (1965).

158. C. P. Poole, Jr., H. E. Swift, and J. F. Itzel, Jr., *J. Chem. Phys.*, **42**, 2576 (1965).
159. K. C. Williams and T. L. Brown, *J. Am. Chem. Soc.*, **88**, 5460 (1966).
160. M. B. Smith, *J. Phys. Chem.*, **71**, 364 (1967).
161. T. Mole and J. R. Surtees, *Australian J. Chem.*, **17**, 310 (1964).
162. T. Mole, *Australian J. Chem.*, **18**, 1183 (1965).
163. N. S. Ham, E. A. Jeffery, T. Mole, and J. K. Saunders, *Australian J. Chem.*, **20**, 2641 (1967).
164. E. A. Jeffery, T. Mole, and J. K. Saunders, *Chem. Commun.*, **1967**, 696.
165. J. P. Maher and D. F. Evans, *Proc. Chem. Soc.*, **1961**, 208.
166. J. P. Maher and D. F. Evans, *J. Chem. Soc.*, **1963**, 5534.
167. A. Ladenburg, *Ber.*, **7**, 387 (1874).
168. W. Ipatiev and B. Dolgow, *Ber.*, **62**, 1220 (1929).
169. B. Dolgov and Y. Volnov, *Zh. Obshch. Khim., Khim Ser. I*, **1931**, 91; *Chem. Abstr.*, **25**, 4535 (1931).
170. P. D. George, L. H. Sommer, and F. C. Whitmore, *J. Am. Chem. Soc.*, **77**, 1677 (1955).
171. G. A. Russell, *J. Am. Chem. Soc.*, **81**, 4815 (1959).
172. F. H. Pollard, G. Nickless, and P. C. Uden, *J. Chromatog.*, **19**, 29 (1965).
173. J. A. Semlyen, G. R. Walker, R. E. Blofeld, and C. S. G. Phillips, *J. Chem. Soc.*, **1964**, 4948.
174. E. J. Bulten and J. G. Noltes, *J. Organometal. Chem.*, **11**, P19 (1968).
175. F. Challenger and L. R. Ridgway, *J. Chem. Soc.*, **121**, 104 (1922).
176. H. K. Hofmeister and J. R. Van Wazer, *J. Inorg. Nucl. Chem.*, **26**, 1209 (1964).
177. G. E. Coates, F. Glockling, and N. D. Huck, *J. Chem. Soc.*, **1952**, 4512.
178. G. E. Coates and A. H. Fishwick, *J. Chem. Soc. (A)*, **1968**, 477.
179. R. Köster, *Angew. Chem.*, **71**, 31 (1959).
180. R. Köster, *Angew. Chem.*, **71**, 520 (1959).
181. L. I. Zakharkin and V. I. Stanko, *Bull. Acad. Sci. USSR, Div. Chem. Sci., English Transl.*, **1960**, 1774.
182. G. Wilke and P. Heimbach, *Ann.*, **652**, 7 (1962).
183. E. W. Abel, W. Gerrard, and M. F. Lappert, *J. Chem. Soc.*, **1958**, 1451.
184. B. M. Mikhailov, L. S. Vasilev, and E. N. Safonova, *Proc. Acad. Sci. USSR, Chem. Sec., English Transl.*, **147**, 1023 (1962).
185. B. M. Mikhailov and L. S. Vasilev, *Bull. Acad. Sci. USSR, Div. Chem. Sci., English Transl.*, **1962**, 1668.
186. B. M. Mikhailov, V. F. Pozdnev, and V. G. Kiselov, *Dokl. English Transl.*, **151**, 571 (1963).
187. P. A. McCusker, *Angew. Chem.*, **69**, 677 (1957).
188. J. Goubeau and H. Keller, *Z. Anorg. Allgem. Chem.*, **267**, 1 (1951).
189. P. A. McCusker, E. C. Ashby, and H. S. Makowski, *J. Am. Chem. Soc.*, **79**, 5179 (1957).
190. G. F. Hennion, P. A. McCusker, E. C. Ashby, and A. J. Rutkowski, *J. Am. Chem. Soc.*, **79**, 5194 (1957).
191. P. A. McCusker and J. H. Bright, *J. Org. Chem.*, **29**, 2093 (1964).
192. A. V. Grosse and J. M. Mavity, *J. Org. Chem.*, **5**, 106 (1940).
193. T. Mole, *Australian J. Chem.*, **19**, 373 (1966).
194. T. Mole, *Australian J. Chem.*, **19**, 381 (1966).
195. V. P. Glushkova and K. A. Kocheshkov, *Dokl. Akad. Nauk, English Transl.*, **116**, 233 (1957).
196. J. W. Ryan, *J. Am. Chem. Soc.*, **84**, 4730 (1962).

197. E. W. Beck, W. H. Daudt, H. J. Fletcher, M. J. Hunter, and A. J. Barry, *J. Am. Chem. Soc.*, **81**, 1256 (1959).
198. M. R. Stober, M. C. Musolf, and J. L. Speier, *J. Org. Chem.*, **30**, 1651 (1965).
199. K. A. Andrianov, V. I. Pakhomov, and V. M. Gelperina, *Proc. Acad. Sci. USSR, Chem. Sec., English Transl.*, **162**, 407 (1965).
200. O. Schmitz-Dumont, *Z. Anorg. Allgem. Chem.*, **248**, 289 (1941).
201. R. F. Chambers and P. C. Scherer, *J. Am. Chem. Soc.*, **48**, 1054 (1926).
202. C. A. Kraus and R. H. Bullard, *J. Am. Chem. Soc.*, **51**, 3605 (1929).
203. W. J. Considine, J. J. Ventura, B. G. Kushlefsky, and A. Ross, *J. Organometal. Chem.*, **1**, 299 (1964).
204. E. Amberger and M. R. Kula, *Chem. Ber.*, **96**, 2562 (1963).
205. D. Seyferth and J. M. Burlitch, *Z. Naturforsch*, **22b**, 1358 (1967).
206. Y. A. Aleksandrov and T. I. Mokeeva, *Tr. Khim. Khim. Tekhnol.*, **4**, 365 (1961); *Chem. Abstr.*, **56**, 493 (1962).
207. W. R. Cullen, *Advan. Organometal. Chem.*, **4**, 145 (1966).
208. H. H. Jaffe and G. O. Doak, *J. Am. Chem. Soc.*, **71**, 602 (1949).
209. G. O. Doak and H. H. Jaffe, *J. Am. Chem. Soc.*, **72**, 3025 (1950).
210. H. H. Jaffe and G. O. Doak, *J. Am. Chem. Soc.*, **72**, 3027 (1950).
211. E. Wiberg and W. Sturm, *Z. Naturforsch.*, **10b**, 112 (1955).
212. E. Wiberg and W. Sturm, *Z. Naturforsch.*, **10b**, 113 (1955).
213. E. C. Ashby, R. Sanders, and J. Carter, *Chem. Commun.*, **1967**, 997.
214. H. Gilman and F. Schulze, *J. Am. Chem. Soc.*, **49**, 2904 (1927).
215. B. J. Wakefield, *Organometal. Chem. Rev.*, **1**, 131 (1966).
216. M. B. Smith and W. E. Becker, *Tetrahedron Letters*, **1965**, 3843.
217. M. B. Smith and W. E. Becker, *Tetrahedron*, **22**, 3027 (1966).
218. M. B. Smith and W. E. Becker, *Tetrahedron*, **23**, 4215 (1967).
219. H. J. Becher, *Z. Anorg. Allgem. Chem.*, **271**, 243 (1953).
220. H. J. Becher, *Z. Anorg. Allgem. Chem.*, **291**, 151 (1957).
221. P. McCusker, G. F. Hennion, and E. C. Ashby, *J. Am. Chem. Soc.*, **79**, 5192 (1957).
222. W. Schabacher and J. Goubeau, *Z. Anorg. Allgem. Chem.*, **294**, 183 (1958).
223. V. W. Buls, O. L. Davis, and R. I. Thomas, *J. Am. Chem. Soc.*, **79**, 337 (1957).
224. E. W. Abel, W. Gerrard, and M. F. Lappert, *J. Chem. Soc.*, **1957**, 5051.
225. B. M. Mikhailov and T. A. Shchegoleva, *J. Gen. Chem. USSR, English Transl.*, **29**, 3404 (1959).
226. K. Niedenzu, D. H. Harrelson, W. George, and J. W. Dawson, *J. Org. Chem.*, **26**, 3707 (1961).
227. C. D. Good and D. M. Ritter, *J. Am. Chem. Soc.*, **84**, 1162 (1962).
228. F. C. Gunderloy and C. E. Erickson, *Inorg. Chem.*, **1**, 349 (1962).
229. L. I. Zakharkin and A. I. Kovredov, *J. Gen. Chem. USSR, English Transl.*, **32**, 1408 (1962).
230. K. Niedenzu, H. Beyer, and J. W. Dawson, *Inorg. Chem.*, **1**, 738 (1962).
231. L. I. Zakharkin and A. I. Kovredov, *Bull. Acad. Sci. USSR, Div. Chem. Sci., English Transl.*, **1962**, 2149.
232. L. I. Zakharkin and A. I. Kovredov, *Bull. Acad. Sci. USSR, Div. Chem. Sci., English Transl.*, **1962**, 357.
233. C. Chambers and A. K. Holliday, *J. Chem. Soc.*, **1965**, 3459.
234. P. A. McCusker and J. H. Bright, *J. Inorg. Nucl. Chem.*, **28**, 2261 (1966).
235. J. P. Tuchagues and J. P. Laurent, *Bull. Soc. Chim. France*, **1967**, 4160.
236. R. Köster and G. Benedikt, *Angew. Chem.*, **75**, 346 (1963).

237. A. I. Kovredov and L. I. Zakharkin, *Bull. Acad. Sci. USSR, Div. Chem. Sci.*, *English Transl.*, **1964**, 42.
238. S. Brownstein, B. C. Smith, G. Ehrlich, and A. W. Laubengayer, *J. Am. Chem. Soc.*, **82**, 1000 (1960).
239. T. Mole, *Australian J. Chem.*, **17**, 1050 (1964).
240. H. Jenker, *Z. Naturforsch.*, **12b**, 809 (1957).
241. K. Ziegler and R. Köster, *Ann.*, **608**, 1 (1957).
242. I. Zakharkin and I. M. Khorlina, *J. Gen. Chem. USSR, English Transl.*, **30**, 1905 (1960).
243. L. I. Zakharkin and V. V. Gavrilenko, *J. Gen. Chem. USSR, English Transl.*, **32**, 688 (1962).
244. G. Pajaro, *Ann. Chim. Rome*, **48**, 193 (1958), *Chem. Abstr.*, **52**, 18195 (1958).
245. T. Mole, *Australian J. Chem.*, **16**, 794 (1963).
246. J. Brandt and E. G. Hoffmann, *Brennstoff-Chem.*, **45**, 201 (1964).
247. P. G. Perkins and M. E. Twentyman, *J. Chem. Soc.*, **1965**, 1038.
248. N. N. Melnikov and G. P. Gracheva, *J. Gen. Chem. USSR*, **5**, 1786 (1935); *Chem. Abstr.*, **30**, 3403 (1936).
249. N. N. Melnikov and M. S. Rokitskaya, *J. Gen. Chem. USSR*, **7**, 1472 (1937); *Chem. Abstr.*, **32**, 127 (1938).
250. D. Sarrach, *Z. Anorg. Allgem. Chem.*, **319**, 16 (1962).
251. C. R. Hart and C. K. Ingold, *J. Chem. Soc.*, **1964**, 4372.
252. A. N. Nesmejanov, A. E. Borisov, I. S. Saveleva, and E. I. Golubeva, *Bull. Acad. Sci. USSR, Div. Chem. Sci., English Transl.*, **1958**, 1490.
253. B. A. Bluestein and H. R. McEntee, *Advan. Chem. Ser.*, **23**, 233. (1961).
254. G. D. Cooper and A. R. Gilbert, *J. Am. Chem. Soc.*, **82**, 5042 (1960).
255. K. A. Andrianov and V. M. Kotov, *Proc. Acad. Sci. USSR, Chem. Sec., English Transl.*, **167**, 375 (1966).
256. G. J. M. Van der Kerk, R. Rijkens, and M. J. Janssen, *Recueil*, **81**, 764 (1962).
257. F. Rijkens and G. J. M. Van der Kerk, *Recueil*, **83**, 723 (1964).
258. F. Rijkens, E. J. Bulten, W. Drenth, and G. J. M. Van der Kerk, *Recueil*, **85**, 1223 (1966).
259. I. Schumann-Ruidisch, V. Lieb, and B. Jutzi-Mebert, *Z. Anorg. Allgem. Chem.*, **355**, 64 (1967).
260. G. M. Burch and J. R. Van Wazer, *J. Chem. Soc. (A)*, **1966**, 586.
261. R. Schwarz and M. Lewinsohn, *Ber.*, **64**, 2352 (1931).
262. R. Schwarz and M. Schmeisser, *Ber.*, **69**, 579 (1936).
263. P. Mazerolles, *Bull. Soc. Chim. France*, **1961**, 1911.
264. M. Lesbre, P. Mazerolles, and G. Manuel, *Compt. Rend.*, **255**, 544 (1962).
265. G. B. Buckten, *Ann.*, **112**, 220 (1859).
266. A. Cahours, *Ann.*, **122**, 48 (1862).
267. K. A. Kocheschkov, *Ber.*, **66**, 1661 (1933).
268. W. J. Pope, *Proc. Roy. Soc. London.*, **72**, 7 (1903).
269. R. M. Kary and K. C. Frisch, *J. Am. Chem. Soc.*, **79**, 2140 (1957).
270. D. Seyferth and F. G. A. Stone, *J. Am. Chem. Soc.*, **79**, 515 (1957).
271. S. D. Rosenberg and A. J. Gibbons, Jr., *J. Am. Chem. Soc.*, **79**, 2138 (1957).
272. W. P. Neumann and G. Burckhardt, *Ann.*, **663**, 11 (1963).
273. H. Gilman and H. W. Melvin, Jr., *J. Am. Chem. Soc.*, **71**, 4050 (1949).
274. K. A. Kocheshkov, *J. Gen. Chem. USSR*, **4**, 1359 (1934); *Chem. Abstr.*, **29**, 3650 (1935).
275. C. R. Dillard, E. H. McNeill, D. E. Simmons, and J. B. Yeldell, *J. Am. Chem. Soc.*, **80**, 3607 (1958).

276. K. V. Vijayaraghavan, *J. Indian Chem. Soc.*, **22**, 135 (1945).
277. A. Saitov, E. G. Rochow, and D. Seyferth, *J. Org. Chem.*, **23**, 116 (1958).
278. R. H. Prince, *J. Chem. Soc.*, **1959**, 1783.
279. G. J. M. Van der Kerk and J. G. A. Luijten, *J. Appl. Chem. London*, **6**, 93 (1956).
280. G. J. M. Van der Kerk and J. G. A. Luijten, *J. Appl. Chem. London*, **6**, 49 (1956).
281. O. H. Johnson and H. E. Fritz, *J. Org. Chem.*, **19**, 74 (1954).
282. D. L. Alleston and A. G. Davies, *J. Chem. Soc.*, **1962**, 2050.
283. C. S. Bobashinkskaya and K. A. Kocheshkov, *J. Gen. Chem. USSR*, **8**, 1850 (1938); *Chem. Abstr.*, **33**, 5820 (1939).
284. K. A. Kocheschkov, *Ber.*, **62**, 996 (1929).
285. H. Gilman and L. A. Gist, Jr., *J. Org. Chem.*, **22**, 368 (1957).
286. H. Zimmer and H. W. Sparmann, *Chem. Ber.*, **87**, 645 (1954).
287. J. D'Ans and H. Zimmer, *Ber.*, **85**, 585 (1952).
288. K. A. Kocheschkov, M. M. Nadj, and A. P. Allesandrow, *Ber.*, **67**, 1348 (1934).
289. G. A. Razuvaev, *Akad. Nauk SSSR, Inst. Organ. Khim. Sintezy Organ. Soedin. Sb.*, **1**, 41 (1950), *Chem. Abstr.*, **47**, 8004 (1953).
290. K. A. Kocheschkov and A. N. Nesmejanow, *Ber.*, **64**, 628 (1931).
291. H. Gilman and S. D. Rosenberg, *J. Am. Chem. Soc.*, **74**, 5580 (1952).
292. V. A. Zasov and K. A. Kocheshkov, *Sb. Statei Obshchei Khim.*, *Akad. Nauk SSSR*, **1**, 278 (1953), *Chem. Abstr.*, **49**, 912 (1955).
293. H. G. Langer, *Tetrahedron Letters.*, **1967**, 43.
294. G. J. M. Van der Kerk and J. G. Luijten, *J. Appl. Chem. London*, **7**, 369 (1957).
295. H. G. Kuivila and O. F. Beumel, Jr., *J. Am. Chem. Soc.*, **80**, 3250 (1958).
296. K. A. Kocheskov and M. M. Nadj, *J. Gen. Chem. USSR, English Transl.*, **5**, 1158 (1935); *Chem. Abstr,*, **30**, 1036 (1936).
297. K. A. Kocheschkov and M. M. Nadj, *Ber.*, **67**, 717 (1934).
298. T. V. Talalaeva, N. A. Saitseva, and K. A. Kocheshkov, *J. Gen. Chem. USSR*, **16**, 901 (1946); *Chem. Abstr*, **41**, 2014 (1947).
299. D. Grant and J. R. Van Wazer, *J. Organometal. Chem.*, **4**, 229 (1965).
300. E. V. Van den Berghe and G. P. Van der Kelen, *J. Organometal. Chem.*, **6**, 522 (1966).
301. A. Ladenburg, *Ber.*, **4**, 19 (1871).
302. A. Ladenburg, *Ann.*, **159**, 251 (1871).
303. M. E. Pavlovskaya and K. A. Kocheshkov, *Compt. Rend. Acad. Sci. USSR*, **49**, 263 (1945); *Chem. Abstr.*, **40**, 5696 (1946).
304. H. G. Kuivila, R. Sommer, and D. C. Green, *J. Org. Chem.*, **33**, 1119 (1968).
305. P. R. Austin, *J. Am. Chem. Soc.*, **54**, 3287 (1932).
306. A Michaelis and V. Paetow, *Ann.*, **233**, 60 (1866).
307. A. Michaelis and H. Loesner, *Ber.*, **27**, 264 (1894).
308. W. J. Pope and E. E. Turner, *J. Chem. Soc.*, **1920**, 1447.
309. W. R. Cullen, *Can. J. Chem.*, **41**, 317 (1963).
310. E. G. Walaschewski, *Chem. Ber.*, **86**, 272 (1953).
311. H. J. Emeleus, R. N. Haszeldine, and E. G. Walaschewski, *J. Chem. Soc.*, **1953**, 1552.
312. L. Maier, D. Seyferth, F. G. A. Stone, and E. G. Rochow, *J. Am. Chem. Soc.*, **79**, 5884 (1957).
313. J. F. Morgan, E. J. Cragoe, Jr., B. Elpern, and C. S. Hamilton, *J. Am. Chem. Soc.* **69**, 932 (1947).
314. A. G. Evans and E. Warhurst, *Trans. Faraday Soc.*, **44**, 189 (1948).

315. H. D. N. Fitzpatrick, S. R. C. Hughes, and E. A. Moelwyn-Hughes, *J. Chem. Soc.*, **1950**, 3542.
316. A. Michaelis, *Ann.*, **233**, 57 (1886).
317. J. Hasenbäumer, *Ber.*, **31**, 1910 (1898).
318. A. Michaelis and A. Günther, *Ber.*, **44**, 2316 (1911).
319. G. T. Morgan, F. M. G. Micklethwait, *J. Chem. Soc.*, **99**, 2286 (1911).
320. P. May, *J. Chem. Soc.*, **101**, 1033 (1912).
321. G. Grüttner and M. Wiernik, *Ber.*, **48**, 1749 (1915).
322. G. Grüttner and M. Wiernik, *Ber.*, **48**, 1759 (1915).
323. A. Marquardt, *Ber.*, **20**, 1516 (1887).
324. A. Marquardt, *Ber.*, **21**, 2038 (1888).
325. A. Michaelis, *Ann.*, **251**, 323 (1889).
326. F. Challenger and C. F. Allpress, *J. Chem. Soc.*, **119**, 913 (1921).
327. N. I. Sheverdina and K. A. Kocheshkov, *J. Gen. Chem. USSR*, **8**, 1839 (1938); *Chem. Abstr.*, **33**, 5819 (1939).
328. Z. M. Manulkin, A. N. Tatarenko, and F. Yusupov, *Sb. Statei Obschei Khim.*, **2**, 1308 (1953); *Chem. Abstr.*, **49**, 5397 (1955).
329. B. M. Mikhailov and L. S. Vasilev, *J. Gen. Chem. USSR, English Transl.*, **35**, 1077 (1965).
330. R. H. Cragg, *J. Inorg. Nucl. Chem.*, **30**, 711 (1968).
331. J. R. Van Wazer and K. Moedritzer, *J. Inorg. Nucl. Chem.*, **26**, 737 (1964).
332. K. Moedritzer and J. R. Van Wazer, *J. Am. Chem. Soc.*, **86**, 802 (1964).
333. J. R. Van Wazer, and S. Norval, *Inorg. Chem.*, **4**, 1294 (1965).
334. K. Moedritzer and J. R. Van Wazer, *Inorg. Chem.*, **3**, 139 (1964).
335. K. Moedritzer and J. R. Van Wazer, *Inorg. Chem.*, **4**, 893 (1965).
336. E. Wiberg and W. Sturm, *Z. Naturforsch.*, **8b**, 689 (1953).
337. E. Wiberg and W. Sturm, *Z. Naturforsch.*, **10b**, 109 (1955).
338. K. Moedritzer and J. R. Van Wazer, *Inorg. Nucl. Chem. Letters.*, **2**, 45 (1966).
339. K. Moedritzer and J. R. Van Wazer, *J. Phys. Chem.*, **70**, 2030 (1966).
340. A. B. Burg and J. Banus, *J. Am. Chem. Soc.*, **76**, 3903 (1954).
341. H. J. Becher, *Z. Anorg. Allgem. Chem.*, **288**, 235 (1966).
342. D. W. Aubrey, M. F. Lappert, and M. K. Majumdar, *J. Chem. Soc.*, **1962**, 4088.
343. K. Niedenzu, H. Beyer, J. W. Dawson, and H. Jenne, *Chem. Ber.*, **96**, 2653 (1963).
344. A. J. Bannister, N. N. Greenwood, B. P. Straughan, and J. Walker, *J. Chem. Soc.*, **1964**, 995.
345. N. N. Greenwood, K. A. Hooten, and J. Walker, *J. Chem. Soc. (A)*, **1966**, 21.
345a. J. Goubeau, M. Rahtz, and H. J. Becher, *Z. Anorg. Allgem. Chem.*, **275**, 161 (1954).
346. K. Moedritzer and J. R. Van Wazer, *Inorg. Chem.*, **3**, 268 (1964).
347. K. Moedritzer and J. R. Van Wazer, *J. Inorg. Nucl. Chem.*, **27**, 1851 (1967).
348. K. Moedritzer and J. R. Van Wazer, *Inorg. Chem.*, **7**, 2105 (1968).
349. H. H. Anderson, *J. Am. Chem. Soc.*, **75**, 1575 (1953).
350. J. R. Van Wazer and K. Moedritzer, *J. Chem. Phys.*, **41**, 3122 (1964).
351. P. Geymayer and E. G. Rochow, *Angew. Chem.*, **77**, 613 (1965).
352. U. Wannagat, *Angew. Chem.*, **77**, 626 (1965).
353. U. Wannagat, E. Bogusch, and F. Höfler, *J. Organometal. Chem.*, **7**, 203 (1967).
354. W. Fink, *Helv. Chim. Acta.*, **51**, 978 (1968).
355. W. Fink, *Helv. Chim. Acta*, **51**, 1011 (1968).
356. W. Sundermeyer, *Chem. Ber.*, **96**, 1293 (1963).
357. W. H. Eisenhuth and J. R. Van Wazer, *Inorg. Chem.*, **7**, 1642 (1968).

358. M. D. Rausch, J. R. Van Wazer, and K. Moedritzer, *J. Am. Chem. Soc.*, **86**, 814 (1964).
359. J. R. Van Wazer, K. Moedritzer, and M. D. Rausch, *J. Chem. Phys.*, **42**, 3302 (1965).
360. T. Colclough, W. Gerrard, and M. F. Lappert, *J. Chem. Soc.*, **1965**, 3006.
361. P. A. McCusker, P. L. Pennartz, R. C. Pilger, Jr., *J. Am. Chem. Soc.*, **84**, 4362 (1962).
362. G. L. O'Connor and H. R. Nace, *J. Am. Chem. Soc.*, **77**, 1578 (1955).
363. M. F. Lappert, *J. Chem. Soc.*, **1958**, 2790.
364. H. K. Hofmeister and J. R. Van Wazer, *J. Inorg. Nucl. Chem.*, **26**, 1201 (1964).
365. J. W. Post and C. H. Hofrichter, Jr., *J. Org. Chem.*, **5**, 572 (1940).
366. D. F. Peppard, W. G. Brown, and W. C. Johnson, *J. Am. Chem. Soc.*, **68**, 77 (1946).
367. F. C. Boye and H. W. Post, *J. Org. Chem.*, **17**, 1389 (1952).
368. D. L. Bailey and A. N. Pines, *Ind. Eng. Chem.*, **46**, 2363 (1954).
369. D. Grant, *J. Inorg. Nucl. Chem.*, **29**, 69 (1967).
370. K. Moedritzer and J. R. Van Wazer, *Makromol. Chem.*, **104**, 148 (1967).
371. D. W. Scott, *J. Am. Chem. Soc.*, **68**, 356 (1946).
372. W. Patnode and D. F. Wilcock, *J. Am. Chem. Soc.*, **68**, 358 (1946).
373. M. J. Hunter, J. F. Hyde, E. L. Warrick, and H. J. Fletcher, *J. Am. Chem. Soc.*, **68**, 667 (1946).
374. D. W. Scott, *J. Am. Chem. Soc.*, **68**, 2294 (1946).
375. F. R. Mayo, *J. Poly. Sci.*, **55**, 57 (1961).
376. W. Simmler, *Makromol. Chem.*, **57**, 12 (1962).
377. M. Morton and E. E. Bostick, *J. Polymer Sci. (A)*, **2**, 523 (1964).
378. E. D. Brown and J. B. Carmichael, *Polymer Letters*, **3**, 473 (1965).
379. J. B. Carmichael and R. Winger, *J. Poly. Sci. (A)*, **3**, 971 (1965).
380. J. B. Carmichael and J. Heffel, *J. Phys. Chem.*, **69**, 2213 (1965).
381. J. B. Carmichael and J. Heffel, *J. Phys. Chem.*, **69**, 2218 (1965).
382. A. J. Barry, W. H. Daudt, J. J. Domicone, and J. W. Gilkey, *J. Am. Chem. Soc.*, **77**, 4248 (1955).
383. K. A. Andrianov, G. A. Kurakov, F. F. Sushchentsova, V. A. Myagkov, and V. A. Abilov, *Doklady Akad. Nauk SSSR English Transl.*, **166**, 151 (1966).
384. J. F. Brown, Jr., L. H. Vogt, Jr., A. Katzmann, J. W. Eustance, K. M. Kiser, and K. W. Krantz, *J. Am. Chem. Soc.*, **82**, 6194 (1960).
385. J. F. Brown, Jr., L. H. Vogt, Jr., and P. I. Prescott, *J. Am. Chem. Soc.*, **86**, 1120 (1964).
386. J. F. Brown, Jr. and L. H. Vogt, Jr., *J. Am. Chem. Soc.*, **87**, 4313 (1965).
387. J. F. Brown, Jr., *J. Am. Chem. Soc.*, **87**, 4317 (1965).
388. K. Moedritzer, *J. Organometal. Chem.*, **5**, 254 (1966).
389. R. C. Mehrotra and G. Chandra, *J. Indian Chem. Soc.*, **39**, 235 (1962).
390. A. G. Davies and P. G. Harrison, *J. Chem. Soc. (C)*, **1967**, 298.
391. W. P. Neumann, *Ann. N. Y. Acad. Sci.*, **159**, Art 1, 56 (1969).
392. M. Komura, T. Tanaka, T. Mukai, and R. Okawara, *Inorg. Nucl. Chem. Letters*, **3**, 17 (1967).
393. D. L. Alleston, A. G. Davies, M. Hancock, and R. F. M. White, *J. Chem. Soc.*, **1963**, 5469.
394. A. G. Davies and P. G. Harrison, *J. Organometal. Chem.*, **7**, P13 (1966).
395. E. Thilo and P. Flögel, *Z. Anorg. Allgem. Chem.*, **329**, 244 (1964).
396. G. Russias, F. Damm, A. Deluzarche, and A. Maillard, *Bull. Soc. Chim. France*, **1966**, 2275.

397. J. R. Van Wazer, K. Moedritzer, and L. C. D. Groenweghe, *J. Organometal. Chem.*, **5**, 420 (1966).
398. K. Moedritzer, J. R. Van Wazer, and C. H. Dungan, *J. Chem. Phys.*, **42**, 2478 (1965).
399. K. Moedritzer and J. R. Van Wazer, *J. Am. Chem. Soc.*, **90**, 1708 (1968).
400. A. G. Davies and P. G. Harrison, *J. Organometal. Chem.*, **8**, P19 (1967).
401. E. Wiberg and W. Ruschmann, *Ber.*, **70**, 1393 (1937).
402. J. D. Edwards, W. Gerrard, and M. F. Lappert, *J. Chem. Soc.*, **1955**, 1470.
403. P. B. Brindley, W. Gerrard, and M. F. Lappert, *J. Chem. Soc.*, **1956**, 824.
404. P. B. Brindley, W. Gerrard, and M. F. Lappert, *J. Chem. Soc.*, **1956**, 1540.
405. T. Colclough, W. Gerrard, and M. F. Lappert, *J. Chem. Soc.*, **1956**, 3006.
406. E. W. Abel, J. D. Edwards, W. Gerrard, and M. F. Lappert, *J. Chem. Soc.*, **1957**, 501.
407. J. A. Blau, W. Gerrard, and M. F. Lappert, *J. Chem. Soc.*, **1957**, 4116.
408. S. H. Dandegaonker, W. Gerrard, and M. F. Lappert, *J. Chem. Soc.*, **1957**, 2872.
409. T. D. Coyle, J. J. Ritter, and T. C. Farrar, *Proc. Chem. Soc.*, **1964**, 25.
410. A. B. Burg, *J. Am. Chem. Soc.*, **62**, 2228 (1940).
411. P. A. McCusker and L. J. Glunz, *J. Am. Chem. Soc.*, **77**, 4253 (1955).
412. E. W. Abel, S. H. Dandegaonker, W. Gerrard, and M. F. Lappert, *J. Chem. Soc.*, **1956**, 4697.
413. P. A. McCusker, E. C. Ashby, and H. S. Makowski, *J. Am. Chem. Soc.*, **79**, 5182 (1957).
414. P. A. McCusker and H. S. Makowski, *J. Am. Chem. Soc.*, **79**, 5185 (1957).
415. B. M. Mikhailov and F. B. Tutorskaya, *Bull. Acad. Sci. USSR, Div. Chem. Sci. English Transl.*, **1959**, 1785.
416. B. M. Mikhailov and T. K. Kozminskaya. *J. Gen. Chem. USSR, English Transl.*, **30**, 3587 (1960).
417. C. Friedel and J. M. Crafts, *Ann.*, **127**, 28 (1863).
418. C. Friedel and J. M. Crafts, *Compt. Rend.*, **56**, 590 (1863).
419. C. Friedel and J. M. Crafts, *Ann. Chim. Phys.*, [4], **9**, 5 (1866).
420. G. S. Forbes and H. H. Anderson, *J. Am. Chem. Soc.*, **66**, 1703 (1944).
421. W. Gerrard and J. U. Jones, *J. Chem. Soc.*, **1952**, 1690.
422. M. Kumada, *J. Inst. Polytech. Osaka City, Univ. Ser. C*, **2**, 139 (1952); *Chem. Abstr.*, **48**, 11303 (1954).
423. J. R. Van Wazer, K. Moedritzer, and L. C. D. Groenweghe, *J. Organometal. Chem.*, **5**, 420 (1966).
424. K. Moedritzer and J. R. Van Wazer, *J. Organometal. Chem.*, **6**, 242 (1966).
425. K. Moedritzer and J. R. Van Wazer, *J. Inorg. Nucl. Chem.*, **29**, 1851 (1967).
426. H. J. Emeleus and H. G. Heal, *J. Chem. Soc.*, **1949**, 1696.
427. M. G. Voronkov and L. M. Chudesova, *Bull. Acad. Sci. USSR, Div. Chem. Sci., English Transl.*, **1957**, 1440.
428. K. A. Andrianov and V. V. Severnyi, *Proc. Acad. Sci. USSR, Chem. Sec. English Transl.*, **134**, 1167 (1960).
429. K. A. Andrianov, V. V. Severnyi, and B. G. Savin, *Bull. Acad. Sci. USSR, Div. Chem. Sci., English Transl.*, **1961**, 1356.
430. K. A. Andrianov and V. V. Severnyi, *Proc. Acad. Sci. USSR, Chem. Sec. English Transl.*, **146**, 828 (1962).
431. K. A. Andrianov, V. V. Severnyi, and B. A. Izmailov, *Bull. Acad. Sci. USSR, Div. Chem. Sci., English Transl.*, **1963**, 256.

432. K. A. Andrianov and V. V. Severnyi, *Bull. Acad. Sci. USSR, Div. Chem. Sci., English Transl.*, **1964**, 1174.
433. K. A. Andrianov and V. V. Severnyi, *Bull. Acad. Sci. USSR, Div. Chem. Sci., English Transl.*, **1964**, 1178.
434. K. A. Andrianov and V. V. Severnyi, *J. Organometal. Chem.*, **1**, 268 (1964).
435. K. Moedritzer, unpublished results, 1968.
436. K. Moedritzer and J. R. Van Wazer, *J. Inorg. Nucl. Chem.*, **29**, 1571 (1967).
437. K. Moedritzer and J. R. Van Wazer, *Inorg. Chem.*, **4**, 1753 (1965).
438. I. P. Goldshtein, N. N. Zemlyanski, O. P. Shamagina, E. N. Guryanova, E. M. Panov, N. A. Slovokhotova, and K. A. Andrianov, *Dok. Akad. Nauk, English Transl.*, **163**, 715 (1965).
439. A. H. Westlake and D. F. Martin, *J. Inorg. Nucl. Chem.*, **27**, 1579 (1965).
440. F. Huber and R. Kaiser, *J. Organometal. Chem.*, **6**, 126 (1966).
441. R. Okawara and M. Wada, *J. Organometal. Chem.*, **1**, 81 (1965).
442. R. L. Barker, E. Booth, W. E. Jones, A. F. Millidge, and F. N. Woodward, *J. Soc. Chem. Ind. London*, **68**, 285 (1949).
443. R. L. Barker, E. Booth, W. E. Jones, A. F. Millidge, and F. N. Woodward, *J. Soc. Chem. Ind. London*, **68**, 289 (1949).
444. R. L. Barker, E. Booth, W. E. Jones, and F. N. Woodward, *J. Soc. Chem. Ind. London*, **68**, 295 (1949).
445. J. R. Van Wazer, K. Moedritzer, and D. W. Matula, *J. Am. Chem. Soc.*, **86**, 807 (1964).
446. B. M. Mikhailov, T. A. Shchegoleva, E. M. Shashkova, and V. D. Sheludyakov, *Bull. Acad. Sci. USSR, Div. Chem. Sci., English Transl.*, **1961**, 1163.
447. E. L. Muetterties, N. R. Miller, K. J. Packer, and H. C. Miller, *Inorg. Chem.*, **3**, 870 (1964).
448. E. Wiberg and W. Sturm, *Z. Naturforsch*, **10b**, 114 (1955).
449. J. Goubeau and H. W. Wittmeier, *Z. Anorg. Allgem. Chem.*, **270**, 16 (1952).
450. D. J. Panckhurst, C. J. Wilkins, and P. W. Craighead, *J. Chem. Soc.*, **1955**, 3395.
451. K. Moedritzer and J. R. Van Wazer, *J. Am. Chem. Soc.*, **87**, 2360 (1965).
452. K. Moedritzer and J. R. Van Wazer, *Inorg. Chem.*, **3**, 943 (1964).
453. H. Landesmann and R. E. Williams, *J. Am. Chem. Soc.*, **83**, 2663 (1961).
454. T. D. Coyle and F. G. A. Stone, *J. Chem. Phys.*, **32**, 1892 (1960).
455. J. C. Lockhart, *J. Chem. Soc.*, **1962**, 1197.
456. A. Finch and J. C. Lockhart, *Chem. Ind. London*, **1964**, 497.
457. N. N. Greenwood and J. Walker, *Inorg. Chem. Letters*, **1**, 65 (1965).
458. F. E. Brinckman and F. G. A. Stone, *J. Am. Chem. Soc.*, **82**, 6235 (1960).
459. H. H. Anderson, *J. Am. Chem. Soc.*, **66**, 934 (1944).
460. H. H. Anderson, *J. Am. Chem. Soc.*, **67**, 223 (1945).
461. H. H. Anderson, *J. Am. Chem. Soc.*, **73**, 5804 (1951).
462. M. Kumada, *J. Inst. Polytech. Osaka City Univ. Ser. C*, **2**, 131 (1952); *Chem. Abstr.*, **48**, 303 (1954).
463. H. H. Anderson, *J. Am. Chem. Soc.*, **75**, 1576 (1953).
464. G. Fritz and D. Kummer, *Z. Anorg. Allgem. Chem.*, **310**, 326 (1961).
465. B. Kanner, D. L. Bailey, and E. J. Pepe, 145th National ACS Meeting, Sept. 1963, Abstracts p. 74Q.
466. K. Moedritzer and J. R. Van Wazer, *J. Inorg. Nucl. Chem.*, **28**, 957 (1966).
467. K. Moedritzer and J. R. Van Wazer, *Inorg. Chem.*, **5**, 547 (1966).
468. R. S. Feinberg and E. G. Rochow, *J. Inorg. Nucl. Chem.*, **24**, 165 (1962).
469. J. J. Burke and P. C. Lauterbur, *J. Am. Chem. Soc.*, **82**, 326 (1961).

470. M. L. Delwaulle, M. B. Bensset, M. Delhaye, *J. Am. Chem. Soc.*, **74**, 5768 (1952).
471. O. Johnson, H. E. Fritz, D. O. Halvorson, and R. L. Evans, *J. Am. Chem. Soc.*, **77**, 5857 (1955).
472. E. V. Van den Berghe, G. P. Van der Kelen, and Z. Eeckhaut, *Bull. Soc. Chim. Belges*, **76**, 79 (1967).
473. C. G. Moreland, M. H. O'Brien, C. E. Douthit, and G. G. Long, *Inorg. Chem.*, **7**, 834 (1968).
474. G. G. Long, C. G. Moreland, G. O. Doak, and M. Miller, *Inorg. Chem.*, **5**, 1358 (1966).
475. A Stock, A. Brandt, and H. Fischer, *Ber.*, **58**, 643 (1925).
476. H. Nöth, P. Fritz, K. H. Hermannsdörfer, W. Meisler, H. Schick, and G. Schmidt, *Angew. Chem.*, **75**, 1114 (1963).
477. M. Abedini and A. G. McDiarmid, *Inorg. Chem.*, **2**, 608 (1963).
478. J. A. Morrison and M. A. Ring, *Inorg. Chem.*, **6**, 100 (1967).
479. S. P. Garrity and M. A. Ring, *Inorg. Nucl. Chem. Letters*, **4**, 77 (1968).
480. W. M. Ingle, E. A. Groschwitz, and M. A. Ring, *Inorg. Chem.*, **6**, 1429 (1967).
481. W. H. Atwell and D. R. Weyenberg, *J. Organometal. Chem.*, **5**, 594 (1966).
482. W. H. Atwell and D. R. Weyenberg, *J. Organometal. Chem.*, **7**, 71 (1967).
483. W. H. Atwell and D. R. Weyenberg, *J. Am. Chem. Soc.*, **90**, 3438 (1968).
484. C. G. Wilkins, *J. Chem. Soc.*, **1953**, 3409.
485. A. Kaczmarczyk and G. Urry, *J. Am. Chem. Soc.*, **82**, 751 (1960).
486. J. V. Urenovitch and A. G. McDiarmid, *J. Am. Chem. Soc.*, **85**, 3372 (1963).
487. J. V. Urenovitch, R. Pejic, and A. G. McDiarmid, *J. Chem. Soc.*, **1963**, 5563.
488. H. Sakurai, K. Tominaga, and M. Kumada, *Bull. Chem. Soc. Japan*, **39**, 1820 (1966).
489. H. Sakurai, K. Tominaga, T. Watanabe, and M. Kumada. *Tetrahedron Letters*, **1966**, 5493.
490. E. Wiberg and H. Behringer, *Z. Anorg. Allgem. Chem.*, **329**, 290 (1964).
491. G. Calingaert, H. Soroos, and H. Shapiro, *J. Am. Chem. Soc.*, **64**, 462 (1942).
492. G. A. Razuvaev, M. S. Fedotov, T. B. Zavarova, and N. N. Bazhenova, *Tr. Khim. Khim. Tekhnol.*, **4**, 622 (1961); *Chem. Abstr.*, **58**, 543 (1963).
493. W. R. Cullen, *Can. J. Chem.*, **41**, 322 (1963).
494. K. Knoll, H. C. Marsmann, and J. R. Van Wazer, *J. Am. Chem. Soc.*, **91**, 4986 (1969).
495. R. Köster and G. Benedikt, private communication.

Reactions of Organotellurium Compounds

KURT J. IRGOLIC AND RALPH A. ZINGARO

Department of Chemistry, Texas A&M University,
College Station, Texas

117

INTRODUCTION

The first organotellurium compound was synthesized by Wöhler in 1840 (W-4). Since then more than 300 papers have been published which deal with various aspects of organotellurium chemistry. In this review an attempt has been made to present the reported reactions of organotellurium compounds. The reader will note that a fair amount of material has been included which cannot be considered to be established knowledge. Many substances reported in the literature are poorly characterized and many contradicting statements have been published concerning the reactions or compounds. Wherever possible, these discrepancies have been noted. It is our hope that some of the problems will be solved by more extensive research in the area of organo-metallic chemistry of tellurium.

Throughout this article R denotes an organic group, aliphatic or aromatic, unless otherwise stated. In each of the various sections the general reactions of a class of organotellurium compounds are discussed. Individual compounds and their reactions are included in the tables. The first section deals with the synthesis of organotellurium compounds with simple inorganic materials serving as the source of tellurium. Sections II through XII describe the reactions of the various types of organotellurium compounds. Heterocyclic compounds are treated separately in Section XIII. Section XIV is concerned with compounds which contain one or two tellurium–metal bonds. Sections XV and XVI discuss optically active tellurium compounds and spectroscopic and structural studies, respectively.

A number of review articles have appeared in the literature which have proven to be helpful in the preparation of this manuscript. These include articles by Rheinboldt (R-10) (general), M. de Moura Campos (M-44) (participation of vicinal groups involving organic tellurium compounds), Schumann and Schmidt (S-11) (R_3M—Te—$M'R_3$ compounds), Petragnani and de Moura Campos (P-13) (new topics in organotellurium chemistry), Abrahams (A-1) (tellurium stereochemistry), Goddard (G-10) (general),

Morgan (M-39) (heterocyclic tellurium compounds), Frommes (F-18) (analytical chemistry of tellurium), Day (D-4) (organotellurium compounds), Dubien (D-18) (organomagnesium compounds containing tellurium), and Cerwenka and Cooper (C-6) (toxicology). The book "The Chemistry of Selenium, Tellurium and Polonium" by K. W. Bagnall (B-1) has a section on organotellurium compounds.

The authors wish to express their appreciation to the United States Atomic Energy Commission for financial support under Contract No. At-(40-1)-2733, and to the Robert A. Welch Foundation of Houston, Texas.

I. THE ROLE OF ELEMENTAL TELLURIUM AND INORGANIC TELLURIUM COMPOUNDS IN ORGANOTELLURIUM CHEMISTRY

A. Elemental Tellurium

Elemental tellurium is relatively unimportant as a starting material in syntheses leading directly to organotellurium compounds when compared with tellurium tetrachloride.

Elemental tellurium reacts with:

1. alkyl iodides to give dialkyltellurium diiodides.
2. organic dihalides and polyhalides to give heterocyclic compounds.
3. organometallic compounds.
   (a) Grignard reagents and alkaliorganic compounds.
   (b) organomercurials.
   (c) organometallic compounds of group IVA.
   (d) organophosphorus compounds.
4. diazonium and iodonium salts.
5. free radicals.
6. miscellaneous reagents.

Figure 1 summarizes these reactions of elemental tellurium.

1. Elemental Tellurium and Alkyl Iodides

The reported reactions of elemental tellurium with simple alkyl halides are concerned almost entirely with alkyl iodides. Demarçay (D-5) stated that dimethyltellurium diiodide, $(CH_3)_2TeI_2$, is obtained by the action of methyl iodide on tellurium. Scott (S-12) confirmed Demarçay's results and mentioned that neither sulfur nor selenium behave in the same manner. With allyl iodide an easily decomposable oil was obtained. Vernon (V-1) followed up these investigations and extended them (V-3) to include ethyl and benzyl iodide. The reaction of alkyl iodides with elemental tellurium is summarized by the equation

$$2 RI + Te \longrightarrow R_2TeI_2$$

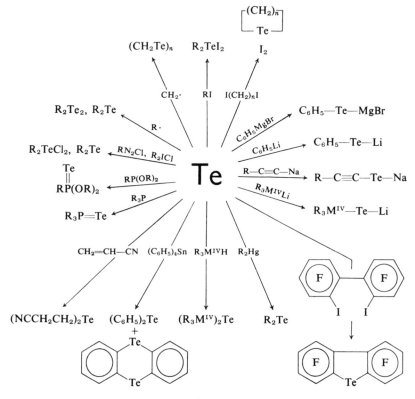

Figure 1

The interaction between methyl iodide and tellurium was observed to be initially very rapid, but it gradually became slow and eventually appeared to stop. Prolonged heating, even up to three weeks, did not result in any significant increase in the yield.

The reaction between tellurium and methyl iodide is reversible at higher temperatures (V-1).

$$Te + 2\ CH_3I \underset{100°}{\overset{80°}{\rightleftharpoons}} (CH_3)_2TeI_2$$

Cohen and coworkers obtained bis(pentafluorophenyl) telluride from tellurium and iodopentafluorobenzene (C-26).

Experimental. Dimethyltellurium Diiodide (V-1):   In a glass bulb, of 150–200 ml capacity, were sealed 50 g (0.39 moles) of amorphous tellurium and 112 g (0.78 moles) of methyl iodide. The sealed container was kept in a water bath at 80° for 36–48 hr. Very little pressure was developed and none was found when the cold tube was opened. The brittle, red mass was powdered and extracted with chloroform. The mixture was filtered through a hot funnel to separate any unreacted tellurium. The yield of dimethyltellurium diiodide, based on the amount of crystalline solid obtained from the chloroform solution, was about 50–55%. It recrystallized from various solvents in shades of red and melted at 127° with decomposition.

2. Elemental Tellurium and Organic Dihalides and Polyhalides

Morgan and Burgess (M-35) and Morgan and Burstall (M-40) were able to prepare heterocyclic tellurium derivatives from elemental tellurium and dihalides. Thus, amorphous tellurium dissolved in tetramethylene diiodide at 130° with quantitative production of telluracyclopentane 1,1-diiodide (M-40). However, pentamethylene dihalides, when refluxed in xylene for 20 hr gave only a 10% yield of the corresponding telluracyclohexane 1,1-dihalide (M-35). In the latter case, when aluminum telluride was substituted for elemental tellurium, the yields were substantially increased (see Section IB-2). Farrar and Gulland (F-1) reinvestigated the reaction between polymethylene dihalides and tellurium. Tetramethylene dibromide and pentamethylene dichloride gave a 60 and 35% yield, respectively, of the cyclic dihalides.

$$n = 1 \text{ or } 2$$

The reactivity of the dihalides seems to decrease with decreasing atomic mass of the halide (see Table I for yields). However, these authors claim better results when sodium telluride is used in place of tellurium. Trimethylene diiodide heated with amorphous tellurium to 120–150° formed a black, high melting product containing unchanged tellurium. The reagents were recovered almost quantitatively from a mixture of ethylene dibromide and tellurium heated to 160–170° for 16.5 hr. β,β'-Diiododiethyl ether gave only 10–13% of 4-tellura-1-oxacyclohexane 4,4-diiodide in impure form. 1-Bromo-2-iodoethane when treated with tellurium at 80° evolved ethene leaving a black substance of undetermined constitution.

The reaction between methylene iodide and tellurium has been discussed

TABLE I

Reaction of Tellurium with Organic Halides

Organic halide	Product	Yield %	m.p., b.p. (torr)	Ref.
CH_3I	$(CH_3)_2TeI_2$	—	—	D-5,C-10
C_2H_5I	$(C_2H_5)_2TeI_2$	50	127	V-1
C_2H_5I	$(C_2H_5)_2TeI_2$	40	57	V-3,S-12
		20		G-4
$C_6H_5CH_2I$	$(C_6H_5CH_2)_2TeI_2$	—	134	V-3
CH_2I_2	$(ICH_2)_2TeI_2$	2–30	127	F-1,D-17
$I—(CH_2)_4—I$	(cyclic) TeI_2	100	149–150	M-40,D-1?
$Br—(CH_2)_4—Br$	(cyclic) $TeBr_2$	60	128–131	F-1
$I—(CH_2)_5—I$	(cyclic) TeI_2	10	135–136	M-35
		63	136	F-1
$Cl—(CH_2)_5—Cl$	(cyclic) $TeCl_2$	35	105–106	F-1
$(ICH_2CH_2)_2O$	(cyclic) O...TeI_2	13	155 (dec.)	F-1
C_6F_5I	$(C_6F_5)_2Te$	10	50–51	C-26
		66	116–119	C-26,C-2?
$Cl_2C{=}CCl—CCl{=}CCl_2$		14	49	M-5
			31 (0.02)	

in a very brief manner by Drew (D-17) who reported the product of this reaction to be bis(iodomethyl)tellurium diiodide with the concurrent formation of a red, amorphous polymer.

Farrar and Gulland (F-1) found that the red polymer was always the main product. The yields of $(ICH_2)_2TeI_2$ fluctuated irregularly with the conditions (4–13 moles CH_2I_2; 110–150°) from 2 to 14%, the best results (30% yield) being obtained with 13 moles CH_2I_2 per mole of tellurium at 80° for 49 hr. An exact repetition of this experiment gave only 3.5% of the desired product.

Mack (M-5) described the preparation of tetrachlorotellurophene from tellurium with hexachloro-1,3-butadiene

The utilization of elemental tellurium for the direct synthesis of fluoroaromatic tellurium derivatives has been described by Cohen and co-workers (C-25,C-26). They heated iodo-perfluoroarenes with tellurium powder in sealed, evacuated tubes and were able to isolate the product in pure form by direct sublimation. The synthesis described by the following equation is illustrative

Table I summarizes the results obtained in reactions of elemental tellurium with organic halides.

Experimental. Telluracyclopentane 1,1-diiodide (M-40):   A mixture of 10 g (0.078 moles) of amorphous tellurium and 63 g (0.20 moles) of tetramethylene diiodide was heated for 5 hr at 130–140° with occasional shaking. The dark solution was filtered hot in order to remove any unreacted tellurium and, on cooling, most of the product crystallized. Addition of petroleum ether (b.p. 40–60°) caused precipitation of an additional quantity of the diiodide. The yield was quantitative.

From acetone, the compound was obtained in the form of shiny, purple plates, whereas hot benzene solutions gave the diiodide in the form of bright red prisms. Both forms melted at 149–150° without decomposition and remained unchanged above 200°. Both appear to be stable at ordinary temperatures.

Tetrachlorotellurophene (M-5):   Finely powdered tellurium (0.3 moles) was shaken with hexachloro-1,3-butadiene (0.6 moles) at 250° for 40 hr. $TeCl_4$ formed and was removed with concentrated hydrochloric acid. Distillation gave tetrachlorotellurophene (13.1 g, 14%) boiling at 31°/0.02 mm. Yellow needles melting at 49° were obtained from methanol.

3. Elemental Tellurium and Organometallic Compounds

a. Grignard Reagents and Organoalkali Compounds. Giua and Cherchi (G-6) claim that theirs was the first investigation concerned with the behavior of tellurium toward organomagnesium compounds. They reported that tellurium reacts with C_6H_5MgBr to give first $C_6H_5TeMgBr$ and then C_6H_5TeH, $(C_6H_5)_2Te$, and H_2Te. It was observed that the amount of H_2Te formed is less than that expected on the basis of the reactions

$$2\ C_6H_5TeMgBr \longrightarrow (C_6H_5)_2Te + (BrMg)_2Te$$
$$(BrMg)_2Te + 2\ HCl \longrightarrow H_2Te + 2\ MgBrCl$$

It is quite obvious that these investigators did not work in an oxygen-free atmosphere because they interpreted the formation of diphenyl telluride in the following manner:

$$2\ C_6H_5TeH \xrightarrow{O_2} C_6H_5Te\text{---}TeC_6H_5$$
$$C_6H_5Te\text{---}TeC_6H_5 \longrightarrow Te + (C_6H_5)_2Te$$

Bowden and Braude (B-14) used $C_6H_5TeMgBr$ as an intermediate for the synthesis of phenyl ethyl telluride. Similar procedures were employed by Pourcelot for phenyl alkynyl tellurides (P-16, P-17) and by Hooton and Allred (H-8) for bis(trimethylsilyl) telluride (see Section XIV).

Petragnani and de Moura Campos (P-10) also utilized the reaction between tellurium and phenylmagnesium bromide in the following synthetic scheme

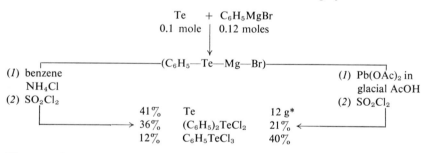

These authors propose two mechanisms to account for the isolated reaction products. The first involves the formation of phenyltelluromagnesium bromide, which immediately reacts with tellurium to give the ditelluride.

$$C_6H_5MgBr + Te \longrightarrow C_6H_5TeMgBr$$
$$2\ C_6H_5TeMgBr + Te \longrightarrow C_6H_5Te\text{---}TeC_6H_5 + Te(MgBr)_2$$
$$C_6H_5Te\text{---}TeC_6H_5 + C_6H_5MgBr \longrightarrow C_6H_5TeC_6H_5 + C_6H_5TeMgBr$$

Sulfuryl chloride will convert the primary reaction products into R_2TeCl_2 and $RTeCl_3$. The absence of any $C_6H_5TeMgBr$ was concluded from the following experimental facts: (*a*) treatment of the reaction mixture with lead

* Crystalline, black, shiny precipitate, which contained Te and Pb.

acetate resulted only in the formation of lead telluride; (b) treatment with ethyl iodide yielded no phenyl ethyl telluride. There exists an obvious contradiction between the results reported here and those obtained by Bowden and Braude (B-14). The workers suggest that this discrepancy may be due to a difference in the manner of preparation of the tellurium used in each case.

In view of the suggestion of Waters (W-1), who proposed a free radical mechanism in the reaction of tellurium with diazonium salts, a free radical mechanism of the following type has been considered,

$$2 \ C_6H_5MgBr \longrightarrow 2 \ C_6H_5 \cdot + 2 \ BrMg \cdot$$
$$2 \ C_6H_5 \cdot + 2 \ Te \longrightarrow 2 \ C_6H_5Te \cdot$$
$$2 \ C_6H_5Te \cdot \longrightarrow C_6H_5TeTeC_6H_5$$
$$2 \ BrMg \cdot + Te \longrightarrow Te(MgBr)_2$$

$$\overline{2 \ C_6H_5MgBr + 3 \ Te \longrightarrow C_6H_5TeTeC_6H_5 + Te(MgBr)_2}$$

followed by

$$C_6H_5TeTeC_6H_5 \longrightarrow 2 \ C_6H_5Te \cdot$$
$$2 \ C_6H_5Te \cdot + 2 \ C_6H_5MgBr \longrightarrow 2 \ C_6H_5TeC_6H_5 + 2 \ BrMg \cdot$$
$$2 \ BrMg \cdot + Te \longrightarrow Te(MgBr)_2$$

$$\overline{C_6H_5TeTeC_6H_5 + 2 \ C_6H_5MgBr + Te \longrightarrow 2 \ (C_6H_5)_2Te + Te(MgBr)_2}$$

The existence of $Te(MgBr)_2$ in the reaction mixture was suggested by the nature of the hydrolytic products whose formation can be described by the following:

$$Te(MgBr)_2 + 2 \ H_2O \longrightarrow H_2Te + 2 \ MgBr(OH)$$
$$H_2Te + \tfrac{1}{2}O_2 \longrightarrow H_2O + Te$$

The free radical mechanism also finds some support in the observations of Farrar (F-2) who suggested, from molecular weight determinations in solution, that ditellurides may be dissociated into free radicals. Similar reactions with p-methoxyphenylmagnesium bromide gave no definite product (P-10).

Amorphous tellurium reacted with phenyllithium in absolute ether under argon atmosphere. After the reaction mixture was subjected to the treatment described for the reaction between tellurium and phenylmagnesium bromide, 27.5% Te, 13% $(C_6H_5)_2TeCl_2$, and 51% $C_6H_5TeCl_3$ were isolated (P-10).

Brandsma, Wijers, and their co-workers (B-15, B-17) and Radchenko, Petrov, and their co-workers (P-14, R-3) reacted sodium acetylides in liquid ammonia with amorphous tellurium.

$$R—C{\equiv}C—Na + Te \longrightarrow R—C{\equiv}C—Te—Na$$
$$R = CH_3 \ (B\text{-}15,B\text{-}17) \qquad R = CH_2{=}CH \ (P\text{-}14) \qquad R = CH_2{=}C(CH_3) \quad (R\text{-}3)$$

The alkynyl sodium telluride was not isolated but reacted with an alkyl halide to give the alkynyl alkyl telluride. For experimental procedures and a listing of the compounds prepared turn to Section XIV.

Experimental. Finely Divided Amorphous Tellurium (P-10): Commercial tellurium was dissolved in concentrated nitric acid. After addition of concentrated hydrochloric acid the solution was evaporated. This operation, with the addition of fresh hydrochloric acid, was repeated until no nitric acid remained.

To the hydrochloric acid solution, which was cooled in ice and stirred vigorously, was added aqueous hydrazine sulfate. The tellurium precipitated slowly as a very fine powder. It was separated by filtration, washed well with water and dried *in vacuo* over P_2O_5.

Reaction between Amorphous Tellurium and Phenylmagnesium Bromide (P-10): A solution of phenylmagnesium bromide was prepared from 2.9 g (0.12 moles) of magnesium and 18.8 g (0.12 moles) of bromobenzene in 80 ml of anhydrous ether. To the phenylmagnesium bromide solution was added, in small portions, 12.7 g (0.1 moles) of amorphous tellurium powder under an atmosphere of nitrogen while stirring.

Following a two-hour period of reflux the tellurium was completely dissolved. The red solution containing a light colored precipitate was utilized as the source of "phenyltelluromagnesium bromide".

b. Organomercurials. The reaction between elemental tellurium and diarylmercury compounds was utilized as early as 1894 (K-5) for the preparation of diaryl tellurides. The reaction is summarized by the equation

$$R_2Hg + 2\,Te \longrightarrow R_2Te + HgTe$$

The reaction involves simple fusion of the components in the absence of any solvent. The temperature must be kept below the decomposition temperature of the mercury compound. Generally, temperatures between 180 and 230° were employed. Stoichiometric amounts of the reactants were kept under nitrogen or CO_2 atmosphere or under vacuum in sealed tubes during the reaction. In this manner Adloff and co-workers (A-2,L-34) prepared [132]Te-$(C_6H_5)_2$ using [132]Te and diphenylmercury.

Dibenzotellurophene, which is otherwise difficult to prepare, and then only in very low yields, has been obtained in 82% yield by Hellwinkel and Fahrbach (H-3) when biphenylylenemercury was heated with tellurium.

Table II lists all the tellurides prepared by this method.

TABLE II
Reactions of Tellurium with Organomercurials

R—Hg—R		R—Te—R		
R	Conditions	Yield %	m.p., b.p. (torr)	Ref.
1-Naphthyl	190–198°/8 hr	53	126.5	L-41
Phenyl	220°/few hr	70	174 (10)	K-5
			182–183 (16.5)	
			312–320 (760) (dec.)	
	220°/15 hr	—	174 (10)	S-18
	—	—	4.2	A-2
	—	—	183–184 (10)	
o-Tolyl	225–235°/12 hr	—	37–38	Z-1,V-6
			202.5 (16.5)	
m-Tolyl	—	—	—	V-6
p-Tolyl	225–230°/15 hr	—	63–64	Z-1,V-6
			210 (16)	
2,5-Dimethylphenyl	—	—	—	V-6
4-Methoxyphenyl	—	—	—	V-6
Pentafluorophenyl	230°	100	50–51	C-26
2,2′-Biphenylylene	—	82	—	H-3

Experimental. Bis(1-naphthyl) Telluride (L-41): A mixture of 4.5 g (0.01 moles) bis(1-naphthyl)mercury and 2.5 g (0.02 moles) powdered tellurium contained in a small flask was kept at 190–198° at 16.5 mm pressure for 8 hr. The cooled mixture was steam distilled to remove any naphthalene which had collected in the neck of the flask. The residue was then extracted with ether. The ether solution was filtered and evaporated as rapidly as possible in order to minimize excessive separation of tellurium. The solid residue was recrystallized from ethanol. Rather large volumes of ethanol were required because of the limited solubility of the compound in this solvent. Bis(1-naphthyl) telluride was obtained in the form of glittering brownish-yellow leaflets which melted at 126.5°. The yield of crude product was 53%.

Excessive heating, that is, beyond the 8-hr period, resulted in the formation of an unidentified yellow solid, which was tellurium free and melted at 190°.

c. *Organometallic Compounds of Group IVA.* The preparation and study of organometallic compounds containing a bond between tellurium and a group IVA element is due largely to the work of Vyazankin (V-7–9,E-3) and co-workers and Schumann, Thom, and Schmidt (S-3,6,7,9,10).

Compounds of the formula $[(C_2H_5)_3M]_2Te$ (M = Ge or Sn) were obtained by heating a mixture of tellurium and $(C_2H_5)_3MH$ in a sealed, evacuated tube (V-7,V-9).

$$2\ (C_2H_5)_3MH + Te \longrightarrow (C_2H_5)_3M\text{—}Te\text{—}M(C_2H_5)_3 + H_2$$

Tetraphenyltin (0.05 moles) and tellurium (0.1 mole) reacted with exchange of phenyl groups above 240°. Reaction time and temperature determined the nature of the end product as given in Table III (S-3).

No intermediates of this phenyl exchange reaction could be isolated. The formation of telluranthrene is attributed to the thermal decomposition of diphenyl telluride.

Schumann and co-workers (for references see Table III) prepared $(C_6H_5)_3M$—Te—Li compounds (M = Ge, Sn, Pb) by addition of tellurium powder to a solution of $(C_6H_5)_3MLi$ in tetrahydrofuran. The reaction proceeds via a polytelluride.

$$R_3MLi + Te_n \longrightarrow R_3M\text{—}Te_n\text{—}Li$$
$$R_3M\text{—}Te_n\text{—}Li + (n\text{-}1)\ R_3MLi \longrightarrow n\ R_3M\text{—}Te\text{—}Li$$

The grayish-black solutions are very unstable, depositing tellurium when in contact with air. These lithium compounds were therefore not isolated, but reacted with $R_3M'X$ (X = halogen, R = C_6H_5, M' = Ge, Sn, Pb).

$$R_3M\text{—}Te\text{—}Li + XM'R_3 \longrightarrow R_3MTeM'R_3 + LiX$$

For individual compounds, $R_3MTeM'R_3$, consult Section XIV.

Experimental. Lithium Triphenylplumbyl Telluride (S-10):   To 0.62 g (0.08 moles) of lithium shavings in 20 ml of dry tetrahydrofuran was added dropwise, with rapid stirring, under an atmosphere of nitrogen, a solution containing 7.55 g (0.016 moles) of triphenyllead chloride in 20 ml of anhydrous THF. The reaction was exothermic. Following three hours of stirring, the dark green solution was filtered through glass wool under an atmosphere of nitrogen in order to remove lithium chloride and any unreacted lithium. Triphenylplumbyl lithium was not actually isolated.

To this solution, still under nitrogen and stirred by means of a magnetic stirrer, were added 2 g (0.016 moles) of tellurium powder. After two hours of stirring, an intense black solution had formed. This solution was used as a source of $(C_6H_5)_3PbTeLi$, (I), which was not isolated. The quantity of I formed is based upon the quantity of Te consumed.

TABLE III

Reactions of Elemental Tellurium with Various Organometallic Compounds of Group IVA Elements

Reagent	Reaction conditions	Product	Yield, %	m.p. or b.p. (mm Hg)	Ref.
$(C_2H_5)_3GeH$	210°/18 hr	$(C_2H_5)_3GeTeGe(C_2H_5)_3$	75.3	112–115 (1)	V-7,V-9
$(C_2H_5)_3SnH$	130°/4 hr	$(C_2H_5)_3SnTeSn(C_2H_5)_3$	63.5	134–135 (1.5)	V-7,V-9
$(C_6H_5)_3GeLi$	THF soln.	$(C_6H_5)_3Ge$—Te—Li	Not isolated		S-6,S-7
$(C_6H_5)_3SnLi$	THF soln.	$(C_6H_5)_3Sn$—Te—Li	Not isolated		S-9
$(C_6H_5)_3PbLi$	THF soln.	$(C_6H_5)_3Pb$—Te—Li	Not isolated		S-10
$(C_6H_5)_4Sn$	230–240°/80 hr	$(C_6H_5)_2Te$	58		S-3
	310°/8 hr	$(C_6H_5)_2Te$	14	174–176 (10)	S-3
		![dibenzo ditellurin structure with Te and Te]	19	188–190	S-3

Bis(triethylgermyl) Telluride (V-9): A mixture of triethylgermane (2.01 g, 0.0125 moles) and tellurium (0.80 g, 0.00625 moles) was ground to a powder and heated for 18 hr at 210°. Hydrogen (70 ml, 50%) was evolved. Distillation in vacuum under nitrogen gave 2.10 g of the compound (75.3%), boiling at 112–115°/1 mm Hg.

d. Organophosphorus Compounds. Tellurium has been shown to undergo nucleophilic attack by phosphines (Z-2,Z-3) and by the diethyl ester of allyl-tellurophosphonous acid (R-5). These compounds containing a phosphorus–tellurium bond decompose in air or under the influence of light or heat.

The phosphine tellurides have a pale, golden yellow color and are obtained in a pure state by the direct reaction between phosphines and elemental tellurium in anhydrous toluene under an atmosphere of nitrogen. Warming of solutions of these compounds in saturated aliphatic hydrocarbons results in the deposition of shiny tellurium mirrors on the surface of the glass containers.

TABLE IV

Phosphine Tellurides

Derivative	Melting point	Per cent composition		Molecular weight	
		Calc'd	Found	Calc'd	Found
Triethyl	76.0–78.0	C: 29.32	29.40	245.6	—[a]
		H: 6.15	6.14		
		P: 12.63	12.48		
Tri-*n*-propyl	41.0–42.0	C: 37.55	37.56	287.8	—[a]
		H: 7.35	7.34		
		P: 10.78	10.01		
Tri-*n*-butyl	35.0–35.5	C: 43.69	43.88	329.9	343 ± 5
		H: 8.25	8.24		
		P: 9.39	9.62		
Tricyclohexyl	184.0–187.0	C: 52.98	53.09	408.1	410 ± 5
		H: 8.15	8.14		
		P: 7.59	7.49		
Phenyldi-*n*-propyl[b]	67.0–71.0	C: 44.78	44.82	321.9	—[a]
		H: 5.95	5.91		
		P: 9.62	9.67		
Phenyldi-*n*-butyl	51.7–53.5	C: 48.05	48.06	349.9	—[a]
		H: 6.63	6.65		
		P: 8.85	8.84		

[a] Not determined because of insufficient solubility and/or decomposition in solution.
[b] Washed with *n*-heptane at Dry Ice acetone temperature to remove excess, unreacted phosphine.

It has been noted that the solubility of tellurium in various phosphines parallels their chemical reactivity toward this element, that is, the solubility decreases as the nucleophilic character of the phosphine decreases. Table IV, taken from ref. (Z-3), lists the phosphine tellurides which have been prepared and their melting points.

Attempts to prepare compounds containing P—Te—C bonds failed (P-13).

Experimental.   Thirty-five grams (0.27 moles) of finely powdered tellurium were added to 120 g (0.75 moles) of freshly distilled tri(n-propyl)phosphine dissolved in 350 ml of sodium-dried and redistilled toluene contained in a one-necked 500 ml round bottomed flask equipped with a reflux condenser. (The amount of tellurium used with different phosphines varied, depending upon the extent to which a particular phosphine reacts. If the tellurium was consumed, more was added.) The mixture was refluxed under nitrogen. After the tellurium was used up after three hours, an additional five grams were added. After a total of 21 hr of reflux the hot solution was filtered rapidly through a sintered glass filter under nitrogen. This removed any unreacted tellurium. The yellow–orange filtrate, evaporated under reduced pressure to about three-fourths of its original volume, was cooled in a Dry Ice Acetone bath under dry nitrogen. An equal volume of sodium-dried, low-boiling petroleum ether, previously cooled in the cold bath, was added to the toluene solution. Soft yellow crystals separated, which were washed with cold petroleum ether and placed immediately in a desiccator under dry nitrogen and stored in a Dry Ice chest. Tri(n-propyl)phosphine telluride melted at 41.0–42.0°.

4. Elemental Tellurium and Diazonium and Iodonium Salts

The reaction of elemental tellurium with diazonium salts has been utilized by Waters (W-1) and Taniyama and co-workers (T-1) to prepare diaryltellurium dichlorides.

Waters used tellurium essentially as a free radical scavenger. When tellurium was permitted to react with benzenediazonium chloride under cold acetone containing chalk, diphenyltellurium dichloride was easily isolated. Sulfur or selenium reacted only when the solution was warmed.

According to Waters these observations confirm the hypothesis that the diazonium chlorides decompose under acetone by the nonionic mechanism

$$(C_6H_5N_2)^+Cl^- \longrightarrow C_6H_5N\!\!=\!\!NCl \longrightarrow C_6H_5\!\cdot\ +\ N_2\ +\ Cl\cdot$$

It is these radicals that react with tellurium in the same manner as those formed in the gas phase (Section I-5).

$$Te\ +\ 2\,C_6H_5\cdot\ +\ 2\,Cl\cdot\ \longrightarrow\ (C_6H_5)_2TeCl_2$$

Taniyama (T-1), essentially following Water's experimental procedures, synthesized the following 15 diaryltellurium dichlorides, $(R-C_6H_4)_2TeCl_2$ (R and m.p. given): o-CH$_3$, 184.0–184.5; m-CH$_3$, 128; p-CH$_3$, 163.5–164.0; o-CH$_3$O, 184.5–185.5; m-CH$_3$O, 163; p-CH$_3$O, 182–183; o-C$_2$H$_5$O, 166; m-C$_2$H$_5$O, 114–115; p-C$_2$H$_5$O, 110; o-Cl, 192; m-Cl, 119–120; p-Cl, 182–183; o-Br, 244–246; m-Br, 143–144; p-Br, 190.

A yield of 11.2% was reported for the bis(o-methylphenyl)tellurium dichloride.

Sandin and co-workers (S-2) reported that when thioglycolic acid dissolved in water was shaken with diphenyliodonium chloride, sodium carbonate, tellurium, and ether at room temperature, diphenyl telluride was formed. In an earlier publication (S-1), Sandin, McClure, and Irwin investigated the decomposition of diphenyliodonium chloride in the presence of tellurium in n-propanol solution. They isolated diphenyltellurium dibromide after addition of bromine to the concentrated reaction mixture.

On the basis of their results they concluded that at least part of the iodonium salt decomposed by way of a nonionic mechanism. The assumption was made that in the iodonium salt the iodine acted as an acceptor for the chloride ion. The subsequent transformations of this donor–acceptor complex were caused by the tendency of the iodine atom to revert from an expanded valence shell to an octet. They suggested the following course for the reaction

$$(C_6H_5)_2I^+Cl \longrightarrow (C_6H_5)_2I \leftarrow Cl$$
$$(C_6H_5)_2I \leftarrow Cl \longrightarrow C_6H_5I + C_6H_5 \cdot + Cl \cdot$$

The formation of an intermediate, undissociated complex involving tellurium, which decomposed at a lower temperature than the iodonium salt would otherwise, was also discussed in this paper. Such a reaction mechanism would not necessarily involve the formation of free radicals and the decompositions with and without a metal could proceed via different mechanisms.

When water was used as a solvent in these reactions, only traces of the organotellurium compound were obtained. This suggests that in this highly polar medium ionic decomposition occurred. It was observed that finely divided tellurium in the presence of water speeded up the decomposition of the iodonium salt.

When an aqueous solution of diphenyliodonium chloride was shaken with tellurium, ether, and aqueous hydrogen sulfide for about an hour at room temperature, diphenyl telluride was again formed.

Di-p-tolyliodonium chloride in place of the phenyl compound, gave the yellow di-p-tolyltellurium dibromide, melting at 203–204°. When a mixture of diphenyliodonium chloride and finely divided tellurium was heated together without a solvent, diphenyltellurium dichloride, melting at 158°, was isolated.

Experimental. Diphenyltellurium Dichloride via Benzenediazonium Chloride (W-1):   Fifteen grams (0.117 moles) of powdered tellurium were added to 20 g (0.142 moles) of benzenediazonium chloride and 20 g of chalk in 300 ml of acetone. The reaction, controlled by cooling in ice, was complete after a one-hour reflux period. After cooling, the solution was filtered and the filtrate evaporated on a water bath. The residue, which contained chloroacetone (caution: *chloroacetone* is highly *lachrymatory*) was extracted with ligroin (b.p. 40–60°). The latter, on cooling, deposited colorless crystals of diphenyltellurium dichloride (1 g, 2.5% yield), m.p. 159°. By using the double salt of benzenediazonium chloride and zinc chloride (from 20 g of aniline) the yield was increased to three grams (7.5% yield).

Diphenyltellurium Dibromide via Diphenyliodonium Chloride (S-1):   Diphenyliodonium chloride was refluxed in *n*-propyl alcohol with finely divided tellurium. After several hours the unchanged tellurium was filtered off and the filtrate evaporated to a small volume. The addition of bromine caused the separation of yellow crystals of diphenyltellurium dibromide, m.p. 199–200°.

5. Elemental Tellurium and Free Radicals in the Gas Phase

In a series of investigations, whose goal was to prove the existence of free radicals and learn more about their properties, tellurium mirrors were employed. Tellurides, R_2Te, ditellurides, R_2Te_2, and telluroformaldehyde, $(CH_2Te)_n$, were formed by attack of various radicals on the metallic tellurium.

Rice and co-workers obtained dimethyl ditelluride, $CH_3TeTeCH_3$, (m.p. $-19.5°$, b.p. 196° dec.) with methyl radicals generated by the thermal decomposition of butane, acetone, or diethyl ether at low pressures in the range of 800–900° (R-14) and by decomposition of methane (R-15). In a similar type of experiment (R-13) methylene radicals generated by the thermal decomposition of diazomethane below 550° carried in a current of ether or butane at low pressures were found to react with tellurium mirrors to form polytelluroformaldehyde, $(CH_2Te)_n$. These investigators reported that tellurium is almost ideal with respect to its ability to combine with free radicals to form compounds which are easily separated and identified.

Glazebrook and Pearson (G-7,G-8) were able to remove tellurium mirrors with free radicals generated from diisopropyl ketone, acetone, methyl ethyl ketone, methyl propyl ketone, and methyl butyl ketone. The radicals from acetone yielded dimethyl telluride in their reaction with tellurium. They interpreted the fission of acetone as follows

$$(CH_3)_2CO \longrightarrow 2\ CH_3\cdot\ +\ CO$$

or

$$(CH_3)_2CO \longrightarrow CH_3\text{---}CO\cdot\ +\ CH_3\cdot \longrightarrow 2\ CH_3\cdot\ +\ CO$$

In subsequent experiments (G-9) the photolysis of acetophenone was found to yield active fragments which react with tellurium mirrors. The products were identified as diphenyl, dimethyl, and phenyl methyl tellurides. The formation of methyl and phenyl radicals from acetophenone was thus clearly indicated. They proposed the following mechanistic scheme for the radical formation.

$$C_6H_5C(O)CH_3 \longrightarrow C_6H_5CO \cdot + \cdot CH_3$$
$$C_6H_5CO \cdot \longrightarrow C_6H_5 \cdot + CO$$
$$2\ C_6H_5 \cdot \longrightarrow C_6H_5\text{---}C_6H_5$$
$$2\ CH_3 \cdot \longrightarrow C_2H_6$$
$$C_6H_5 \cdot + CH_3 \cdot \longrightarrow C_6H_5\text{---}CH_3$$

The phenyl and methyl radicals also combined with tellurium to form all the possible tellurides.

Similar experiments have been carried out by Pearson and Purcell (P-1), who removed tellurium mirrors with propyl radicals generated from dipropyl ketone, and by Belchetz and Rideal (B-7). The latter generated both methyl and methylene radicals by the thermal decomposition of methane. Finally, Williams and Dunbar (W-2) have furnished the most convincing evidence that tritelluroformaldehyde, $(CH_2Te)_3$, is formed when methylene radicals from diazomethane react with tellurium mirrors.

6. Miscellaneous Reactions of the Element

Lowy and Dunbrook (L-39) unsuccessfully attempted the preparation of dyes analogous to the sulfur dyes by reacting dinitrophenol, diphenylamine, and p-phenylenediamine with elemental tellurium in sodium hydroxide under a variety of conditions.

Cullinane, Rees, and Plummer (C-29) were able to prepare dibenzotellurophene by heating together a mixture of biphenylylene sulfone and tellurium

Kraft and Vorster (K-4) failed, however, in their attempt to obtain diphenyl telluride from diphenyl sulfone and tellurium.

Electrolysis of a suspension of tellurium in $1N$ Na_2SO_4 at 24–26° and vinyl cyanide gave bis(β-cyanoethyl) telluride melting at 58–59°. This reaction very probably proceeds via the intermediate hydrogen telluride.

Experimental. Dibenzotellurophene (C-29): Biphenylylene sulfone (8 g, 0.037 moles) and tellurium (6 g, 0.047 moles) were mixed thoroughly and heated in an atmosphere of carbon dioxide until the evolution of sulfur dioxide indicated that the reaction had begun. The temperature was adjusted so that the evolution of gas was continuous. The sublimated sulfone must be melted down periodically. After 36 hr the mixture was extracted with boiling acetone. The extract was evaporated to dryness. The solid residue was washed with cold alcohol dissolving the telluride only. The product remaining after evaporation of the alcohol was steam distilled and recrystallized from low boiling petroleum ether. The telluride was obtained in the form of colorless needles (1 g, 10% yield) melting at 93°.

B. Metallic Tellurides, Diphosphorus Pentatelluride, Hydrogen Telluride, Tellurium Dioxide, Tetramethoxy- and Hexamethoxytellurium

Tellurides, especially alkali metal tellurides, have found application for the synthesis of diorgano tellurides, aliphatic as well as aromatic, and for the preparation of heterocycles containing tellurium. Aluminum telluride is the starting material for the synthesis of tellurols. Magnesium telluride, phosphorus telluride, tellurium dioxide, and esters of the acids H_4TeO_4 and H_6TeO_6 are only of secondary importance. Alkali tellurites are methylated biologically to dimethyl telluride (C-7–C-9).

1. Alkali Metal Tellurides

Lithium, sodium, and potassium telluride have been employed in the preparations of various tellurides. These compounds are listed in Table V together with pertinent references.

The synthesis of the alkali metal tellurides, M_2Te, was accomplished through five different methods.

1. Tschugaeff and Chlopin (T-2) introduced sodium formaldehyde sulfoxylate, also known as "Rongalite-C," as a reagent for the preparation of sodium telluride, Na_2Te. The inexpensive compound, Rongalite-C, is commercially available. Its preparation involves the reaction of sodium hyposulfite, $NaHSO_2$, with formaldehyde. Sodium hyposulfite is itself formed by hydrolysis of sodium dithionite, $Na_2S_2O_4$.

$$NaO\!-\!\overset{\overset{O}{\uparrow}}{S}\!-\!\overset{\overset{O}{\uparrow}}{S}\!-\!ONa + H_2O \longrightarrow NaHSO_3 + H\!-\!\overset{\overset{O}{\uparrow}}{S}\!-\!ONa$$

$$\overset{H}{\underset{H}{>}}C\!=\!O + H\!-\!\overset{\overset{O}{\uparrow}}{S}\!-\!ONa + 2\,H_2O \longrightarrow HO\!-\!CH_2\!-\!\overset{\overset{O}{\uparrow}}{S}\!-\!ONa \cdot 2\,H_2O$$

Rongalite—C

Tellurium metal is added to solutions of sodium hydroxide containing sodium formaldehyde sulfoxylate, dissolving in the base with formation of sodium polytellurides and sodium tellurite.

$$3\,Te + 6\,NaOH \longrightarrow 2\,Na_2Te + Na_2TeO_3 + 3\,H_2O$$
$$Na_2Te + Te \longrightarrow Na_2Te_2$$
$$Na_2Te_2 + HOCH_2S(O)ONa + 3\,NaOH \longrightarrow H_2CO + 2\,Na_2Te + Na_2SO_3 + 2\,H_2O$$

TABLE V
Reactions of Alkali Metal Tellurides

M_2Te_n M	n	Method of preparation of M_2Te_n[a]	Reagent	Product	mp., b.p. (torr)	Yield, %	Ref.
Na	1	1	CH_3I	$(CH_3)_2Te$	93.5 (749)	—	B-12
Na	2	2	CH_3I	$(CH_3)_2Te_2$	68 (vacuum)	38	C-20
K	1	5	$(CH_3OSO_3)_2Ba$	$(CH_3)_2Te$	82 (—)	—	W-6
		5	CH_3OSO_3K	$(CH_3)_2Te$	—	—	H-2
K	1	5	$(C_2H_5OSO_3)_2Ba$	$(C_2H_5)_2Te$	—	—	W-4,M-7
		5	$C_2H_5OSO_3K$	$(C_2H_5)_2Te$	137–138 (760)	—	H-2,M-9,W-5
Na	1	2	$i\text{-}C_3H_7Br$	$(i\text{-}C_3H_7)_2Te$	49 (14)	78	B-16
Na	1	1	C_4H_9Br	$(C_4H_9)_2Te$	109–112 (12)	60	B-2
Na	2	1	C_4H_9Br	$(C_4H_9)_2Te_2$	112–115 (1.5)	23	B-8
K	1	5	$(C_5H_{11}OSO_3)_2Ca$	$(C_5H_{11})_2Te$	198 (—) dec.[b]	—	W-7
Na	1	5	$C_5H_{11}Br$	$(C_5H_{11})_2Te$	138–140 (18) 99–100 (2)	55	B-3
Na	1	3	$C_6H_5CH_2Cl$	$(C_6H_5CH_2)_2Te$	50–52	90	S-16
		1,3	$C_6H_5CH_2Cl$ $[(CH_3)_2(C_6H_5)C_6H_5CH_2N]Cl$	$(C_6H_5CH_2)_2Te$	53	—	T-2
Na	1	2	$C_{16}H_{33}Cl$	$(C_{16}H_{33})_2Te$	43–45	—	D-6
Li	1	2	$[IC(C_6H_5){=}C(C_6H_5)]_2$	tetraphenyltellurophene	239	82	B-18,H-10
Na	1	—[c]	$HC{\equiv}C{-}C{\equiv}CH$	tellurophene	148 (714)	69	M-6
Na	1	—[c]	$(C{\equiv}C{-}CH_2OH)_2$	2,5-bis(hydroxymethyl)-tellurophene	107	25	M-6

Alkali metal	Equiv.	Reactant	Product	M.p. (b.p.) °C	Yield %	Reference
	—c	[C≡C—C(CH₃)₂OH]₂	2,5-bis(dimethylhydroxy-methyl)tellurophene	94	100	M-6
Na	1	(C≡C—C₆H₅)₂	2,5-diphenyltellurophene	225	55	M-6
	1	Br—(CH₂)₄—Br	tetrahydrotellurophene[d]	128–131[d]	70	F-1
Na	1	Cl—(CH₂)₅—Cl	telluracyclohexane[d]	106–107[d]	33	F-1
	1	(ClCH₂CH₂)₂O	1-oxa-4-telluracyclo-hexane[e]	179.5 (dec.)[e]	45	F-1
Na	1	(ClCH₂CH₂)₂S	1-thia-4-telluracyclo-hexane	69.5	6.6	M-1
	1	[ICH₂CH₂CH(CH₂I)]₂	3,3′-bis(telluracyclo-pentyl)	145	62	B-20,B-21
Na	1	o-BrCH₂CH₂—C₆H₄—CH₂Br	telluroisochroman	64	52	H-7
Na	2	o-carboxyphenyl-diazonium chloride	bis(o-carboxyphenyl)telluride[f]	116–126 (0.3)	—	M-8
	1			215	—	M-11[f]
K	1	bz-1-halobenz-anthrone	bz-1,bz-1′-dibenzanthronyl ditelluride	278 (dec.)	22	F-2[f] F-2
Na	1			—	—	P-3

a See text for discussion.
b Impure product.
c Not given.
d Isolated as dibromide after addition of bromine; melting point of dibromide given.
e Isolated as dichloride after addition of chlorine; melting point of dichloride given.
f Identity of compound doubtful; for discussion of these investigations see Section IV–B–1.

The reaction mixture first acquires a permanganate-like color caused by the polytelluride (the ditelluride has been used in the above equations for the sake of simplicity) which then converts to pale pink. According to Tschugaeff and Chlopin the best results were obtained when six grams Rongalite and 40 ml of a 10% solution of sodium hydroxide were used per gram of tellurium. In more recent investigations a smaller Rongalite: tellurium ratio was used (M-1,B-2,B-3,B-12). In order to minimize atmospheric oxidation of the very sensitive telluride to tellurium, the reaction is run under an atmosphere of nitrogen. The sodium telluride is not isolated.

2. Sodium telluride (B-16) and lithium telluride (B-18) have also been prepared by the reaction of elemental tellurium with solutions of the metals in liquid ammonia

$$2n\ Na + Te_n \xrightarrow{\ NH_3\ } n\ Na_2Te$$

The telluride is generally not isolated. The alkyl halide is added to the liquid ammonia containing the telluride.

3. Sodium telluride is also formed from tellurium in alkaline aqueous solution of sodium dithionite (S-16,T-2). The following mechanism was discussed by Tschugaeff and Chlopin (T-2).

The polytellurides formed from tellurium and sodium hydroxide are converted to the telluride according to the equation

$$Na_2Te_2 + Na_2S_2O_4 + 4\ NaOH \longrightarrow 2\ Na_2SO_3 + 2\ Na_2Te + 2\ H_2O$$

It is possible that the sulfoxylate group is the active agent.

$$Na_2S_2O_4 + 2\ NaOH \longrightarrow Na_2SO_3 + Na_2SO_2 + H_2O$$

The preparation of sodium telluride using $Na_2S_2O_4$ is inferior to the method employing Rongalite-C (T-2).

4. Farrar (F-2) dissolved tellurium (3 g, 0.0235 moles) in a solution of KOH (15 g, 0.27 moles) in water (27 ml) and heated the mixture to 100°. Then zinc filings (5 g, 0.077 moles) were cautiously added giving a deep purple solution, which was stirred into a cold mixture of sodium bicarbonate (30 g) and water (100 ml). The resulting liquid was then added as rapidly as possible to the diazonium salt prepared from o-aminobenzoic acid.

5. The last method for the preparation of tellurides used by Wöhler and co-workers is only of historic interest. They heated tellurium with carbon

obtained from the mono-potassium salt of d-tartaric acid to synthesize K_2Te, and tellurium or a tellurium–bismuth alloy with carbon and sodium carbonate to obtain Na_2Te. Under these conditions ditellurides, Na_2Te_2, were also formed which gave dialkyl ditellurides upon treatment with alkyl halides.

The aqueous sodium telluride solution obtained by the Rongalite-C procedure is then treated with the desired organic halide generally dissolved in ethanol. Thus, dialkyl tellurides and a variety of heterocyclic ring systems using dihaloalkanes have been synthesized employing this method.

Mann and Holliman, for instance, (H-7,M-8) have prepared telluroisochroman in good yield by the condensation of o-(β-bromoethyl)benzyl bromide with sodium telluride.

Mazza and Melchionna (M-11) claim to have prepared bis(o-carboxyphenyl) telluride by the reaction between the diazonium salt made from o-aminobenzoic acid and sodium telluride prepared by method 2. Farrar (F-2), however, was not able to reproduce these results employing sodium telluride made by method 1; using K_2Te (method 4) he isolated what he claimed to be the authentic bis(o-carboxyphenyl) telluride (F-2).

The alkylation of sodium telluride, prepared from the elements in liquid ammonia, has also been shown to proceed very smoothly. Bis(isopropyl) telluride, for example, was prepared in yields of $> 80\%$ by treatment of sodium telluride with isopropyl bromide in liquid ammonia. $tert$-Butyl bromide, however, gave only traces of the expected telluride (B-16), and chloroacetic acid precipitated elemental tellurium (B-8).

Mack (M-6) synthesized tellurophene and 2,5-disubstituted tellurophenes by addition of the calculated amounts of diacetylenes to a methanolic solution of sodium telluride at room temperature. After the reaction was completed, most of the methanol was distilled off. The residue was shaken with water. The tellurophene was extracted with ether.

$$R = H, CH_2OH, C(CH_3)_2OH, C_6H_5$$

Although the author did not propose a mechanism for this reaction, it seems likely that the sodium telluride added to the conjugated system forming an intermediate 3,4-disodiotellurophene which was then hydrolyzed.

This mechanism is highly speculative. Mack's reaction, however, differs from the other reactions discussed in this section, insofar as it cannot proceed via a condensation.

For individual compounds prepared by these methods consult Table V.

It is interesting to note that when tellurium is added to Rongalite-C (2:1 molar ratio), the ditelluride, Na_2Te_2, appears to be formed together with other polytellurides and Na_2Te. The only illustration of this potentially interesting observation, from the point of view of ditelluride synthesis, is that of Bergson (B-8). He found that such a solution reacted readily with butyl bromide to give dibutyl ditelluride. Chloroacetic acid treated similarly yielded only elemental tellurium (B-8). Chen and George (C-21) obtained dimethyl ditelluride in 38% yield from Na_2Te_2 synthesized from 0.14 moles each of sodium metal and tellurium metal in liquid ammonia at $-78°$. Perkins (P-3) claimed to have prepared bz-1, bz-1'-dibenzanthronyl ditelluride in a similar reaction.

Experimental. Telluroisochroman (H-7):   An aqueous solution of sodium telluride was prepared by adding Rongalite (34 g) to a solution of sodium hydroxide (26 g) in 100 ml of water contained in a flask equipped with a reflux condenser and an inlet for methane, which replaced the air. Finely divided, freshly precipitated tellurium (15 g) (see Section I-A-3-a) was added and the mixture heated on a water bath for 30 min with frequent shaking. A permanganate-colored solution was formed. Ethanol (100 ml) was added followed by a solution of o-(β-bromoethyl)benzyl bromide (25 g) in hot ethanol (150 ml). The mixture formed two layers. The heating was continued for an additional two hours with occasional vigorous shaking. The alcohol was then distilled off. The methane atmosphere was maintained throughout the course of the preparation until the distillation was complete.

The residue was broken up, cooled, and then shaken vigorously with a mixture of water (300 ml) and ether (300 ml). Any undissolved material was removed by filtration. The ethereal extract was separated, washed with water, dried over sodium sulfate, and distilled. The telluroisochroman was collected as a pale yellow liquid (8.5 g, 37%), b.p. 116–126°/0.3 mm which readily crystallized on cooling. A second crop (3.8 g) was obtained when the undissolved residue was extracted with 50 ml of hot alcohol and the extract was filtered and cooled. The combined crystals were recrystallized from methanol under an atmosphere of carbon dioxide. Telluroisochroman was obtained in the form of silky needles, m.p. 63.5–64°.

Dibutyl Ditelluride (B-8):   In a 500-ml, three-necked flask 9.24 g of Rongalite-C and 9.0 g of sodium hydroxide were dissolved in 200 ml of water. The flask was flushed thoroughly with nitrogen and 15.32 g of finely powdered tellurium was added. The mixture was vigorously stirred under reflux for five hours. After that time 16.4 g of butyl bromide were added dropwise. The mixture was allowed to cool and the stirring was continued for 30 min. The content of the flask was extracted with two 150-ml portions of ether. The ether extract was dried over $CaCl_2$ and the ether was distilled under a nitrogen atmosphere. The residue, distilled under reduced pressure, gave the ditelluride boiling between 112–115°/1.5 mm as a dark red oil with a pungent smell. The yield was 23%.

Dimethyl Ditelluride (C-21):   Clean sodium metal (3.2 g, 0.14 moles) was added to 100 ml of anhydrous ammonia at $-78°$ in a 500-ml flask equipped with reflux condenser, stirrer, and dropping funnel. After stirring for one hour, high-purity, powdered tellurium (18.2 g, 0.14 moles) was added in 0.5-g portions. Methyl iodide (24 g, 0.17 moles) was then added dropwise over a 20-min period with stirring. After removal of the cooling bath and the evaporation of ammonia, water was added to the residue and the mixture extracted four times with 50-ml portions of ether. The deep red ether extract was dried over night with $CaCl_2$. The solvent was removed and the product distilled. Dimethyl ditelluride (38% yield) was collected at 68° under vacuum.

2. Aluminum Telluride and Other Inorganic Tellurium Compounds

Aluminum telluride, Al_2Te_3, reacted with methanol at 320–335° and gave dimethyl telluride in good yield.

$$Al_2Te_3 + 6\,CH_3OH \longrightarrow 3\,(CH_3)_2Te + Al_2O_3 + 3\,H_2O$$

With ethanol at 240–260° a substance with a boiling point close to that of ethanol was produced, which is referred to as ethyltelluromercaptan (N-1). Baroni (B-4), by dropping alcohols onto Al_2Te_3 at 300–350° in a hydrogen atmosphere, isolated upon distillation the following tellurols (boiling points given): CH_3TeH (57°), C_2H_5TeH (90°), C_3H_7TeH (121°), and C_4H_9TeH (151°).

An attempt to prepare tellurophene and tetrahydrotellurophene from Al_2Te_3 and sodium succinate and tetramethylene glycol, respectively, failed

M-4). Al_2Te_3 heated with acetylene over bauxite yielded an unidentified tellurium-containing organic compound (M-4).

Aluminum telluride and polymethylene dihalides combine to form telluracycloalkanes (M-40).

$$Al_2Te_3 + 3 \text{ X}-(CH_2)_n-\text{X} \longrightarrow 3 \left[\begin{array}{c} -(CH_2)_{\overline{n}}- \\ \\ -Te- \end{array} \right] + 2 \text{ AlX}_3$$

$n = 5$ (M—40)
X = Cl, Br, I

$n = 6$ (M—35)

A mixture of Al_2Te_3 (0.062 moles) and pentamethylene dibromide (0.275 moles) (molar ratio 1:4.5; 50% excess bromide) began to react at 165°. The pentamethylene dichloride reacted less vigorously than the bromide. The diiodide combined with Al_2Te_3 at 135–145° (M-35). Tetramethylene dibromide behaved similarly (M-40). The telluracycloalkanes, however, were not obtained as such, because excess halide added to the cyclic tellurides forming telluronium halides (for details see Section XIII-A). When magnesium telluride was employed in the condensation with pentamethylene dibromide, the reaction commenced only at 200°. The yield of organotellurium compound was small (M-35).

It is claimed that P_2Te_5 and saturated aliphatic carboxylic acids containing at least seven carbon atoms gave compounds of the type $(RCTe)_2Te$, which were investigated as additives to mineral lubricating oils to improve their extreme pressure properties (F-11).

Telluroketones, R_2CTe, were obtained by Lyons and Scudder (L-42) by passing H_2Te through a mixture of a ketone with concentrated HCl. The telluroketones were isolated and purified by vacuum distillation. The following compounds were prepared (boiling points given): $(CH_3)_2CTe$ (55–58°/ 10–13 mm); $(CH_3)(C_2H_5)CTe$ (63–66°/9–10 mm); $(C_2H_5)_2CTe$ (69–72°/8–11 mm); and $(C_3H_7)_2CTe$, which could not be purified. The analytical results for tellurium deviate ∼4% from the calculated values. Benzil and benzophenone did not react.

For the reaction involving the electrolysis of a mixture of tellurium and vinyl cyanide in Na_2SO_4 solution and probably proceeding via hydrogen telluride, see Section I-A-6. Tellurobenzoic acid, $C_6H_5C(Te)OH$, was not formed in the reaction between H_2Te and the pyridine adduct of benzoyl chloride. The analogous selenium compound was successfully prepared using this method (L-33).

Tellurium dioxide was inert toward m-cresol (M-37).

Tetramethoxytellurium and 2,2′-dilithiobiphenyl gave 48% bis(biphenylylene)tellurium and 23% biphenylylenebiphenylyltelluronium hydroxide (m.p. 176°) (H-3).

Hexamethoxytellurium with the same lithium compound produced only bis(biphenylylene)tellurium in 55% yield, no tris(biphenylylene)tellurium was detected (H-3).

C. Tellurium Tetrahalides

Tellurium tetrachloride is employed almost exclusively. The few experiments with tellurium tetrabromide and tetraiodide are not conclusive with respect to their reactivity as compared with that of tellurium tetrachloride. With Grignard reagents all of the tetrahalides give equally good yields of the tellurides. However, condensation reactions seem to proceed satisfactorily only with tellurium tetrachloride.

1. Tellurium tetrachloride reacts with 1,3-diketones to yield as the main products 1-tellura-3,5-cyclohexanedione 1,1-dichlorides. In some cases, noncyclic compounds of the type

$$R-\overset{O}{\underset{\|}{C}}-CH=\overset{OH}{\underset{|}{C}}-CH_2-TeCl_3, \quad (R-\overset{O}{\underset{\|}{C}}-CH=\overset{OH}{\underset{|}{C}}-CH_2)_2TeCl_2,$$

and

$$R-\overset{O}{\underset{\|}{C}}-CH=\overset{OC_2H_5}{\underset{|}{C}}-CH_2-TeCl_3$$

have also been isolated.

2. Tellurium tetrachloride combines with ketones to form bis(organo)-tellurium dihalides or organotellurium trihalides.

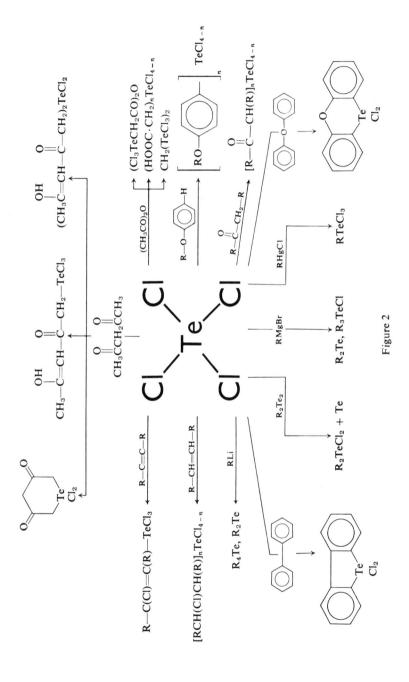

Figure 2

146

3. Carboxylic acid anhydrides give bis(organo)tellurium dihalides and organotellurium trihalides.

4. Aromatic compounds with a ring-hydrogen atom activated by an RO—, RS—, or R_2N— group condense with $TeCl_4$ with the elimination of HCl to give organotellurium trichlorides or bis(organo)tellurium dichlorides.

5. Tellurium tetrachloride adds to olefins with the formation of bis(organo)-tellurium dichlorides and organotellurium trichlorides.

6. Tellurium tetrachloride reacts with biphenyl and diphenyl ether to give dibenzotellurophene and phenoxtellurine, respectively.

7. With Grignard reagents tris(organo)tellurium chlorides and diaryl tellurides are obtained. Other organometallic reagents used in the synthesis of organotellurium compounds include lithium, mercury, and zinc organo-metallics.

8. Tellurium tetrahalides react with ditellurides with the formation of bis(organo)tellurium dihalides and elemental tellurium.

Figure 2 is a schematic representation of the various reactions of tellurium tetrachloride.

1. Tellurium Tetrachloride and 1,3-Diketones

1,3-Diketones react with tellurium tetrachloride to form cyclic and linear condensation products (1-3).

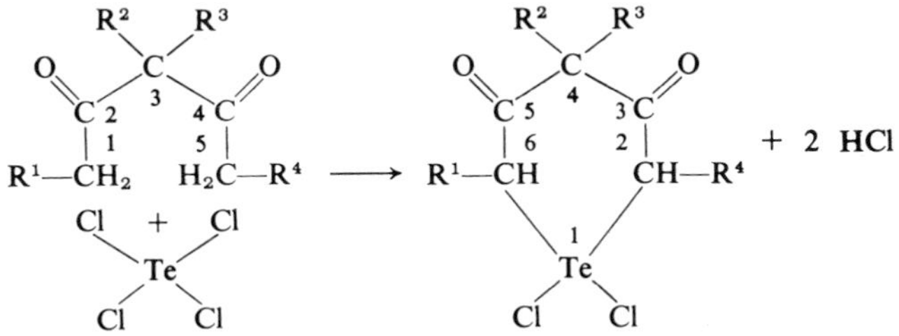

(1)

1-tellura-3,5-cyclohexanedione 1,1-dichloride

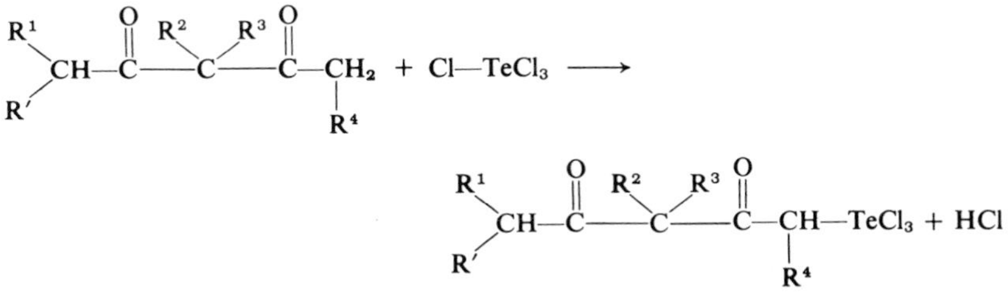

(2)

acetylacetonyltellurium trichloride

$$2 \quad \begin{matrix} R^1 \\ \diagdown \\ CH-C \\ \diagup \\ R' \end{matrix} \begin{matrix} O \\ \| \\ \\ \end{matrix} \begin{matrix} R^2 \\ | \\ C \\ \end{matrix} \begin{matrix} R^3 \\ | \\ C \\ \end{matrix} \begin{matrix} O \\ \| \\ -C-CH_2 \\ | \\ R^4 \end{matrix} + Cl_2TeCl_2 \longrightarrow$$

$$\left[\begin{matrix} R^1 \\ \diagdown \\ CH-C \\ \diagup \\ R' \end{matrix} \begin{matrix} O \\ \| \\ \\ \end{matrix} \begin{matrix} R^2 \\ | \\ C \\ \end{matrix} \begin{matrix} R^3 \\ | \\ C \\ \end{matrix} \begin{matrix} O \\ \| \\ -C-CH \\ | \\ R^4 \end{matrix} \right]_2 TeCl_2 + 2 HCl$$

(3)

bis(acetylacetonyl)tellurium dichloride

The nature of the substituents R^1, R', R^2, R^3 and R^4 determine which compounds will be formed in the reaction with tellurium tetrachloride. 1-Tellura-3,5-cyclohexanedione-1,1-dichlorides were isolated, when

1. There were at least two hydrogen atoms in each of the positions, 1 and 5, of the ketone. Therefore, dibenzoylmethane (M-17) acetylbenzoylmethane (M-16), 1,1,1-trimethylacetylacetone (M-17), 1,1,5,5-tetramethylacetylacetone (M-27), acetylcamphor (M-21), and acetylmesityloxide (M-21) did not yield cyclic compounds. Morgan and Drew (M-21), however, reported that definite products were not obtained from 3-allylacetylacetone, 3,3'-benzylidenacetyl-acetone, 2-acetylcyclohexanone, and 1-phenylacetylacetone, although the requirement of two hydrogen atoms each in the 1 and 5 positions is fulfilled. 1-Isopropylacetylacetone gave only linear condensation products.

2. R^2 and R^3 represent hydrogen atoms and/or alkyl groups. One phenyl group in the 1- or 3-position of the ketone was found to inhibit the condensation reaction. 3-Benzylacetylacetone and 3,3-dibenzylacetylacetone gave the expected cyclic product (M-27), while 3,3-bis(*p*-nitrobenzyl)acetylacetone (M-27) did not react.

Morgan and co-workers (M-29) proposed the following mechanism for the formation of the cyclic tellurium dichlorides.

$$R^1-CH_2-\overset{\overset{O}{\|}}{C}\cdots\overset{R^2}{\underset{|}{C}}\cdots\overset{R^3}{\underset{|}{C}}\cdots\overset{\overset{O}{\|}}{C}-CH_2-R^4 \rightleftharpoons R^1-CH=\overset{\overset{HO}{|}}{C}\cdots\overset{R^2}{\underset{|}{C}}\cdots\overset{R^3}{\underset{|}{C}}\cdots\overset{\overset{OH}{|}}{C}=CH-R^4$$

$$\bigg\downarrow TeCl_4$$

Since cyclic compounds were obtained from diketones with two alkyl groups

(R^2 and R^3) in the 3 position of the ketone, enolization using hydrogen atoms attached to the median carbon atom cannot be required for this reaction. If addition of $TeCl_4$ across the carbon–carbon double bond created by enolization really occurs, then enolization from the median carbon atom followed by addition of $TeCl_4$ would lead to a linear product which could not undergo ring formation.

$$CH_3-\overset{\overset{\displaystyle O}{\|}}{C}-CH_2-\overset{\overset{\displaystyle O}{\|}}{C}-CH_3 \rightleftharpoons CH_3-\overset{\overset{\displaystyle O}{\|}}{C}-CH=\overset{\overset{\displaystyle OH}{|}}{C}-CH_3$$

$$\downarrow TeCl_4$$

$$CH_3-\overset{\overset{\displaystyle O}{\|}}{C}-\underset{\underset{\displaystyle TeCl_3}{|}}{CH}-\overset{\overset{\displaystyle O}{\|}}{C}-CH_3 \longleftarrow CH_3-\overset{\overset{\displaystyle O}{\|}}{C}-\underset{\underset{\displaystyle TeCl_3}{|}}{CH}-\overset{\overset{\displaystyle OH}{|}}{\underset{}{C}}\overset{Cl}{\diagdown\diagup}CH_3$$

Such a product was obtained from a mixture of dibenzoylmethane and $TeCl_4$.

The tellurium trichloride formed, $(C_6H_5-\overset{\overset{\displaystyle O}{\|}}{C})_2CH-TeCl_3$, was very unstable and decomposed in air to $TeOCl_2$ and the ketone (M-17).

Tellurium tetrachloride is known to add to carbon–carbon double bonds (see Section I-C-5).

The lengthening of the chain in the 1 and 5 positions of the ketone does not inhibit cyclization. Experimental difficulties, however, were encountered in the isolation and purification because of the high solubilities of these compounds.

The condensation reactions were carried out in boiling chloroform using a 1:2 molar ratio of $TeCl_4$ to ketone. Table VI lists the cyclic compounds prepared, their melting points, and yields.

Morgan and co-workers, in their earlier publications about these cyclic derivatives, formulated these compounds as 1-oxa-2-tellura-5-cyclohexen-4-ones. The formation of such a ring system requires the presence of at least one hydrogen atom on the carbon atom connecting the two carbonyl groups. The discovery that 3,3-diethylacetylacetone also gave a cyclic compound (M-21) led to the formulation of a 1-tellura-3,5-cyclohexanedione structure, which was supported by the formation of oximes with hydroxylamine (M-24).

$$CH_3-\overset{\overset{\displaystyle O}{\diagup\diagdown}}{\underset{\overset{\displaystyle \|}{}}{C}_6} \quad {}_2Te$$

3-methyl-1-tellura-2-oxa-3-cyclohexen-5-one
6-methyl-1-oxa-2-tellura-5-cyclohexen-4-one

TABLE VI

Reactions of Tellurium Tetrachloride with 1,3-Diketones

Reactant: $R^1-CH_2-C(OR^2)(R^3)=C-\overset{O}{\overset{\|}{C}}-CH_2-R^4$

R1	R2	R3	R4	1-Tellura-3,5-cyclohexanedione 1,1-dichloride m.p.	Yield	Ref.	$O=C(R^1-CH_2)-C(R^2)(OR^5)=C-C(R^4)-CH-TeCl_3$ R5	m.p.	Yield	$O=C(R^1-CH_2)-C(R^2)(OR^5)=C-C(R^4)-CH_2TeCl_2$ R5	m.p.	Yield
H	H	H	H	155–173 (dec.)	62	M-15	—	—	—	—	—	—
H	H	H	H	—	20	M-16	C2H5	106–107	40	H	115 (dec.)	18
H	CH3	H	H	180–190 (dec.)	30	M-17	—	—	—	—	—	—
H	C2H5	H	H	185–190 (dec.)	68	M-17	—	—	—	—	—	—
H	C3H7	H	H	185–190 (dec.)	80	M-31	—	—	—	—	—	—
H	i-C3H7	H	H	~180	88	M-22	—	—	—	—	—	—
H	C4H9	H	H	~150		M-22	—	—	—	—	—	—
H	i-C4H9	H	H	155	36	M-23	—	—	—	—	—	—
H	sec-C4H9	H	H	142	37	M-29	—	—	—	—	—	—
H	C5H11[a]	H	H	168–169	35	M-29	—	—	—	—	—	—
H	C6H5CH2	H	H	162	52	M-29	—	—	—	—	—	—
H	CH3	CH3	H	180 (dec.)	60	M-27	—	—	—	—	—	—
H	CH3	C2H5	H	194 (dec.)	—[d]	M-24	—	—	—	—	—	—
H	C2H5	C2H5	H	—	23	M-29	—	—	—	—	—	—
H	C6H5CH2	C6H5CH2	H	178–180 (dec.)	23	M-21	—	—	—	—	—	—
H	Cl	H	H	189–190 (dec.)	5	M-27	—	—	—	—	—	—
CH3	H	H	H	161–162 (dec.)	65	M-17	—	—	—	H	131–132 (dec.)	16
CH3	CH3	H	H	175–185 (dec.)	80	M-19	—	—	—	—	—	—
CH3	CH3	H	CH3	166–167 (dec.)	68	M-21	—	—	—	—	—	—
CH3	C2H5	H	CH3	180 (dec.)	76	M-27	—	—	—	—	—	—
				182 (dec.)		M-21						

R¹	R²	R³	R⁴	m.p.	%		R	m.p.	%	R	m.p.	%
CH₃	C₅H₁₁[a]	CH₃	H	oily	—	M-29	—	—	—	—	—	—
CH₃	C₂H₅	H	H	167 (dec.)	63	M-22	—	—	—	—	—	—
CH₃	C₃H₇	H	H	145–150 (dec.)	62	M-29	—	—	—	—	—	—
CH₃	i-C₃H₇	H	H	168–173 (dec.)	—	M-29	—	—	—	—	—	—
CH₃	C₄H₉	H	H	103	30	M-23	—	—	—	—	—	—
CH₃	C₆H₅CH₂	H	H	168	63	M-29	—	—	—	—	—	—
C₂H₅	H	H	H	153–154	58	M-21	—	—	—	—	—	—
C₂H₅	C₂H₅	H	H	140 (dec.)	54	M-21	C₂H₅	105.5–106.5	28	—	—	—
C₃H₇	H	H	H	125	50	M-23	—	—	—	—	—	—
i-C₃H₇	H	H	H	—	—	M-21	C₂H₅	100 (dec.)[b]	42	C₂H₅	98 (dec.)	12
C₄H₉	H	H	H	102	—	M-23	—	—	—	—	—	—
C₅H₁₁	H	H	H	87	15	M-21	H	117 (dec.)	30	—	—	—
C₆H₁₃	H	H	H	80	23	M-27	—	—	—	—	—	—
C₇H₁₅	H	H	H	89	26	M-27	H	114–115 (dec.)	16	—	—	—
C₆H₁₃	H	H	H	80	23	M-27	—	—	—	—	—	—
C₇H₁₅	H	H	H	89	25	M-27	H	114–115 (dec.)	16	—	—	—
C₈H₁₇	H	H	H	49	—	M-23	—	—	—	—	—	—
C₁₀H₂₁	H	H	H	89	21	M-29	—	—	—	—	—	—
C₆H₅CH₂	H	H	H	not cryst.	—	M-29	—	—	—	—	—	—
CH₃	CH₃	CH₃	H	157–162 (dec.)	78	M-21	C₂H₅	110–111 (dec.)	7	—	—	—
CH₃	CH₃	CH₃	H	—	—	M-27	—	—	—	—	—	—
CH₃	C₂H₅	C₂H₅	H	150–156 (dec.)[c]	19	M-21	—	—	—	—	—	—
C₂H₅	C₂H₅	C₂H₅	H	138/148[c]	—	M-22	H	87–88	—	—	—	—

[a] 2-Methylbutyl.
[b] Forms a glass at 88–92° which then melts at 100–101°.
[c] Two polymorphic forms.
[d] Not isolated, but was detected in reaction mixture by reduction to cyclic telluride.

The ring system is stable in the diketo form. Enolization would introduce destabilizing ring strains.

The structure of 1-tellura-3,5-cyclohexanedione has been confirmed by NMR and IR spectroscopy (D-7). For the compound, in $CDCl_3$, derived from acetylacetone and tellurium tetrachloride two singlets were recorded with a chemical shift of 207 and 234 cps (Varian HR—60, TMS) and a relative intensity of 1:2. The IR spectrum revealed typical carbonyl absorptions.

The linear condensation products, compounds of the constitution given by formulas **1** and **2**, were usually isolated as reaction byproducts from the mother liquors which remained after separation of the cyclic compounds. If cyclization was not possible, due to the foregoing reasons, the linear compounds were the only products isolated. These linear derivatives—not forced into the diketo configuration by ring-strain considerations—can still exhibit keto-enol tautomerism as long as the middle methylene group bears at least one hydrogen atom. The enol forms, **4** and **5**, are capable of being intramolecularly hydrogen bonded.

$n = 1$ or 2

R^1, R', R'', R''', R^4 = organic groups

The NMR spectrum of bis(3-chloroacetylacetonyl)tellurium dichloride showed three signals at 140, 292, and 879 cps relative to TMS with an intensity ratio of 3:2:1. A carbonyl absorption was not observed in the IR spectrum. These findings are consistent with a hydrogen-bonded cyclic enol-structure (D-7).

If the potential function of the hydrogen atom forming the hydrogen bond has only one minimum, the difference between the hydrogen-bonded cyclic forms **4** and **5** ceases to exist.

Morgan and co-workers (for references see Tables VI and VII) isolated O–ethyl derivatives of the type

$$
\begin{array}{c}
R'' \quad\quad O \quad\quad\quad OC_2H_5 \quad R^4 \\
\diagdown \quad\quad \parallel \quad\quad\quad\quad | \quad\quad\quad \diagup \\
R'\!\!-\!\!-\!\!C\!-\!C\!-\!CH\!\!=\!\!C\!\!-\!\!-\!\!-\!\!C\!\!-\!\!-\!\!-\!TeCl_3 \\
\diagup \quad\quad\quad\quad\quad\quad\quad\quad\quad\quad \diagdown \\
R^1 \quad\quad\quad\quad\quad\quad\quad\quad\quad R'''
\end{array}
$$

In these compounds there exists the possibility for isomerism. The ethyl group can be attached to the oxygen atom closest to or furthest from the tellurium atom. Morgan used both formulations (M-21,M-27). These ethyl ethers, upon treatment with alkali, lose the tellurium atom with the formation of the O–ethyl compound of the enol form of the parent diketone. A decision as to the site of attachment of the ethyl group in the tellurium compound could be made by careful examination of the hydrolytic products of nonsymmetric acetylacetone derivatives (M-16).

$$
\begin{array}{c}
OC_2H_5 \quad\quad O \\
| \quad\quad\quad\quad \parallel \\
R\!-\!CH_2\!-\!C\!\!=\!\!CH\!-\!C\!-\!CH_2\!-\!TeCl_3 + 4\,KOH \longrightarrow
\end{array}
$$

$$
\begin{array}{c}
OC_2H_5 \quad\quad O \\
| \quad\quad\quad\quad \parallel \\
R\!-\!CH_2\!-\!C\!\!=\!\!CH\!-\!C\!-\!CH_3 + 3\,KCl + KHTeO_3 + H_2O
\end{array}
$$

The ethyl group comes from the ethanol which was present as an impurity in the chloroform used as a solvent in these condensation reactions. Purified chloroform did not give the O–ethyl products (M-21,M-22,M-27). O–Ethyl derivatives of bis(acetylacetonyl)tellurium dichlorides were never isolated.

Dibenzoylmethane gave a very unstable tellurium trichloride of the formula $(C_6H_5CO)_2CH\!-\!TeCl_3$ (M-17). Bis(acetyl)acetone, and dimethyldihydro-resorcinol in the presence of $TeCl_4$ condensed to pyrones, which subsequently formed salts with the hexachlorotellurite anion (M-21). 1,1,2,2-Tetrakis-(acetyl)ethane probably gave a 2-(5-methyl-3,4-diacetylfuryl)methyltellurium trichloride (M-21).

Although these linear tellurium compounds have been obtained in only a few instances from the residues left after separation of the cyclic products, it is probable that they were formed in other condensation reactions, but owing to their instability and great solubility were not isolated (M-29).

For specific compounds and literature references refer to Tables VI and VII.

Experimental. 1-Tellura-3,5-cyclohexanedione 1,1-dichloride (M-15):   Sub-limed $TeCl_4$ (10.3 g, 0.038 moles) was mixed with acetylacetone (7.6 g, 0.076 moles) in 55 ml of dry $CHCl_3$. The orange solution was refluxed on a water bath for two hours. After the evolution of hydrogen chloride had ceased, the

TABLE VII
Noncyclic Condensation Products from TeCl$_4$ and 1,3-Diketones

Ketone	Product	m.p.	Yield, %	Ref.
(CH$_3$)$_2$CH—CO—CH$_2$—CO—CH$_3$	(CH$_3$)$_2$CH—CO—CH=C(OC$_2$H$_5$)—CH$_2$—TeCl$_3$	103 (dec.)	30	M-21
(CH$_3$)$_2$CH—CO—CH$_2$—CO—CH(CH$_3$)$_2$	(CH$_3$)$_2$CH—C(OC$_2$H$_5$)=CH—CO—C(CH$_3$)$_2$—TeCl$_3$	87–89 (dec.)	0.4	M-27
(CH$_3$)$_3$C—CO—CH$_2$—CO—CH$_3$	(CH$_3$)$_3$C—CO—CH=C(OC$_2$H$_5$)—CH$_2$—TeCl$_3$	116	80	M-17
	[(CH$_3$)$_3$C—CO—CH=C(OH)—CH$_2$]$_2$TeCl$_2$	133 (dec.)	23	M-17
(CH$_3$)$_2$CHCH$_2$—CO—CH$_2$—CO—CH$_3$	(CH$_3$)$_2$CH—CH$_2$—CO—CH=C(OC$_2$H$_5$)—CH$_2$—TeCl$_3$	100–101 (dec.)	42	M-21
	[(CH$_3$)$_2$CH—CH$_2$—CO—CH=C(OH)—CH$_2$]$_2$TeCl$_2$	97–98	12	M-21
(phenyl)—C(=O)—CH$_2$—C(=O)—CH$_3$	(phenyl)—C(=O)—CH=C(OC$_2$H$_5$)—CH$_2$—TeCl$_3$	138–140 (dec.)	80	M-16
(phenyl)—C(=O)—CH$_2$—C(=O)—(phenyl)	[(phenyl)—C(=O)—CH=C(OH)—CH$_2$]$_2$ TeCl$_2$	148 (dec.)	13	M-17
	[(phenyl)—C(=O)—CH—]$_2$ CH—TeCl$_3$	—	—	M-17
(phenyl)—NH—N=C(CO—CH$_3$)$_2$	Not isolated	—	—	M-21

154

$CH_3—CO—CH_2—CO—CH_2—CO—CH_3$	$\left[\begin{array}{c} \text{H}_3\text{C} \\ \text{C}{=}\overset{\oplus}{\text{O}}\cdot\text{H} \\ \text{H}_3\text{C} \end{array}\right]_2 TeCl_6^{2-}$	—	43	M-21
dimedone (5,5-dimethyl-1,3-cyclohexanedione)	$\left[\begin{array}{c} \text{H}_3\text{C} \\ \text{H}_3\text{C} \\ \text{C}{=}\overset{\oplus}{\text{O}}\cdot\text{H} \\ \text{OH} \end{array}\right]_2 TeCl_6^{2-}$	160–165 (dec.)[a]	32	M-21
$(CH_3—CO)_2CHCH(CO—CH_3)_2$	$CH_3—CO,\ COCH_3,\ H_3C—CO,\ CH_2—TeCl_3$ furan	—	—	M-21
$CH_3—CO—CH_2—CO—OC_2H_5$	$C_2H_5O—CO—CH{=}C(OC_2H_5)—CH_2—TeCl_3$	90–92	32	M-21

[a] Resinified at 130°.

solution was filtered and concentrated over lime in a desiccator. Crystals separated. The concentrated filtrates yielded additional crystals. After re-crystallization from ethanol, benzene, or acetone the substance darkened at 155–160° and melted with decomposition between 169 and 173° (yield: 7 g; 62%).

O-Ethylacetylacetonyltellurium Trichloride (M-16): Acetylacetone (7.8 g, 0.078 moles) dissolved in 30 ml of $CHCl_3$, dried on $CaCl_2$ for several days and fractionated, was added to $TeCl_4$ (10.4 g, 0.038 moles) and the clear solution remained for 14 hr at room temperature. A dense yellow oil separated. Hydrogen chloride was evolved after refluxing this mixture for 1.5 hr. The resulting liquor was decanted from the small amount of a dark, heavy oil and elemental tellurium. The major portion of the chloroform was removed under reduced pressure and 2.3 g of the cyclic dichloride separated. The mother liquid, further concentrated, left an oily residue which was spread on a porous tile to remove any unchanged diketone. The solid residue (5.7 g, 40% yield) was recrystallized from chloroform or benzene/light petroleum ether. The crystals melted at 106–107°.

1-Pentylacetylacetonyltellurium Trichloride (M-21): Tellurium tetra-chloride (3.2 g, 0.012 moles) added to *n*-heptoylacetone (3.2 g, 0.019 moles) in 20 ml of ethanol-free $CHCl_3$ gave a brown solution, which evolved hydro-gen chloride while refluxing for two hours. After separation from 0.22 g of tellurium, the filtrate, left in a vacuum desiccator, deposited crystals which were washed with CCl_4 (1.4 g, 30% yield). No further solid could be isolated from the mother liquor. The trichloride melted with blackening at 117°.

Bis(acetylacetonyl)tellurium Dichloride (M-17): Tellurium tetrachloride (8 g, 0.03 moles) and acetylacetone (6 g, 0.06 moles) dissolved in 90 ml of $CHCl_3$, purified by the salicylide process, were refluxed for 45 min. The orange solution, after removal of elemental tellurium, was concentrated to 40 ml and finally evaporated *in vacuo*. The remaining yellow syrup was stirred with a small volume of $CHCl_3$. Crystals of the cyclic product (2.7 g, 30% yield) separated.

The filtrate was shaken with a few milliliters of water, the TeO_2 was re-moved, and the $CHCl_3$ solution evaporated. The semi-solid residue, spread on porous tile, yielded two grams (18%) of the linear dichloride, which melted with decomposition at 115° following recrystallization from benzene.

2. Tellurium Tetrachloride and Monoketones

The reactions between aliphatic and aliphatic-aromatic monoketones and tellurium tetrachloride were investigated by Morgan and Elvins (M-30) and by Rohrbaech (R-17) and Rust (R-18). The tetrachloride easily combines with the ketones in an inert solvent such as ether or chloroform with the evolution of hydrogen chloride to form bis(organo)tellurium dichlorides and organotellurium trichlorides. Chloroform is preferred as a solvent for the reaction, since ether forms an adduct with TeCl$_4$, which is present as impurity in the final product.

The solubilities of the di- and trichlorides vary irregularly within the homologous series of the ketones. Therefore, only the less soluble compound was isolated (M-30), although it is probable that both types of compounds were formed (see Table VIII). *tert*-Butyl methyl ketone gave both products. The low yields of purely aliphatic compounds are due to their hydrolytic instability.

When 1,1,3-trimethylacetone was allowed to react with tellurium tetrachloride, hydrogen chloride was evolved. Only an oily, easily hydrolyzable product was formed, which was not further characterized. 1,1,3,3-Tetramethylacetone did not react with tellurium tetrachloride. The reactants were recovered quantitatively (M-30). The electron-releasing tendency or steric requirements of the methyl groups may explain these observations.

Bromomethyl phenyl ketone did not condense with tellurium tetrachloride (R-18). Morgan and Elvins proposed the following mechanism for the formation of these condensation products (M-30).

The tellurium trichloride can then react with another enolized acetone molecule to form the dichloride. The loss in the acidic character of the hydrogen atoms in position α- to the carbonyl group due to the presence of electron releasing groups might prevent enolization which is required for the addition of TeCl$_4$.

Experimental. Bis(acetonyl)tellurium Dichloride (M-30):   Redistilled acetone (2 g, 0.039 moles) was refluxed with TeCl$_4$ (4.7 g, 0.018 moles) in 30 ml of dry CHCl$_3$. The turbid yellow solution rapidly evolved HCl. After 40 min

TABLE VIII

Reaction of $TeCl_4$ with Monoketones

$R—CH_2—\overset{\overset{O}{\|}}{C}—R'$		$(R'—\overset{\overset{O}{\|}}{C}—\overset{\overset{R}{\|}}{C}H—)_n TeCl_{4-n}$			
R	R'	n	m.p.	Yield, %	Ref.
H	$CH_3—$	2	126–128	16.7	M-30
H	$C_2H_5—$	1	101	75	M-30
H	$C_3H_7—$	2	92–93	14	M-30
H	i-$C_3H_7—$	2	90	52	M-30
H	$C_4H_9—$	2	62	28	M-30
	i-$C_4H_9—$	2	95	45.5	M-30
H	$tert$-$C_4H_9—$	2	191–192	26	M-30
		1	114–115	21	M-30
H	$C_6H_5—$	2	186–187	—	M-30
		2	186–187	—	R-18
H	p-Tolyl	2	200	—	R-18
H	p-Methoxyphenyl	2	190	—	R-18
H	p-Ethoxyphenyl	2	212–213[b]	—	R-17
H	2,4-Dimethylphenyl	2	180	—	R-17
H	2,4,5-Trimethylphenyl	2	188	—	R-17
H	4-Isopropylphenyl	2	183	—	R-17
H	1-Naphthyl	2	203–204	—	R-17
$CH_3—$	$C_2H_5—$	1	77–78	84.6	M-30
$CH_3—$	i-$C_3H_7—$	Oily product[a]	—	—	M-30
$CH_3—$	$C_6H_5—$	1	114–115	61	M-30
$C_2H_5—$	$C_3H_7—$	1	70	77	M-30
$C_2H_5—$	$C_6H_5—$	1	128–129	71	M-30
$C_6H_5—$	$C_6H_5—$	1	142–143	60	M-30

[a] Not characterized.
[b] Blackened at 184°.

the filtrate, which was separated from 0.1 g of tellurium, was concentrated to a brown syrup in a vacuum desiccator. The syrup was seeded with crystals obtained by extracting a small portion with petroleum ether, dissolving the residue in CCl_4, and diluting the solution with petroleum ether. The crystals, drained on a porous tile and recrystallized from $CHCl_3/CCl_4$ (1:1), melted at 126–128° (0.9 g, 16.7% yield).

1-(1-Benzoylethyl)tellurium Trichloride (M-30): $TeCl_4$ (3 g, 0.11 moles) and the ketone (3 g, 0.022 moles) were refluxed for 45 min in 25 ml of $CHCl_3$. A solid separated on concentration which, when recrystallized from CCl_4 in a dry atmosphere, melted at 114–115° (2.5 g, 61% yield).

3. Tellurium Tetrachloride and Carboxylic Acid Anhydrides

Acetic anhydride condenses with tellurium tetrachloride to give products, the nature of which is dependent on the molar ratio of the reagents employed (M-26).

A large excess of the anhydride and $TeCl_4$ (6:1) formed bis(carboxymethyl)tellurium dichloride, the cyclic product, 4-tellura-1-oxa-2,6-cyclohexanedione 4,4-dichloride, was probably formed as an intermediate.

When only a 1¾ to 2-fold molar excess of the anhydride was used in the condensation, a grey solid, a mixture of carboxymethyltellurium trichloride and its anhydride was formed, which was reduced with $K_2S_2O_5$ to bis(carboxymethyl)ditelluride. From the final mother liquors of this condensation, colorless crystals of methylenebis(tellurium trichloride) were isolated. The formation of this methylene compound is unexpected. Morgan and Drew (M-26) discussed possible mechanisms leading to this product.

The tendency of the anhydrides to undergo a condensation reaction decreases rapidly as the homologous series is ascended (M-32). Propionic acid anhydride and $TeCl_4$ in a 3:1 molar ratio yielded methylcarboxymethyltellurium trichloride, melting between 145 and 150°, in 72% yield.

Experimental. Bis(carboxymethyl)tellurium Dichloride (M-26): $TeCl_4$ (20 g, 0.073 moles) and acetic acid anhydride (48 g, 0.47 moles) dissolved in 150 ml of dry $CHCl_3$ were refluxed for 4 hr. The liquid was decanted from tellurium and diluted with $CHCl_3$. On concentration several crops of crystals were deposited which initially were deliquescent, but became stable when stored in a vacuum desiccator occasionally opened to the atmosphere. The dichloride crystallized from ethyl acetate/$CHCl_3$ upon addition of petroleum ether and melted at 160–161° (56% yield).

Methylenebis(tellurium trichloride) (M-26): When $TeCl_4$ condensed with acetic anhydride in a molar ratio of 1:2 in $CHCl_3$ (4 ml of $CHCl_3$ per gram of $TeCl_4$) a grey solid (32% yield) was obtained which was reduced to bis(carboxymethyl)ditelluride by $K_2S_2O_5$. The chloroform filtrates, when concentrated and allowed to remain in a vacuum desiccator, deposited methylenebis(tellurium trichloride) (14% yield). Recrystallized from dry $CHCl_3$ it melted at 173° with effervescence.

4. Tellurium Tetrachloride and Aromatic Compounds

Aromatic compounds combine with tellurium tetrachloride with the evolution of hydrogen chloride when a ring hydrogen atom is activated by an R—O, HO—, R_2N—, RS—, R·CONH—, 2-quinolyl or a 9-acridinyl group.

When the reactions were performed under relatively mild conditions, for example, refluxing the reactants in dry chloroform or carbon tetrachloride for several hours, or heating them in the absence of solvents on the water bath, the organotellurium trichloride was isolated exclusively, or as the main product. From the reaction of 0.089 moles of $TeCl_4$ with 0.11 moles of diphenyl ether a small amount of the bis(p-phenoxyphenyl)tellurium dichloride was recovered.

(X = activating group)

The yields of the dichlorides were not increased when a twofold excess of the aromatic component was employed. This is probably due to the insolubility of the trichlorides which precipitate during the reaction. Only resorcinol gave the dichloride upon refluxing tellurium tetrachloride with the dihydroxy compound in diethyl ether for $\frac{1}{2}$ hr (R-18).

A large excess of the aromatic component and/or prolonged heating at higher temperatures, in most cases in the absence of any solvent, yielded the bis(organo)tellurium dichlorides. Methoxybenzene or ethoxybenzene and

TeCl$_4$ employed in a molar ratio of 5:1 and 4:1, respectively, and heated for 24 hr at 150–180° formed the dichlorides in good yields (M-32,M-38).

Dimethylaminobenzene and methylphenylaminobenzene form adducts with TeCl$_4$ in dry diethyl ether. The dimethylaminobenzene adduct could be converted to the dichloride by boiling it in H$_2$O. From the mother liquor of the methylphenylaminobenzene adduct a small amount of dichloride was recovered. This adduct did not give the dichloride (M-35) when boiled in H$_2$O.

Additional compounds of tellurium tetrachloride with the following substances were reported, but not further characterized (R-8): tetraphenylpyrrole, indole, 2-phenylindole, 4-phenylindole, carbazole, benzoylcarbazole, phenylcarbazoacridine, pyridine, 2-phenylpyridine, 4-phenylpyridine, quinoline, 8-phenylquinoline, 8-hydroxyquinoline, 7-hydroxy-6-methoxyquinoline, 2-hydroxy-4-carboxyquinoline, its acid chloride, amide and diethylamide, fluorenylquinoline, acridine and trypaflavine. Aminophenols, diaminophenols, acylaminophenols (R-6) and p-nitroanisole (M-32) did not form tellurium compounds. Aniline and other free amines gave only tarry products (R-6). Tellurium tetrabromide and chloride adducts of various amines were prepared by Montignie (M-13), Lenher (L-31), and Lowy and Dunbrook (L-39).

Crystalline substances could not be obtained from naphthyl alkyl ethers (M-32,R-8) naphthols (R-18), monoethyl ethers of hydroquinone and resorcinol, and the monomethyl ether of hydroquinone (M-32).

Purely aliphatic ethers, for example, diethyl ether gave only adducts (M-28). Ethyl phenyl sulfide (R-17) yielded only elemental tellurium (R-17), while thiophenol formed tellurium, hydrogen chloride, and diphenyl disulfide (R-17).

Montignie (M-13,M-14) reported that TeBr$_4$ reacted with benzoquinone, anthraquinone, naphthoquinone, and pyrrole to give substances which, according to elemental analysis, might be organotellurium tribromides.

Phenetole and TeBr$_4$ heated at 180–190° gave only bromophenetoles and metallic tellurium (M-38). Although detailed investigations as to the dependence of reaction rates and the nature of the products isolated upon the substituents in the aromatic ring are lacking, the rules of aromatic substitutions seem to apply. Two o,p-directing substituents in positions m- to each other facilitate the reaction; the reactivity seems to be reduced when they are in o-positions (M-28). Thus m-dimethoxybenzene easily formed bis(2,4-dimethoxyphenyl)tellurium dichloride. An attempt to prepare 2,4-dimethoxy-1,5-bis(trichlorotelluro)benzene failed. 1-Hydroxy-2-methoxybenzene is reported to react less smoothly (M-28). Excess 1-methoxy-2-methylbenzene, when heated with TeCl$_4$ at 170°, did not form a dichloride, although the trichloride was easily obtained. The m- and p- isomers reacted with greater difficulty (M-32). An unsuccessful attempt was made to synthesize p,p'-bis(trichlorotelluro)diphenyl ether (D-11).

Rust (R-18) discovered, in 1897, that anisole and phenetole react with tellurium tetrachloride. His analyses for these derivatives agree with the for-mulations, bis(p-methoxyphenyl)- and bis(p-ethoxyphenyl)tellurium dichlo-ride. However, Morgan and Drew (M-28) and Reichel and Kirschbaum (R-6) were unable to reproduce Rust's work. They obtained, using Rust's reaction conditions, the organotellurium trichlorides. Unfortunately, the melting points for the anisole derivatives are very close, the bis(p-methoxyphenyl)-tellurium dichloride being reported to melt from 181 to 185° (M-32) while the p-methoxyphenyltellurium trichloride was found to have melting points of 182° (R-6) and 192° (P-5), the latter value probably being more reliable. Rust reported 190° as the m.p. of his yellow product.

According to the melting point of 185° of the phenetole derivative, Rust had obtained the trichloride, although his analyses again agree with the values calculated for the dichloride. Melting points of 179–184° (P-5), (R-6), and 108–111° (M-38) are reported for the trichloride and dichloride, respectively.

Rust (R-18) prepared, from materials that he considered to be the dihy-droxide, dibromide, and dinitrate, derivatives of bis(p-methoxyphenyl)- and bis(p-ethoxyphenyl) telluride. He also obtained bis(p-methoxyphenyl)-tellurium diiodide and a platinum salt of the alleged formula $(CH_3OC_6H_4)_2TeCl_2 \cdot PtCl_4$.

These "dichlorides" were reduced to red "tellurides" by Lederer (L-13) and Rohrbaech (R-17). Rohrbaech converted the "bis(p-methoxyphenyl) telluride" into the dichloride and prepared sulfur derivatives of the two tellurides.

The melting points of the anisole derivatives agree fairly well with more recently reported literature values, while Rust's bis(p-ethoxyphenyl)tellurium dibromide melted 64° higher than the value of 116–117° reported by Lederer (L-25).

The red to orange color of the reduction products of Rust's compounds is characteristic of ditellurides. The melting point of the bis(p-methoxyphenyl)-ditelluride is reported as 56–60°, while the telluride melts between 52 and 56°. Rohrbaech (R-17) and Lederer (L-13) gave 50° as the melting point of their product. The bis(p-ethoxyphenyl)ditelluride fuses at 109° (R-6), the telluride at 64° (P-6). Lederer (L-13) and Rohrbaech reported 64°. It is, however, very surprising that Rust's and Rohrbaech's analyses all agree with the structure of bis(organo)tellurium derivatives. All of the results obtained with Rust's two compounds are of doubtful value. The compounds prepared by these authors, therefore, will not be discussed in the following sections. Among these investigations, in addition to the ones already mentioned, are Lederer's experiments with "bis(p-methoxyphenyl) telluride" and methyl iodide (L-13) and his preparations of mercuric halide adducts of "bis(p-methoxyphenyl) telluride" (L-11).

Morgan and Burgess (M-37) investigated the reaction of a "tellurium oxychloride" (prepared by dissolving tellurium in HNO_3 ($D = 1.2$), evaporation of the resultant solution with concentrated HCl and the removal of all volatile materials under reduced pressure on a water bath) with the isomeric cresols.

p-Cresol gave three products.

m.p. 229–230° (dec.)   m.p. 244–245° (dec.)   (m.p. 213–214° (dec.)

These three substances were isolated by successive treatment of the reaction mixture with benzene, chloroform and alcohol.

o-Cresol yielded only a 3-methyl-4-hydroxyphenyltellurium trichloride (m.p. 208–209° dec.) and two isomeric dichlorides melting at 177–178°, and 197–198°, respectively.

m-Cresol gave the bis(2-methyl-4-hydroxyphenyl)tellurium dichloride only, which easily converted to tris(2-methyl-4-hydroxyphenyl)telluronium chloride, melting at 184–185°, upon boiling in aqueous ethanol or on acidification of its sodium carbonate solution. The o-cresol derivative was not easily transformed into the telluronium salt under the same conditions.

Tellurium dioxide did not react with m-cresol. Tellurium tetrachloride and m-cresol heated on a water bath gave only a very small yield of the telluride (melting 143–144°) on reduction of the oil obtained in this condensation with $K_2S_2O_5$. The small yield was due, according to Morgan and Burgess, not to tellurium tetrachloride, but to the presence of a small amount of basic tellurium chloride formed by hydrolysis.

The position of the chlorotellurium group is not certain in all cases. The products obtained from $TeCl_4$ and aromatic compounds are listed in Table IX.

Experimental. p-Ethoxyphenyltellurium Trichloride (R-6): Tellurium tetra-chloride (12 g, 0.044 moles), phenetole (6 g, 0.049 moles) and 75 ml of dry CCl_4 were refluxed for two hours in a dry atmosphere. Hydrogen chloride was given off. Yellow crystals were deposited, which were separated and washed with dry CCl_4 (14.5 g, 93% yield). After recrystallization from glacial acetic acid the substance melted at 184°.

Bis(p-methoxyphenyl)tellurium Dichloride (M-32): Tellurium tetrachloride (15 g, 0.056 moles) and anisole (30 g, 0.28 moles) were heated for 24 hr at 150–180°, the solution then being allowed to solidify *in vacuo*. The dichloride,

TABLE IX

Reactions of TeCl$_4$ with Aromatic Compounds

Aromatic compound[a]	R$_n$TeCl$_{4-n}$	m.p.	Yield, %	Ref.
4-Methoxybenzene	1	190	100	M-32
	1	182	96	R-6
	1	—[b]	—[b]	L-13
	?[c]	183–184	—	R-17
	?[c]	190	—	R-18
	2	181–182	95	M-32
4-Ethoxybenzene	?[c]	185	—	R-17,R-18
	1	182–183	92	M-28
	2	108	—	M-38
	?[c]	—[b]	—[b]	L-13
	1	184	93	R-6
3-Methyl-4-methoxybenzene	1	232–233	—	M-32
2-Methyl-4-methoxybenzene[d]	1	154	—	M-32
2-Methoxy-5-methylbenzene	1	135	—	M-32
3-Hydroxy-4-methoxybenzene	1	157–159 (dec.)[e]	—	M-28
2,4-Dimethoxybenzene	1	155–156	67	M-28
	2	204–205	~60	M-28

p-Bromo-methoxybenzene[f]	1	184	—	M-32
4-Phenoxybenzene	2	157–158	0.1	D-11
	1	156–157	90	D-11
4-Thiophenoxybenzene	1	164.5–165.5	42	P-5
4-Dimethylaminobenzene	2	188–189[g]	35[h]	M-35
4-(N-Methyl-N-phenylamino)benzene	2	170–172	5	M-35
4-Acetamidobenzene	1	—[i]	—[i]	R-6
2,4-Dihydroxybenzene	2	188–189	—	R-18
4-Hydroxybenzene	1	213	~100	R-6
4-(4'-Carboxyquinolyl-2')benzene	1	237	47.6	R-8
4-(4'-Aminoquinolyl-2')benzene	1	243 (dec.)	45.1	R-8
4-(Acridinyl-9')benzene	1	254	63.0	R-8
4-(5,6-Benzo-4-carboxyquinolyl-2')-benzene	1	276–280 (dec.)	38.0	R-8

[a] The aromatic rings are numbered in such a way that the chlorotellurium group will always be linked to the 1-position.
[b] Not isolated; reduced with NaHSO₃. For discussion see text.
[c] See discussion in text.
[d] Could also be 2-methoxy-4-methylbenzene.
[e] Trichloride obtained in impure form only.
[f] Position of TeCl₃ group unknown.
[g] Turned blue at 181°.
[h] Based on adduct.
[i] Trichloride not purified but reduced with K₂S₂O₅ to the ditelluride, m.p. 172°, 7% yield.

crystallized from benzene/petroleum ether (40–60°), melted at 181–182° (95% yield).

5. Tellurium Tetrachloride and Olefins

Tellurium tetrachloride yields tellurium trichlorides and dichlorides in addition reactions with ethylenes and acetylenes.

Cyclohexene reacted violently. In CCl_4 the reaction proceeded more smoothly and in better yields than in experiments without solvent (M-41). The 2-chloro-cyclohexyltellurium trichloride melted at 111–114° (dec.).

Funk and Weiss (F-19) prepared bis(2-chlorocyclohexyl)tellurium dichloride from $TeCl_4$ and cyclohexene (1:2 molar ratio) in CCl_4 in 65% yield by warming the mixture to 40°. It is a crystalline substance melting at 132° (dec.). Bis(2-chloroethyl)tellurium dichloride, melting 166° (dec.), was similarly prepared from ethylene in 50% yield.

There are, however, statements in the literature, which claim that $TeCl_4$ does not combine with ethylene (F-1) and that 2-chlorocyclohexyltellurium trichloride is stable in excess cyclohexene (M-45).

Diisobutylene, styrene, and 1,4-diphenyl-1,3-butadiene gave only unidentified viscous products and tellurium. Stilbene did not react, even at higher temperatures (M-45).

2,2-Diphenyl-4-pentenoic acid and tellurium tetrachloride in dioxane in a

2:1 molar ratio gave a 100% yield of the tellurium dichloride (M-45) which melts at 195–198°. For a more detailed mechanistic presentation see Section III.

Phenylacetylene or diphenylacetylene and TeCl$_4$ in equimolar proportions gave the tellurium trichlorides.

R = H: 94% yield, m.p. 205–215° (dec.)
R = C$_6$H$_5$: 77% yield, m.p. 125–128°

The stereochemistry of the products was not investigated (M-46). A corresponding dichloride was not formed when an excess of the acetylenic component was employed.

A definite compound was not obtained from TeBr$_4$ and ethylene (F-1). Attempts to prepare bis(2-haloethyl)tellurium dihalides by action of TeCl$_4$ or TeBr$_4$ on ethylene in CCl$_4$, aqueous HCl, or HBr were unsuccessful (F-1).

Carbethoxymethylene triphenylphosphorane when reacted with TeBr$_4$ formed a phosphonium salt (P-11).

m.p. 212–215° (dec.)

Experimental. Bis(2-chlorocyclohexyl)tellurium Dichloride (F-19):   Cyclohexene (1.6 g, 0.02 moles) was added to tellurium tetrachloride (2.7 g, 0.01 moles) suspended in 50 ml of dry CCl$_4$. The flask, sealed with a CaCl$_2$ tube, was warmed to 40°. The solution was then evaporated at reduced pressure and the residue was triturated with petroleum ether (30–40°). The solid which separated was recrystallized from ethanol and melted at 132° (dec.) (65% yield).

2-Phenyl-2-chloroethenyltellurium Trichloride (M-46): Tellurium tetrachloride (2.70 g, 0.01 moles) was triturated in a mortar with a solution of phenylacetylene (1.20 g, 0.01 moles) in 10 ml of CCl$_4$. A yellow solution formed with the evolution of heat. The trichloride separated immediately. Recrystallized from benzene it melted at 205–215° (dec.) (94% yield).

6. Tellurium Tetrachloride as a Starting Material for Heterocyclic Compounds

Heterocyclic ring systems containing a tellurium atom are the products of reactions between biphenyl (C-27) and diphenyl ether (C-4,D-11,D-13) and tellurium tetrahalides.

Courtot and Bastani (C-27) obtained dibenzotellurophene dichloride by heating biphenyl and tellurium tetrachloride at 140–165°. The product decomposed at 200°. Hellwinkel and Fahrbach (H-3), however, report a melting point of 333–335° for the same compound. Similarly (C-27), $TeBr_4$ and biphenyl, at 225–240°, gave a small yield of the dibromide (dec. 210–220°) which largely decomposed to 4,4′-dibromobiphenyl at the reaction temperature.

Diphenyl ether, when condensed with $TeCl_4$ at 200° gave phenoxtellurine 10,10-dichloride (D-11,D-13).

2-Nitrodiphenyl ether and $TeCl_4$ yielded only phenolic decomposition products (D-13). The preparation of 2-chloro-8-methylphenoxtellurine 10,10-dichloride required slightly different conditions than those employed for phenoxtellurine dichloride (C-4).

The same compound was probably obtained by Drew and Thomason (D-13) in their synthesis of phenoxtellurine due to the presence of phenyl *p*-tolyl ether as an impurity in their diphenyl ether. Tellurium tetrachloride in this case functioned also as a chlorinating agent, introducing a chlorine atom into the 4-position of the parent ether or into the 2-position of the phenoxtellurine system. This compound, however, was isolated only as an adduct with phenoxtellurine following the reduction of the mixture of dichlorides.

Experimental. Phenoxtellurine 10,10-Dichloride (D-11): Tellurium tetrachloride (11.1 g, 0.041 moles) and diphenyl ether (7.0 g, 0.043 moles) were heated in a flask equipped with a condenser and a drying tube. The temperature was raised slowly from 100 to 240° over a period of 13 hr. Hydrogen chloride was evolved at about 120° and again near 200°. The cooled melt was ground, the powder stirred with ether, the remaining solid dissolved in acetone, and the solution separated from the tellurium by filtration. Evaporation of the acetone left 9.4 g of the dichloride (62% yield). After treating this product with toluene in a Soxhlet, the crystals separating from the solvent melted at 265° (48% yield of purified product).

2-Chloro-8-methylphenoxtellurine 10,10-Dichloride (C-4): Tellurium tetrachloride (27 g, 0.1 moles) and 4-methyl-4'-chlorodiphenyl ether (22 g, 0.1 moles) was kept at 160° for three hours and then at 200–240° for eight hours. The solid obtained on cooling was powdered and freed from unchanged starting materials by extraction with ether. The residue (40 g) was crystallized from $CHCl_3$. Following separation from precipitated tellurium, six grams (14% yield) of the reddish brown microcrystalline dichloride, melting at 284°, were isolated.

7. Tellurium Tetrachloride and Organometallic Compounds

a. Grignard Reagents. The interaction of tellurium tetrachloride with Grignard reagents proceeds in a rather complex way. Elemental tellurium, bis(organo)tellurium dihalides, bis(organo) tellurides, and tris(organo)-telluronium halides have been observed as products. The mechanisms, by which such a variety of compounds is formed, remain largely unknown.

The following reaction steps were postulated by Lederer. Reduction of $TeCl_4$ (L-30):

$$TeCl_4 + 4\,RMgBr \longrightarrow Te + 4\,MgClBr + 2\,R_2$$

Formation of R_3TeX (L-30):

$$TeCl_4 + 3\,RMgBr \longrightarrow R_3TeCl \text{ and/or } R_3TeBr + 3\,MgClBr$$

In this reaction the telluronium chlorides are the primary products, which reaction with bromide ions to form the more difficultly soluble telluronium bromides (L-4).

Reduction of the telluronium halide to the telluride (L-4,L-30):

$$R_3TeX + RMgBr \longrightarrow R_2Te + R_2 + MgXBr$$

Petragnani and de Moura Campos (P-8,P-13), however, stated that telluronium salts tend to be inert toward arylmagnesium halides. Lederer (L-4) claimed that phenylmagnesium bromide and triphenyltelluronium chloride gave diphenyl telluride and diphenyl.

Formation of R_2TeX_2 (L-4):

$$TeCl_4 + 2\,RMgBr \longrightarrow R_2TeX_2 + 2\,MgXBr$$

TABLE X

TeCl$_4$ and Grignard Reagents

R—MgBr	Moles RMgBr / Moles TeCl$_4$	R$_m$TeX$_n$			Yield, %	m.p.	Ref.
R		m	n	X			
C$_4$H$_9$—	2	2	0[a]			b$_{99}$132–135	B-22
Phenyl—	5	3	1	Br		247–248 (dec.)	E-4
	5	2	0	I[b]		—	L-3
		3	1	Br		203–204	
	5	2	2	I[b]	11	247–249	L-4
		3	1	Br	36		
		2	0		12		
		2	2	Br/I			
	5	2	2	Cl[c]	88.3	162–163	R-12
	6	2	2	Cl[c]	91.7	162–163	R-12
	5	3	1	Cl[d]	16	242–243	N-4
	5	2	0		80		N-4
p-Tolyl	5	3	1	I[b]	32	232–233 (dec.)	L-4
		2	0		14		
	6	3	1	Cl[d]	25		N-7
	6	2	0				N-7

m-Tolyl	5	3	1	I[b]	26	160–161	L-19
o-Tolyl	5	3	1	I[b]		195–196	L-4
2,4-Dimethylphenyl	5	3	1	I[b]	12	208–209	L-19
2,5-Dimethylphenyl	5	3	1	I[b]	20	187	L-19
2,4,6-Trimethylphenyl	5	3	1	I[b]	16	169–170	L-19
p-Methoxyphenyl	5	3	1	I[b]	15	160	L-19
	5	3	1				N-6
	10	2	0		40		N-6
o-Methoxyphenyl	5	3	1	I[b]	43	191–192	L-30
	5	2	2	Br	32		
p-Ethoxyphenyl	5	3	1	I[b]	46	208–209	L-21
o-Ethoxyphenyl	5	3	1	I[b]	44	225	L-21
1-Naphthyl	6	2	2	Cl[c]	80.3	265–266 (dec.)	R-12
	5	3	1	I[b]			N-5
	10	2	0				N-5
Phenylacetylenyl	5	2	2	I[e]	33	120–125 (dec.)	M-46

[a] After reducing the reaction mixture with Zn/glac. acetic acid.
[b] Product treated with KI to produce R_3TeI.
[c] Product treated with SO_2Cl_2.
[d] Starting material $TeCl_4$ and $PoCl_4$; products converted to R_3TeI, which with $AgCl$ gave R_3TeCl, yield of Te compound given.
[e] I_2 added to purified reaction mixture.

Organotellurium trihalides were never observed among the products of reactions between $TeCl_4$ and Grignard reagents. The above-mentioned authors (P-8,P-13) reported that the reactions described by the following equations do *not* take place.

$$TeCl_4 + RMgBr \longrightarrow RTeCl_3 + MgBrCl$$
$$RTeCl_3 + RMgBr \longrightarrow R_2TeCl_2 + MgBrCl$$

On the other hand, diaryltellurium dihalides (L-3,L-4,L-30) and dibutyltellurium dihalide (which was reduced to the telluride without being isolated) (B-22) were obtained in such Grignard reactions. If dihalides, as such, are formed, they could be reduced by excess Grignard to the tellurides (see Section VI). The compounds which have been isolated in reactions between tellurium tetrachloride and Grignard reagents are listed in Table X. A reference is given for each separate experiment. If more than one compound was obtained in a reaction, all the isolated or mentioned products are listed without repeating the reference.

The elemental tellurium formed during these reactions will also combine with the Grignard reagents (see Section I-A-3-a for mechanistic details). Rheinboldt and Petragnani (R-12) found that the reduction of the tellurium tetrahalides by Grignard compounds was largely inhibited when the ethereal solution, diluted with twice the volume of benzene, is added slowly to an ice-cold suspension of the halide in ether. The bis(organo) tellurides were obtained in excellent yields by refluxing the combined solutions for two hours. The tellurides were usually not isolated as such but converted to the dichlorides or diiodides. Byproducts were not observed in these reactions.

If the reaction mixture as prepared above was not refluxed, but kept cold and then hydrolyzed, tris(organo)telluronium halides, which were converted to the more insoluble iodides by addition of KI, were isolated in low to moderate yields, together with tellurides and bis(organo)tellurium dihalides (for references see Table X). The preferred molar ratio of Grignard reagent to tellurium tetrachloride is 5:1.

Tellurium tetrabromide and tellurium tetraiodide refluxed with phenylmagnesium bromide gave a 90.8 and 96% yield, respectively, of diphenyltellurium dichloride (after treatment of the mixture with SO_2Cl_2) as compared with 88% in the case of tellurium tetrachloride (R-12).

Among unsaturated Grignard reagents, only two compounds were investigated. Phenylacetylenylmagnesium bromide gave the expected telluride, while β-styrylmagnesium bromide yielded 1,4-diphenyl-1,3-butadiene and an unidentified tellurium compound (M-46).

Butylmagnesium bromide and $TeCl_4$ (2:1) formed the dichloride which was reduced by zinc in glacial acetic acid to the telluride (B-22). Lederer reported (L-4) that ethyl- and methylmagnesium iodides and $TeCl_4$ did not yield telluronium salts.

Experimental. Diphenyltellurium Dichloride (R-12): To a suspension of vacuum-dried, finely powdered $TeCl_4$ (13.5 g, 0.05 moles) was added, over a 40-min period with ice–water cooling and vigorous stirring, a Grignard solution (0.25 moles) prepared from 6.1 g of Mg and 39.25 g of bromobenzene in 100 ml of ether and diluted with 200 ml of dry benzene. The initially vigorous reaction mixture was then refluxed for two hours. The precipitated tellurium redissolved and magnesium halides precipitated. The yellow solution, cooled with ice water, was treated with dilute NH_4Cl solution. The organic layer was washed three times with a small amount of water, filtered and dried over Na_2SO_4, and concentrated in vacuum until an orange-yellow oil remained. The latter was dissolved in CCl_4 and it yielded the dichloride upon treatment with SO_2Cl_2 in CCl_4. Addition of petroleum ether completely precipitated the substance, which was washed with petroleum ether and recrystallized from benzene/petroleum ether. It melted at 162–163° (88.3% yield).

Tris(p-ethoxyphenyl)telluronium Iodide (L-21): *p*-Ethoxyphenylmagnesium bromide (0.188 moles), prepared from 37.3 g of *p*-ethoxybromobenzene and 4.6 g of Mg, was poured, with vigorous stirring, into 250 ml of absolute ether containing $TeCl_4$ (10 g, 0.037 moles). The solution was immediately hydrolyzed with 25 ml of ice water, the precipitate separated, extracted with ethanol and then with $CHCl_3$ and the combined extracts concentrated. The residue was dissolved in hot H_2O and then treated with KI to give tris(*p*-phenetyl)telluronium iodide, which, recrystallized from ethanol, melts at 208–209° (10.5 g, 46% yield).

b. Organomercuric Chlorides. Arylmercuric chlorides are useful intermediates in the preparation of aromatic tellurium compounds which, because of the absence of an active site, cannot be obtained by direct condensation of an aromatic compound with tellurium tetrachloride. Campbell and Turner (C-4) introduced this method in 1938. The mercuric chloride, prepared from the diazonium salt, was refluxed for several hours in dry dioxane with an equimolar amount of tellurium tetrachloride. Chloroform and acetonitrile (C-4) were also used as solvents. Dioxane, however, has the advantage of precipitating the $HgCl_2$ formed as a dioxane adduct. The reaction between $TeCl_4$ and 2-chloromercuri-4'-carboxydiphenyl ether did not take place in

$$R—HgCl + TeCl_4 \longrightarrow R—TeCl_3 + HgCl_2$$

$CHCl_3$, benzene or toluene due to the insolubility of the starting material in these solvents.

The following compounds were prepared from $TeCl_4$ and the respective organomercuric chlorides (melting point and yield given): $C_6H_5TeCl_3$ (215–218°/66%) (F-2); *p*-$CH_3C_6H_4TeCl_3$ (181–182°/—) (F-2); 1-$C_{10}H_7TeCl_3$ (175–180°, dec./96%) (P-5); *o*-thiophenoxyphenyltellurium trichloride

$(213–215°/80.6\%)$ (P-5); $2\text{-}C_{10}H_7TeCl_3$ $(200–202°,$ dec./$85.2\%)$ (R-11); 4′-methyl-2-trichlorotellurodiphenyl ether $(180–185°,$ dec./$80\%)$ (C-4) and 4′-carboxy-2-trichlorotellurodiphenyl ether $(205–206°,$ dec./—) (C-4). De Moura Campos and Petragnani (M-45) isolated the corresponding bis-(organo)tellurium dichloride from a mixture containing $TeCl_4$ and 2,2-diphenyl-5-chloromercuri-4-pentanolactone in a 1:2 molar ratio.

$$2 \quad \begin{array}{c} C_6H_5 \\ \diagdown \\ C \\ \diagup \\ C_6H_5 \end{array} \begin{array}{c} CH_2\text{---}CH\text{---}CH_2\text{---}HgCl \\ | \qquad\qquad | \\ \\ O \\ \diagdown \diagup \\ C \\ \| \\ O \end{array} \quad + \ TeCl_4 \ \longrightarrow \ (R\text{---}CH_2)_2TeCl_2 + 2\ HgCl_2$$
$$\text{m.p. } 195–198°;\ 100\% \text{ yield}$$

c. Alkaliorganic Compounds and Diethylzinc. Aromatic and aliphatic lithium compounds have served as reagents in the preparation of tellurides, tetraorganotellurium compounds, and heterocyclic tellurium derivatives.

Tellurium tetrachloride and pentafluorophenyllithium in ether/hexane, warmed from $-78°$ to $20°$ and stirred overnight gave a mixture of bis-(pentafluorophenyl) telluride (m.p. 50–51°, 30% yield), tetrakis(pentafluorophenyl)tellurium, which decomposed without melting at $\sim 210°$, and perfluorobiphenyl. Similarly, 2,2′-dilithioperfluorobiphenyl, gave 17% perfluorodibenzotellurophene, melting at 116–119° (C-26).

Braye and co-workers (B-18) prepared in 56% yield, tetraphenyltellurophene, melting at 239.0–239.5°, from 1,4-dilithiotetraphenylbutadiene and tellurium tetrachloride.

In these reactions the primary products would be the organotellurium chlorides and tetraorganotellurium compounds.

$$n\ RLi + TeCl_4 \longrightarrow \begin{cases} RTeCl_3 \\ R_2TeCl_2 \\ R_3TeCl \\ R_4Te \end{cases}$$

In the experiments described so far only R_4Te (C-26) or R_2Te (B-18) were isolated. The tellurides are probably formed by decomposition of R_4Te, giving a hydrocarbon as the accompanying product (see Section VIII).

Wittig and Fritz (W-3) synthesized tetraphenyltellurium by the reaction between $TeCl_4$ and LiBr-free phenyllithium solution in ether in a molar ratio 1:5.6. Phenyllithium solution was added until the Gilman test was positive. The excess of reagent which was found to be necessary was taken as an indication that a complex $[(C_6H_5)_5Te]^-Li^+$, which could not be obtained in crystalline form, was formed as an intermediate. When a benzene solution of phenyllithium containing an equivalent amount of lithium bromide was employed, tetraphenyltellurium could not be isolated (P-8,P-13).

Hellwinkel and Fahrbach (H-3), starting with $TeCl_4$ and 2,2'-dilithiobiphenyl, prepared in 30–40% yield bis(biphenylylene)tellurium, a yellow crystalline substance with a melting point of 214°. 2,2'-Biphenylylene-2-biphenylyltelluronium chloride was observed as a byproduct. The authors formulated the following reaction sequence.

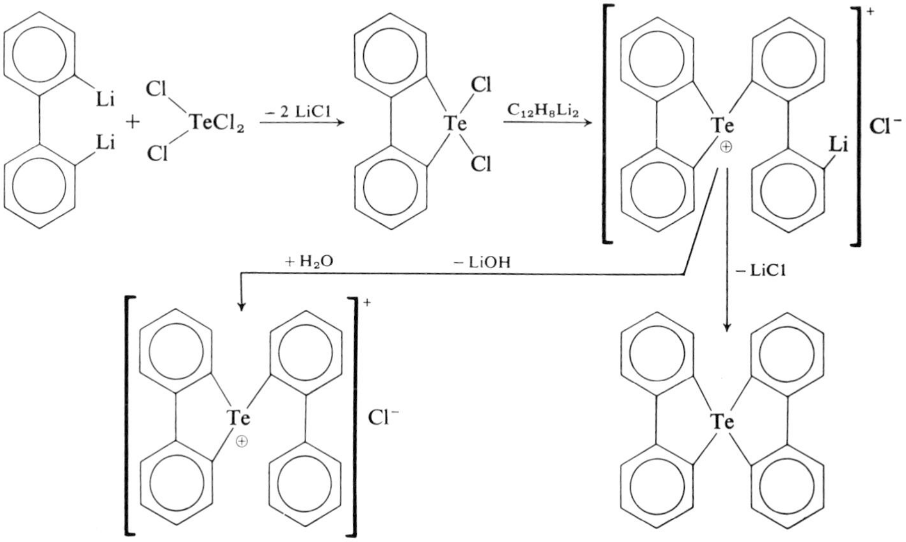

Tetrabutyl- and tetramethyltellurium which were formed from $TeCl_4$ and the lithioalkane when brought together in a 1:4 molar ratio, are rather unstable (H-4). The solution containing $(CH_3)_4Te$ upon hydrolysis with iodide containing ice water deposited trimethyltelluronium iodide. An attempt to distill $(C_4H_9)_4Te$ gave hydrocarbon and dibutyl telluride.

In the first attempt to prepare a tetrakis(organo)tellurium compound, Marquardt and Michaelis (M-9) mixed $TeCl_4$ and diethylzinc in ethereal solution and isolated a zinc containing salt upon evaporation of the ether. When dissolved in HCl and treated with Na_2CO_3 for the removal of zinc only triethyltelluronium chloride, melting at 174°, was obtained.

They also reported that $TeCl_4$ in ether, together with bromo- or chlorobenzene, gave a vigorous reaction upon addition of sodium metal. However, only tellurium and a tarry mass was isolated.

Experimental. 2-Naphthyltellurium Trichloride (R-11): 2-Chloromercurinaphthalene (14.4 g, 0.04 moles) and $TeCl_4$ (10.8 g, 0.04 moles) in 80 ml of dry dioxane were refluxed for two hours. After cooling to 10° the dioxane·$HgCl_2$

complex precipitated. The residue, triturated with CCl_4, was filtered and recrystallized from benzene (85.2% yield, m.p. 200–202° dec.).

Tetraphenyltellurium (W-3):   An ethereal solution of phenyllithium (50 ml, 1.04*N*, 0.052 moles), free from LiBr, was dropped into a solution of $TeCl_4$ (25 g, 0.0093 moles) in 25 ml of absolute ether at $-70°$. The reaction mixture was shaken and slowly warmed. The precipitated product was recrystallized from ether (m.p. 100–102°; softening from 60°). All operations must be performed in an inert atmosphere.

Tetraphenyltellurophene (B-18): $TeCl_4$ (5.3 g, 0.0197 moles) dissolved in 150 ml of dry ether was added at room temperature within 15 min to a suspension of 1,4-dilithiotetraphenylbutadiene in 100 ml of ether, prepared from 10 g (0.056 moles) of diphenylacetylene and an excess of clean lithium shavings. The reaction mixture was poured into CH_2Cl_2 and H_2O. After filtration on Kieselguhr, the organic layer was dried over Na_2SO_4 and evaporated. The residue, triturated with ether, left 5.35 g (56% yield) of crystals. Treatment with charcoal and recrystallization from CH_2Cl_2/ethanol gave a product melting at 239.0–239.5°.

8. Tellurium Tetrahalides and Diaryl Ditellurides

Petragnani and de Moura Campos (P-6) refluxed an equimolar mixture of the tellurium tetrahalides TeX_4 (X = Cl, Br, I) and bis(*p*-methoxyphenyl) and bis(*p*-ethoxyphenyl) ditellurides in toluene. The reaction proceeded according to the equation

$$2 \text{ R—Te—Te—R} + \text{TeX}_4 \longrightarrow 2 \text{ R}_2\text{TeX}_2 + 3 \text{ Te}$$

It can be seen in Table XI, that the reactivity of the tellurium tetrahalides tends to decrease in the sequence $TeCl_4$, $TeBr_4$, TeI_4.

TABLE XI
Tellurium Tetrahalides and Diaryl Ditellurides

R_2Te_2	TeX_4	Reflux time,	R_2TeX_2		Te	
R	X	hr	Yield, %	m.p.	Yield, %	Ref.
p-Methoxyphenyl	Cl	1	83.1	183–184	95.4	P-6
	Br	2	78.0	198–200	96.8	
	I	3	85.8	167–168	91.6	
p-Ethoxyphenyl	Cl	1	81.7	110–111	100	P-6
	Br	2	86.6	124–125	96.8	
	I	3	80.0	138–140	80.0	

D. Tellurium Dihalides

"Tellurium dihalides" were employed extensively by Lederer, along with Grignard reagents, to prepare aromatic tellurides.

$$TeX_2 + 2\,RMgX \longrightarrow R_2Te + 2\,MgX_2$$

The reaction, however, does not proceed according to this simple equation. The mixtures obtained after hydrolysis contain telluride, ditelluride, elemental tellurium, and diaryl.

Generally, an ethereal solution of the Grignard reagent was added to the tellurium dihalide. In most cases, a fourfold excess of the Grignard reagent was employed. After refluxing for 2–3 hr, hydrolysis with ice water resulted in the formation of a black residue, which was extracted with ether. Ether, aromatic hydrocarbon, and the starting bromide were removed by distillation in a CO_2 atmosphere. The residue was then distilled at low pressure. Frequently it was treated with Cu powder at 280° to transform the ditellurides into tellurides before distillation. The telluride was almost always converted into the dibromide by treatment with bromine for purposes of purification.

It is almost certain that the investigators, who used tellurium dihalides as starting materials, were working with mixtures of elemental tellurium and tellurium tetrahalides. Aynsley (A-5), Damiens (D-2,D-3), and Aynsley and Watson (A-6) synthesized tellurium dichloride and dibromide, and found that these substances are not stable in ether solution. At room temperature tellurium is slowly deposited, and the decomposition is accelerated by heat. A trace of moisture causes the dihalides to instantaneously disproportionate:

$$2\,TeX_2 \longrightarrow Te + TeX_4$$

Tellurium diiodide does not exist. A preparation of this composition consists of a solid solution of tellurium in tellurium tetraiodide (D-1). The reaction schemes devised by Lederer (L-10) are therefore not applicable. The Grignard reagent will react with the tellurium tetrahalide as well as with the elemental tellurium. The following equations explain the formation of the observed products

$$TeX_4 + 2\,RMgX \longrightarrow R_2TeX_2 + 2\,MgX_2$$
$$TeX_4 + 3\,RMgX \longrightarrow R_3TeX + 3\,MgX_2$$
$$R_2TeX_2 + 2\,RMgX \longrightarrow R_2Te + R_2 + 2\,MgX_2$$
$$R_3TeX + RMgX \longrightarrow R_2Te + R_2 + MgX_2$$
$$2\,R\!-\!MgBr + 3\,Te \longrightarrow R_2Te_2 + Te(MgBr)_2$$

For a more detailed discussion of these reactions consult Sections I-C-7, VI-1, and I-A-3-a. There exists also the possibility that the ditellurides react with the Grignard reagent to give tellurides (see Section V).

$$R_2Te_2 + R\!-\!MgBr \longrightarrow R_2Te + RTeMgBr$$

This may account for the small amounts of ditelluride isolated.

TABLE XII
"Tellurium Dihalides" and Grignard Reagents

R—MgBr	TeX$_2$	Ratio Moles RMgX / Moles TeX$_2$	Product R$_m$Te$_n$X$_p$						
R	X		m	n	p	X	Yield	m.p.	Ref.
C$_6$H$_5$—	Br	4	2	1	2	Br[a]	60	—	L-10
			2	2	0		14[b]	53–54	L-10
	Cl	4	2	1	2	Br[a]	60		L-10
			2	2	0		8[b]		
	I	4[c]	2	1	2	Br[a]	60		L-10
			2	2	0		13[b]		
o-Methylphenyl	Cl	2	2	1	2	Br[a]	36	—	L-13
			2	2[d]	0		—		
	Br	4	2	1	2	Br[a]	64	—	L-13
			2	2[d]	0		—		
m-Methylphenyl	Br	4[e]	2	1	2	Br[a]	56	165–166	L-16
p-Methylphenyl	Br	4	2	1	2	Br[a]	55	—	L-13
	I	4	2	1	2	Br[a]	55	—	L-13
			2	2[d]	0		—		

2,5-Dimethylphenyl	Br	4[e]	2	1	0[f]		72	52	L-14
2,4-Dimethylphenyl	Br	4[e]	2	1	2	Br[a]	57	200–201	L-14
2,4,6-Trimethylphenyl	Br	4[e]	2	1	0[f]		71	129	L-15
p-Methoxyphenyl	Br	4[e]	2	1	2	Br[a]	62	—	L-17
m-Methoxyphenyl	Br	2.2[e]	2	1	2	Br[a]	48	185–186	L-26
o-Methoxyphenyl	Br	4	2	1	2	Br[a]	30	195–196	L-27
p-Etinoxyphenyl	Br	3[e]	2	1	2	I[a]	41	134–135	L-25
	Br	2	2	1	0		—	64	L-40
o-Ethoxyphenyl	Br	2.5[e]	2	1	2	Br[a]	33	183–184	L-23
p-Chlorophenyl	Br	3	2	1	2	Br[a]	66	184–185	L-22
p-Bromophenyl	Br	3	2	1	2	Br[a]	27	192–193	L-22
1-$C_{10}H_7$	Br	4[e]	2	1	2	Br[a]	62	—	L-24
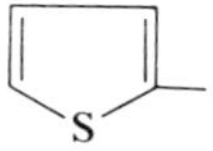 (thienyl structure)	Br	3.6	2	1	2	Br[a]	62	190.5	K-7

[a] Initial product R_2Te; converted to dihalide for purification.

[b] Yield calculated according to following reaction scheme: $2\ TeCl_2 \longrightarrow TeCl_4 + Te$; $2\ RMgBr + 3\ Te \longrightarrow R_2Te_2 + Te(MgBr)_2$.

[c] With C_6H_5MgI using the same molar amounts a 70% yield of R_2TeBr_2 was obtained.

[d] Only in tarry form, not dealt with further.

[e] Crude product heated with Cu powder, distilled and converted into dibromide.

[f] Crude product heated with Cu powder, distilled and recrystallized.

The yields of tellurides in the reactions employing "tellurium dihalides" range between 27 and 72% (see Table XII), probably because the amounts of Grignard reagent used in the reactions were not sufficient to convert the available tellurium into tellurides.

Rogoz (R-16) synthesized methyl phenyl telluride (b_{22} 118–122°) in 39.3% yield and methyl p-biphenylyl telluride (m.p. 84°) in 28% yield by slowly adding a mixture of 1 mole each of methylmagnesium bromide and the aromatic Grignard reagent to 1 mole of $TeBr_2$ in anhydrous ether.

Hellwinkel and Fahrbach (H-3) used "tellurium dichloride" together with 2,2′-dilithiobiphenyl in their synthesis of dibenzotellurophene (52% yield). Finally, tetraphenyltellurophene (53.5% yield, m.p. 240–241°) was obtained from $TeBr_2$ and 1,4-dilithio-1,2,3,4-tetraphenyl-1,3-butadiene (H-10). According to Krafft and Lyons (K-5), diphenylmercury and $TeCl_2$ gave chlorobenzene.

$$(C_6H_5)_2Hg + TeCl_2 \longrightarrow 2\,C_6H_5Cl + HgTe$$

II. REACTIONS OF TELLUROLS, R—Te—H

Tellurols have the general formula R—Te—H. Our knowledge about these compounds is quite limited. Only four alkanetellurols have been prepared (B-4,N-1). Although aromatic tellurols have been postulated as reaction intermediates, they have never been isolated as pure compounds (G-6,L-10,P-9). The sparsity of literature which is concerned with the tellurols may be in some way related to the exceedingly unpleasant odor of these compounds, especially of the alkyl derivatives. However, their extreme sensitivity toward oxygen is probably a more important factor.

Few reactions of tellurols are known. Baroni (B-4) reported that alkanetellurols are easily and rapidly oxidized to tellurides. However, one would expect ditellurides (R_2Te_2) to be formed as the initial products.

$$2\,R\text{—}TeH + \tfrac{1}{2}\,O_2 \longrightarrow R_2Te_2 + H_2O$$

Giua and Cherchi (G-6) studied the reaction of elemental tellurium with phenylmagnesium bromide and isolated diphenyl telluride as a product. They postulated the following reaction sequence

$$C_6H_5MgBr + Te \longrightarrow C_6H_5Te\text{—}MgBr$$
$$2\,C_6H_5Te\text{—}MgBr \longrightarrow (C_6H_5)_2Te + Te(MgBr)_2$$
$$Te(MgBr)_2 + 2\,HCl \longrightarrow H_2Te + 2\,MgClBr$$

The amount of H_2Te collected upon hydrolysis of the reaction mixture was not sufficient to account for all of the telluride formation according to the given equations. As an alternative, they discuss the formation of benzenetellurol by hydrolysis of phenyltelluromagnesium bromide, followed by oxidation of the tellurol to the ditelluride, R_2Te_2, which, upon elimination of tellurium, gave diphenyl telluride. Although this finding might explain the observation of Baroni (B-4) that tellurols are oxidized to tellurides, more direct experimental evidence is needed to ascertain the reaction sequence with confidence.

Alcoholic solutions of Hg, Ag, and Pb salts when combined with the aliphatic tellurols did not produce the corresponding metal salts of the tellurol (B-4). However, Lederer (L-10) reported that benzenetellurol reacts with mercuric chloride to yield C_6H_5—Te—HgCl. The higher reactivity of this aromatic tellurol might be caused by the greater acidity of this aromatic derivative.

Miller (M-12) was able to prepare monocrystals of photoconductive cadmium by treatment of zone-refined cadmium with tellurols containing alkyl groups with less than four carbon atoms at temperatures between 750 and 1200° and at pressures ranging from moderate vacuum to 50 atm.

Experimental. Phenyltelluromercuric Chloride: Diphenyl ditelluride (3 g) was dissolved in 250 ml of ethanol. Sodium (5 g) was added with cooling under a dry nitrogen atmosphere. A small amount of benzenetellurol was formed, with the main products being tellurium and diphenyl telluride. The solution was diluted with water, extracted with ether, filtered, and acidified with dilute hydrochloric acid. A slight turbidity was observed. After extraction with ether, the solution, when treated with $HgCl_2$, produced a yellow, amorphous precipitate, which was purified by recrystallization from hot methanol. The compound began to decompose at 90°.

III. REACTIONS OF TELLURENYL COMPOUNDS, R—Te—X

Tellurenyl compounds are known only in the aromatic series. The phenyltellurenyl compounds, C_6H_5—Te—X, where X represents Cl^-, Br^-, ClO_4^-, and NO_3^-, have been isolated only as complexes with thiourea (F-5–F-8). *p*-

Methoxyphenyltellurenyl derivatives, where X is —$S_2O_3^-$, C_6H_5—$\overset{\overset{O}{\uparrow}}{\underset{\downarrow}{S}}$—S⁻—,

CH_3—$\overset{\overset{O}{\uparrow}}{\underset{\downarrow}{S}}$—S⁻—, and CH_3O—$\overset{\overset{S}{\parallel}}{C}$—S⁻—, have also been prepared (F-3). *β*-

Naphthyltellurenyl iodide was synthesized from the ditelluride and iodine (V-5).

p-Methoxyphenyltellurenyl methanethiosulfonate, obtained from *p*-methoxyphenyltellurium trichloride and sodium methanethiosulfonate, reacts with potassium benzenethiosulfonate in methanol with displacement of the methanethiosulfonate group

TABLE XIII

Reactions of p-Methoxyphenyltellurenyl Compounds

The product was obtained in the form of stable, light orange-red crystals. The thiosulfonates react with sodium thiosulfate in aqueous solution or with a suspension of potassium methyl xanthate in warm ether to give the products listed in Table XIII (F-3).

Foss and co-workers have carried out complete crystallographic studies of some of these molecules which showed that the sulfur atom is directly co-ordinated to the tellurium atom (F-5–F-8).

2-Naphthyltellurenyl iodide is a convenient starting material for the synthesis of asymmetric tellurides, which are obtained in good yields by the reaction of the iodide with alkyl, alkenyl, alkynyl- or arylmagnesium bromides (see Table XIV).

When phenylmercuric chloride was used in place of the Grignard reagent the product isolated was the mercuric chloride adduct of the asymmetric telluride, which was reduced quantitatively to the telluride by molten $Na_2S \cdot 9H_2O$ (V-5).

2-Naphthyltellurenyl iodide and allyldiphenylacetic acid react in $CHCl_3$ to give α,α-diphenyl-δ-[2-naphthyltelluro]-γ-valerolactone

TABLE XIV

Asymmetric Tellurides Te—R from 2-Naphthyltellurenyl Iodide

R	Reagent	Yield, %	m.p./b.p.	Ref.
Phenyl[a]	C_6H_5HgCl	100	49–50	V-5
Phenyl	C_6H_5MgBr	80[b]	—	V-5
Cyclohexyl	$C_6H_{11}MgBr$	91	190 (0.024)	V-5
Ethyl	C_2H_5MgBr	75.4	143 (0.03)	V-5
		59[c]	80–95[c]	M-46
		95	88–89	M-46
		61[d]	—	M-42

[a] Reaction product is $2\text{-}C_{10}H_7TeC_6H_5\cdot HgCl_2$, which is subsequently reduced with $Na_2S\cdot 9H_2O$.

[b] The telluride was converted by means of SO_2Cl_2 to the dichloride. Yield based on dichloride.

[c] Yield based on the diiodide. The telluride was isolated in an impure state only.

[d] Yield based on diiodide. The telluride was not isolated.

Hydrogen iodide was oxidized to iodine under reaction conditions which converted the telluride into the diiodide. The treatment of the naphthyltellurenyl iodide with 10% NaOH resulted in the formation of the ditelluride and naphthalenetellurinic acid anhydride (V-5).

With 2-naphthyltellurenyl iodide triphenylphosphine gave a 1:1 adduct when a 100% excess of the phosphine was employed. The adduct could be recrystallized from a $CHCl_3$/benzene mixture in the presence of triphenylphosphine, but in the absence of the phosphine it decomposed to the ditelluride. The filtered solutions, upon addition of water, yielded triphenylphosphine oxide (P-12).

$$2\text{-}C_{10}H_7Te\text{—}I + (C_6H_5)_3P \longrightarrow 2\text{-}C_{10}H_7Te\text{—}I\cdot(C_6H_5)_3P$$
$$2\ 2\text{-}C_{10}H_7TeI\cdot(C_6H_5)_3P \longrightarrow (2\text{-}C_{10}H_7)_2Te_2 + \text{phosphorus product}$$
$$\downarrow H_2O$$
$$(C_6H_5)_3PO$$

Triethylamine and tri-isopropyl phosphite also converted the tellurenyl iodide to the ditelluride. The following reaction sequence was postulated for the reaction with phosphite (P-12).

$$2\ 2\text{-}C_{10}H_7\text{—}Te\text{—}I + (i\text{-}C_3H_7O)_3P \longrightarrow (2\text{-}C_{10}H_7)_2Te_2 + (i\text{-}C_3H_7O)_2P\overset{O}{\underset{I}{\nearrow}} + i\text{-}C_3H_7I$$

The ditelluride was obtained in good yield by employing a 100% excess of phosphite. The di-isopropyl iodophosphate was isolated as the di-isopropyl N-phenylamidophosphate.

$$(i\text{-}C_3H_7O)_2P\overset{O}{\underset{I}{\nearrow}} + 2\ C_6H_5NH_2 \longrightarrow (i\text{-}C_3H_7O)_2P\overset{O}{\underset{NH\text{—}C_6H_5}{\nearrow}} + C_6H_5NH_2\cdot HI$$

Experimental. Phenyl 2-Naphthyl Telluride (V-5):

a. From Phenylmercuric Chloride: 2-Naphthyltellurenyl iodide (0.35 g, 0.001 mole) and phenylmercuric chloride (0.31 g, 0.001 mole) were heated together in dioxane for 30 min. After cooling the solution it was poured, while stirring, into 50 ml of cold water. The yellow precipitate was filtered and mixed with 3.6 g of molten $Na_2S\cdot9H_2O$. After keeping the mixture at 90–95°, ice water was added. The yellow oil solidified and the solid was recrystallized from methanol/water. The yield was 0.4 g (100%); m.p. 49–50°.

b. From Phenylmagnesium Bromide: 2-Naphthyltellurenyl iodide (1.9 g) was slowly added to phenylmagnesium bromide, prepared from 0.9 g of bromobenzene and 0.2 g of magnesium in 15 ml of anhydrous ether. The iodide dissolved immediately. After the reaction was complete, 30 ml of 1.5% HCl were added. The ether layer was decanted and the aqueous layer was extracted with ether. The combined ethereal solutions were washed with water and then dried over Na_2SO_4. After filtration, SO_2Cl_2 was added with cooling until the reaction mixture became colorless. The addition of petroleum ether caused the separation of 1.6 g of white crystals of the diaryltellurium dichloride; yield: 80%; m.p. 171.5–172.5° (from benzene and petroleum ether). The dichloride was easily reduced to the telluride with $Na_2S \cdot 9H_2O$.

α,α-Diphenyl-δ-[2-naphthyl-diiodotelluro]-γ-valerolactone (M-42): 2-Naphthyltellurenyl iodide (0.76 g, 0.002 mole) was refluxed for 12 hr with diphenylallylacetic acid (0.50 g, 0.002 mole) in 40 ml of $CHCl_3$. Following evaporation of the solvent at room temperature and washing of the residue with CCl_4, 0.66 g of a reddish solid was isolated. To the CCl_4 extract a solution of iodine in the same solvent was added until the brown color persisted. This yielded an additional 0.28 g of product. The red crystals melted at 174–176° after recrystallization from benzene or $CHCl_3$/petroleum ether (yield: 61%).

Reaction of 2-Naphthyltellurenyl Iodide and Triphenylphosphine (M-42): A solution of 1.02 g (0.002 mole) di-2-naphthyl ditelluride and 2.10 g (0.008 mole) triphenylphosphine in 40 ml of benzene was treated dropwise while stirring with a solution of 0.51 g (0.002 moles) of I_2 in 40 ml of benzene. After two hours a yellow precipitate was isolated and washed with benzene. The product (2.45 g, 95% yield) melted with decomposition at 155–160°. The adduct could be recrystallized from $CHCl_3$/benzene mixtures containing an equal amount of triphenylphosphine. Treatment with boiling ethanol, benzene, chloroform, acetone, and glacial acetic acid converted the adduct to the ditelluride.

The adduct (1.2 g, 0.0019 mole) was boiled in 50 ml of ethanol until dissolution was complete. On cooling, the ditelluride (0.30 g, 64% yield, m.p. 120–122°) separated in the form of red needles. Triphenylphosphine oxide was isolated from the filtered solution upon the addition of water.

IV. REACTIONS OF TELLURIDES, R—Te—R

Tellurides are easily prepared by reduction of bis(organo)tellurium dihalides (see Section VI). Such compounds among others containing the ^{125m}Te isotope have been obtained by decay of a variety of ^{125}Sb compounds (N-2). The tellurides undergo a variety of reactions. The organic groups, however, participate only if a functional group is present in the organic moiety at the time the tellurium–carbon bond is formed. The chemical changes take place preferentially on the tellurium atom or affect the tellurium–carbon linkages. The following reactions of tellurides have been investigated.

1. Reactions which increase the valency of the tellurium atom:

a. The addition of organic halides to form tris(organo)telluronium halides.

b. The addition of two halogen atoms in reactions with elemental halogen, sulfuryl halides, thionyl chloride, transition metal chlorides, vic. organic dibromides and tris(*p*-tolyl)bismuth difluoride to yield the respective bis-(organo)tellurium dihalides.

c. The oxidation of tellurides by nitric acid to bis(organo)tellurium dinitrates, by oxygen and hydrogen peroxide to bis(organo) telluroxides, to tellurinic acids or a mixture of oxides and acids and by benzoyl peroxide to bis(organo)tellurium dibenzoates.

d. The formation of addition compounds with transition metal salts, especially mercuric halides, with hydrochloric and picric acids and with trimethylaluminum, trimethylgallium, and boron tribromide.

2. Reactions affecting the tellurium–carbon bond:

a. The oxidation of tellurides to tellurinic acids by oxygen and hydrogen peroxide.

b. The exchange of organic groups.

c. The exchange of the tellurium atom for a sulfur or selenium atom.

d. The decomposition of tellurides, especially of those which contain a benzyl group and the cleavage of Te—C bonds by transition metal complexes and triethylmetal hydrides.

3. Reactions which modify the organic moiety.

Figure 3 summarizes the reactions of the tellurides.

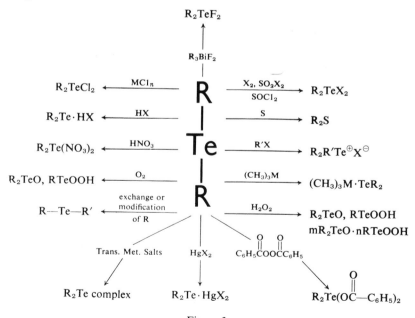

Figure 3

A. Reactions Increasing the Valency of the Tellurium Atom in the Tellurides

1. Tellurides and Organic Halides

In general, organic halides add easily to aromatic and aliphatic tellurides with the formation of tris(organo)telluronium halides.

$$R\text{—Te—}R' + R''X \longrightarrow \left[\begin{array}{c} R' \\ | \\ R\text{—Te}^{\oplus} \\ | \\ R'' \end{array} \right]^{+} X^{-}$$

The reactions are carried out by mixing the telluride and the halide. The combination proceeds quite vigorously with lower aliphatic tellurides and methyl iodide (B-3,C-1,C-2,M-9,V-2). In the case of aromatic and higher aliphatic tellurides, the mixture was heated so as to speed up the reaction (B-2,L-8,L-12,L-13,L-23,M-43). Several investigators used methanol or ether (B-2,B-12,L-12) as solvents. In many instances an excess of the organic halide was employed as the reaction medium. The telluronium halides precipitated as crystalline substances upon mixing of the components. Aromatic tellurides reacted more slowly and required days for complete conversion. Compounds containing a free carboxylic acid group or a menthyl group tended to separate as viscous, uncrystallizable products. Some of the telluronium salts were claimed to be amorphous.

Table XV lists all of the telluronium salts prepared from tellurides and gives their physical characteristics together with pertinent literature references.

Lederer reported that when p-tolyl telluride or phenyl telluride was permitted to react with excess methyl iodide, the telluronium salt which formed incorporated methyl iodide as solvate into the crystal (L-6,L-14). The solvate molecule was not held very strongly. In contact with ether methyl iodide was lost and crystals of the unsolvated salt remained in pure form. The diphenyl methyltelluronium iodide solvate lost methyl iodide upon heating to 60°. Bis(p-tolyl) carboxymethyltelluronium bromide loosely binds one additional molecule of bromoacetic acid, which can be removed by washing the solid with chloroform.

$$Ar_2Te + RI \longrightarrow [Ar_2RTe]^+I^- + RI \longrightarrow [Ar_2RTe \cdot RI]^+I^-$$

$$\underset{\underset{-RI}{\text{ether, CHCl}_3}}{\underbrace{\hspace{4cm}}}$$

Lederer (L-13) investigated the reaction of bis(p-methoxyphenyl) telluride with methyl iodide and reported the formation of a red-brown product. Addition of picric acid to a hot aqueous solution of this substance caused yellow needles to form. The analytical data for carbon and hydrogen correspond to p-methoxyphenyl dimethyltelluronium picrate. Morgan and Drew

TABLE XV

Tris(organo)telluronium Halides from Tellurides and Alkyl Halides

| R—Te—R' | | R'X | RR'R"Te⁺X⁻ | | |
R	R'		m.p.	Yield, %	Ref.
CH₃—	CH₃—	CH₃I	—	—	C-1–C-3, V-2
		Br—CH₂CO₂C₂H₅	137.5	—	B-12
		BrCH₂C(O)C₆H₅	90–91	—	B-12
		C₆H₅CH₂Cl[a]	121[a]	—	B-12
		C₂H₅I	92	—	B-6,M-9
	C₂H₅—	CH₃I	159 (dec.)	83	B-2
	C₄H₉—	CH₃I	94–95	—	H-4
		C₆H₅C(O)CH₂Br	87	—	B-2
		BrCH₂COOH	viscous	—	B-2
		BrCH₂CO₂C₂H₅	62	87	B-2
		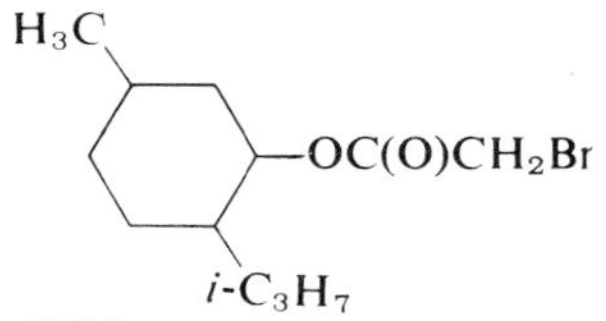 (cyclohexyl)—OC(O)CH₂Br	90	79	B-2
C₄H₉—	—CH₂CO₂C₂H₅	CH₃I	Not crystalline	—	B-2
C₄H₉—	(menthyl)—CH₂COO	BrCH₂C(O)C₆H₅	89	4	B-2
	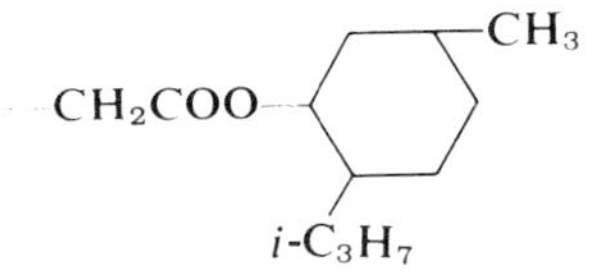	CH₃I	Not crystalline	—	B-2
		BrCH₂C(O)C₆H₅	106	65	B-2

(continued)

189

TABLE XV (*Continued*)

R	R'	R"X	m.p	Yield, %	Ref.
C_5H_{11}—	C_5H_{11}—	CH_3I	70	—	B-3
		$BrCH_2CO_2C_2H_5$	50	—	B-3
		$BrCH_2C(O)C_6H_5$	84	—	B-3
		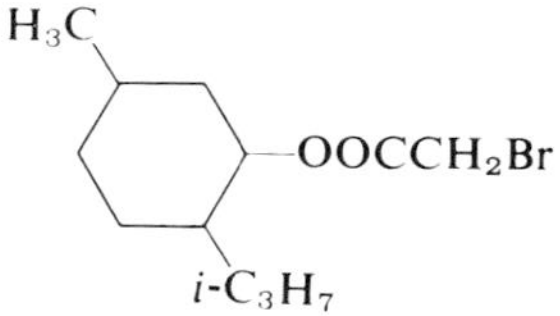 (menthyl $OOCCH_2Br$ structure: H_3C—cyclohexyl, i-C_3H_7)	Pasty	—	B-3
—CH_2CH_2CN	—CH_2CH_2CN	C_2H_5I	58–59	—	K-1
C_6H_5—	C_6H_5—	CH_3I	123–124	100	L-6
		C_2H_5I[b]	—	—	L-6
		$ICH_2CO_2C_2H_5$	110	36	L-12
		$BrCH_2CO_2C_2H_5$	63–64 (dec.)	100	L-7
		$BrCH_2CO_2CH_3$	99–100	91	L-7
		$BrCH_2CO_2H$	Not solid	—	L-7
		$BrCH(C_2H_5)CO_2H$	84–85[c]	Very small	L-12
		$BrCH(C_2H_5)CO_2C_2H_5$	142–143[d,e]	—	L-12
		$BrC(CH_3)_2CO_2C_2H_5$	~130[e]	—	L-12
		$CH_3CH(Br)CO_2CH_3$	~110	50	L-8
		$CH_3CH(Br)CO_2C_2H_5$	~125	43	L-8
		$CH_3CH(Br)CO_2C_3H_7$	~99[e]	9	L-8
		$(CH_3)_2C(Br)CO_2CH_3$	~116[e]	9	L-8
C_6H_5—	C_6H_5—	$C_6H_5CH_2Br$	90–91	—	L-10
		Br—CH_2CH_2—Br	259–263[f]	74	M-43
p-Tolyl	C_6H_5—	CH_3I	73–74	—	L-20
o-Tolyl		CH_3I	119–120	—	L-29

190

p-Methoxyphenyl	p-Methoxyphenyl	CH_3I	108–109	—	L-17[b]
p-Methoxyphenyl	CH_3—	C_2H_5I	168	100	R-6
o-Methoxyphenyl	o-Methoxyphenyl	CH_3I	124–125	—	L-27
m-Methoxyphenyl	m-Methoxyphenyl	CH_3I	Not crystalline	—	L-26
p-Ethoxyphenyl	p-Ethoxyphenyl	CH_3I	~69[g]	9	L-25
o-Ethoxyphenyl	o-Ethoxyphenyl	CH_3I	140	—	L-23
p-Tolyl	p-Tolyl	CH_3I	85–86 (dec.)	100	L-6
p-Tolyl		$BrCH_2CO_2H$	No pure product obtained	—	L-12
		$BrCH_2CO_2CH_3$	92–93	56	L-12
		$BrCH_2CO_2C_2H_5$	102–103	20	L-12
m-Tolyl	m-Tolyl	CH_3I	121–122	100	L-12
o-Tolyl	o-Tolyl	CH_3I	125–126	100	L-6
2,5-Dimethylphenyl	2,5-Dimethylphenyl	CH_3I	137	100	L-14
2,4,6-Trimethylphenyl	2,4,6-Trimethylphenyl	CH_3I	168–169[b]	—	L-15
[2,2'-biphenyldiyl structure]		CH_3I	102–104 (dec.)	87	H-4
1-Naphthyl	1-Naphthyl	CH_3I	146	10	L-24
[2-thienyl structure]	[2-methyl-thienyl structure]	CH_3I	106–115	—	K-8

[a] The chloride was not isolated. Addition of picric acid to the solution of $R_2R'TeCl$ gave the picrate, the melting point of which is reported.
[b] See discussion in text.
[c] No analytical data, identity doubtful.
[d] %C 1.3% too low.
[e] Amorphous.
[f] Product $[(C_6H_5)_2TeCH_2CH_2CH_2Te(C_6H_5)_2]^{2+}\,2Br^-$.
[g] Analysis for I 10% too low.

191

(M-28), however, claim that Lederer dealt with bis(*p*-methoxyphenyl) ditelluride (see Section IC-4).

Aromatic tellurides are less reactive than the aliphatics. This is borne out by the longer reaction times needed and in the failure to isolate telluronium salts from mixtures of diphenyl telluride with ethyl iodide (L-6) and butyl bromide (M-43). It has been reported that no reaction takes place between diphenyl telluride and butyl bromide when the components are heated for two days in a sealed tube at 180°. Bis(2,4-dimethylphenyl) telluride did not combine with methyl iodide (L-14). A mixture of bis(2,4,6-trimethylphenyl) telluride and methyl iodide deposited only a very small amount of a crystalline material melting at 168–169° after standing for four weeks at room temperature (L-15). Boiling the reaction mixture for two days and heating it in a sealed tube for 15 hr at 100° left the telluride unchanged. Although a detailed investigation of *o*-substituted tellurides is lacking, it seems to be reasonable to conclude that steric hindrance is responsible for the unreactive nature of the bis(2,4,6-trimethylphenyl) telluride. Aromatic tellurides bearing a substituent in the 2-position react normally. Only *o*-tolyl telluride gave no product with the methyl ester of bromoacetic acid (L-12). Especially interesting in this context would be 2,6-dialkylphenyl tellurides. No literature reports could be found, however, concerning these compounds.

Experimental. Dibutyl Methyltelluronium Iodide (B-2):   A mixture of 12.1 g dibutyl telluride and 8.0 g of methyl iodide became warm and turned rapidly into a solid mass, which was washed with dry ether and recrystallized from chloroform. Yield: 16 g (83%), m.p. 159° (dec.). The salt is soluble in water. Hellwinkel and Fahrbach (H-4) reported a m.p. of 94–95° (from ethanol) for this compound.

Carbethoxymethyl Dimethyltelluronium Bromide (B-12):   The telluronium salt was rapidly precipitated in the form of small white crystals upon the addition of an ethereal or ethanolic solution of ethyl bromoacetate to dimethyl telluride. The product was very soluble in water or ethanol, but insoluble in ether, m.p. 137.5°.

Diphenyl Methyltelluronium Iodide (L-6):   Diphenyl telluride (5 g) dissolved in 10–15 ml of methyl iodide was allowed to stand for 48 hr. After 10–15 min crystals began to form. After two days absolute ether was added; the separated crystals were then stored for 12 hr under ether. Yield: 7.15 g (100%), m.p. 123–124°.

If the crystals are dried at room temperature and the treatment with ether eliminated, a solvate having the composition $[(C_6H_5)_2CH_3Te \cdot CH_3I]^+I^-$ is obtained. The additional molecule of methyl iodide is lost at 60°.

2. Tellurides and Elemental Halogens, Sulfuryl Halides, Thionyl Chloride, and Metal Chlorides

Tellurides combine with elemental halogens and yield crystalline bis-(organo)tellurium dihalides.

$$R_2Te + X_2 \longrightarrow R_2TeX_2$$

This reaction has been frequently used to purify tellurides. The halides are recrystallized and reduced back to the telluride (L-10,L-29).

Dichlorides precipitate when a stream of dry chlorine is passed through a solution of the tellurides in ether (L-5,L-14–L-17,L-19,L-20,L-22–L-24, L-26,L-36), benzene (N-6,R-17), petroleum ether (J-2), CCl_4 (M-38), or $CHCl_3$ (M-26).

The bromides are synthesized by the addition of a slight excess or, more often, by the addition of an equimolar amount of bromine, sometimes dissolved in petroleum ether (J-2), $CHCl_3$ (M-32,R-11), benzene (R-11) or ether (W-3) to the tellurides in the same solvent. To the reaction mixture in benzene, petroleum ether is usually added to precipitate the dibromides (R-11) which are analytically pure in most of the cases. Recrystallization can be carried out from chloroform, chloroform–ethanol, benzene, CS_2, or benzene–petroleum ether mixtures.

The diiodides are formed by mixing stoichiometric amounts of the telluride and iodine in chloroform (P-7,R-11,T-1), CCl_4 (M-42), benzene (N-5,R-11, V-4), CS_2 (R-11), or petroleum ether (V-4,V-5), but the preferred solvent is diethyl ether (B-2,C-25,E-4,L-5,L-14–L-17,L-20,L-23,L-24,L-26,L-27,L-29, R-7).

Tables XVI and XVII list the compounds prepared, their melting points and yields together with references.

Bis(p-ethoxyphenyl)tellurium dichloride and dibromide are difficult to purify. Very pure tellurides were necessary as starting material to successfully synthesize these dihalides (L-25). The diiodide was found to be more stable.

A crystalline substance could not be isolated from the viscous products formed by addition of iodine dissolved in chloroform or CCl_4 to $(C_5H_{11})_2Te$ or C_5H_{11}—Te—$CH_2CO_2C_2H_5$ (B-3).

A mixture of bis(p-methoxyphenyl) telluride and polonide in benzene solution was converted to the dichlorides by elemental chlorine. The products were separated by paper chromatography (N-6). The p-tolyl derivatives, treated similarly in $CHCl_3$, gave the corresponding dichlorides (N-7). The diphenyl polonide was preferentially transformed into the dichloride upon stepwise addition of a solution of chlorine in benzene (N-4). Difficulties are reported in the purification of bis(p-tolyl)tellurium and polonium dibromides and diiodides prepared in the conventional manner (N-7). Benzyl naphthyl and benzyl p-methoxyphenyl tellurides cannot be converted to the respective

TABLE XVI

Bis(organo)tellurium Dihalides from Symmetric Tellurides and Elemental Halogen

R₂Te	R₂TeCl₂			R₂TeBr₂			R₂TeI₂		
R	m.p.	Yield, %	Ref.	m.p.	yield, %	Ref.	m.p.	Yield, %	Ref.
CH_3—							127–130		R-7
—CH_2COOH	160–161	—	M-26	105–106	—	K-1			
—CH_2CH_2CN									
C_4H_9—	160						61	70	B-2
C_6H_5—	160	—	L-10,L-41	204	100	K-5,L-29, R-7	237–238	—	L-5
p-Tolyl	166–167	—	A-2,J-2, L-5	198–200	—	G-6,W-3	238–239	—	E-4,R-7
				200–201	91	R-9,Z-1	218–219	—	L-5
m-Tolyl	131–132	—	L-16	201	100	L-13,R-7 J-2			
				203–204	—				
				165–166	—	L-16	164	—	L-16

Substituent	m.p.	Yield (%)	Ref.	m.p.	Yield (%)	Ref.	m.p.	Yield (%)	Ref.
o-Tolyl				182	—	Z-1, L-13	175–176	—	L-5
2,4-Dimethylphenyl	187–188	—	L-14	200–201	—	L-14	181–182	—	L-14
2,5-Dimethylphenyl	197–198	100	L-14	189–190	—	L-14	161–162	—	L-14
2,4,6-Trimethylphenyl	178–179	—	L-15	205–206	—	L-15	111	—	L-15
p-Bromophenyl				195	—	T-1	231	—	T-1
p-Methoxyphenyl	183–184	—	L-17	195–196	100	L-17	166–167	—	L-17
				198	—	T-1			
				190	—	M-32			
m-Methoxyphenyl	162–163	—	L-26	185–186	—	L-26	167–168	—	L-26
o-Methoxyphenyl	184–185	—	L-27	195–196	—	L-27	199–200	—	L-27
p-Ethoxyphenyl	125	—	L-25	127	—	M-38	134–135	—	L-25
o-Ethoxyphenyl	164–165	—	M-38	116–117	69	L-25	144	—	M-38
1-Naphthyl	265	—	L-23	183–184	—	L-23	214–215	—	L-23
2-Naphthyl		—	L-41	244(dec.)	—	L-41	184–186	100	L-24
				250–253	100	R-1	216–219	94.8	R-11
(thienyl structure with S)	185.5	—	K-7	190.5	—	K-7	125	—	K-7
2-Chloro-2-phenylvinyl							168–169	100	M-46

TABLE XVII

Bis(organo)tellurium Dihalides from Unsymmetric Tellurides and Elemental Halogen

R—Te—R'		RR'TeCl₂			RR'TeBr₂			RR'TeI₂		
R	R'	m.p.	Yield, %	Ref.	m.p.	Yield, %	Ref.	m.p.	Yield, %	Ref.
C₄H₉—	—CH₂CO₂C₂H₅	135–136	100	L-20				60	—	B-2
C₆H₅—	p-Tolyl	179–180	—	L-29	175–176	—	L-20	195	—	L-20
	o-Tolyl				154–155	—	L-29	172–173	—	L-29
	p-Methoxy-phenyl				148.5–149.5	100	R-11	166–167	100	R-11
	p-Phenoxy-phenyl				156.5–158	100	R-11	184.5	100	R-11
	1-Naphthyl				180.5–182.0	100	R-11	212	100	R-11
	2-Naphthyl				186–187	100	R-11	182	100	R-11
p-Methoxy-phenyl	CH₃—							109	—	R-7
	C₂H₅—				107–108	100	V-4	135	100	V-4
	1-Naphthyl				220–222	100	R-11	189	100	R-11
	2-Naphthyl				163–164	100	R-11	173–173.5	100	R-11

p-Phenoxy-	1-Naphthyl	189–190	100	R-11	183–184	100	R-11
phenyl	2-Naphthyl	146–147	100	R-11	204–206	100	R-11
p-Dimethyl-	p-Methoxy-				129–130	100	P-7
amino-	phenyl						
phenyl	p-Ethoxyphenyl				96–97 (dec.)	100	P-7
	p-Phenoxyphenyl				135–136 (dec.)	100	P-7
2-Naphthyl	C_2H_5—	141–142	100	V-5	131 (102)[a]	100	V-5
	Cyclohexyl	157	100	V-4	139	100	V-4
	1-Naphthyl	215–217	100	R-11	205	100	R-11
	$(C_6H_5)_2C$				174–176		M-42

[a] Compound exists in two polymorphic modifications.

197

dihalides by treatment with bromine or iodine. Cleavage of the tellurium–carbon bond occurred with elimination of the benzyl group. (For details see Section IV-B4).

Rheinboldt and Vicentini (R-11) introduced sulfuryl chloride and thionyl chloride as reagents for the conversion of tellurides into the dichlorides. They found these chlorides to be better suited than elemental chlorine. The reaction with thionyl chloride proceeded more slowly than with sulfuryl chloride, but the products were obtained as well-formed crystals. Petragnani (P-7) showed that sulfuryl bromide is equally suited for bromination reactions. The yields are reported to be quantitative. Table XVIII presents the compounds prepared in this way.

Nefedov and co-workers (N-5–N-7) synthesized difluorides of bis(1-naphthyl), bis(p-methoxyphenyl) and bis(p-tolyl) telluride and polonide by adding equivalent amounts of tris(p-tolyl)bismuth difluoride to the mixtures of telluride and polonide. The products were separated chromatographically.

The ease with which bis(organo)tellurium dihalides are formed, led to the investigation of the tellurides as dehalogenation agents. De Moura Campos and co-workers (M-43) investigated the reaction of vic-dibromides with aromatic tellurides and obtained the tellurium dibromides and the olefin in good yields. Ethylene bromide and diphenyl telluride, gently refluxed for 30 min in ethylene bromide as a solvent, formed the following telluronium salt in 75% yield.

$$Br-CH_2-CH_2-Br + 2 \; \langle\bigcirc\rangle-Te-\langle\bigcirc\rangle \longrightarrow$$

The telluronium salt converted to the diphenyltellurium dibromide and ethylene upon recrystallization from glacial acetic acid or vigorous refluxing in ethylene bromide. This finding suggests that the dehalogenation proceeds via the telluronium salt.

Metal chlorides, for example, $FeCl_3$, $CuCl_2$, and $HgCl_2$, when heated in a sealed tube in glacial acetic acid with bis(p-tolyl) telluride were reduced to the lower chlorides. The bis(p-tolyl)tellurium dichloride was isolated from these

Bis(organo)tellurium Dihalides from Tellurides and SO_2X_2 or SOX_2

R	R'	m.p.	Yield, %	Ref. SO_2Cl_2	Ref. $SOCl_2$	m.p.	Yield, %	Ref. SO_2Br_2
		RR'TeCl2				RR'TeBr2		
C6H5—	C6H5—	161–162	100	V-6				
	p-Methoxyphenyl	114–115	100	N-4[a]				
	p-Phenoxyphenyl	128.5–129.5	100	R-11	R-11			
	1-Naphthyl	202.5–203.5	100	P-9,R-11	R-11			
	2-Naphthyl	171.5–172.5	100	R-11,V-5	R-11			
p-Tolyl	p-Tolyl		100	N-7[a],V-6				
m-Tolyl	m-Tolyl			V-6				
p-Methoxyphenyl	C2H5—	80–81	100	V-4	V-4			
	p-Methoxyphenyl	182–184	100	P-6				
	1-Naphthyl	219–220	100	R-11	R-11			
	2-Naphthyl	129.5–130.5	100	R-11	R-11			
p-Ethoxyphenyl	p-Ethoxyphenyl	110–111	100	P-6				
	1-Naphthyl	193–194	100	P-9				
	2-Naphthyl	175–176	100	R-11				
p-Phenoxyphenyl	p-Phenoxyphenyl	138–139	100	R-11	R-11			
	1-Naphthyl			R-11	R-11			
	2-Naphthyl			R-11	R-11			
p-Dimethyl-aminophenyl	p-Methoxyphenyl	170–172	100	P-7		183–184 (dec.)		P-7
	p-Ethoxyphenyl	153–154		P-7		121–123 (dec.)		P-7
	p-Phenoxyphenyl	194–195 (dec.)		P-7		188–189 (dec.)		P-7
1-Naphthyl	1-Naphthyl	265–266	80.3	N-5[a],R-12				
2-Naphthyl	1-Naphthyl	232–233	100	R-11	R-11			
	2-Naphthyl	245–247	100	R-11	R-11			
	C2H5—	138–139	100	V-5				
	Cyclohexyl—	140–141	100	V-4,V-5	V-4			

[a] Mixture of telluride and polonide.

TABLE XIX

Tellurides as Dehalogenation Agents

R—Te—R'						
R	R'	Reagent	Solvent	$RR'TeX_2$ Yield, %	Te-free compound	Yield	Ref.
2-Naphthyl	p-Methoxy-phenyl	$C_2H_4Br_2$	$(CH_2)_2Br_2$[a]	66			M-43
1-Naphthyl	p-Methoxy-phenyl	$C_2H_4Br_2$	$(CH_2)_2Br_2$[a]	70			M-43
C_6H_5—	C_6H_5—	$C_2H_4Br_2$	$(CH_2)_2Br_2$[a]	86[b]			M-43
		$(C_6H_5CHBr)_2$	—[c]	85	C_6H_5—CH=CH—C_6H_5 (trans)	94	M-43
		1,2-Dibromo-cyclohexane	c-$C_6H_{10}Br_2$[a]	84			M-43
		Dibromocholesterol	Xylene[a]	77	Cholesterol	58	M-43
			—[d]	71	Cholesterol	93	M-43
		$C_6H_5(CHBr)_2COOH$	—[e]	74	C_6H_5—CH=CH—COOH	100	M-43
		CH_2=CH—CH_2—Br[f]	CH_2=CH—CH_2—Br[f]	Yield not reported	$(CH_2$=CH—$CH_2)_2$[g]		M-43

p-Tolyl	p-Tolyl	FeCl₃	Glac. AcOH[h]	83[l]	FeCl₂	83	M-47
		CuCl₂	Glac. AcOH[h]	83	CuCl	83	M-47
		HgCl₂	Glac. AcOH[h]	52	Hg₂Cl₂	35	M-47
		AgCl	Glac. AcOH[j]		No reaction		M-47
		FeCl₃ anhydrous	Dry toluene[a]	90	FeCl₂		M-47
		CuSO₄ hydrated	Glac. AcOH[k]		No reaction		M-47
		Fe₂(SO₄)₃	Glac. AcOH[k]	Elem. Te only	FeSO₄	73	M-47
		C₆H₅COCl	—[k]	86[l]			M-47

[a] Reflux temperature.
[b] m.p. 199–202°.
[c] Temperature kept at 150°.
[d] Temperature at 95°.
[e] Temperature at 100°.
[f] Temperature at 180°/3 hr.
[g] Olefin not identified.
[h] Temperature at 100°/10 hr and 140°/1hr.
[i] m.p. 170–172°.
[j] Temperature at 160°/168 hr.
[k] Temperature kept at 150° for 6 hr.
[l] Bis(p-tolyl)tellurium dibenzoate was also isolated, when the reaction mixture was in contact with air.

201

reactions in yields ranging from 52 to 90% (M-47,S-17). Silver chloride was reduced only slightly under the same conditions and most of the telluride was recovered unchanged. When hydrated ferric chloride was employed in this reaction with toluene as the solvent the anhydride of the bis(p-tolyl) tellurium hydroxychloride was obtained. Anhydrous $FeCl_3$ did not yield this anhydride.

$$2 \; FeCl_3 + R_2Te \longrightarrow R_2TeCl_2 + 2 \; FeCl_2$$
$$R_2TeCl_2 + H_2O \longrightarrow R_2Te(OH)Cl + HCl$$

$$2 \; R_2Te(OH)Cl \longrightarrow R_2Te \underset{O}{\overset{Cl \quad Cl}{\diagup \diagdown}} TeR_2 + H_2O$$

Ferric sulfate in glacial acetic acid was reduced by the telluride to ferrous sulfate. A crystalline tellurium compound was not formed in this reaction. The principal product was elemental tellurium.

Although allyl bromide (M-43) was transformed into 1,5-hexadiene by diphenyl telluride, benzoyl chloride was not converted into benzil in the reaction with bis(p-tolyl) telluride. Bis(p-tolyl)tellurium dichloride and dibenzoate were the products obtained. A more detailed discussion of this reaction is given in Section IV-A-3.

Table XIX summarizes the dehalogenation reactions of tellurides.

Experimental. Bis(2,5-dimethylphenyl)tellurium Dichloride (L-14): The telluride (5 g) was dissolved in ether and dry chlorine was passed into the solution. The white dichloride precipitated immediately. For purification the solid was dissolved in $CHCl_3$ and reprecipitated with CH_3OH. Yield: 6 g (100%), m.p. 197–198°. The compound is soluble in benzene, toluene, xylene, chloroform, and CS_2, slightly soluble in methanol and ethanol, and insoluble in petroleum ether.

Bis(2-naphthyl)tellurium Dibromide (R-11): The telluride (0.38 g) was dissolved in 10 ml of benzene. This solution was mixed with 0.16 g of bromine in 5 ml of benzene with ice water cooling. The addition of 20 ml of petroleum ether caused precipitation of the dibromide. The crude product was dissolved in $CHCl_3$. After addition of a small volume of petroleum ether light-yellow prisms separated on cooling. These melt at 253–255°.

p-Dimethylaminophenyl p-Methoxyphenyltellurium Diiodide (P-7): To the ice-cooled solution of the telluride (0.35 g, 0.001 mole) in $CHCl_3$ was added dropwise an equimolar quantity of iodine in $CHCl_3$. The red diiodide precipitated quantitatively upon addition of petroleum ether (50–70°). Recrystallized from benzene/petroleum ether the compound melted at 129–130° (dec.).

Bis(organo)tellurium Dichlorides from Tellurides and SO$_2$Cl$_2$ (R-11):   A rather concentrated solution of the telluride in benzene or chloroform was shaken with ice water cooling with a solution which contained a small excess of SO$_2$Cl$_2$ in the same solvent. After the SO$_2$ evolution had ceased, the product was precipitated quantitatively by the addition of petroleum ether.

Bis(p-tolyl)tellurium Dichloride (M-47):   The telluride (1.62 g, 0.005 moles) and FeCl$_3$ (1.62 g, 0.01 moles) and 2 ml of glacial acetic acid were heated in a sealed tube at 100° for 10 hr and then at 140° for one hour. The mixture was treated with water and the crystalline solid washed with petroleum ether. The dichloride (1.65 g, 83% yield) melted at 170–172°. The aqueous solution was titrated with K$_2$Cr$_2$O$_7$ showing an 83% reduction of FeCl$_3$.

3. The Oxidation of Tellurides by HNO$_3$, H$_2$O$_2$, O$_2$, and Benzoyl Peroxide

Tellurides, especially the aliphatic derivatives, are easily oxidized by a variety of oxidizing reagents. Nitric acid, ranging in concentration from dilute to 6N, reacts with tellurides on gentle heating with the formation of the bis(organo)tellurium dinitrate and evolution of nitric oxide (L-18).

$$3 \text{ R}_2\text{Te} + 8 \text{ HNO}_3 \longrightarrow 3 \text{ R}_2\text{Te(NO}_3)_2 + 2 \text{ NO} + 4 \text{ H}_2\text{O}$$

It is likely that the telluride is first oxidized to the oxide, R$_2$TeO, which afterward combines with an excess of nitric acid to form the dinitrate, but no attempt has been made to isolate these oxides from the reaction mixture.

Lederer (L-18) employing nitric acid (d 1.2) was not able to prepare the dinitrate of bis(p-tolyl) telluride, but isolated instead the hydroxynitrate R$_2$Te(OH)NO$_3$. The phenyl and o-tolyl compounds reacted as expected. The hydroxy nitrate was also obtained from butyl carbomenthoxymethyl telluride and 6N nitric acid and dibutyl telluride and dilute acid (B-2). Additional experimental data are needed to explain this behavior. Table XX summarizes the reactions of tellurides with nitric acid.

The oxidation of tellurides by oxygen should yield, as the primary product, the bis(organo) telluroxide.

$$2 \text{ R}_2\text{Te} + \text{O}_2 \longrightarrow 2 \text{ R}_2\text{TeO}$$

In the presence of water the oxide is readily hydrated and forms the dihydroxide, R$_2$Te(OH)$_2$. These simple substances, however, have been isolated in a few cases only (L-5,M-7,M-26,V-2). Analytical results are generally not given and the purity and exact composition of these compounds is doubtful.

H$_2$O$_2$ is reported to react similarly. Vernon (V-2) obtained a basic solution containing dimethyltellurium dihydroxide upon addition of enough hydrogen peroxide to dimethyl telluride to effect complete dissolution. A dihydroxide

TABLE XX
The Oxidation of Tellurides by HNO$_3$

| R—Te—R' | | Product[a] | m.p. | Yield, % | Reagent | Ref. |
R	R'					
CH$_3$—	CH$_3$—		142		HNO$_3$	M-48,V-3
					Mod. conc. HNO$_3$	H-2,W-6
					Mod. conc. HNO$_3$	H-2
C$_2$H$_5$—	C$_2$H$_5$—				HNO$_3$	M-7,W-4,W-5
C$_4$H$_9$—	C$_4$H$_9$—	R$_2$Te(OH)NO$_3$	84	75	Dil. HNO$_3$	B-2
C$_4$H$_9$—	—CH$_2$—CO$_2$—(4-CH$_3$, 2-i-C$_3$H$_7$-cyclohexyl)	RR'Te(OH)NO$_3$	42	100	6N HNO$_3$	B-2
C$_5$H$_{11}$—	C$_5$H$_{11}$—		40		Mod. conc. HNO$_3$	W-7
C$_6$H$_5$—	C$_6$H$_5$—		160		HNO$_3$ (D = 1.2)	L-18
p-Tolyl	p-Tolyl	R$_2$Te(OH)NO$_3$	237–238		HNO$_3$ (D = 1.2)	L-18
o-Tolyl	o-Tolyl				HNO$_3$ (D = 1.2)	L-18

[a] The product obtained has the formula RR'Te(NO$_3$)$_2$ unless otherwise stated.

was also formed by the oxidation of dibutyl telluride with alkaline hydrogen peroxide. This substance was recrystallized from ethanol. It lost water at 120°. The analytical data are in good agreement with the proposed structure (B-2).

Two papers by Balfe and co-workers (B-2,B-3) deal exclusively with the oxidation of tellurides by oxygen and hydrogen peroxide. Generally, the products obtained have the formula $mR_2TeO \cdot nRTeOOH$ when neutral solutions of the tellurides were oxidized. The compounds actually isolated and the values of m and n can be found in Tables XXI and XXII. However, only alkanetellurinic acids were formed from the pentyl tellurides and from butyl carbomenthoxymethyl telluride and atmospheric oxygen as given in Table XXI (B-2,B-3).

The formation of tellurinic acids, RTeOOH, in these oxidation processes, was attributed by Balfe (B-2) to the instability of the primary oxidation product. Telluroxide, R_2TeO, in its enol-form for example,

$$\begin{array}{c} C_3H_7\!\!-\!\!CH \\ \diagdown \\ Te\!\!-\!\!OH \\ \diagup \\ C_4H_9 \end{array}$$

is further oxidized to the tellurinic acid and a carboxylic acid. In fact, butyric acid was isolated in the oxidation of dibutyl telluride with hydrogen peroxide. The stability of the $mR_2TeO \cdot nRTeOOH$ compounds is attributed to the association of the oxide with the tellurinic acid, possibly through a Te—O—Te linkage or a hydrogen bond, which would prevent enolization. There is, however, no firm experimental evidence for the existence of the oxide in the enol form.

It is interesting that such a mixed compound could not be isolated from the reaction of bis(carbomenthoxymethyl) telluride and hydrogen peroxide (B-2). The substance obtained oxidized hydriodic acid and rapidly decolorized $KMnO_4$. The analytical data agree with the values calculated for a hydrated tellurone $R_2TeO_2 \cdot H_2O$. However, Balfe preferred to formulate this product as the hydrogen peroxide adduct of the oxide, $R_2Te\diagup^{OH}_{}\diagdown_{OOH}$. In such a compound enolization is not possible and therefore further oxidation to the tellurinic acid cannot occur.

When tellurides were treated with oxygen in the presence of hydrohalic acids, the dihydroxides initially formed reacted with the acid and gave bis-(organo)tellurium dihalides in fair yields (L-5,M-47).

Vernon (V-2) reported that acidic solutions of $KMnO_4$ absorb and oxidize dimethyl telluride. Concentrated H_2SO_4 acquires a delicate pink color on bubbling dimethyl telluride through this acid.

TABLE XXI
Reaction of Tellurides with Oxygen

R	R'	Product	m.p.	Yield, %	Reagent	Ref.
CH_3-	CH_3-	$(CH_3)_2Te(OH)_2$			Air/H_2O	V-2
C_2H_5-	C_2H_5-	White substance			Air	W-4
		$(C_2H_5)_2TeO$			Air	M-7
$-CH_2COOH$	$-CH_2COOH$	Amorphous oxide			Air	M-26
C_4H_9-	C_4H_9-	$(C_4H_9)_2TeO \cdot 3C_4H_9TeOOH$	180 (dec.)	13	Air/6 months	B-2
	$-CH_2CO_2C_2H_5$	$(C_4H_9)(C_2H_5O_2CCH_2)TeO \cdot 2C_4H_9TeOOH$	180 (dec.)	26	Air/2 months	B-2
	(menthyl structure: cyclohexane ring bearing H_3C, $i\text{-}C_3H_7$, and $-O_2C \cdot CH_2-$)	C_4H_9TeOOH	250 (dec.)	16	Air/8 months	B-2

R—Te—R′

C$_5$H$_{11}$—	C$_5$H$_{11}$TeOOH	200–220 (dec.)		Air	B-3
—CH$_2$CO$_2$C$_2$H$_5$	C$_5$H$_{11}$TeOOH	200–220 (dec.)		Air	B-3
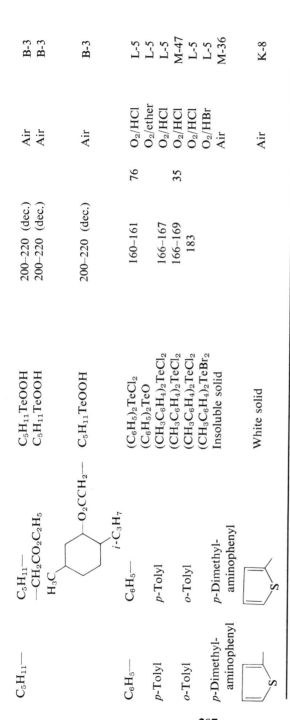	C$_5$H$_{11}$TeOOH	200–220 (dec.)		Air	B-3
C$_6$H$_5$—	(C$_6$H$_5$)$_2$TeCl$_2$	160–161	76	O$_2$/HCl	L-5
	(C$_6$H$_5$)$_2$TeO			O$_2$/ether	L-5
p-Tolyl	(CH$_3$C$_6$H$_4$)$_2$TeCl$_2$	166–167		O$_2$/HCl	L-5
o-Tolyl	(CH$_3$C$_6$H$_4$)$_2$TeCl$_2$	166–169	35	O$_2$/HCl	M-47
	(CH$_3$C$_6$H$_4$)$_2$TeCl$_2$	183		O$_2$/HCl	L-5
	(CH$_3$C$_6$H$_4$)$_2$TeBr$_2$			O$_2$/HBr	L-5
p-Dimethyl-aminophenyl	Insoluble solid			Air	M-36
p-Dimethyl-aminophenyl	White solid			Air	K-8

Structure label: H$_3$C — (cyclohexane) —O$_2$CCH$_2$— with i-C$_3$H$_7$

Thiophene structures (lower left entries).

TABLE XXII

The Reaction of Tellurides with H_2O_2

R—Te—R'						
R	R'	Product	m.p.	Yield, %	Reagent	Ref.
CH_3—	CH_3—	$R_2Te(OH)_2$			H_2O_2	V-2
		$R_3Te^+RTeOO^-$ [a]			Excess H_2O_2	V-2
		R_2TeO_2			Excess H_2O_2 boiling	V-2
C_4H_9—		$2\ R_2TeO\cdot RTeOOH$	180 (dec.)	61	H_2O_2 (30%)	B-2
		$RTe(OH)_2$		65	Alkaline H_2O_2 (20%)	B-2
C_4H_9—	$—CH_2CO_2C_2H_5$	$RR'TeO\cdot 2\ RTeOOH$	180 (dec.)		H_2O_2 (30%)	B-2
$—CH_2—CO_2$ (menthyl: CH_3, $i\text{-}C_3H_7$)	C_4H_9—	$R'TeOOH$		30	$H_2O_2\cdot$ urea	B-2
		$RR'TeO\cdot R'TeOOH$	220 (dec.)		H_2O_2 (30%)	B-2
$—CH_2CO_2$ (menthyl: CH_3, $i\text{-}C_3H_7$)		$R_2Te(OH)(OOH)$	90		H_2O_2 (30%)	B-2
C_5H_{11}—		$3\ R_2TeO\cdot RTeOOH$	144 (152) (dec.)	36	H_2O_2	B-3
$—CH_2CO_2C_2H_5$	(methylthiophene)	$RR'TeO\cdot 2\ RTeOOH$	200 (dec.)	100	H_2O_2	B-3
(methylthiophene)		White solid			3% H_2O_2	K-8

[a] $(CH_3)_2Te$ boiled with excess H_2O_2 for a short time produces "Vernon's β-base." See Section VI.

Benzoyl peroxide transformed pentyl carbethoxymethyl and butyl carbo-menthoxymethyl telluride into the dibenzoate upon refluxing the reactants in $CHCl_3$ for one hour. The dibenzoates melt at 77–78° (B-3) and 133° (87% yield) (B-2), respectively.

De Moura Campos (M-47), in an attempt to obtain benzil from benzoyl chloride and bis(p-tolyl) telluride, isolated bis(p-tolyl)tellurium dichloride and dibenzoate (m.p. 232–235°) in 86 and 46% yields, respectively. No benzil was ever detected. The following reaction sequence satisfactorily explains the experimental results (M-47).

$$2 C_6H_5COCl + 2 H_2O \longrightarrow 2 HCl + 2 C_6H_5CO_2H$$
$$2 R_2Te + O_2 \longrightarrow 2 R_2TeO$$
$$R_2TeO + 2 HCl \longrightarrow R_2TeCl_2 + H_2O$$
$$R_2TeO + 2 C_6H_5CO_2H \longrightarrow R_2Te(C_6H_5CO_2)_2 + H_2O$$
$$\overline{2 C_6H_5COCl + 2 R_2Te + O_2 \longrightarrow R_2TeCl_2 + R_2Te(C_6H_5CO_2)_2}$$

It should be possible to obtain the observed products with only a catalytic amount of water present, since it is regenerated in the formation of the tetra-valent tellurium compounds. The primary reaction involves the oxidation of the telluride to the bis(organo) telluroxide, or, in the presence of sufficient water, to the dihydroxide. This assumption is supported by the observation that the telluride was recovered unchanged when the reaction was performed under nitrogen. These substances then react with hydrogen chloride formed by the hydrolysis of benzoyl chloride.

Additional proof for the proposed reaction sequence was obtained by isolating the dibenzoate (4.1% yield) from a reaction mixture containing the telluride and benzoic acid.

Experimental. Diphenyltellurium Dinitrate (L-18): Diphenyl telluride (5 g) was covered with 20 ml of halogen-free nitric acid ($D = 1.2$). The oxidation began without heating. After heating for 10 min on a water bath, excess nitric acid was removed in a vacuum desiccator over KOH. Recrystallized from $CHCl_3$ the compound melted at 160°.

Diphenyltellurium Dichloride (L-5): To a solution of the telluride (10 g) in 300 ml of ether was added 100 ml of concentrated HCl. Oxygen was bubbled through the solution. After a few minutes crystals separated and the reaction was complete in three to four hours. Following recrystallization from absolute ethanol, the dichloride melted at 160–161° (100% yield).

$2(C_4H_9)_2TeO \cdot C_4H_9TeOOH$ (B-2): To a cooled solution of the telluride (5.8 g) in 15 ml of acetone was added 30% hydrogen peroxide until precipitation ceased. The white solid was washed with water and dried *in vacuo*. The substance decomposed at 180°. The mother liquor contained butyric acid.

Butyl Carbomenthoxymethyltellurium Dibenzoate (B-2): Benzoyl peroxide (2.4 g) in 25 ml of $CHCl_3$ was added to the telluride (3.8 g) in 25 ml of $CHCl_3$. The mixture was heated under reflux for one hour. The $CHCl_3$ was then evaporated at room temperature. The product (5.4 g, 87% yield) washed with water and dried *in vacuo* melted at 133°.

4. Addition Compounds Formed by Tellurides

Most of the mercuric halide adducts of the tellurides were prepared by Lederer. These crystalline complexes found application in the determination and isolation of dimethyl telluride formed by the action of various organisms, for example, *Penicillium brevicaule*, *P. chrysogenum*, and a mold related to *P. notatum*, upon inorganic tellurites (B-12). The tellurides arising from the reaction between organic free radicals and tellurium mirrors were also identified as the mercuric halide adducts (G-9).

The preparation of these substances is very simple. The tellurides dissolved in ether were shaken with an aqueous solution of the mercuric halide, or solutions of the components in acetone or ethanol were combined. Most of the adducts are insoluble in these solvents and precipitate as crystalline solids or amorphous powders (see Table XXIII).

$$R_2Te + HgX_2 \longrightarrow R_2Te \cdot HgX_2$$

The products were recrystallized from glacial acetic acid, ethanol, acetone, or benzene. The melting points of the pure products are not very sharp. Softening was observed 5–10 degrees before the melting points were reached. A number of derivatives slowly converted to viscous oils on heating (L-16, L-17,L-20,L-23). While the dimethyl telluride adducts are crystalline, the $HgCl_2$ and HgI_2 derivatives of dipentyl telluride, the only other aliphatic telluride investigated, could not be obtained in crystalline form. Bis(*p*-methoxyphenyl) telluride reacted with $HgCl_2$ and $HgBr_2$ (L-11). The product obtained is a brown amorphous substance, which is reported to have no melting point.

Bis(2,4,6-trimethylphenyl) telluride did not combine with mercuric halides (L-15). Lederer (L-11) was unable to prepare adducts between tellurides and basic mercuric cyanide, thiocyanate and sulfate. Only the basic mercuric nitrate, $Hg(OH)NO_3$, when dissolved in water and shaken with an ethereal solution of bis(*o*-tolyl) telluride gave a white precipitate of the composition $R_2Te \cdot Hg(OH)NO_3$ which melts at 98–99°. The corresponding diphenyl telluride compound was isolated as an oil.

Other metal halides which have been found to form complexes with tellurides are AgI, CdI_2 (C-24), $PtCl_2$ (F-13,F-14,G-6,L-9) and $AuCl_3$ (G-6,L-9). Coates (C-24) prepared the adducts $[(CH_3)_2Te]_2 \cdot AgI$ (78% yield) and $(CH_3)_2Te \cdot (AgI)_2$ (88% yield) as white solids melting at 73–74° and 137–138°

Mercuric Halide Adducts of Tellurides, RR'Te·HgX_2

R	R'	X = Cl m.p.	X = Cl Ref.	X = Br m.p.	X = Br Ref.	X = I m.p.	X = I Ref.
CH_3—	CH_3—	179 (dec.)	B-12,C-5	160-161	C-5	108	C-5,G-9
C_5H_{11}—	C_5H_{11}—	132	B-3[a]	88	B-3	89-90	B-3[b]
C_6H_5—	CH_3—		G-9	124-125	G-9	146[e]	G-9
C_6H_5—	C_6H_5—	160-161[c]	G-6[d],L-9	148	L-11	133-134	G-9,L-11
	o-Tolyl	91[f]	L-20	54	L-20		L-29
	p-Tolyl					74	L-20
p-Tolyl	p-Tolyl	135-136[g]	L-9	85	L-11	165	L-11
m-Tolyl	m-Tolyl	116-117	L-16	53[h]	L-16		L-16[i]
o-Tolyl	o-Tolyl	212	L-9	199-200	L-11	142-143	L-11
2,4-Dimethylphenyl	2,4-Dimethylphenyl	106	L-14	99	L-14	107-108	L-14
2,5-Dimethylphenyl	2,5-Dimethylphenyl	179-180	L-14	169-170	L-14	166-167	L-14
p-Methoxyphenyl	p-Methoxyphenyl	90[f]	L-11[j],L-17	77-78	L-11[j],L-17	63	L-17
m-Methoxyphenyl	m-Methoxyphenyl	89	L-26	114-115 (dec.)	L-26	122-123	L-26
o-Methoxyphenyl	o-Methoxyphenyl	143-144	L-27	84	L-27	80-81	L-27
p-Ethoxyphenyl	p-Ethoxyphenyl	150-151	L-25	155-156	L-25	123-124	L-25
o-Ethoxyphenyl	o-Ethoxyphenyl	174-175	L-23	162-163	L-23	90	L-23
1-Naphthyl	1-Naphthyl	187-188	L-24	178-179	L-24	152-153 (dec.)	L-24
Thienyl	Thienyl	Dec.	K-8				

[a] No pure product obtained.
[b] Viscous product.
[c] From glacial AcOH.
[d] m.p. reported 85–90 (dec.).
[e] From EtOH.
[f] Amorphous.
[g] $Ar_2Te \cdot HgCl_2 \cdot 6$ EtOH.
[h] Softens at 53°.
[i] No definite m.p.
[j] Starting materials not well defined. See Section I-B-3.

(dec.), respectively, by mixing the telluride in acetone solution with AgI in nearly saturated KI solution in the required stoichiometry. An acetone solution of $[(CH_3)_2Te]_2AgI$ gave a precipitate of AgI on addition of $AgNO_3$, suggesting a complex of the type $[(CH_3)_2Te{\rightarrow}Ag{\leftarrow}Te(CH_3)_2]^+I^-$. Cadmium iodide also combined with this telluride. The adduct, however, was not analyzed (C-24).

Diphenyl telluride was observed to react with platinum(II) chloride and gold chloride. The latter adduct was very unstable and decomposed, even in absence of light or oxygen (L-9). Giua and Cherchi (G-6), however, isolated a diphenyl telluride–$AuCl_3$ complex by shaking an ethereal solution of diphenyl telluride with a 10% aqueous solution of $AuCl_3$. The gray solid melted at 154–156°. Complexes between Pt(II) and diphenyl and diethyl telluride were prepared by Jensen (J-1). Fritzman (F-16,F-17) obtained a dibenzyl telluride adduct.

Chatt and co-workers (C-12–C-20) in their investigations of inductive and mesomeric effects in platinum(II) and palladium(II) complexes prepared three types of compounds with the general formulas L_2MCl_2, $L_2M_2Cl_4$ and L,amMCl_2, where M represents Pt(II) or Pd(II), am piperidine or p-toluidine, and L an aliphatic telluride. The following reaction scheme outlines the preparative steps.

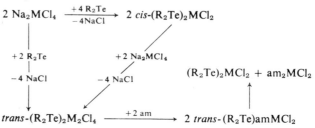

The complexes prepared are listed in Table XXIV.

The mononuclear complexes, synthesized by shaking two equivalents of the ligand with one equivalent of Na_2PtCl_4 in ethanol or $(NH_4)_2PdCl_4$ in aqueous solution at room temperature, isomerize spontaneously to the chloro-bridged dimers. It was therefore possible to prepare $L_2M_2Cl_4$ directly from an equimolar mixture of the ligand and Na_2MCl_4 (C-13,C-17). The bridged complexes reacted with amines with formation of the mixed complexes trans-L,amMCl_2. These complexes, however, disproportionate spontaneously into L_2MCl_2 and am$_2MCl_2$ (C-14,C-18). Only the compound trans-$[(C_2H_5)_2Te$, piperidine, $PtCl_2]$ was stable enough to be obtained pure (C-14). The solubility of this product in nonpolar solvents was taken as indicative of the trans-configuration. The existence of the complexes $[(C_3H_7)_2Te$, p-toluidine, $PtCl_2]$ and $[(C_2H_5)_2Te$, piperidine, $PdCl_2]$ in CCl_4 solution was demonstrated by the use of infrared spectroscopy using the different locations of the N—H

TABLE XXIV

Transition Metal Complexes with Tellurides, R_2Te

Complex	Method	Solvent	m.p.	Yield	Ref.
cis-[(C₂H₅)₂Te]₂PtCl₂	K₂PtCl₄ + 2 R₂Te	Ethanol or ethanol/H₂O	—	—	C-13,J-1
cis-[(C₃H₇)₂Te]₂PtCl₂	Na₂PtCl₄ + 2 R₂Te	ethanol/H₂O	—	—	C-13
cis-[(C₆H₅)₂Te]₂PtCl₂	K₂PtCl₄ + 2 R₂Te	H₂O/C₂H₅OH	~200° (dec.)	—	J-1
cis-[(C₆H₅CH₂)₂Te]₂PtCl₂	[(C₃H₇)₃NH]₂PtCl₄ + 2 R₂Te	C₂H₅OH	—	—	F-17
	K₂PtCl₄ + R₂Te	H₂O/C₂H₅OH			
cis-[(C₂H₅)₂Te]₂PdCl₂	(NH₄)₂PdCl₄ + R₂Te	H₂O	98.5-99.0	28	C-17
trans-(C₂H₅)₂Te, piperidine, PtCl₂	trans-(R₂Te)₂Pt₂Cl₄ + piperidine	Acetone	60.5-61.5	30	C-14
trans-[(C₂H₅)₂Te]₂Pt₂Cl₄	cis-(R₂Te)₂PtCl₂ + Na₂PtCl₄	C₂H₅OH	142 (dec.)	40.5	C-13
trans-[(C₃H₇)₂Te]₂Pt₂Cl₄	R₂Te + Na₂PtCl₄	C₂H₅OH	120-131 (dec.)	79	C-13
trans-[(C₂H₅)₂Te]₂Pd₂Cl₄	(R₂Te₂)PdCl₂ + Na₂PdCl₄	C₂H₅OH	110-125 (dec.)	23	C-17
	R₂Te + Na₂PdCl₄	C₂H₅OH	110-125 (dec.)	70	C-17
trans-[(C₃H₇)₂Te]₂Pd₂Cl₄	R₂Te + Na₂PdCl₄	C₂H₅OH	131.5-132.0 (dec.)	30	C-17
[Mn(CO)₄TeC₆H₅]₂	[Mn(CO)₅]₂ + 2 Te(C₆H₅)₂	p-Xylol	—	—	H-6
{Mn(CO)₄[Te(C₆H₅)₂]₂}⁺ [Cr(SCN)₄(NH₃)₂]⁻	Mn(CO)₃(TeR₂)₂Cl + CO		85 (dec.)	35	K-9
{Mn(CO)₃[Te(C₆H₅)₂]₂}Br	Mn(CO)₅Br + TeR₂	Et₂O/36°	—	—	H-6
{Mn(CO)₃[Te(C₆H₅)₂]₂}I	Mn(CO)₅I + TeR₂	Et₂O/36°	—	—	H-6
{Mn(CO)₃[Te(C₆H₅)₂]₂}Cl	Mn(CO)₅Cl + 2 TeR₂	Et₂O/36°	—	—	H-6
{Mn(NO)₃[Te(C₆H₅)₂]}	2 Mn(CO)₃(TeR₂)Cl + 6 NO	Benzene 20°	Not pure	—	H-6
Fe(CO)₄Te(C₆H₅)₂	[Fe(CO)₄]₃ + 3 TeR₂	Cyclohexane 80°	—	~25	H-6
Fe(CO)₃Te(C₆H₅)₂Br₂	Fe(CO)₄Br₂ + TeR₂	CH₂Cl₂/−38°	—	—	H-6
Fe(CO)₃Te(C₆H₅)₂I₂	Fe(CO)₄I₂ + R₂Te	Et₂O/10°	—	—	H-6
Fe(NO)₂COTe(C₆H₅)₂	Fe(NO)₂(CO)₂ + TeR₂	Ether/25°	—	—	H-6
[Co₂(CO)₅Te]ₙ	[Co(CO)₄]₂ + TeR₂	Benzene or acetone/20°	—	—	H-6

stretching frequency bands in the mixed and simple complexes (C-16,C-18)
The platinum complexes $L_2Pt_2Cl_4$ are more stable than the corresponding
palladium complexes. The relative stability of telluride-containing complexes
within a series having ligands drived from a group VI element is shown by the
following series:

$$L_2Pt_2Cl_4:R_2O \lll R_2S \gg R_2Se < R_2Te \quad (C\text{-}16,C\text{-}17)$$
$$L_2Pd_2Cl_4:R_2S > R_2Se > R_2Te \quad (C\text{-}17)$$

From a study of the N—H stretching frequencies in complexes [L,amPtCl$_2$]
it was concluded that the telluride ligand influences the N—H bond mainly
through an inductive effect, back-bonding from the metal to the ligand being
unimportant (C-16).

Hieber and Kruck (H-6) and Kruck and Hoefler (K-9) investigated the
reaction of diphenyl telluride with transition metal carbonyls. Bis(pentacar-
bonylmanganese) reacted only slowly with the telluride at 130° in *p*-xylene
forming di-μ-phenyltelluro-bis(tetracarbonylmanganese) (H-6).

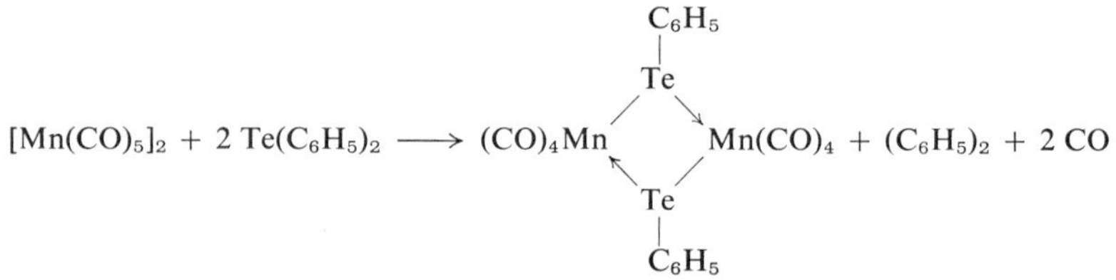

$$[Mn(CO)_5]_2 + 2\ Te(C_6H_5)_2 \longrightarrow (CO)_4Mn \overset{\displaystyle C_6H_5}{\underset{\displaystyle C_6H_5}{\overset{\displaystyle |}{\underset{\displaystyle |}{\overset{\displaystyle Te}{\underset{\displaystyle Te}{\diagup\diagdown}}}}}} Mn(CO)_4 + (C_6H_5)_2 + 2\ CO$$

One of the tellurium–carbon bonds is broken during this reaction. Molecular
mass determinations in benzene and the observed diamagnetism support the
dinuclear structure of the complex.

Halopentacarbonylmanganese was found to react much more easily than
the pentacarbonyl. In boiling diethyl ether two CO groups were exchanged.

$$Mn(CO)_5X + 2\ Te(C_6H_5)_2 \longrightarrow Mn(CO)_3[Te(C_6H_5)_2]_2X + 2\ CO$$
$$(X = Cl, Br, I)$$

The reactivity of $Mn(CO)_5X$ decreases in the sequence $Cl > Br > I$. These
compounds are non electrolytes, thermally stable, insensitive to moisture and
atmospheric oxygen, and diamagnetic. The telluride ligand causes color-
deepening. Infrared spectroscopic investigations did not allow for conclusions
to be drawn concerning the positions of the ligands in the complex. The strong
trans-effect of the CO group, however, makes the following structure plausible.

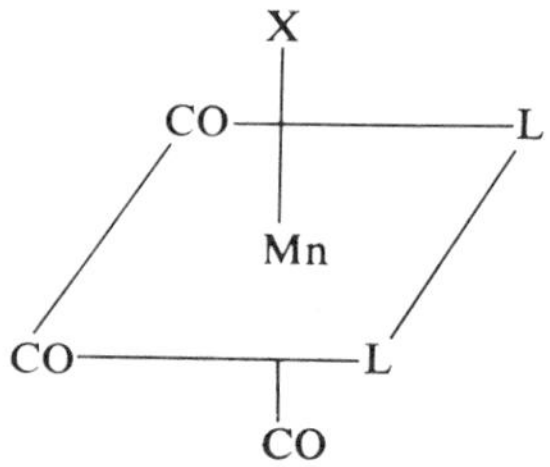

(L = diphenyl telluride)

This complex ($X = Cl$) reacted with NO and formed $Mn(NO)_3[Te(C_6H_5)_2]Cl$, which, because of similar solubility behavior of reactants and product, could not be obtained in pure form.

$$2 Mn(CO)_3[Te(C_6H_5)_2]_2Cl + 6 NO \longrightarrow$$
$$2 Mn(NO)_3Te(C_6H_5)_2Cl + 6 CO + Te(C_6H_5)_2 + (C_6H_5)_2TeCl_2$$

It was not possible to isolate monosubstituted complexes $Mn(CO)_4LX$ or compounds in which more than two CO ligands were replaced by diphenyl telluride.

Di-n-butyl telluride was recovered unchanged after boiling with $Mn(CO)_5Cl$ in ether. The manganese complex, however, dimerized quantitatively with loss of one CO ligand. The following reaction sequence is postulated.

$$2 Mn(CO)_5Cl + 2 Te(C_4H_9)_2 \rightleftharpoons 2 Mn(CO)_4Te(C_4H_9)_2Cl + 2 CO$$
$$2 Mn(CO)_4Te(C_4H_9)_2Cl \longrightarrow [Mn(CO)_4Cl]_2 + Te(C_4H_9)_2$$

This catalytic conversion of the mononuclear into the dinuclear species is believed to be caused by the instability of the manganese–dialkyl telluride complex. While aryl telluride complexes are stabilized by d_π–d_π backbonding made possible by the electron-withdrawing phenyl groups, the alkyl groups in the dialkyl telluride increase the electron density on the tellurium atom thus preventing or decreasing the d_π–d_π interactions and weakening the Mn—Te bond to such an extent as to give the dialkyl telluride complex only transitory existence.

Tris(irontetracarbonyl) and diphenyl telluride in boiling cyclohexane yielded $Fe(CO)_4Te(C_6H_5)_2$ with a trigonal-bipyramidal structure.

$$[Fe(CO)_4]_3 + 3 Te(C_6H_5)_2 \longrightarrow 3 Fe(CO)_4Te(C_6H_5)_2$$

Diphenyl telluride has been reported to replace only one CO ligand in $Fe(CO)_4X_2$ at room temperature.

$$Fe(CO)_4X_2 + Te(C_6H_5)_2 \longrightarrow Fe(CO)_3Te(C_6H_5)_2X_2 + CO$$
$$(X = Br, I)$$

The positions of the ligands are uncertain. Dipole moment measurements, however, have confirmed that the two iodine atoms in $Fe(CO)_4I_2$ are in *cis*-positions. From this fact and the *trans*-effect of the CO group the following structure was proposed for the complexes $Fe(CO)_3Te(C_6H_5)_2X_2$:

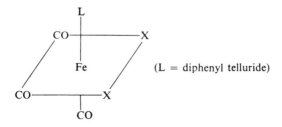

(L = diphenyl telluride)

Fe(NO)$_2$(CO)$_2$ gave, with diphenyl telluride in ether, at room temperature the complex Fe(NO)$_2$(CO)Te(C$_6$H$_5$)$_2$:

$$Fe(NO)_2(CO)_2 + Te(C_6H_5)_2 \longrightarrow Fe(NO)_2(CO)Te(C_6H_5)_2 + CO$$

The complex containing the telluride has greater stability than Fe(NO)$_2$(CO)$_2$.

Both Te—C bonds in diphenyl telluride were cleaved in the reaction with bis(tetracarbonylcobalt) forming the paramagnetic compound [Co$_2$Te(CO)$_5$]. Chromium carbonyl did not react with the telluride when kept at 125° for several days. Diphenyl telluride heated with nickel carbonyl gave only elemental nickel and tellurium.

The C—O and N—O stretching frequencies, magnetic susceptibilities and dipole moments of the complexes prepared are tabulated (H-6). The CO exchange rate of the complexes Mn(CO)$_3$[Te(C$_6$H$_5$)$_2$]$_2$X (X = Cl, Br, I) was determined in experiments with ^{14}CO in benzene at 30° and ratios of 15:7:1 were found (H-5) for the halogens in that order.

The cationic complex {Mn(CO)$_4$[Te(C$_6$H$_5$)$_2$]$_2$}$^+$[Cr(SCN)$_4$(NH$_3$)$_2$]$^-$ was synthesized from Mn(CO)$_3$[Te(C$_6$H$_5$)$_2$]Cl and CO in benzene solution in the presence of AlCl$_3$ (K-9).

One of the lone electron pairs of the tellurium atom in dimethyl telluride was found to form a coordinate-covalent bond to trimethylgallium and trimethylaluminum (C-24).

$$(CH_3)_2Te + (CH_3)_3M \longrightarrow (CH_3)_2Te \longrightarrow M(CH_3)_3$$
$$(M = Al, Ga)$$

The gallium adduct had a melting point of −32.5 to −32.0° and boiled at 122°, This compound is 90% dissociated in the vapor phase at 100°. The heat of the dissociation was calculated to be ∼8 kcal/mole from a plot of the equilibrium constant log K_p versus T^{-1}. The trimethylaluminum adduct, which boiled at 146°, dissociated almost completely at temperatures between 100 and 150° making the measurement of the heat of dissociation impossible. The adduct (CH$_3$)$_2$Te·BBr$_3$ was prepared by Chen and George (C-21).

Tellurides are reported to give complexes with acidic substances. Lederer (L-3) mentioned that diphenyl telluride combined with HCl to give a compound (C$_6$H$_5$)$_2$Te·HCl, which melts at 233–234°. Hellwinkel and Fahrbach (H-4) prepared the picric acid adduct of biphenylylene telluride upon addition of the acid to the ethanolic solution of the product obtained from trimethyltelluronium iodide and 2,2-dilithiobiphenyl. p-Methoxyphenyl methyltelluride also formed a picric acid adduct (M-28).

The transition metal complexes formed by tellurides are listed in Table XXIV.

Experimental. Bis(o-phenoxyphenyl) Telluride Mercuric Chloride Adduct (L-23):   HgCl$_2$ (3 g) was dissolved in hot ethanol and poured into a hot ethanolic solution of the telluride (1.5 g). The clear solution (hot ethanol must be added to dissolve any precipitate formed on mixing) deposited, upon cooling, the pure adduct, which melts at 174–175°.

Bis(dimethyl telluride)silver Iodide (C-24):   Dimethyl telluride (3.05 g, 0.02 moles) in 20 ml of acetone was added to a solution of AgI (2.35 g, 0.01 moles) in 10 ml of nearly saturated KI solution. A white precipitate appeared, which dissolved on warming. Long colorless needles were deposited upon cooling which, after washing with acetone and recrystallization from acetone, softened at 60–162° and melted at 73–74°. The product smelled of dimethyl telluride and lost the telluride quantitatively in vacuum at 180°. The yield was 78%. Similarly $(CH_3)_2Te \cdot (AgI)_2$ was obtained upon mixing the reactants in the required stoichiometric ratio in acetone solution.

Picric Acid Adduct of 2,2′-Biphenylylene Telluride (H-4):   The telluride was dissolved in warm ethanol and mixed with an ethanolic solution of picric acid. Upon cooling and concentrating the solution, the picric acid adduct was obtained in 37% yield.

Complexes cis-L_2PtCl_2 *(L = telluride)* (C-13):   Two equivalents of the ligand were shaken with one equivalent of K_2PtCl_4 dissolved in water. With higher alkyl tellurides better results were obtained by treating an ethanolic solution of Na_2PtCl_4 at room temperature, or a dilute acetic acid solution of K_2PtCl_4 at boiling temperature, with the ligand.

Dichlorobis(diethyl telluride)palladium(II) (C-17):   Ammonium chloropalladate(II) (3.2 g) in 20 ml of water was treated with diethyl telluride (3.8 g) in a nitrogen atmosphere. After shaking a dark-orange precipitate was formed, which was filtered, washed with water, ethanol, and hot ether and dried. Yield 40.5%.

Dichloro-μμ′-dichlorobis(dipropyl   telluride)dipalladium(II) (C-17):   The telluride was added to a solution of Na_2PdCl_4 in ethanol. After three days the solid product was filtered, washed with water, ethanol, and ether and recrystallized from methanol. Yield: 30%.

B. Reactions Affecting the Tellurium–Carbon Bond

Reactions affecting the tellurium–carbon bond are not well understood and examples are few in number. Presently they are rather unimportant in preparative organotellurium chemistry.

1. Reactions with Hydrogen Peroxide and Oxygen

Balfe and co-workers (B-2,B-3) isolated tellurinic acids as oxidation products of tellurides. The oxidizing reagents employed were atmospheric oxygen

(B-2,B-3) and alkaline hydrogen peroxide (B-2). The following equation may describe the overall reaction.

$$(RCH_2)_2Te + 2 O_2 \longrightarrow R—CH_2—TeOOH + R \cdot COOH$$

A possible reaction mechanism, which would lead to the oxidative cleavage of one Te—C bond is discussed in Section IV-A-3. For individual compounds isolated in this way consult Tables XXI and XXII.

2. The Exchange of Organic Groups

There are only two literature reports claiming the exchange of an organic group in a telluride. Mazza and Melchionna (M-11) studied the reaction of bis(o-carboxyphenyl) telluride with powdered zinc in hot 10% KOH. They filtered the resulting mixture into the calculated amount of sodium chloroacetate. Upon acidification with H_2SO_4, o-carboxyphenyl carboxymethyl telluride, m.p. 195°, was obtained. It is likely that a tellurol was formed as an intermediate and that the reaction proceeded according to the equations

$$R_2Te + Zn + 3 KOH + H_2O \longrightarrow RTeK + RH + K_2Zn(OH)_4$$
$$RTeK + Cl—CH_2COONa \longrightarrow RTeCH_2COONa + KCl$$

The results obtained by Mazza and Melchionna, however, are of doubtful value. Farrar (F-2) was unable to repeat the synthesis of Mazza's starting telluride. Farrar's analytical results did not agree with the structure of bis(o-carboxyphenyl) telluride. Therefore all compounds derived from Mazza's telluride are of an uncertain constitution (see Sections IV-C, XIII-D).

Bowden and Braude (B-14) observed that solutions of phenyl ethyl telluride in cyclohexane and ethanol exhibit a rapid change with respect to their absorption characteristics, probably as a result of disproportionation to the symmetric tellurides.

$$2 C_6H_5—Te—C_2H_5 \longrightarrow (C_6H_5)_2Te + (C_2H_5)_2Te$$

3. The Exchange of the Tellurium Atom for a Sulfur or Selenium Atom

The tellurium atom in tellurides can be replaced by a sulfur or selenium atom. The reaction between diphenyl telluride and elemental sulfur is one of the known examples of such a replacement (K-6).

In this particular experiment 1.28 g of the telluride and 0.14 g of sulfur were heated under an atmosphere of CO_2 at 220° for 10 hr. The reaction mixture, after cooling, was treated with ether. Of the total tellurium present 58.4% was isolated in elemental form. Unchanged telluride, diphenyl sulfide, and diphenyl disulfide were obtained from the ethereal solution.

Bis(pentafluorophenyl) telluride heated with selenium at 320° gave a quantitative yield of the corresponding selenide (C-26).

4. The Decomposition of Tellurides and Cleavage of Tellurium–Carbon Bonds by Transition Metal Complexes and Triethylmetal Hydrides

Although the Te—C bond in tellurides is generally not cleaved under conditions encountered in the various reactions given by tellurides, there are compounds which decompose under these relatively mild conditions. Bis-(carboxymethyl) telluride is reported to blacken on mixing with HCl (M-26).

Tellurides containing a benzyl group are rather unstable toward water and atmospheric oxygen. Dibenzyl telluride, stable for some time in an inert atmosphere (T-2), decomposes quickly to elemental tellurium when exposed to air (S-16,T-2).

2-Naphthyl benzyl and p-methoxyphenyl benzyl tellurides, upon treatment with SO_2Cl_2, $SOCl_2$, Br_2 or I_2, give aryltellurium trihalides and benzyl halide (V-4).

$$\text{[phenyl]}-CH_2-Te-R + 2\,Br_2 \longrightarrow \text{[phenyl]}-CH_2-Br + R-TeBr_3$$

These benzyl tellurides decompose also, especially in the impure state, when exposed to atmospheric agents and wet solvents. Benzaldehyde, which was identified as the dinitrophenylhydrazone, and diaryl ditelluride were formed (V-4).

$$2\,\text{[phenyl]}-CH_2-Te-R + O_2 \xrightarrow{H_2O} R-Te-Te-R + 2\,\text{[phenyl]}-C\!\!\begin{array}{c}O\\ \\H\end{array} + H_2O$$

For a discussion of reactions between diphenyl telluride and $[Mn(CO)_5]_2$ and $[Co(CO)_4]_2$ in which Te—C bonds were broken, see Section I-A-4.

Diphenyl telluride, when treated in liquid ammonia with two equivalents of KNH_2, gave benzene, dipotassium triimidotellurite, and dipotassium tritelluride (S-4).

$$7\,Te(C_6H_5)_2 + 10\,KNH_2 + 2\,NH_3 \longrightarrow 4\,K_2[Te(NH)_3] + K_2Te_3 + 14\,C_6H_6$$

Cleavage of tellurium–carbon bonds also occurred in reactions between diethyl telluride and trialkylmetal hydrides, R_3MH, where M represents Si, Ge, and Sn, and R cyclohexyl and C_2H_5-. The reactions proceeded at temperatures between 20 and 200° and gave compounds of the type C_2H_5—Te—MR_3 and R_3M—Te—MR_3. Ethane was liberated during the reactions (V-8, V-9). $(C_2H_5)_3SiH$ and $(C_2H_5)_3GeH$ successively replaced the ethyl groups in diethyl telluride (V-8,V-9) yielding the symmetric telluride, $(C_2H_5)_3M$—Te—$M(C_2H_5)_3$ and the unsymmetric telluride, $(C_2H_5)_3M$—Te—C_2H_5.

$$(C_2H_5)_3MH + (C_2H_5)_2Te \longrightarrow (C_2H_5)_3M\text{—Te—}C_2H_5 + C_2H_6$$
$$(C_2H_5)_3M\text{—Te—}M(C_2H_5)_3 + C_2H_6 \xleftarrow{\quad\quad} \underset{(C_2H_5)_3MH}{\mid}$$

Triethyltin hydride is more reactive than the Si and Ge analogs. The reaction begins spontaneously following an induction period. No unsymmetric telluride was formed. The only products isolated were $[(C_2H_5)_3Sn]_2Te$ and a small amount of $[(C_2H_5)_3Sn]_2$. The reactivity of the group IV metal hydrides toward diethyl telluride increases in the series $(C_2H_5)_3SiH < (C_2H_5)_3GeH \ll (C_2H_5)_3SnH$ (V-9). The symmetric tellurides probably form from a triethyl-metal ethyl telluride. However, the possibility of disproportionation of the unsymmetric compound must not be overlooked (V-9).

$$2 (C_2H_5)_3M\text{—Te—}C_2H_5 \longrightarrow [(C_2H_5)_3M]_2Te + (C_2H_5)_2Te$$

For individual compounds prepared and for reaction conditions see Table XXV.

The reaction between phenyltelluromagnesium bromide and trimethyl-chlorosilane gave bis(trimethylsilyl) telluride (H-8) (see Section XIV).

Experimental. Triethylgermyl Tellurides (V-9): A mixture of triethyl-germane (8.00 g, 0.05 moles) and diethyl telluride (7.86 g, 0.042 moles) was heated in a sealed, evacuated tube to 140° for 7 hr. 1100 ml. (98.5%) of ethane were evolved. By fractionation under a current of nitrogen $(C_2H_5)_3GeTeC_2H_5$ (5.22 g, 39%) and $[(C_2H_5)_3Ge]_2Te$ (6.45 g, 57.9%) boiling at 61 and 113–116°, respectively, at 1 mm pressure were obtained.

C. Reactions Modifying the Organic Moiety

A fair number of reactions are known in which the organic moiety in the tellurides is modified.

o-Carboxyphenyl carboxymethyl telluride refluxed for two hours with sodium acetate and acetic anhydride, the latter distilled off and the residue boiled with 10% NaOH yielded benzo-3-hydroxytellurophene.

This heterocyclic compound, a yellowish white solid, melted at 160° and decomposed at 200°. It was soluble in basic solutions (M-11). The results obtained in this investigation are of doubtful value (see Section IV-B-2).

Bis(2-cyanoethyl) telluride when treated with 28% hydrochloric acid gave bis(2-carboxyethyl) telluride, melting at 158.5–159.5° (K-1).

TABLE XXV
Tellurium–Carbon Bond Cleavage by Triethylmetal Hydrides

C_2H_5—Te—Y	R_3MH		Reaction		Product	Yield %	b.p. (mm Hg)	n_D^{20}	Ref.
Y	M	R	Temp., °C	Time/hr					
C_2H_5—	Si	C_2H_5—	200	7	R—Te—SiR$_3$	11.5	52–53 (1)	1.5340	V-8,V-9
	Si	C_2H_5—	200	7	R$_3$Si—Te—SiR$_3$	55.4	97–99 (1)	1.5339	V-8
	Ge	C_2H_5—	140	7	R—Te—GeR$_3$	39	61 (1)	1.5458	V-8,V-9
	Ge	C_2H_5—	140	7	R$_3$Ge—Te—GeR$_3$	57.9	113–116 (1)	1.5610	V-9
	Ge	c-C_6H_{11}—	200	3.5	R$_3$Ge—Te—GeR$_3$	77.7	mp. 128–129	—	V-9
C_2H_5—	Sn	C_2H_5—	20	1	R$_3$Sn—Te—SnR$_3$	91.4	119–121 (1)	1.5972	V-8,V-9
$(C_2H_5)_3Si$—	Sn	C_2H_5—	20	1	R$_3$Si—Te—SnR$_3$	90.8	109–112 (1)	1.5680	V-9
$(C_2H_5)_3Ge$—	Sn	C_2H_5—	20	1	R$_3$Ge—Te—SnR$_3$	61.9	126–128 (1)	1.5723	V-9

Pourcelot (P-17) isolated propargylic and allenic compounds from the reaction of phenyltelluromagnesium bromide with propargyl bromide

$$C_6H_5\text{—}Te\text{—}MgBr + Br\text{—}CH(R)\text{—}C\equiv CH \longrightarrow$$

$$C_6H_5\text{—}Te\text{—}CH(R)\text{—}C\equiv CH + C_6H_5\text{—}Te\text{—}C(R)\text{=}C\text{=}CH_2$$

	R = H	25	75
proportions in %	R = CH$_3$	12	88

For the allenic derivative with R = H a boiling point of 65–75°/10^{-3}mm Hg and a yield of 27% is reported. The isomerization from the acetylenic to the allenic systems by migration of a hydrogen ion must have taken place after the propargylic telluride was formed.

The allenic isomer (4%) present in Br—CH(CH$_3$)—C≡CH did not take part in the reaction. Strong bases have been observed to facilitate the isomerization of corresponding ethers, sulfides, and selenides (P-16,P-17).

Rogoz (R-16) prepared a variety of tellurides starting with phenyl methyl and biphenylyl methyl telluride. The bromoacetyl group was introduced into the p-position of the phenyl group in a reaction employing bromoacetyl chloride and anhydrous aluminum chloride in dry chloroform at −10°.

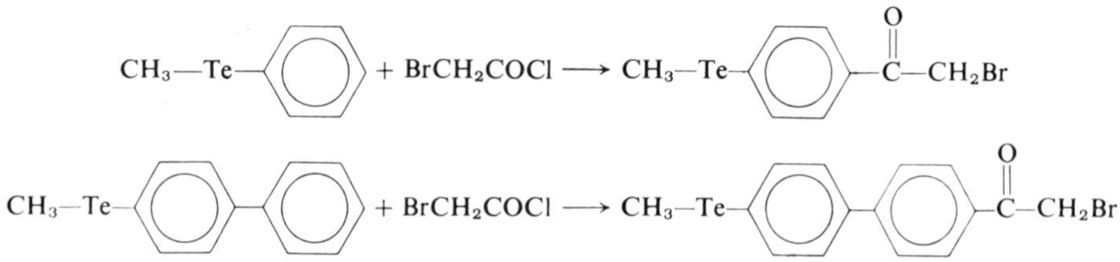

The functional groups in these tellurides made possible the synthesis of a number of new tellurides. The following scheme summarizes these reactions using the phenyl methyl telluride as an example.

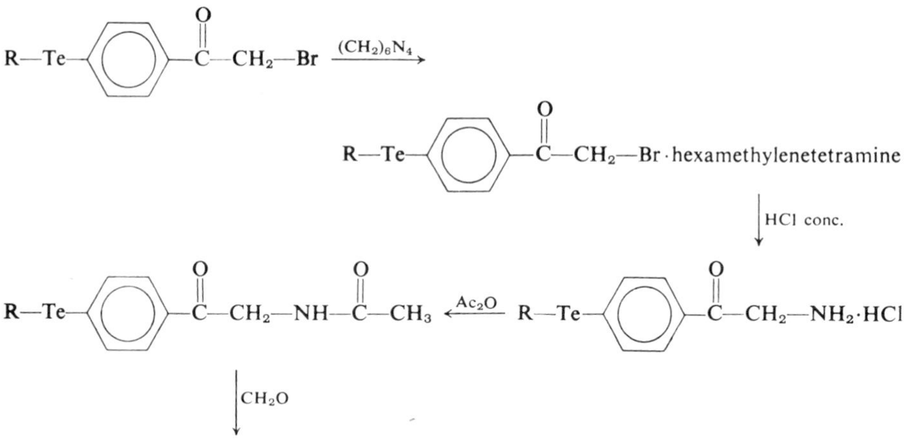

$$-Te-\langle\bigcirc\rangle-\overset{O}{\overset{\|}{C}}-\underset{\underset{CH_2OH}{|}}{CH}-NH-\overset{O}{\overset{\|}{C}}-CH_3 \xrightarrow[i\text{-PrOH}]{(i\text{-PrO})_3Al} R-Te-\langle\bigcirc\rangle-\underset{\underset{OH}{|}}{CH}-\underset{\underset{CH_2OH}{|}}{CH}-NH-\overset{O}{\overset{\|}{C}}-CH_3$$

$$\downarrow \text{HCl, then } NH_3$$

$$-Te-\langle\bigcirc\rangle-\underset{\underset{CH_2OH}{|}}{\overset{\overset{OH}{|}}{CH}}-CH-NH\cdot\overset{O}{\overset{\|}{C}}-CHCl_2 \xleftarrow[CaCO_3/NaCN]{CCl_3CH(OH)_2} R-Te-\langle\bigcirc\rangle-\underset{\underset{CH_2OH}{|}}{\overset{\overset{OH}{|}}{CH}}-CH-NH_2$$

(R = CH$_3$)

Table XXVI lists all the compounds prepared by Rogoz.

TABLE XXVI
Tellurides prepared by F. Rogoz (R-16)

R'	CH$_3$—Te—⬡—R'		CH$_3$—Te—⬡—⬡—R'	
	Yield	m.p.	Yield	m.p.
—CH$_2$Br	52	55–6	22.3	118
—CH$_2$·Br·HMTA[a]	88.1	107 (dec.)		
—CH$_2$—NH$_2$·HCl	30.2	233	57.3	243
—CH$_2$·NH·$\overset{O}{\overset{\|}{C}}$—CH$_3$	78.9	163	86	190
$\overset{CH_2OH}{\overset{\|}{—CH}}$—NH·$\overset{O}{\overset{\|}{C}}$—CH$_3$	45.8	155	52	161–163
H $\overset{CH_2OH}{\overset{\|}{H—CH}}$—NH·$\overset{O}{\overset{\|}{C}}$—CH$_3$	79.5	128	89	147–148
H $\overset{CH_2OH}{\overset{\|}{H—CH}}$—NH$_2$	83.3	108	91.6	131
H $\overset{CH_2OH}{\overset{\|}{H—CH}}$—NH—$\overset{O}{\overset{\|}{C}}$—CHCl$_2$	92.9	114	71.6	146

HMTA = hexamethylenetetramine.

Benzonitrile-*N*-oxide (R-1), *N*-α-diphenylnitrone (R-1), and 1,3-diphenyl-nitrilimine (R-2) react with alkylvinylacetylenyl tellurides. The vinyl groups add possibly via a 1,3-dipolar addition reaction to the unsaturated, nitrogen-containing reagents.

R = CH₃: m.p. 127–128°

5-(alkyltelluroethynyl)-1,3-diphenyl-pyrazoline

R = C₂H₅: m.p. 115.5–116.0
Yields: ~ 70%

5-(methyltelluroethynyl)-3-phenyl-2-isoxazoline
m.p. 66–67°, 89% yield

5-(ethyltelluroethynyl)-2,3-diphenyl-isoxazolidine
m.p. 89.5–90.5°, 19.8% yield

V. REACTIONS OF DITELLURIDES, R_2Te_2

Both diaryl and dialkyl ditellurides, R—Te—Te—R, where R is an alkyl or aryl group, are known, but the aryl compounds are much more common. The colors of all of the known ditellurides range from orange to dark red. The odor of the alkyl derivatives is exceedingly obnoxious and they are very sensitive toward atmospheric oxygen. Both of these objectionable properties are diminished in the aryl series and this accounts for their more extensive study.

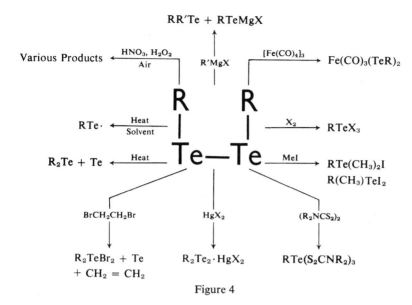

Figure 4

As can be seen in Figure 4, most of the reactions given by the ditellurides proceed either by the elimination of one tellurium atom with the formation of tellurides or their derivatives, or via fission of the tellurium–tellurium bond to yield a monoorganotellurium compound.

The following reactions of ditellurides have been investigated.

1. Reactions with Elemental Halogens and Bis(thiocarbamoyl) Disulfides

Fission of the Te—Te bond is brought about by halogenolysis. Quantitative conversion of diaryl ditellurides to the aryltellurium trihalides by chlorine, bromine, and iodine has been reported by several investigators (M-32,P-5, V-5).

$$R_2Te + 3 X_2 \longrightarrow 2 RTeX_3$$

The ditelluride in chloroform, CCl_4 or benzene solution is treated with the stoichiometric amount of the halogen. The organotellurium trihalide is precipitated quantitatively.

Because of the extreme paucity of our knowledge concerning tellurenyl derivatives, the preparation of 2-naphthyltellurium iodide (V-5) by the addition of iodine to a solution of bis(2-naphthyl) ditelluride in cold benzene in a 1:1 molar ratio should be noted. Attempts to prepare other tellurenyl compounds gave only indefinite products (P-5). Foss (F-4) was able to prepare an interesting series of aryltellurium tris(dithiocarbamates) by treatment of diaryl ditellurides with bis(thiocarbamoyl) disulfides in chloroform

$$R_2Te_2 + 3 (R_2NCS_2)_2 \longrightarrow 2 RTe(S_2CNR_2)_3$$

TABLE XXVII

Reactions of Bis(organo) Ditellurides with Halogen and Bis(thiocarbamoyl) Disulfides

| R_2Te_2 | | $RTeX_n$ | | | | |
R or complete formula	Reagent	X or complete formula	n	m.p.	Yield, %	Ref.
CH_3—	Br_2	Br	3	140 (dec.)	100	C-21
$(CH_2Te_2)_n$	Cl_2	$CH_2(TeCl_3)_2$	3	173 (dec.)	—	M-26
	Br_2	$CH_2(TeBr_3)_2$	3	214 (dec.)	—	M-26
—CH_2COOH	Cl_2	Cl	3	—	—	M-26
	Br	Br	3	—	—	M-26
—$CH(CH_3)COOH$	Br	Br	3	139–141	—	M-32
C_6H_5—	Br_2	Br	3	246 (dec.)	—	P-4
	I_2	I	3	179 (dec.)	—	P-4
4-Methoxyphenyl	$(R_2NCS_2)_2$	R_2NCS_2	3	90	—	F-4
	Cl_2	Cl	3	192	—	M-32
	Br_2	Br	3	188–190 (dec.)	—	M-32,P-5
	I_2	I	3	131–133 (dec.)	93	P-5
	$[(CH_3)_2NCS_2]_2$	$(CH_3)_2NCS_2$	3	184 (dec.)	90	F-4

	$[(C_2H_5)_2NCS_2]_2$	$(C_2H_5)_2NCS_2$				
(piperidine–N–CS_2)$_2$			3	156 (dec.)	90	F-4
piperidine–N–CS_2			3	178 (dec.)	90	F-4
4-Ethoxyphenyl	Br_2	Br	3	195–205 (dec.)	100	P-5
	I_2	I	3	133–134 (dec.)	100	P-5
4-Phenoxyphenyl	Br_2	Br	3	207	100	P-5
	I_2	I	3	161 (dec.)	100	P-5
4-Thiophenoxyphenyl	Br_2	Br	3	207	100	P-5
	I_2	I	3	180 (dec.)	100	P-5
3-Methyl-4-methoxyphenyl	Cl_2	Cl	3	—	—	M-32
	Br_2	Br	3	—	—	M-32
2-Methoxy-5-bromophenyl	Br_2	Br	3	153	—	M-32
1-Naphthyl	Br_2	Br	3	160 (dec.)	100	P-5
	I_2	I	3	133 (dec.)	100	P-5
2-Naphthyl	Br_2	Br	3	212–215 (dec.)	100	V-5
	I_2	I	1	130°/220–222[a] (dec.)	86	V-5

[a] Melts at 130°, solidifies and remelts at 220°.

Mixing the reagents in the amounts required by the above equation in $CHCl_3$, heating and adding ethanol produced the desired compounds on further heating in 90% yields. Sodium dithiocarbamates and tellurium trichlorides give the same products. Preliminary experiments have shown that bis(2-benzothiazolyl) disulfide gave a similar reaction with bis(4-methoxyphenyl) ditelluride.

Chen and George observed that dimethyl ditelluride formed some $TeBr_4$ and methyl bromide upon treatment with bromine in concentrated ether solution or in the absence of any solvent. In benzene solution methyltellurium tribromide was obtained quantitatively (C-21).

Table XXVII lists the compounds prepared together with pertinent references.

Experimental. 4-Methoxyphenyltellurium Trichloride (P-5): To a cold solution of bis(4-methoxyphenyl) ditelluride (0.47 g, 0.001 mole) in CCl_4 was added dropwise with stirring bromine (0.48 g, 0.003 moles) dissolved in a few ml of CCl_4. The tribromide separated quantitatively as an orange solid.

2-Naphthyltellurenyl Iodide (V-5): Iodine (2.6 g, ~0.01 mole) in 80 ml of benzene was added slowly, with cooling and stirring, to a suspension of bis(2-naphthyl) ditelluride (5.1 g, 0.01 mole) in 80 ml of benzene. After the ditelluride dissolved, naphthyltellurenyl iodide began to precipitate. The bluish-black powder (6.6 g, 86% yield) melted at 130°, solidified and melted again at 220–222° with decomposition.

2. Reaction with Methyl Iodide

Morgan and Drew (M-28) showed that bis(4-methoxyphenyl) ditelluride, when warmed with methyl iodide, undergoes a smooth decomposition to give stoichiometric amounts of the iodo- derivatives

This reaction was confirmed by Reichel and Kirschbaum (R-6). That this method is not general is illustrated by the fact that the *p*-ethoxyphenyl derivative decomposes when treated with methyl iodide.

Experimental. 4-Methoxyphenyl Dimethyltelluronium Iodide from the Ditelluride (M-28): To 78 g of pure methyl iodide was added 5 g of pure bis(4-

methoxyphenyl) ditelluride. The solution was refluxed for $\frac{1}{2}$ hr after which time 4.6 g of salmon colored dimethyl p-methoxyphenyltelluronium iodide separated. This compound melted at 170–172° and was insoluble in acetone, benzene, and ordinary organic solvents, but was decomposed by boiling alcohol.

The methyl iodide mother liquor from the previous reaction contained 4.7 g of p-methoxyphenyl methyltellurium diiodide. This is a red, crystalline solid, m.p. 109°, and is soluble in ordinary organic solvents.

3. Reactions of Ditellurides with Halogen-Containing Compounds Yielding Bis(organo)tellurium Dihalides

Petragnani and de Moura Campos (P-6) investigated the reactions of bis(4-methoxyphenyl) and bis(4-ethoxyphenyl) ditelluride with vic-dibromides, with tellurium tetrahalides, with diphenylselenium dibromide, diaryltellurium dibromide and aryltellurium trichloride with or without solvent and obtained diaryltellurium dihalides, tellurium and a halogen-free compound.

The reaction of diaryl ditellurides with vic-dibromides was reported to proceed according to the following equation

$$\text{RTeTeR} + \text{RCHBr—CHBrR}' \longrightarrow \text{R}_2\text{TeBr}_2 + \text{RCH}{=}\text{CHR}' + \text{Te}$$

The yields of the alkenes were reported to be excellent (P-6). Because *trans*-olefins were obtained from *meso*-dibromides a mechanism involving a *trans*-elimination was proposed.

The stoichiometries of the other reactions are summarized by the following equations

$$\text{R}_2\text{Te}_2 + (\text{C}_6\text{H}_5)_2\text{SeBr}_2 \longrightarrow \text{R}_2\text{TeBr}_2 + (\text{C}_6\text{H}_5)_2\text{Se} + \text{Te}$$
$$2\,\text{R}_2\text{Te}_2 + 2\,\text{RTeCl}_3 \longrightarrow 3\,\text{R}_2\text{TeCl}_2 + 3\,\text{Te}$$
$$2\,\text{R}_2\text{Te}_2 + \text{TeX}_4 \longrightarrow 2\,\text{R}_2\text{TeX}_2 + 3\,\text{Te}$$
$$\text{R}_2\text{Te}_2 + \text{R}_2\text{TeBr}_2 \longrightarrow \text{R}_2\text{TeBr}_2 + \text{R}_2\text{Te} + \text{Te}$$

It is interesting to note that the bis(organo)tellurium dibromide removed one tellurium atom from the ditelluride. The authors (P-6) postulate the following mechanism

Attempts to isolate such intermediates failed. The reactions actually performed

TABLE XXVIII

Reaction of Bis(4-methoxyphenyl) (a) and Bis(4-ethoxyphenyl) Ditelluride (b) with Halogen Compounds (P-6)

Halogen compound	Solvent (reflux time, hr)	Diaryl tellurium dihalide Yield, %	m.p.	Halogen-free reaction product Formula	Yield, %	m.p.
a $BrCH_2-CH_2-Br$	Ethylene bromide (2½)	90.0	198–200	$CH_2{=}CH_2$		
b	Toluene (1)	95.0	124–125		90	
a $C_6H_5-CH(Br)-CH(Br)-C_6H_5$		75.0	198–200	trans-$C_6H_5CH{=}CHC_6H_5$		121–124
b (meso, m.p. 237°)	Toluene (4)	76.0	124–125		100	
a $C_6H_5-CH(Br)-CH(Br)-COOH$		75.0	198–200	$C_6H_5CH{=}CHCOOH$	67.5	
b (meso, m.p. 203°)	Glacial acetic acid (1)	85.2	124–125		84.4	128–132
a $HOOC-CH(Br)-CH(Br)-COOH$		81.0	198–200	trans-$HOOCH{=}CHCOOH$	73.2	280 dec.

	Solvent					
b (*meso*, m.p. 256° dec.)		78.1	124–125		86.2	
a $(C_6H_5)_2SeBr_2$	Toluene	84.0	198–200	$(C_6H_5)_2Se$ (isolated as the dichloride)	100.0	
b (m.p. 140° dec)					83.0	155–160
a $TeCl_4$	Toluene (1)	83.1	183–184		95.4	
b		81.7	110–111	Te	100.0	
a $TeBr_4$	Toluene (2)	78.0	198–200		96.8	
b		86.6	124–125	Te	96.8	
a TeI_4	Toluene (3)	85.8	167–168		91.6	
b		80.0	138–140	Te	80.0	
a $p\text{-}CH_3O\text{—}C_6H_4TeCl_3$	Toluene (2)	95.1	182–184		94.4	
b $p\text{-}C_2H_5O\text{—}C_6H_4TeCl_3$	Toluene	96.6	110–111		94.4	
a $(p\text{-}CH_3O\text{—}C_6H_4)_2TeBr_2$	(4 ½)	97.0	198–200	$(p\text{-}CH_3O\text{—}C_6H_4)_2Te$	67.4	52–53
b $(p\text{-}C_2H_5O\text{—}C_6H_4)_2TeBr_2$		92.4	124–125	$(p\text{-}C_2H_5O\text{—}C_6H_4)_2Te$	67.7	63–64

231

employed ditellurides and tellurium dibromides with the same R-group, for example, bis(4-methoxyphenyl) ditelluride and bis(4-methoxyphenyl)tellurium dibromide. It is therefore impossible to decide whether the bromine atoms in the end product are attached to the telluride formed from the ditelluride by elimination of a tellurium atom, or remained with the telluride to which it originally belonged. The latter possibility implies that the catalytic decomposition of the ditelluride to the telluride occurred with the dibromide acting as catalyst. This question could easily be settled by further experiments.

In Table XXVIII are listed the data reported by Petragnani and de Moura Campos for the reactions of diaryl ditellurides with various halogen containing compounds.

Experimental. Bis(4-ethoxyphenyl)tellurium Dibromide (P-6): Equimolecular quantities (0.001 mole) of bis(4-ethoxyphenyl) ditelluride and *meso-α, β*-dibromo-*β*-phenylpropionic acid (m.p. 203°) were refluxed for four hours in anhydrous toluene. Tellurium, which is precipitated quantitatively, is removed by filtration. Addition of petroleum ether (b.p. 60–70°) brought about separation of bis(4-ethoxyphenyl)tellurium dibromide, m.p. 124–125°, yield 85.2%, and fumaric acid, m.p. 128–132°. Inasmuch as the fumaric acid is also insoluble in petroleum ether, separation of the two components was accomplished by recrystallization from ethanol.

4. Reactions with Organometallic Compounds

The reaction of diaryl ditellurides with a three to five-fold excess of an aryl Grignard reagent has been found to result in scission of the Te—Te bond (P-9) to form the unsymmetric telluride in 100% yield.

$$R_2Te_2 + R'MgBr \longrightarrow R—Te—R' + RTeMgBr$$

Upon addition of petroleum ether to the ethereal reaction mixture, both the telluride and the telluromagnesium bromide precipitated. The unsymmetric tellurides, which were obtained in oily form, were converted to the dichlorides by treatment with SO_2Cl_2. The following compounds were prepared (compound, melting point, ditelluride, Grignard reagent given):

$(C_6H_5)(4-CH_3OC_6H_4)Te$, 60.5–61.5°, $(4-CH_3OC_6H_4)_2Te_2$, C_6H_5MgBr;
$(4-C_2H_5OC_6H_4)(1-C_{10}H_7)TeCl_2$, 193–194°, $(4-C_2H_5OC_6H_4)_2Te_2$, $1-C_{10}H_7MgBr$;
$(C_6H_5)(4-C_6H_5OC_6H_4)TeCl_2$, 128.5–129.5°, $(4-C_6H_5OC_6H_4)_2Te_2$, C_6H_5MgBr.

In this same study it was reported that alkylmagnesium halides do not cleave the ditelluride link. The reaction of ditellurides with phenyllithium yielded a difficultly separable mixture of tellurides.

Experimental. Phenyl 4-Methoxyphenyl Telluride (P-9): Into a solution of phenylmagnesium bromide (from 0.84 g, 0.035 moles, of Mg and 5.5 g, 0.035 moles, bromobenzene in 15 ml of absolute ether) was dropped, with cooling and stirring, under a nitrogen atmosphere, a solution of bis(4-methoxyphenyl) ditelluride (4.7 g, 0.01 mole) in 100 ml of ether. After addition of 250 ml of petroleum ether (30–50°) a precipitate was formed which was allowed to remain in the reaction mixture for two hours. The filtrate was hydrolyzed with aqueous NH_4Cl. The organic layer, dried over Na_2SO_4, was concentrated. The remaining oil crystallized after addition of methanol and gave quantitatively phenyl 4-methoxyphenyl telluride melting at 60.5–61.5°.

5. Thermal Decomposition and Dissociation into Radicals

The thermal decomposition of diaryl ditellurides has been studied by Farrar (F-2) and Petragnani and de Moura Campos (P-6). Pyrolysis is reported to begin at a temperature of 250° with concurrent formation of the diaryl telluride and elemental tellurium. Good yields were obtained only above 300°.

$$(C_6H_5)_2Te_2 \xrightarrow{\Delta} (C_6H_5)_2Te + Te$$

Heating bis(p-methoxyphenyl) and bis(p-ethoxyphenyl) ditelluride in xylene for four hours to 137–140° was without effect (P-6).

Miller (M-12) prepared mono-crystals of photoconductive cadmium compounds by treating zone refined cadmium at 800–1000° with vapors of ditellurides having alkyl groups with less than four carbon atoms. Methylene ditelluride is reported to explode in air at temperatures higher than 40°C.

Farrar (F-2), from measurements of the molecular mass of diphenyl ditelluride in camphor, concluded that this compound is dissociated into radicals to the extent of 30% at 160°. As evidence of free radical formation he reported a marked increase in the intensity of the red color of a solution in ethanol on warming from −80° to 80° and the absorption of oxygen by such a solution to give a colorless, gelatinous substance, presumably a peroxide. Earlier investigators (M-28,M-32,R-6) have also observed that the molecular masses of ditellurides in organic solvents are less than those predicted from their molecular formulas. Such a dissociation is known to occur in disulfides (S-5).

6. Reactions with Nitric Acid, Hydrogen Peroxide and Air, and Reduction by Sodium Metal

Concentrated nitric acid is reported to decompose bis(p-phenoxyphenyl) ditelluride (D-11) and bis(p-ethoxyphenyl) ditelluride (M-28) into p,p'-dinitrodiphenyl ether and p-nitroethoxybenzene, respectively. Lederer observed that diphenyl ditelluride is not affected by cold, halogen-free, 65% nitric acid. However, on heating, it is oxidized to the nitric acid ester of benzenetellurinic acid (L-10).

Solutions of ditellurides exposed to air form an insoluble, white amorphous oxide, which can be reduced back to the ditelluride (D-11,M-26,M-28,M-32). Hydrogen peroxide decolorizes ethanol solutions of ditellurides. Upon evaporation of the solution a white amorphous product was obtained (M-32). In ethanolic solution, treatment with sodium metal was reported to result in the formation of the tellurol (L-10).

$$RTeTeR \xrightarrow[C_2H_5OH]{Na} 2\ RTeH$$

7. Ditellurides as Ligands

Addition compounds which form between diaryl ditellurides and mercuric chloride have been reported (G-9,M-32). However, these are very few in number and they include $(C_6H_5)_2Te_2 \cdot HgI_2$ (G-9) and $(p\text{-}CH_3OC_6H_5)_2Te_2 \cdot HgCl_2$ (M-32).

Hieber and Kruck (H-6) investigated the reaction between various metal carbonyls and diaryl ditellurides. With $(p\text{-}CH_3OC_6H_5Te)_2$, $[Fe(CO)_4]_3$ was found to yield the complex $[p\text{-}CH_3OC_6H_4TeFe(CO)_3]_2$. This red-brown compound is unusually stable. Its diamagnetism can be interpreted only through the existence of an Fe—Fe bond with electron spins paired. If each TeC_6-H_4OCH_3-p moiety forms one coordinate bond and a covalent bond to each iron atom, each of the latter achieves the krypton configuration. The presence of four ν(C—O) frequencies as well as the large electric dipole moment, $\mu_D = 4.06D$, requires that the molecule does not possess a center of symmetry. By using the similar thio-derivative, $[Fe(CO)_3SC_2H_5]_2$, as a model, a nonplanar arrangement of the type

$$\begin{array}{c} \text{Te} \\ \diagup \quad \diagdown \\ \text{Fe} \qquad \text{Fe} \\ \diagdown \quad \diagup \\ \text{Te} \end{array}$$

is suggested. For the tellurium atoms, a trigonal-pyramidal arrangement is proposed.

Experimental. Bis(tricarbonyl-μ-p-methoxyphenyltelluroiron) (H-6): In a closed system equipped with a reflux condenser, 2.0 g of freshly extracted iron tetracarbonyl (0.004 moles) were mixed with 2.8 g (0.006 mole) of bis(p-methoxyphenyl) ditelluride and the whole refluxed for five hours. The carbon monoxide was permitted to escape by way of a mercury valve. On cooling, the small amount of tellurium formed was separated from the red-browm solution by filtration and the filtrate was concentrated under reduced pressure. Crystallization of the oily liquid, which still contained some of the unreacted ditelluride, was accomplished by slow digestion with petroleum ether at 0°. The product, $[Fe(CO)_3TeC_6H_4CH_3$-$p]_2$, was recrystallized from a hexane–petroleum ether mixture.

VI. REACTIONS OF BIS(ORGANO)TELLURIUM DIHALIDES AND THEIR DERIVATIVES

The derivatives formed by bis(organo)tellurium dihalides are shown in Figure 5. Their reactions will be discussed in separate sections following the discussions of the dihalides.

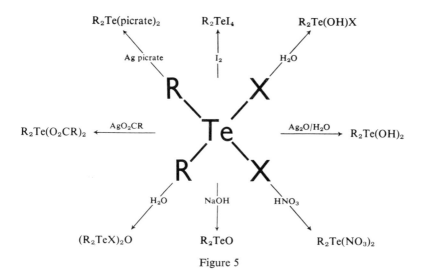

Figure 5

A. Reactions of Bis(organo)tellurium Dihalides

Bis(organo)tellurium dihalides undergo the following reactions.

1. Reduction to the tellurides by $Na_2S \cdot 9H_2O$, $NaHSO_3$, Na_2SO_3, $K_2S_2O_5$, Zn dust, and hydrazine sulfate.

2. Formation of ditellurides by the action of Na_2S or $K_2S_2O_5$ on certain bis(organo)tellurium dihalides.

3. Reaction with organometallic reagents.

4. Halogen exchange.

5. Hydrolysis by various agents.

6. Reactions with ditellurides.

7. Formation of addition compounds.

8. Reactions affecting the organic part of the molecule.

9. Tellurium–carbon bond cleavage and elimination of the tellurium atom.

Figure 6 summarizes the important reactions of bis(organo)tellurium dihalides.

The structure of bis(organo)tellurium dihalides can be described as a

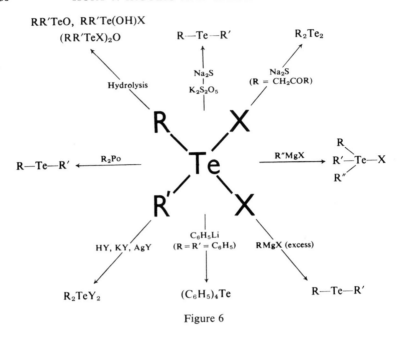

Figure 6

slightly distorted trigonal bipyramid with one equatorial position vacant (C-11,C-22,C-23,G-1) (see Section XVI).

Vernon (V-1,V-2) discovered that dimethyltellurium dihalides can form two types of compounds having differing physical properties but the same analytical composition. The physiological activity of these two compounds has been investigated (C-28,K-2). The following reaction sequence gives the relationship between these two series.

$$Te + 2 CH_3I \underset{>180°}{\overset{80°}{\rightleftarrows}} (CH_3)_2TeI_2 \underset{HI}{\overset{Ag_2O}{\rightleftarrows}} (CH_3)_2Te(OH)_2$$
$$\alpha\text{-diiodide} \qquad\qquad \alpha\text{-base}$$

$$\downarrow \text{Evaporated to dryness, heating } 100°/15 \text{ torr}$$

$$\beta\text{-dihalides} \underset{Ag_2O}{\overset{HX}{\rightleftarrows}} \beta\text{-base}$$

Gilbert and Lowry (G-4) found that diethyltellurium diiodide reacted similarly.

The *cis–trans* isomerism proposed by Vernon (V-2) was proved to be incorrect by X-ray investigations (C-23). Drew (D-15) and then Lowry and Gilbert (L-36) proved, through chemical methods, that the β-series of these compounds consists of complex substances formed in a 1:1 molar ratio from mono-organotellurium and tris(organo)tellurium derivatives. The following scheme summarizes the results.

$$R_2TeI_2 \underset{HI}{\overset{Ag_2O}{\rightleftarrows}} R_2Te(OH)_2 \overset{Heat}{\longrightarrow} R_3TeO\overset{O}{\overset{\uparrow}{—}}TeR$$

$$\downarrow HI$$

$$\begin{array}{c} —R_3TeI + RTeOOH \\ \downarrow HI \end{array}$$

$$R_3TeI \cdot RTeI_3 \longleftarrow \underline{\hspace{5cm}} RTeI_3$$

Vernon's β-base, therefore, is the mixed anhydride of trimethyltelluronium hydroxide and methanetellurinic acid, the β-diiodide a complex formed from one mole each of trimethyltelluronium iodide and methyltellurium triiodide. The exact structure of this complex is uncertain (D-15,L-36)*. All the steps in these reactions with the exception of the conversion of "α-base" to "β-base" are reversible. This conversion requires that one methyl group migrates from one tellurium atom to another. The mechanism of this migration is not known.

Such a migration has also been observed by Morgan and Burgess (M-37). They obtained tris(2-methyl-4-hydroxyphenyl)telluronium chloride upon treatment of the corresponding dichloride with boiling 95% ethanol or with hot Na_2CO_3 solution followed by treatment with HCl.

1. Reduction of Bis(organo)tellurium Dihalides to Tellurides

Bis(organo)tellurium dihalides can be reduced to the corresponding tellurides by a variety of reducing agents: Na_2SO_3, $NaHSO_3$, $K_2S_2O_5$, zinc in $CHCl_3$, benzene or glacial acetic acid, tin(II) chloride in hydrochloric acid, $Na_2S_2O_3$, molten $Na_2S \cdot 9H_2O$, and lithium aluminum hydride. The reduction of dihalides by Grignard reagents is discussed in Section VI-3.

Sodium sulfide nonahydrate is the most widely used reducing agent. The dihalide is added to the molten sulfide. Addition of water precipitates the telluride in very good yields.

The reductions with Na_2SO_3, $NaHSO_3$, $Na_2S_2O_3$ and $K_2S_2O_5$ are performed in aqueous solutions. To increase the solubility of the halide (by hydrolysis to the hydroxyhalide or dihydroxide) and to remove the hydrohalic acid formed, the solutions are made basic with sodium hydroxide or sodium carbonate. These reducing agents, however, do not possess general applicability. Bis(2,4-dimethoxyphenyl)tellurium dichloride was resistant to the reducing action of aqueous alkali bisulfite. Bis(p-methoxyphenyl)tellurium dichloride was not reduced by $K_2S_2O_5$ (M-32). The bis(organo)tellurium

* That the structure has been determined crystallographically has been brought to the authors' attention (F. Einstein, J. Trotter, and C. Williston, *J. Chem. Soc.*, B, **1967**, 2018). It corresponds to $[(CH_3)_3Te]^+[CH_3TeI_4]^-$.

dichlorides derived from acetylacetone (M-17), acetone, 1,1,1-trimethylacetone and acetophenone (M-30) (see Sections I-C-1 and I-C-2) were decomposed to elemental tellurium by $KHSO_3$, $K_2S_2O_5$ or Na_2SO_3. The dichloride, $[(CH_3)_2CHCH_2C(O)CH=C(OH)CH_2]_2TeCl_2$, produced an oily liquid upon treatment with cold aqueous $K_2S_2O_5$ solution, which rapidly decomposed (M-21). p-Ethoxyphenyl 2-chlorocyclohexyltellurium dichloride gave only tellurium when reduction with NaHS, $NaHSO_3$, or hydrazine sulfate was attempted.

Dimethyltellurium diiodide, reduced by Na_2SO_3 in presence of methyl iodide, gave trimethyltelluronium iodide (S-12). Similarly, the corresponding p-methoxyphenyl derivative, treated with $K_2S_2O_5$ in the presence of methyl or ethyl iodide, formed the expected telluronium salts (R-6).

The following equations describe these reduction processes

$$
\begin{array}{llll}
R & + Na_2SO_3 + H_2O \longrightarrow & R & + 2\ HX + Na_2SO_4 \\
| \diagdown & \\
 X & + Zn \longrightarrow & | & + ZnX_2 \\
Te \diagup & & Te \\
| \diagup & \\
 X & + Na_2S \longrightarrow & | & + 2\ NaX + S \\
R & + \tfrac{1}{2}\ LiAlH_4 \longrightarrow & R & + H_2 + \tfrac{1}{2}\ LiX + \tfrac{1}{2}\ AlX_3
\end{array}
$$

Certain mixed tellurium dihalides, $RR'TeX_2$, which are reduced to ditellurides are discussed in the following Section VI-A-2. Table XXIX lists the dihalides reduced and gives physical characteristics of the tellurides formed.

Experimental. 4-Dimethylaminophenyl 4-Methoxyphenyl Telluride (Na_2S Method; P-7): 4-Dimethylaminophenyl 4-methoxyphenyltellurium dichloride (0.42 g; 0.001 moles) was added to 3.6 g (0.015 mole) hydrated sodium sulfide at 100°. The mixture was kept at this temperature for 10–15 min. The telluride, a yellow-orange oil, solidified after dilution of the mixture with water. The yield is quantitative (0.35 g). The telluride recrystallized from ethanol in the form of colorless needles, m.p. 96–97°.

p-Methoxyphenyl Methyl Telluride ($K_2S_2O_5$ Method; R-6): p-Methoxyphenyl methyltellurium diiodide was dissolved in 300 ml of H_2O containing 7 g of NaOH. To this solution was added 20 g of $K_2S_2O_5$ dissolved in 150 ml of H_2O. A red oil, which was extracted with CCl_4, precipitated upon addition of hydrochloric acid while stirring and cooling. Distillation of the CCl_4 extract dried over NaOH gave 67% telluride, boiling at 150–152° at 20 torr.

Dimethyl Telluride (Na_2SO_3 Method; V-2): To a solution of hydrated Na_2SO_3 (13 g) and Na_2CO_3 (6 g) in 200 ml of H_2O contained in a 500 ml distilling flask equipped with a long condenser was added dimethyltellurium diiodide (20 g). The flask was gently warmed. The iodide dissolved rapidly. Upon raising the temperature slowly the telluride distilled.

TABLE XXIX
The Reduction of Bis(organo)tellurium Dihalides to Tellurides

	RR¹TeX₂		Reducing	R—Te—R¹		
R¹	Rᵃ	X	agent	m.p., b.p. (torr)	Yield, %	Ref.
CH₃—	CH₃—	I	Na₂SO₃	94 (770)	Good	B-12,S-12, V-2
ICH₂—	ICH₂—	I	K₂S₂O₅	180–185 (dec.)	100	F-1
—CH₂—COOH	—CH₂—COOH	Cl	K₂S₂O₅	140–141	100	M-26
2-Chloro-2-phenylvinyl	2-Chloro-2-phenylvinyl	Cl	Na₂S	59–61	100	M-46
C₆H₅—	C₆H₅—	Cl	NaHSO₃	182–183 (16.5)	100	A-2,L-34
		Br	Na₂S	182–183 (16)	100	R-7
			NaHSO₃	182 (14)	92	L-10
			LiAlH₄			L-2
4-Bromophenyl	4-Bromophenyl	Cl	Na₂S	121		T-1
4-Methylphenyl	4-Methylphenyl	Br	Na₂S	210 (16)	100	R-7
			Na₂SO₃	70	—	L-13
2-Methylphenyl	2-Methylphenyl	Br	NaHSO₃	202–203 (15)		L-13
				38		
4-Methylphenyl	Phenyl	Br	NaHSO₃	207–208 (16)	87	L-20
				84		
4-Methoxyphenyl	4-Methoxyphenyl	Cl	Zn + glac. AcOH	54	90	B-22
						L-20
			NaHSO₃	50		L-13
			Na₂S	56	93	T-1
		Br	NaHSO₃	54		P-6

(continued)

239

TABLE XXIX (Continued)

R¹	Rᵃ	X	Reducing agent	R–Te–R¹ m.p., b.p. (torr)	Yield, %	Ref.
	Phenyl	Cl	Na₂S	60.5–61.5	87[a]	R-11
	1-Naphthyl	Cl	Na₂S	72	95[a]	R-11
	2-Naphthyl	Cl	Na₂S	87–88	83.5	R-11
4-Methoxyphenyl	4-Dimethylaminophenyl	Cl	Na₂S	96–97	100	P-7
		Br	Na₂S	96–97	100	P-7
	C_2H_5—	Cl	Na₂S	120 (1)	30[a]	V-4
	CH_3—	I	$K_2S_2O_5$	150–152 (20)	67	R-6
	Benzyl	Cl	Na₂S	36–38	100	V-4
	4-Ethoxyphenyl	Cl	Zn in glac. AcOH	45		M-38
4-Ethoxyphenyl	4-Ethoxyphenyl	Cl	Zn in glac. AcOH	64		M-38
		Cl	NaHSO₃	64		L-13
		Br	NaHSO₃	64		P-6
	4-Dimethylaminophenyl	Cl	Na₂S	126.7	100	P-7
		Br	Na₂S	126.7	100	P-7
	2-Naphthyl	Cl	Na₂S	65–66	100	P-9
	—CH₂ CH—CH₂C(C₆H₅)₂CO | O	Cl	NaHSO₃		93	M-45

		X	Reagent	mp or bp (°C (mm))	Yield (%)	Ref.
4-Phenoxyphenyl	Phenyl	Cl	Na_2S	177 (0.03)	70[a]	R-11
	1-Naphthyl	Cl	Na_2S	71–73	70[a]	R-11
	2-Naphthyl	Cl	Na_2S	69–70	87.2[a]	R-11
4-Phenoxyphenyl	4-Dimethylaminophenyl	Cl	Na_2S	75.5–76.5	100	P-7
	4-Dimethylaminophenyl	Br	Na_2S	75.5–76.5	100	P-7
4-Dimethylaminophenyl	4-Dimethylaminophenyl	Cl	$K_2S_2O_5$	128–130		M-36
2-Thienyl	2-Thienyl	Br	$SnCl_2/HCl$			K-7
		I	$Na_2S_2O_3$			K-8
2-Naphthyl	2-Naphthyl	Cl	Na_2S	144–145	93[a]	R-11
2-Naphthyl	Benzyl	Cl	Na_2S	58–59	100	V-4
	$CH_2\!-\!CH\!-\!CH_2\!-\!C(C_6H_5)_2CO$ (O bridge)	Cl	Na_2S			M-42
	Cyclohexyl	Cl	Na_2S	190 (0.024)	59[a]	V-4
	Phenyl	Cl	Na_2S	49–50	75.5[a]	R-11
1-Naphthyl	1-Naphthyl	Cl	Na_2S	93–94	91.5[a]	R-11
1-Naphthyl	Phenyl	Cl	Na_2S	147–148 (0.01)	80.3[a]	R-11

[a] The dichloride, obtained from $ArTeX_3$ and $RHgX$, was not isolated in pure form, but treated with Na_2S. The yield of the overall process is given.

The oil, separated from the water, was dried over Na_2SO_4. The telluride distilled at 94°/770 torr.

Bis(p-methoxyphenyl)tellurium Dichloride (Zn Process; M-32): Bis(*p*-methoxyphenyl)tellurium dichloride (10 g) and 5 g of zinc dust in 150 ml of benzene were heated at reflux temperature for 24 hr. The insoluble residue formed was separated by filtration and the filtrate was concentrated. The white crystals which separated still contained chlorine. More zinc dust was added. The mixture was refluxed further until all of the halogen was eliminated. The product, bis(*p*-methoxyphenyl) telluride, recrystallized from dilute alcohol melted at 53–54°. The yield was 37%.

2. Formation of Ditellurides by the Action of Na_2S or $K_2S_2O_5$ upon Certain Bis(organo)tellurium Dihalides

Petragnani (P-7) observed that arylacetonyl, arylphenacyl, and aryl 2,4-dihydroxyphenyltellurium dichlorides when treated with excess $Na_2S \cdot 9H_2O$ at 95–100°, sodium bisulfite, zinc in chloroform or hydrazine sulfate gave ditellurides.

The ditellurides melted at 59–60° (R = CH_3), 107° (R = C_2H_5) and 88° (R = C_6H_5).

2,2-Diphenyl-5-(4-ethoxyphenyldiiodotelluro)-4-pentanolactone   reduced with $NaHSO_3$ at room temperature gave the telluride while with sodium sulfide at 100° bis(4-ethoxyphenyl) ditelluride was formed (M-42).

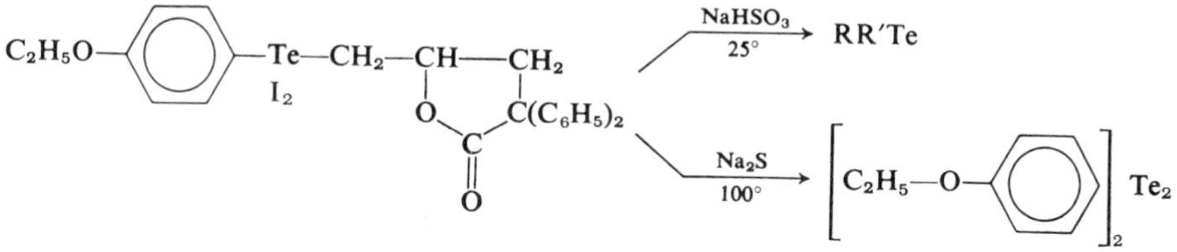

4-Methoxyphenyl methyltellurium diiodide, when reduced with alkaline $K_2S_2O_5$, produced the unsymmetric telluride (R-6). The diiodide shaken with an aqueous solution of $K_2S_2O_5$ in the presence of ether, however, gave, besides the expected telluride, as the main product, bis(4-methoxyphenyl) ditelluride (M-28). The authors propose that the diiodide is first partially reduced and then loses methyl iodide intramolecularly.

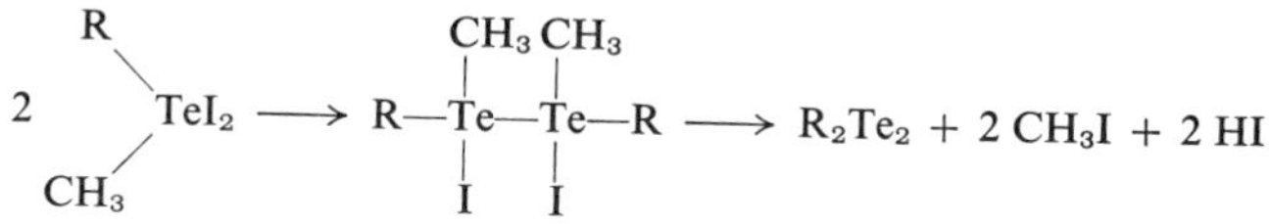

Rheinboldt (R-10,p.1069) however, pointed out that cleavage of 4-methoxyphenyl methyl telluride by HI to methyl iodide and 4-methoxybenzenetellurol with subsequent oxidation of the tellurol to the ditelluride is a more likely mechanism.

3. Reactions with Organometallic Reagents

Bis(organo)tellurium dihalides are known to react with Grignard reagents, with phenyllithium, with diphenylmercury, and with diaryl polonides.

The reactions with Grignard reagents are the most important for preparative purposes. The following scheme identifies the possible products:

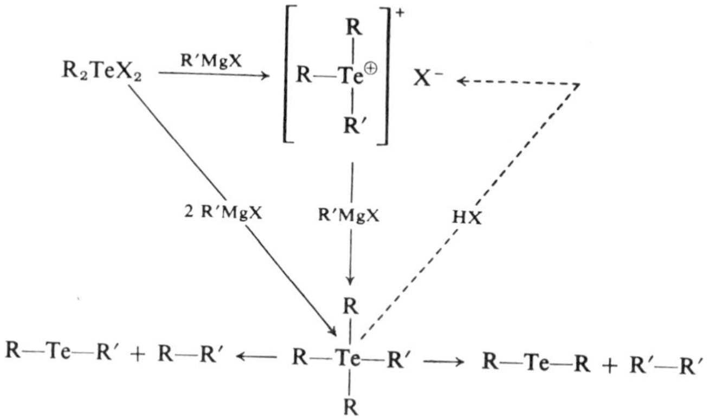

Lederer (L-20,L-28) has shown that triaryltelluronium halides were produced when a diaryltellurium dichloride dissolved in toluene was rapidly poured into a solution containing a two- to threefold molar excess of the Grignard reagent. After shaking, the reaction mixture was rapidly hydrolyzed by dilute hydrochloric acid. The telluronium iodide was isolated following the addition of potassium iodide to aqueous solutions of the crude products.

These telluronium salts (Table XXX lists the compounds prepared by this method) are not necessarily formed according to the equation

$$R_2TeX_2 + R'MgX \longrightarrow R_2R'TeX + MgX_2$$

TABLE XXX
Bis(organo)tellurium Dichlorides + Organomagnesium Bromide $\longrightarrow$
Tris(organo)telluronium Halide

R₂TeCl₂	R¹MgBr	R₂R¹TeI[a]		
R	R¹	m.p.	Yield, %	Ref.
Phenyl	Phenyl	247–248	70	L-28
	4-Methylphenyl	219–220	63	L-28
	3-Methylphenyl	202	60	L-28
	2-Methylphenyl	175–176	36	L-28
	3,4-Dimethylphenyl	114–115	22	L-28
	2,4-Dimethylphenyl	103	40	L-28
	2,5-Dimethylphenyl	213–214	36	L-28
	2,4,6-Trimethylphenyl	153–154	10	L-28
	4-Methoxyphenyl		70	L-28
	3-Methoxyphenyl[b]	95	6	L-28
	2-Methoxyphenyl	225–227	68	L-28
	4-Ethoxyphenyl	131		L-28
	2-Ethoxyphenyl	247–248 (dec.)	54	L-28
	1-Naphthyl	148	77	L-28
4-Methylphenyl	Phenyl	210	75	L-20
2-Thienyl[c]	2-Thienyl	244.5		K-7

[a] The crude telluronium chloride and bromide mixture was converted to the less soluble iodide by treatment with KI.
[b] 3-Methoxyphenylmagnesium iodide was used.
[c] Dibromide was used.

Since enough Grignard reagent is available for the formation of a tetraaryltellurium compound, it is a possible intermediate on the way to the telluronium salt. Hellwinkel and Fahrbach (H-3,H-4) and Wittig and Fritz (W-3) have demonstrated that tetraorganotellurium compounds are easily hydrolyzed to telluronium salts (see also Sections VIII and VII-8).

When a three- to fivefold molar excess of methylmagnesium iodide was employed and when the reaction mixture was heated on the water bath for several hours, the tellurium dihalides were reduced to the tellurides.

$$R_2TeX_2 + 2 CH_3MgI \longrightarrow R_2Te + C_2H_6 + MgX_2 + MgI_2$$

This reaction could again proceed via a tetraorganotellurium intermediate. Hellwinkel and Fahrbach have shown (H-4) that tetraalkyltellurium compounds decompose to hydrocarbon and telluride. Such a reaction mechanism would also explain the observation by Lederer that diphenyltellurium dibromide and 2-methylphenylmagnesium bromide (L-29) and bis(2-methylphenyl)tellurium dibromide and phenylmagnesium bromide (L-29) gave the symmetric and unsymmetric tellurides. From the reaction between bis(4-methylphenyl)tellurium dichloride and phenylmagnesium bromide only the mixed telluride was isolated (L-20). A fivefold molar excess of arylmagnesium bromide was employed in these reductions.

The bis(organo)tellurium dihalides reduced by Grignard reagents are tabulated in Table XXXI together with the yields and the melting and boiling points of the resulting tellurides.

Nefedov and co-workers (N-4,N-7) found that diphenyltellurium dichloride and bis(4-methylphenyl)tellurium dihalides are reduced to the tellurides by the respective bis(aryl) polonides.

$$R_2TeX_2 + R_2Po \longrightarrow R_2Te + R_2PoX_2$$

Diethyltellurium diiodide reacted with diphenylmercury in $CHCl_3$ yielding the adduct $(C_2H_5)_2TeI_2 \cdot Hg(C_6H_5)_2$ with a melting point of 94°. No reaction was observed to take place between bis(4-methylphenyl)mercury and dimethyl- and diethyltellurium diiodides (G-4).

Finally diphenyltellurium dichloride and phenyllithium (free from lithium bromide) in absolute ether gave tetraphenyltellurium in 51.8% yield (W-3).

Experimental. Bis(4-methoxyphenyl) Phenyltelluronium Iodide (L-20):   Bis-(4-methoxyphenyl)tellurium dichloride (10 g, 0.026 moles) dissolved in 250 ml of toluene was rapidly poured into a Grignard solution prepared from bromobenzene (12.4 g, 0.08 moles) and magnesium (1.96 g, 0.08 moles) in 60 ml of absolute ether. The mixture was stirred and immediately hydrolyzed with 25 ml of water. The white precipitate which formed was separated by filtration and extracted with $CHCl_3$ and ethanol. The filtrate and the extracts were combined and evaporated. The residue, dissolved in a large amount of hot water and treated with KI, gave 75% of the telluronium iodide melting at 210°.

The hydrolysis can also be carried out with dilute hydrochloric acid. The precipitate thus formed was dissolved in water and treated with KI (L-28).

TABLE XXXI
Reduction of Bis(organo)tellurium Dihalides by Grignard Reagents

RR'TeX₂				Product			
R	R'	X	R"MgX	Formula	m.p., b.p. (mmHg)	Yield, %	Ref.
Phenyl	R	Br	2-CH₃C₆H₄MgBr	$R_2Te + RR''Te$	212–213 (22)		L-29
Phenyl	2-Methylphenyl	Br	CH₃MgI	$RR'Te$		77	L-29
4-Methylphenyl	R	Br	C₆H₅MgBr	$RR''Te$	64		L-20
3-Methylphenyl	R	Br	CH₃MgI	R_2Te	205–206 (18)	89	L-16
2-Methylphenyl	R	Br	C₆H₅MgBr	$R_2Te + RR''Te$			L-29
2,4-Dimethylphenyl	R	Br	CH₃MgI	R_2Te	202–203 (10)	92	L-14
4-Methoxyphenyl	R	Br	CH₃MgI	R_2Te	237–243 (14) 57		L-17
3-Methoxyphenyl	R	Br	CH₃MgI	R_2Te	247–252 (34–6)		L-26
2-Methoxyphenyl	R	Br	CH₃MgI	R_2Te	248–251 (30) 74		L-27
4-Ethoxyphenyl	R	I	CH₃MgI	R_2Te	235–240 (18) 63		L-25
2-Ethoxyphenyl	R	Br	CH₃MgI	R_2Te	245 (18)	60	L-23
1-Naphthyl	R	Br	C₂H₅MgI	R_2Te	126.5		L-24
2-Thienyl	R	I	CH₃MgI	R_2Te			K-8

246

4-Methylphenyl Phenyl Telluride from Bis(4-methylphenyl)tellurium Dibromide and Phenylmagnesium Bromide (L-20): Bis(4-methylphenyl)tellurium dibromide (90 g, 0.19 moles) was added slowly to a Grignard solution prepared from bromobenzene (157 g, 1 mole) and magnesium (25 g, ~1 mole) in 500 ml of absolute ether. After boiling for three hours on the water bath, the mixture was hydrolyzed with ice water and slightly acidified with hydrochloric acid. The solids were filtered off and discarded. The ether layer, shaken with water, dried over KOH, and distilled under an atmosphere of CO_2 gave a residue which was freed from water in vacuum. The telluride was converted to the dibromide.

Bis(4-ethoxyphenyl) Telluride from the Dibromide and Methylmagnesium Iodide (L-23): Methyl iodide (80.5 g, 0.56 moles) and magnesium (14.2 g, 0.58 moles) were combined in absolute ether. After all the magnesium had dissolved, the mixture was boiled for one hour with fresh magnesium to ensure that all the methyl iodide was consumed (to avoid telluronium salt formation). The dibromide (50 g, 0.094 moles) was slowly dropped into the Grignard solution. If the reaction is too vigorous, the mixture should be cooled. After refluxing for two hours the mixture was hydrolyzed with dilute hydrochloric acid. The ether layer, washed with water and dried over KOH, was distilled under a CO_2 atmosphere. The telluride distilled between 244 and 244.5°/18 mm (60% yield).

Tetraphenyltellurium (W-3): To a suspension of diphenyltellurium dichloride (3.53 g, 0.01 moles) in 30 ml of absolute ether in a two-necked flask, was added, under an atmosphere of nitrogen, 25 ml of a 1.04N lithium bromide-free solution of phenyllithium in ether (a total of 0.026 moles). A positive Gilman test was obtained after the addition of 2.6 moles of phenyllithium per mole of dichloride because of the formation of the intermediate $[(C_6H_5)_5Te]Li$.

The yellowish crystals which precipitated from ether at $-10°$ were handled in a closed system. Final purification was accomplished under an atmosphere of nitrogen by dissolving 6.8 g of $(C_6H_5)_4Te$ in 20 ml of anhydrous benzene. Any undissolved residue was removed by filtration through a porous glass filter. The benzene was removed *in vacuo* and the remaining solid was recrystallized from 100 ml of absolute ether in a sealed tube. The yellowish, monoclinic crystals melted at 104–106°, with decomposition.

4. Halogen Exchange and Replacement Reactions

The insolubility of the silver halides has been utilized for halogen exchange reactions. Thus, bis(organo)tellurium difluorides and dibromides were obtained by the reaction between diiodides and AgF(E-4) and AgBr in acetone (V-4), respectively. De Moura Campos and co-workers (M-47) synthesized

TABLE XXXII

Exchange and Replacement of the Halogen Atoms in Bis(organo)tellurium Dihalides

RR¹TeX₂				RR¹TeY₂			
R	R¹	X	Reagent	Y	m.p.	Yield, %	Ref.
CH₃—	R	I	AgCN	CN	90		L-36
			HNO₃	NO₃	142		V-1,V-3
			Ag picrate	Picrate			V-1
			Ag benzoate	Benzoate	154		V-1
			AgF	F	84		E-4
C₂H₅—	R	Cl	HI	I			W-5
C₆H₅—	R	I	AgF	F	154		E-4
	R	Br	AgF	F	154		E-4
C₆H₅—	4-Methoxyphenyl	Cl	HBr	Br	148.5–149.5		R-11
	1-Naphthyl	Cl	HBr	Br	180.5–182.0		R-11
	2-Naphthyl	Cl	HBr	Br	186.7		R-11
	4-Phenoxyphenyl	Cl	HBr	Br	156.5–158.0		R-11
4-Methylphenyl	R	I	Ag benzoate	Benzoate	230–238	96	M-47
			Ag butyrate	Butyrate	125–127		M-47
			Ag caprilate	Caprilate	85–86		M-47
			Ag laurate	Laurate	66–68		M-47
C₆H₅CH₂—	4-Methoxyphenyl	Cl	KI	I	133 (dec.)	76	V-4
		I	AgBr/HBr	Br	153 (dec.)	77	V-4
	2-Naphthyl	Cl	KI	I	144 (dec.)	100	V-4

C₆H₅CH₂—

R	Aryl	X	Reagent	X'	M.p.	%	Ref.
	2-Naphthyl	I	AgBr	Br	149 (dec.)	80	V-4
4-Methoxyphenyl	2,4-Dihdroxyphenyl	Cl	KI	I	130 (dec).	100	P-7
			HBr	Br	179–180	100	P-7
	4-Dimethylaminophenyl	Cl	HBr	Br	183–184 (dec.)	100	P-7
		Br	HCl	Cl	170–172 (dec.)	100	P-7
4-Ethoxyphenyl	1-Naphthyl	Cl	HBr	Br	220–222		R-11
	2-Naphthyl	Cl	HBr	Br	163–164		R-11
	2,4-Dihydroxyphenyl	Cl	KI	I	137ª	100	P-7
		Br	HBr	Br	190–191 (dec.)	100	P-7
	4-Dimethylaminophenyl	Br	HCl	Cl	121–122 (dec.)	100	P-7
		Br	HCl	Cl	153–154 (dec.)	100	P-7
4-Phenoxyphenyl	2,4-Dihydroxyphenyl	Cl	KI	I	122–124 (dec.)	100	P-7
		Br	HBr	Br	181–182 (dec.)	100	P-7
	4-Dimethylaminophenyl	Cl	HBr	Br	188–189 (dec.)	100	P-7
	1-Naphthyl	Cl	HBr	Br	189–190		R-11
	2-Naphthyl	Cl	HBr	Br	146–147		R-11
4-Dimethylaminophenyl	R	Cl	NaI	I			M-36
2-Naphthyl	1-Naphthyl	Cl	HBr	Br	215–217		R-11
2-Phenyl-2-chlorovinyl	R	Cl	KI	I	168–169 (dec.)	98	M-46

ª The other two modifications melted at 105° and 127°.

some bis(4-methylphenyl)tellurium dicarboxylates from the dichloride and silver carboxylates in dioxane.

$$R_2TeCl_2 + 2\ R'CO_2Ag \longrightarrow R_2Te(O_2CR')_2 + 2\ AgCl$$

When R' was C_5H_{11} and C_9H_{19} hydroxymonocarboxylates, melting at 136–137° and 96–99°, respectively, were formed.

Halogen interconversion has also been accomplished by boiling the dihalide with an excess of concentrated hydrohalic acid (P-7,R-11). Potassium iodide in aqueous solution and tellurium dichlorides in ethanol or methanol (M-46,V-4) gave rise to the diiodides.

Table XXXII and the following reaction scheme will show the extent to which halogen exchange and replacement have been investigated. The reactions with silver oxide are discussed in Section VI-A-5.

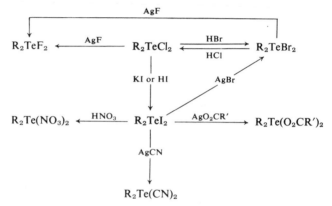

Experimental. 4-Dimethylaminophenyl 4-Phenoxyphenyltellurium Dibromide (P-7): The dichloride (0.16 g, 0.00033 moles) was treated with 20 ml of boiling 40% hydrobromic acid for three minutes. Ammonium hydroxide was added until the color of the precipitate changed to orange. The filtered solid, washed with water, was once more treated with 10 ml of boiling 40% HBr. After addition of ammonia a quantitative yield of the dibromide was obtained. Recrystallized from benzene/petroleum ether it melted at 188–189° (dec.).

Benzyl 4-Methoxyphenyltellurium Diiodide (V-4): A solution of the dichloride (1 g, 0.0025 moles) in a minimum amount of ethanol was combined with a concentrated aqueous solution of KI. The diiodide was extracted with benzene at room temperature from the filtered precipitate. Addition of petroleum ether to the benzene extract gave the diiodide (75.8% yield) which decomposed at 133°.

Bis(4-methylphenyl)tellurium Dibenzoate (M-47): A mixture of the dichloride (1.00 g, 0.0025 moles) and silver benzoate (1.22 g, 0.0053 moles) in 25 ml of dioxane was refluxed for four hours under an inert atmosphere in the dark. The organic solution filtered from the silver chloride gave, upon evaporation, the dibenzoate, which recrystallized from benzene/petroleum ether melted at 230–238° (96% yield).

5. Hydrolysis of Dihalides

Bis(organo)tellurium dihalides undergo hydrolysis upon treatment with water, ammonia, alkali metal hydroxides, and carbonates. The nature of the products isolated depends upon the hydrolyzing agent and the tellurium dihalide being hydrolyzed. The following scheme describes the reactions involved.

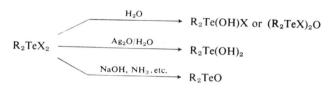

The hydrolysis with water to the bis(organo) hydroxytelluronium chloride proceeded only with the dichlorides and dibromides. The diiodides were found to be inert toward water (L-5,V-1).

The hydroxy halides formed have been isolated. In solution they dissociate into halide ions and the bis(organo) hydroxytelluronium ion (G-4,N-9). In the case of a number of tellurium dihalides, hydrolysis by water produced only the anhydrides of the hydroxy halides (see Table XXXIII for individual compounds). The question as to whether the isolated compound is the hydroxide or the anhydride cannot be answered on the basis of elemental analyses alone for high molecular mass compounds because the calculated values for the two structures lie too closely together. Some of the formulas in Table XXXIII, therefore, are not too dependable.

When moist silver oxide is brought in contact with the dihalides complete hydrolysis occurs. The halide ions are removed as insoluble silver halides. The tellurium compound in such a solution has the formula of a dihydroxide. The dihydroxide, however, cannot be isolated by evaporation of the solvent. Water is lost from the solid thus obtained and the dihydroxide is converted to the oxide during this process (L-5).

Bis(organo) telluroxides were obtained as white precipitates when dihalides were permitted to react with ammonia or sodium hydroxide. Lederer, however, reported that when bis(3-methylphenyl)tellurium dibromide was heated on the water-bath with sodium hydroxide he obtained a substance, insoluble in benzene, to which he assigned the formula $R_2Te(ONa)_2$. The oxides are soluble in a variety of organic solvents.

TABLE XXXIII
Hydrolytic Reactions of Bis(organo)tellurium Dihalides

RR¹TeX₂			Reagent	Product			Ref.
R	R¹	X		Formula	Yield, %	m.p.	
CH_3-	R	I	Ag_2O/H_2O	$R_2Te(OH)_2$			V-1
			H_2O	$R_2Te(OH)_2$			V-1
			NH_3	$(R_2TeI)_2O$		115° (dec.)	V-3
			NH_3, $R_2Te(OH)_2$	$R_2Te(OTeIR_2)_2$[a]		152° (dec.)	V-3
			NH_3, NaOH, K_2CO_3, $R_2Te(OH)_2$	$(ITeR_2)_2O$		112–125	V-3
		Br	NH_3	$(R_2TeBr)_2O$			H-2
		Cl	NH_3	$(R_2TeCl)_2O$			H-2,W-6
C_2H_5-	R	I	Ag_2O/H_2O	$R_2Te(OH)_2$			G-4,V-3
			NH_3	$(R_2TeI)_2O$			H-2,W-5
			NH_3 con.	$(R_2TeI)_2O$		107	G-4
		Br	$R_2Te(OH)_2$	$R_2Te(OH)Br$			G-4
		Br	NH_3	$(R_2TeBr)_2O$			H-2,W-5
		Cl	NH_3, KOH, NaOH	$(R_2TeCl)_2O$			H-2,W-5
			Ag_2O/H_2O	R_2TeO			H-2,M-7
$C_5H_{11}-$	R	Cl	Ag_2O/H_2O	$R_2Te(OH)_2$			W-7
C_6H_5-	R	Br	5% NaOH	R_2TeO	90	191, 185	K-5,L-5, R-9
4-Methylphenyl	R		H_2O	$R_2Te(OH)Br$		264–265	L-5
		Cl	H_2O	$R_2Te(OH)Cl$	88	233–234, 250	L-5,T-1
		Br	5% NaOH	R_2TeO	84	167	L-5,R-9
			H_2O	$R_2Te(OH)Br$		269	L-5
		Cl	H_2O	$R_2Te(OH)Cl$		261–263	L-5

		X	Reagent	Product	Yield	mp (°C)	Ref.
	Phenyl	Br	NaOH	R₂TeO		154–155	L-20
		Cl	H₂O boiling	(RR¹TeCl)₂O		243–244	L-20
3-Methylphenyl	R	Br	NH₃	R₂TeO		163–164	L-16
		Cl	H₂O	R₂Te(OH)Cl		87	L-16
2-Methylphenyl	R	Br	H₂O	(R₂TeBr)₂O		224.5 (dec.)	L-5
			5% NaOH	R₂TeO		205–206 (dec.)	L-5
			R₂TeO	(R₂TeBr)₂O		224–225 (dec.)	L-5
		Cl	H₂O	(R₂TeCl)₂O		220–222	L-5
2,4-Dimethyl-phenyl	Phenyl	Br	NH₃	R₂TeO		217	L-29
	R	Br	NaOH	R₂TeO		216–217	L-14
2,5-Dimethyl-phenyl		Cl	H₂O	(R₂TeCl)₂O	100	239–240	L-14
	R	Br	NaOH	R₂TeO		225–226	L-14
2,4,6-Trimethyl-phenyl		Br	H₂O	R₂Te(OH)Br		223–227	L-14
		Cl	H₂O	R₂Te(OH)Cl			L-14
	R	Br	NaOH, NH₃	R₂TeO	100	204–205	L-15
4-Methoxyphenyl		Cl	H₂O boiling	(R₂TeCl)₂O		237	L-15
	R	Br	NaOH	R₂TeO		190–191	L-17
		Cl	NaOH	R₂TeO			M-32
			H₂O	(R₂TeCl)₂O		250	T-1
			H₂O	R₂Te(OH)Cl		197	L-17
—CH₂C(O)CH₃		Cl	H₂O	RTeOCl		225–235 (dec.)	P-7
—CH₂C(O)C₆H₅		Cl	H₂O boiling	RTeOCl		225–235 (dec.)	P-7
		Br	H₂O boiling	RTeOBr			P-7

253

(continued)

TABLE XXXIII (*Continued*)

| RR¹TeX₂ | | | | Product | | | |
R	R¹	X	Reagent	Formula	Yield, %	m.p.	Ref.
3-Methoxyphenyl	R	Br	NH₃	R_2TeO		90[b]	L-26
2-Methoxyphenyl	R	Br	NH₃	R_2TeO		205–206	L-27
4-Ethoxyphenyl	R	Br	2N NaOH	R_2TeO		181 (dec.)	M-38
		Cl	2N NaOH	R_2TeO		181 (dec.)	M-38
			H₂O boiling	$R_2(TeCl)_2O$		193 (dec.)	M-38
	—CH₂C(O)CH₃	Cl	H₂O	RTeOCl			P-7
	—CH₂C(O)C₆H₅	Cl	H₂O	RTeOCl			P-7
2-Ethoxyphenyl	R	Br	NH₃ con.	R_2TeO		205–206	L-23
4-Phenoxyphenyl	—CH₂C(O)C₆H₅	Cl	H₂O boiling	RTeOCl			P-7
1-Naphthyl	R	Br	NH₃	R_2TeO		224–225 (dec.)	L-24

[a] The hydrate $(CH_3)_2Te[OTeI(CH_3)_2]_2 \cdot 1/2\ H_2O$ melted at 145° (V-3).
[b] Amorphous.

254

Vernon (V-3) found that the hydrolytic products of dimethyltellurium diiodide vary with the amount of reagent employed. When an excess of ammonia was employed, the compound isolated had the formula $(CH_3)_2Te$-$[OTe(I)(CH_3)_2]_2$, melting at 152°. Avoiding a large excess, $[(CH_3)_2TeI]_2O$ was formed. The reaction of the diiodide with the dihydroxide in the proper proportions gave the same compounds as those obtained with ammonia.

Morgan and Drew (M-28) observed a rather surprising difference in the resistance of various diaryltellurium dichlorides toward alkaline hydrolysis. While bis(4-ethoxyphenyl)tellurium dichloride was found to undergo attack by boiling potassium carbonate only very slowly with the liberation of phenetole, bis(2,4-dimethoxyphenyl)tellurium dichloride was hydrolyzed much more rapidly under the same conditions with the liberation of resorcinol dimethyl ether. Bis(4-phenoxyphenyl)tellurium dichloride was unaffected by a cold solution of K_2CO_3. It was decomposed by the boiling reagent with the formation of diphenyl ether.

Bis(organo)tellurium dihalides of the general formula

$$RO-\underset{X_2}{\underset{|}{C_6H_4}}-Te-CH_2-\overset{O}{\overset{\|}{C}}-R'$$

$(R = CH_3, C_2H_5, C_6H_5; R' = CH_3, C_6H_5)$

are unstable toward water. The tellurium–carbon bond is cleaved by cold water in the case of the acetonyl derivatives and by boiling water in the case of the phenacyl derivatives (P-7).

$$R-O-\underset{X_2}{\underset{|}{C_6H_4}}-Te-CH_2\overset{O}{\overset{\|}{C}}-R \xrightarrow{H_2O} RO-C_6H_4-Te\overset{O}{\underset{X}{<}} + X-CH_2\overset{O}{\overset{\|}{C}}-R$$

Experimental. Diphenyl Hydroxytelluronium Chloride (L-5):   Finely powdered diphenyltellurium dichloride (5 g) was boiled in 100 ml of water. The remaining residue was brought into solution with an additional amount of water. The hydroxy chloride precipitated on cooling. The product was recrystallized twice from water and melted at 233–234° (88% yield).

Bis(2,5-dimethylphenyl) Telluroxide (L-14):   The finely powdered dibromide (5 g) was mixed with a sodium hydroxide solution prepared by dissolving 16 g of NaOH in 100 ml of water. The mixture was heated for one hour on the water bath with constant stirring. The oxide was then filtered, washed with cold water, and dried. Recrystallized from benzene, the oxide melted at 225–226° (100% yield).

6. Reactions with Ditellurides

The reactions of bis(organo)tellurium dihalides with ditellurides have been discussed in Section V-3.

7. Formation of Addition Compounds

A whole series of unstable compounds of the type $(CH_3)_2TeI_2 \cdot nNH_3$, where $n = 1 \ldots 6$, have been investigated by Vernon (V-3). The iodide $(CH_3)_2TeI_2$ is a bright red salt, but it rapidly becomes decolorized and is transformed into a milky, semi-fluid mass by the action of dry ammonia. The composition reaches a limiting stoichiometry which corresponds to the formation of $Te(CH_3)_2I_2 \cdot 6NH_3$. Vernon postulated the existence of a whole series of compounds $Te(CH_3)_2I_2 \cdot nNH_3$. The red color of the ammonia-free compounds diminishes with increasing value of n and disappears completely for $n = 3$. The compounds remain solid up to $n = 3$, which has the appearance of white porcelain, but become increasingly fluid as n increases. The process is entirely reversible and the iodide can be recovered unchanged.

$$Te(CH_3)_2I_2 \cdot nNH_3 \longrightarrow Te(CH_3)_2I_2 + nNH_3$$

The stability of the compounds is approximately inversely proportional to n; when $n = 2$, ammonia is lost very slowly. The action of aqueous ammonia upon the iodide is totally different from that of the dry gas and appears to be much more complicated (see Section VI-A-5).

Vernon (V-3), on combining equimolar quantities of dimethyltellurium diiodide and iodine in ethyl acetate, obtained a precipitate of bluish-black crystals of the formula $(CH_3)_2TeI_4$, which were not very stable and released traces of iodine.

Gilbert and Lowry (G-4) prepared the corresponding ethyl derivative, $(C_2H_5)_2TeI_4$ by refluxing the components dissolved in $CHCl_3$ for one minute. Some decomposition was observed during this reaction. The tetraiodide melted at 98°.

A green platinum salt was recovered from a mixture of chloroplatinic acid and dimethyltellurium diiodide (V-1).

For complexes formed with tris(organo)telluronium compounds see Section VII-5.

8. Reactions Affecting the Organic Part of the Molecule

Only a few reactions are known which take place in the organic part of the molecule without effect on the tellurium atom. Petragnani (P-7) treated aryl 2,4-dihydroxyphenyltellurium dichlorides with acetic anhydride and a drop of H_2SO_4 and isolated upon the addition of water, the corresponding diacetates.

$$R-O-\bigcirc-Te-\bigcirc-OH + 2\ CH_3COOH \xrightarrow{H_2SO_4}$$

(with Cl_2 on the Te and OH on the second ring)

$$R-O-\bigcirc-Te-\bigcirc-O\overset{O}{\overset{\|}{C}}-CH_3 + H_2O$$

(with Cl_2 on the Te, and $O-\overset{\|}{\underset{O}{C}}-CH_3$ on the second ring)

R = CH₃, m.p. 166–167°, 88% yield
R = C₂H₅, m.p. 162–168°, 88% yield
R = C₆H₅, m.p. 172–173°, 100% yield

Bis(4-dimethylaminophenyl)tellurium dichloride formed a hydrochloride (m.p. 136–137°). It underwent nitration and its nitro-derivative was reducible to a diazotizable base without elimination of tellurium (M-36).

9. Tellurium–Carbon Bond Cleavage and Elimination of the Tellurium Atom

Tellurium–carbon bond cleavage, which occurs with certain sensitive bis(organo)tellurium dihalides, has already been discussed in this section. Specifically, the reader is referred to the introduction to Section VI-A, where the conversion of Vernon's "α-base" into the "β-base" is dealt with, to Sections VI-A-1 and 2 (bond cleavage by reducing agents) and to Section VI-A-5 (hydrolytic bond cleavage).

In addition, dibutyltellurium diiodide in acetone solution treated with hydriodic acid, followed by evaporation of the acetone at room temperature gave butyltellurium triiodide. A tellurenyl compound is claimed to be a possible intermediate (B-2).

$$R_2TeI_2 \xrightarrow{HI} RTeI + RI$$
$$RTeI + I_2 \longrightarrow RTeI_3$$

The iodine might be formed by atmospheric oxidation of HI.

Bis(4-dimethylaminophenyl)tellurium dichloride suspended in dilute hydrochloric acid at −5° treated with an excess of aqueous sodium nitrite, the temperature being finally raised to +5°, gave an 80% yield of TeO_2 and 4-nitrosodimethylaniline (M-36).

Bis(4-ethoxyphenyl)tellurium dihalides were decomposed by concentrated nitric acid with the formation of varying proportions of o-halogenated p-nitrophenetoles (M-38).

B. Reactions of Bis(organo)tellurium Dihydroxides

Bis(organo)tellurium dihydroxides seem to exist only in aqueous solutions. Attempts to isolate them by evaporation of the solvent lead to the formation of the corresponding oxides through loss of one molecule of water (L-5).

$$R_2Te(OH)_2 \longrightarrow R_2TeO + H_2O$$

Upon titration of diethyltellurium dihydroxide with $1N$ hydrochloric acid the end point corresponded to the formation of $(C_2H_5)_2Te(OH)Cl$. There was no sign of the second hydroxyl group being neutralized by addition of excess acid (G-4).

Hydrohalic acids of higher concentration convert the dihydroxides into dihalides.

$$R_2Te(OH)_2 + 2 HX \longrightarrow R_2TeX_2 + 2 H_2O$$
$$R = CH_3 \text{ (V-1,V-2,W-6)}$$
$$R = C_2H_5 \text{ (V-3)}$$

The dialkyltellurium dihydroxides, $R_2Te(OH)_2$ gave, with nitric acid (V-1, V-3) and with $AgNO_3$ (V-1), the dinitrate ($R = CH_3$), with H_2SO_4 the disulfate ($R = CH_3$, W-6) and with $HClO_4$ (L-36), the diperchlorate ($R = CH_3$, m.p. 37°).

An aqueous solution of dimethyltellurium dihydroxide when evaporated with hydrogen peroxide gave white, insoluble dimethyl tellurone (L-36,V-2). Diethyl tellurone was obtained by evaporating to dryness the dihydroxide solution at 100°/15 torr and allowing the residue to cool in contact with air (G-4). Similarly, diphenyl tellurone was isolated (L-5). The transformation of the dihydroxides into Vernon's "β-bases" was discussed in Section VI-A.

The dihydroxides are reducible to the tellurides by SO_2 (W-6).

C. Reactions of Compounds with the Formula $R_2Te(OR')_2$

Only a small number of compounds having the general formula $R_2Te(OR')_2$ have been prepared. Their chemical properties have not been investigated in detail. The following reactions are reported: Attempts to isolate hydrolysis products after treatment of pentyl carbethoxymethyltellurium dibenzoate with alkali were unsuccessful (B-3). Diphenyltellurium diacetate, however, is reported to be converted to the oxide by 5% NaOH solution (L-5). Reduction of bis(4-methylphenyl)tellurium dibenzoate with hydrated sodium sulfide gave the telluride (M-47).

Vernon (V-3) hydrolyzed the compound $(CH_3)_2Te[OTe(I)(CH_3)_2]_2$ with excess of water to $(CH_3)_2TeI_2$ and $(CH_3)_2Te(OH)_2$; with an amount of water insufficient for complete hydrolysis, $[(CH_3)_2TeI]_2O$ and $(CH_3)_2Te(OH)_2$ were obtained. Silver oxide in water produced the dihydroxide.

D. Reactions of Bis(organo)tellurium Dinitrates

Bis(organo)tellurium dinitrates are known to be reduced to the telluride by sulfurous acid, to give dihalides when treated with concentrated hydrohalic acids, and to hydrolyze to the anhydride of bis(organo) hydroxytelluronium nitrate. Heeren (H-2) and Mallet (M-7) investigated the reduction of diethyl-tellurium dinitrate by H_2SO_3. The dinitrates $R_2Te(NO_3)_2$ (R = CH_3, C_2H_5 and C_5H_{11}) gave the dihalides with concentrated hydrohalic acids (H-2,V-3, W-6,W-7).

Lederer (L-18) observed the hydrolysis of diphenyl- and bis(2-methyl-phenyl)tellurium dinitrate by water.

$$2 R_2Te(NO_3)_2 + H_2O \longrightarrow (R_2TeNO_3)_2O + 2 HNO_3$$

E. Reactions of Bis(organo) Hydroxytelluronium Compounds and Their Anhydrides

It often is difficult to establish with certainty whether the authors were dealing with hydroxy compounds $R_2Te(OH)X$ or their anhydrides, $(R_2TeX)_2O$. This is especially true for the early investigations in this field, for example, by Wöhler, Mallet, and Heeren.

Lederer (L-5) found that the diphenyl hydroxy- and the bis(4-methyl-phenyl) hydroxytelluronium halides lose water upon heating: the chlorides at 150°, the bromides at 170°, and the iodides at 180°. Diphenyl hydroxy-telluronium bromide dissolved in glacial acetic acid produced the dibromide and the diacetate, which was converted into the oxide by treatment with 5% NaOH solution (L-5). Bis(4-methylphenyl) hydroxytelluronium iodide and the 2-isomer disproportionated in methanol above 50° into the diiodide and the oxide (L-5). The chloride ion in the hydroxy chlorides can be exchanged for a bromide or iodide ion by treatment with the respective potassium halides.

$$R_2Te(OH)Cl + KX \longrightarrow R_2Te(OH)X + KCl$$
$$R = C_6H_5, X = I; \text{m.p. } 214–215° \text{ (L-5)}$$
$$R = 4\text{-}CH_3C_6H_4, X = I; \text{m.p. } 203–204° \text{ (L-5)}$$

Similarly the anhydrides are capable of halogen exchange.

$$(RR'TeCl)_2O + 2 KX \longrightarrow (RR'TeX)_2O + 2 KCl$$
$$R = C_6H_5, R' = 4\text{-methylphenyl}, X = I, \text{m.p. } 201° \quad \text{(L-20)}$$
$$X = Br, \text{m.p. } 260 \quad \text{(L-20)}$$
$$R = R' = 2,5\text{-dimethylphenyl}, X = I, \text{m.p. } 70° \quad \text{(L-14)}$$

Heeren (H-2) and Wöhler (W-5,W-6) prepared compounds of the dimethyl and diethyl series, which probably have the formula $(R_2TeX)_2O$. The reaction of these substances with hydrohalic acids yielded the corresponding dihalides. The anhydrides, when shaken with silver sulfate, silver oxalate, silver cyanide, lead formate or silver acetate or by treatment with CO_2 or phosphoric acid, gave compounds containing the respective anions.

The hydroxy halides and their derivatives are reduced by SO_2 and Na_2S (H-2,M-47).

Butyl carbomethoxymethyl hydroxytelluronium nitrate and diethylamine in ether deposited a white solid which analyzed for $(C_4H_9)(CH_2CO_2H)TeO\cdot 2C_4H_9TeOOH$ (B-2).

F. Reactions of Bis(organo) Telluroxides

Bis(organo) telluroxides are very sensitive toward air. Lederer (L-5) reported that bis(4-methylphenyl) telluroxide in presence of air formed a product melting at 205–206°.

The oxides dissolve in water giving a basic solution, probably forming the dihydroxide (L-5).

$$R_2TeO \underset{-H_2O}{\overset{+H_2O}{\rightleftharpoons}} R_2Te(OH)_2$$

Diphenyl telluroxide gave the dinitrate upon treatment with nitric acid (L-18).

Heeren (H-2) and Mallet (M-7) reported that dialkyl telluroxides, R_2TeO ($R = CH_3$, C_2H_5), can be reduced to the telluride and converted to dihalides when acted upon by hydrohalic acids.

VII. REACTIONS OF TRIS(ORGANO)TELLURONIUM COMPOUNDS

Tris(organo)telluronium compounds, R_3TeX, are known where X is F^-, Cl^-, Br^-, I^-, OH^-, NO_3^-, $TeOCl_3^-$, SiF_6^{2-}, picrate$^-$, CrO_4^{2-}, $Cr_2O_7^{2-}$, and various complex transition metal anions. Triphenyltelluronium iodide was investigated as a possible reagent for the determination of bismuth and cobalt (P-15). Triphenyltelluronium chloride was not found to be suited as a reagent for various anions due to the high solubility of the salts (M-10). The polarographic behavior of the triphenyl- (M-10) and the triethyltelluronium chloride (S-13) was determined. Bis(dodecyl) ethyltelluronium chloride was treated with purified Bentonite in water to give a coating composition with excellent suspension and thickening properties (R-4).

Doering and Hoffman (D-9) studied the deuterium exchange between trimethyltelluronium iodide and D_2O. The complex salt tris(2-hydroxy-5-methylphenyl)telluronium trichlorooxotellurite decomposed in ethanol to TeO_2 and the telluronium chloride (M-37). Aside from these isolated investigations the following reactions of telluronium compounds have received attention:

1. Halogen exchange employing KBr, KI, AgCl, AgBr, $AgNO_3$, and Ag_2O/H_2O.

2. Formation of insoluble picrates.

3. Formation of mercuric halide adducts

4. Formation of salts with CrO_4^{2-}, $Cr_2O_7^{2-}$, $AuCl_3$, $ZnCl_2$, H_2PtCl_6, $CuCl_2$, $SnCl_2$, $FeCl_3$, and $SnBr_4$.

5. Reactions with CH_3TeX_3 and $(CH_3)_2TeI_2$. (X = Br, I)

6. Reactions with organolithium and organozinc compounds.

7. Reactions in which a carbon–tellurium bond is cleaved.

8. Reactions of tris(organo)telluronium hydroxides.

Yields for these reactions are generally not given.

1. Halogen Exchange

The halogen exchange in telluronium salts can easily be accomplished with potassium bromide and iodide as well as with silver chloride and bromide.

The crude reaction product obtained during the synthesis of a telluronium salt is usually converted into the less soluble iodide by the addition of potassium iodide to the reaction mixture. The solubility differences between the silver halides are utilized for the preparation of the telluronium bromides and chlorides from the iodides.

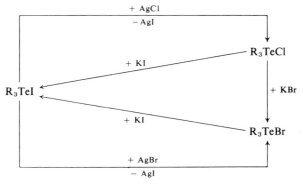

Since the solubility of the telluronium halides decreases in the order chloride > bromide > iodide, the chlorides can be converted into the bromides and iodides, and the bromides into the iodides by treatment with the proper potassium halide. These metathetical reactions are carried out in aqueous medium.

Lederer (L-6) converted diphenyl methyltelluronium iodide and bis(4-methylphenyl) methyltelluronium iodide into the corresponding nitrates by treatment with a slight excess of $AgNO_3$ in aqueous solution. The products crystallized from water and melted at 168–169° and 155–157°, respectively.

Boiling the telluronium halides with freshly precipitated silver oxide in water and evaporating the resultant solution, gave white solids, which have the formula R_3TeOH.

$$2\ R_3TeX + Ag_2O + H_2O \longrightarrow 2\ R_3Te(OH) + 2\ AgX\ (X = Cl, Br, I)$$

TABLE XXXIV
Reactions of Tris(organo)telluronium Compounds

R	R'	X	$R_2R'TeX$ + KBr → $R_2R'TeBr$		$R_2R'TeX$ + KI → $R_2R'TeI$		$R_2R'TeX$ + AgCl → $R_2R'TeCl$		$R_2R'TeX$ + AgBr → $R_2R'TeBr$		$R_2R'TeX$ + Ag_2O/H_2O → $R_2R'TeOH$		$R_2R'TeX$ + HgX_2 → $R_2R'TeX·HgX_2$		$R_2R'TeX$ + picric acid → picrate	
			m.p.	Ref.	m.p.	Ref.	m.p.	Ref.	m.p.	Ref.	m.p.	Ref.	m.p.	Ref.	m.p.	Ref.
CH_3-	CH_3-	I					n	C-3			n	C-1, E-4				
CH_3-	$-C_6H_4OCH_3$	I													127	M-28
C_2H_5-	C_2H_5-	I							162	M-9	ni	B-6				
		Cl									n	C-1, M-9				
C_6H_5-	C_6H_5-	I					245	L-3, L-4, N-4	260	L-3, L-4	G	L-12	178	L-28		
		Br									n	E-4	144	L-4		
		Cl											139	L-28, N-4	n	L-4
C_6H_5-	$p\text{-}CH_3C_6H_4-$	I					n	L-28					223	L-28	133	L-28
		Cl	229	L-28									G	L-28		
C_6H_5-	$m\text{-}CH_3C_6H_4-$	I											133	L-28		
		Br	203	L-28									n	L-28		
		Cl					G	L-28					G	L-28		
C_6H_5-	$o\text{-}CH_3C_6H_4-$	I											a 184	L-28	106	L-28
		Cl	203	L-28			G	L-28					211	L-28		
C_6H_5-	$2,4\text{-}(CH_3)_2C_6H_3-$	I											202	L-28		
C_6H_5-	$2,5\text{-}(CH_3)_2C_6H_3-$	I					211	L-28					~110	L-28	128	L-28
		Br											175	L-28		
C_6H_5-	$2,4,6\text{-}(CH_3)_3C_6H_2-$	Cl	221	L-28									176	L-28		
		I	i	L-28			i	L-28					94	L-28	171	L-28

R_1	R_2	X							
C_6H_5—	$p\text{-}CH_3OC_6H_4$—	I	o L-28		o L-28			a 90 L-28	127 L-28
		Br						o	127 L-28
C_6H_5—	$o\text{-}CH_3OC_6H_4$—	Cl	221 L-28		G L-28			o	127 L-28
		I						219 L-28	166 L-28
		G						166 L-28
C_6H_5—	$p\text{-}C_2H_5OC_6H_4$—	Cl	179 L-28		o L-28			77 L-28	
C_6H_5—	$o\text{-}C_2H_5OC_6H_4$—	I			i L-28			184 L-28	
C_6H_5—	$1\text{-}C_{10}H_7$—	I	i L-28					126 L-28	o L-28
C_6H_5—	$C_6H_5CH_2$—	Cl, Br			116 L-7		o L-10		
C_6H_5—	—$CH_2CO_2CH_3$	Br						36 L-7	145 L-7
C_6H_5—	—CH_2CO_2H	Cl			130 (dec.) L-6	138 L-6	118 L-7		
C_6H_5—	CH_3—	Br, I	138 L-6	124 L-6	261 L-4, N-7	266 L-4	n L-6	136 L-6	94^v L-6
							L-6		
$p\text{-}CH_3C_6H_4$—	$p\text{-}CH_3C_6H_4$—	Cl, I	266 L-4		261 L-4, N-7	231 L-20	110 L-12	n L-4, N-7	194 L-4, L-12
$p\text{-}CH_3C_6H_4$—	C_6H_5—	Br			G L-6				133 L-20
$p\text{-}CH_3C_6H_4$—	CH_3—	I, Cl	74 L-6		G L-6		n L-6	150 L-6	158 L-6
$m\text{-}CH_3C_6H_4$—	$m\text{-}CH_3C_6H_4$—	Cl	a L-19	161 L-19	nc L-19			160 L-19	153 L-19
$m\text{-}CH_3C_6H_4$—	CH_3—	I	a L-16		ni L-16			G L-16	115^z L-16
									n L-4
$o\text{-}CH_3C_6H_4$—	$o\text{-}CH_3C_6H_4$—	Cl	198 L-12		176 L-12	n L-4	G L-12		
$o\text{-}CH_3C_6H_4$—	CH_3—	I, Cl	135 L-6		93^w L-12	135 L-6	n L-6	135 L-6	144 L-6
$2,4\text{-}(CH_3)_2C_6H_3$—		I			164 L-19				139 L-19
$2,4,6\text{-}(CH_3)_3C_6H_2$—									

(continued)

263

TABLE XXXIV (Continued)

R₂R'TeX			R₂R'TeX + KBr → R₂R'TeBr		R₂R'TeX + KI → R₂R'TeI		R₂R'TeX + AgCl → R₂R'TeCl		R₂R'TeX + AgBr → R₂R'TeBr		R₂R'TeX + Ag₂O/H₂O → R₂R'TeOH		R₂R'TeX + HgX₂ → R₂R'TeX·HgX₂		R₂R'TeX + picric acid → picrate	
R	R'	X	m.p.	Ref.	m.p.	Ref.	m.p.	Ref.	m.p.	Ref.	m.p.	Ref.	m.p.	Ref.	m.p.	Ref.
2,5-(CH₃)₂C₆H₃—	CH₃—	I													170	L-14
p-CH₃OC₆H₄	p-CH₃OC₆H₄—	I					n	N-6	n	N-6					160	L-19
p-CH₃OC₆H₄—	CH₃—	I													154	L-17
o-CH₃OC₆H₄—	o-CH₃OC₆H₄—	I					112ˣ	L-30					235	L-30	170	L-30
		Br	112ʸ	L-30									218	L-30		
		Cl											245	L-30		
p-C₂H₅OC₆H₄—	p-C₂H₅OC₆H₄—	I							218	L-21					179	L-21
		Br														
o-C₂H₅OC₆H₄—	o-C₂H₅OC₆H₄—	I							203	L-21					165	L-21
		Br														
1-C₁₀H₇—		I					n	N-5	n	N-5						
		Br			n	K-7										

n = Not reported.

G = Gummy.

a = Amorphous.

i = Impure.

o = Oily.

nc = Not crystalline.

ni = Not isolated.

ᵛ = Crystallizes with 1 H₂O.

ʷ = Crystallizes with 2 H₂O; solidifies and remelts at 148°.

ˣ = Crystallizes with 4 H₂O.

ʸ = Crystallizes with 2 H₂O; solidifies and remelts at 195°.

ᶻ = From ethanol with 5 ROH.

The physical properties of these hydroxides remain largely uninvestigated. For individual compounds prepared see Table XXXIV. Morgan and Burgess (M-37) collected colorless crystals of the sodium salt of tris(2-methyl-4-hydroxyphenyl)telluronium hydroxide upon cooling of a solution of the telluronium chloride in warm $2N$ Na_2CO_3 or $4N$ NaOH. The crystals, after washing with ethanol and drying in air, melted at 137–138° and solidified on further heating.

$$\left[\begin{array}{c} OH \\ \bigcirc \\ CH_3 \end{array} \right]_3 TeCl + 4\ NaOH \longrightarrow \left[\begin{array}{c} ONa \\ \bigcirc \\ CH_3 \end{array} \right]_3 Te(OH) + NaCl + 3\ H_2O$$

Diphenyl carbomethoxymethyltelluronium bromide and the respective carbethoxy compound both gave, upon shaking for three days with silver oxide in water, diphenyl carbomethoxymethyltelluronium hydroxide. The ester was hydrolyzed to the acid during this process (L-7).

$$2\ [(C_6H_5)_2TeCH_2CO_2R]Br + Ag_2O + 3\ H_2O \longrightarrow$$
$$2\ [(C_6H_5)_2TeCH_2COOH](OH) + 2\ AgBr + 2\ ROH$$
$$(R = CH_3, C_2H_5)$$
$$(mp.\ 111–112°)$$

2. Formation of Insoluble Picrates

The telluronium picrates, crystalline, yellow substances, are precipitated upon addition of picric acid or its sodium salt to aqueous solutions of the telluronium halides. They are insoluble in water and can be used for the characterization of the telluronium compounds.

$$[R_3Te]^+X^- + picrate^- \longrightarrow [R_3Te]^+picrate^- + X^-$$

The reported picrates are listed in Table XXXIV.

3. Formation of Mercuric Halide Adducts

Lederer prepared almost all of the known addition compounds between telluronium halides and mercuric halides.

$$R_3TeX + HgX_2 \longrightarrow R_3TeX \cdot HgX_2 \qquad (X = Cl, Br, I)$$

The chlorides were prepared in aqueous medium, while the bromides and iodides were formed by mixing ethanolic solutions of the components. The adducts are not all crystalline. A few of these compounds were obtained only as amorphous solids, as oils or gummy products. The purification was accomplished by recrystallization from a suitable solvent.

The coordination number of mercury in these adducts is not known. The analytical data agree with the formulation $R_3TeX \cdot HgX_2$. Mercury(II) can accommodate four or six ligands. In view of this the formula of $[R_3Te]^+$-$[HgX_3]^-$ is unlikely. A compound $[R_3Te]_2^+[HgX_4]^{2-}$ would satisfy mercury with respect to the required number of ligands, but its composition does not agree with the experimental values. There may exist the possibility of a direct tellurium–mercury bond.

The known adducts, together with their melting points, are listed in Table XXXIV.

4. Salt Formation with Chromate and Dichromate Ions and with Metal Halides

Telluronium halides, R_3TeX, where X usually stands for Cl, form precipitates when dilute aqueous solutions of the telluronium chloride are mixed with aqueous solutions of K_2CrO_4, $K_2Cr_2O_7$, $AuCl_3$, H_2PtCl_6, $ZnCl_2$, and $SnCl_2$. Most of these precipitates are insoluble in cold water. Some of these products crystallize from hot aqueous solutions, while others could not be recrystallized. With the exception of the bis(p-tolyl) methyltelluronium derivative, the salts were found to be rather stable (L-6). $CuCl_2$ formed complex salts with telluronium compounds which, however, were not isolated due to their solubility in H_2O (L-6).

According to analytical data (L-6), the yellow-brownish chloroplatinates, the yellow chromates, and orange dichromates have the composition $(R_3Te)_2^+PtCl_6^{2-}$, $(R_3Te)_2^+CrO_4^{2-}$ and $(R_3Te)_2^+Cr_2O_7^{2-}$, respectively. The gold chloride and tin(II) chloride salts were never analyzed. Hot aqueous solutions of $ZnCl_2$ and bis(2-methylphenyl) methyltelluronium chloride deposited, on cooling, a crystalline substance of the formula $(2-CH_3C_6H_4)_2CH_3TeCl \cdot Zn(OH)Cl$. Since zinc (II) prefers to be tetra-coordinated, arguments similar to those put forth for Hg(II) might apply for these compounds as well.

For individual compounds turn to Table XXXV.

Drew (D-15) reported that trimethyltelluronium bromide combined with $FeCl_3$ to yield a complex salt which crystallized in the form of salmon colored needles. It also gave a compound with $SnBr_4$.

5. Complex Formation with Organotellurium Compounds

Drew (D-15) during his investigation of Vernon's "isomeric" dimethyltellurium dihalides (see Section VI-A) found that the isomers of Vernon's β-series are complexes between trimethyltelluronium halides and methyltellurium trihalides.

Drew prepared a number of these complexes by dissolving the components in warm acetone. Addition of $CHCl_3$ caused the precipitation of deeply colored crystalline solids.

$$R_3TeX + RTeX_3 \longrightarrow R_3TeX \cdot RTeX_3 \qquad (R = CH_3)$$

TABLE XXXV

The Reactions of Tris(organo)telluronium Halides with Various Ionic Salts

$[R_3Te]^+X^-$	Melting points of products**							Ref.
	K_2CrO_4	$K_2Cr_2O_7$	$AuCl_3$	$CuCl_2$	H_2PtCl_6	$ZnCl_2$	$SnCl_2$	
$(C_6H_5)_3TeCl$	N		N		N		N	L-4
$(C_6H_5)_2CH_3TeCl$	151	153 (dec.)		S	158	149–150		L-6
$(C_6H_5)_2(CH_2CO_2CH_3)TeCl$	73	115	O		60 (dec.)	G		L-7
$(p\text{-}CH_3C_6H_4)_3TeCl$	np		G		N		N	L-4
$(p\text{-}CH_3C_6H_4)_2CH_3TeCl$	52(i)	54–55(i)	N	S	104–105(i)	G		L-6
$(m\text{-}CH_3C_6H_4)_2CH_3TeCl$			35–36(i)		155 (dec.)			L-16
$(o\text{-}CH_3C_6H_4)_2CH_3TeCl$	162 (dec.)	172 (dec.)	186					L-6
$(C_6H_5)_2(CH_2CO_2H)TeCl$					i	*		L-6
$(C_2H_5)_3TeCl$					n			L-7
$(CH_3)_3TeI$					n			B-6
								C-3

N = Reagents gave a precipitate; product not further investigated.
O = Product oily.
S = Product did not precipitate.
G = Gummy.
* = See text for discussion.
** = For formulas see text.
np = No precipitate.
i = Impure according to analysis.
n = m.p. not reported.

The following compounds were obtained in a similar manner:

$R_3TeI \cdot RTeI_3$	purple black, with green luster	blackened 83°
$R_3TeBr \cdot RTeI_3$	purple red, with green luster	blackened 90°
$R_3TeI \cdot RTeBr_3$	orange-brown	m. 120° (dec.)
$R_3TeBr \cdot RTeBr_3$	yellow	m. 142° (dec.)

The characteristic color of these complexes developed even by grinding together the components in the absence of a solvent.

A complex $R_3TeI \cdot 2R_2TeI_2$ was obtained by treating the filtrate of the reaction mixture obtained by warming $R_3TeI \cdot RTeI_3$ with aqueous Na_2CO_3, with HI. A similar compound $R_3TeI \cdot R_2TeI_2 \cdot R_2TeI_4$ crystallized when the components were combined in the presence of HI containing iodine. A complex was not formed between TeI_4 and R_3TeI.

Trimethyltelluronium iodide was found to combine with iodine to form R_3TeI_3, a purple, crystalline solid, melting at 76.5° (S-12).

6. Reactions with Organometallic Compounds

Telluronium halides react with organolithium and organozinc compounds. The following systems were investigated:

$$2\,(C_2H_5)_3TeX + (C_2H_5)_2Zn \xrightarrow[]{\text{room temp.}} \text{no reaction}$$
$$(X = Cl, I) \xrightarrow{100°} 2\,(C_2H_5)_2Te + 2\,C_4H_{10} + ZnX_2 \quad (M\text{-}9)$$

$$(CH_3)_3TeI + CH_3Li \longrightarrow (CH_3)_4Te + LiI \xrightarrow[NaI]{H_2O} (CH_3)_3TeI \quad (H\text{-}4)$$

$$(C_4H_9)_3TeI + C_4H_9Li \longrightarrow (C_4H_9)_4Te \xrightarrow{\text{distill.}} (C_4H_9)_2Te + C_8H_{18} \quad (H\text{-}4)$$
$$(C_6H_5)_3TeX + 4\,C_4H_9Li \longrightarrow (C_4H_9)_4Te + LiX + 3\,(C_6H_5)Li \quad (F\text{-}9, H\text{-}4)$$
$$(C_6H_5)_3TeCl + C_6H_5Li \longrightarrow (C_6H_5)_4Te + LiCl \quad (W\text{-}3)$$

$$(CH_3)_3TeI + \text{(dilithiobiphenyl)} \longrightarrow \text{(dibenzotellurophene)Te} + C_2H_6 + LiCH_3 + LiI \quad (H\text{-}4)$$

In all of these reactions a tetraorganotellurium derivative or even pentavalent species are postulated as intermediates, which, upon hydrolysis or thermal decomposition, give the various products. Only tetraphenyltellurium was actually isolated from the reaction between phenyllithium and triphenyltelluronium chloride (W-3). (See also Section VIII).

7. Reactions in which a Carbon–Tellurium Bond is Cleaved

A few reactions of telluronium halides are known, in which a carbon–tellurium bond is cleaved, converting Te(IV) into Te(II) compounds. Boiling ethanol regenerated the components from bis(2-methylphenyl) methyl-telluronium iodide (L-6), diphenyl methyltelluronium bromide (L-6) and diphenyl carbomethoxymethyltelluronium bromide (L-7). 4-Methoxyphenyl dimethyltelluronium iodide is reported to be hydrolyzed by ethanol (M-28). Boiling water was found to cleave diphenyl carbethoxymethyltelluronium bromide into its constituents (L-7).

Although no detailed studies have been carried out, it is likely that the reactions proceed in the following manner:

$$[R_2TeCH_3]^+X^- \xrightarrow{\text{solvent}} R_2Te + CH_3X$$

Dibutyl carbethoxymethyl- and dibutyl carbomenthoxymethyltelluronium bromide both decomposed upon distillation according to the following equations (B-2):

$$[R_2TeCH_2CO_2C_2H_5]^+Br^- \longrightarrow RBr + R\!-\!Te\!-\!CH_2\!-\!CO_2C_2H_5$$
$$R = C_4H_9; \text{ b.p. } 135\text{–}138°/21 \text{ mm}; 80\% \text{ yield}$$
$$R = C_5H_{11}; \text{ b.p. } 140\text{–}150°/17 \text{ mm},$$

$$[R_2TeCH_2CO_2\text{-}l\text{-menthyl}]^+Br^- \xrightarrow{100°/20 \text{ mm}}$$
$$RBr + R\!-\!Te\!-\!CH_2\!-\!CO_2\text{-}l\text{-menthyl} + \text{ b.p. } 120\text{–}125°/0.1 \text{ mm}; 61\% \text{ yield}$$
$$(R = C_4H_9) + (l\text{-menthyl } O_2CCH_2)_2Te \text{ m.p.}58°; 50\% \text{ yield})$$

Diphenyl methyltelluronium iodide upon heating in dimethylaniline formed diphenyl telluride and phenyl trimethylammonium iodide (L-6). Bis(diphenyl methylenetelluronium) dibromide gave the following reactions (M-43):

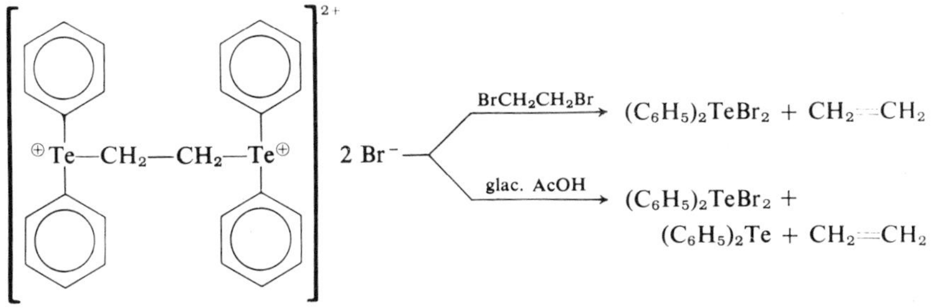

Heal (H-1) and Emeleus and Heal (E-4) discussed the thermal stability of trimethyltelluronium compounds. McDaniel attributed the relative stability of $(CH_3)_3TeI$ to hyperconjugation between the H atoms of the methyl group and low lying vacant tellurium orbitals (M-2). Concerning the reactions of telluronium halides with Grignard reagents, see Section I-C-7.

Reichel and Kirschbaum (R-7) treated triphenyltelluronium iodide and 4-methoxyphenyl dimethyltelluronium iodide with a 15-fold molar excess of $Na_2S \cdot 9H_2O$ at 95–100° and observed reduction to the telluride.

$$(C_6H_5)_3Te^+I^- + NaSH \longrightarrow (C_6H_5)_2Te + C_6H_6 + NaI + S$$

With the 4-methoxyphenyl dimethyltelluronium iodide, dimethyl telluride, 4-methoxyphenyl methyl telluride and anisole were isolated. The mixed telluride is thought to be a product of the reversible dissociation of the telluronium salt.

Experimental. Triphenyltelluronium Chloride (L-4): The telluronium iodide was boiled with AgCl suspended in water. The residue remaining after evaporation of the solvent was dissolved in absolute ethanol. Addition of ether caused the precipitation of the chloride in needlelike crystals. The substance melted at 244–245°.

$(C_6H_5)_2CH_3TeCl \cdot HgCl_2$ (L-6): The combined aqueous solutions of the telluronium chloride and mercuric chloride deposited a white precipitate, which dissolved in hot H_2O. On cooling white needles of the adduct were formed melting at 135–136°.

Diphenyl Methyltelluronium Hydroxide (L-6): The finely powdered tellu-ronium iodide (2 g) and Ag_2O (0.4 g) and 15 ml of H_2O were shaken for one hour. The filtrate, concentrated in a vacuum desiccator, yielded the hydroxide as a clear syrup, which became solid in vacuum.

8. Reactions of Tris(organo)telluronium Hydroxides

Tris(organo)telluronium hydroxides are ill-characterized substances. Generally, they are viscous liquids (L-28) which solidify when kept in vacuum. Analytical data are scarce. Diphenyl methyltelluronium hydroxide is reported to be hygroscopic (L-6). The aqueous solutions give an alkaline reaction and precipitate metal hydroxides from metal ion solutions. The hydroxides liberate NH_3 from its solution, but do not absorb CO_2 from the air. With potassium bromide and iodide, hydrohalic acids, picric acid, H_2SiF_6, and H_2PtCl_6, the corresponding salts were formed. Table XXXVI lists the known reactions. Melting points of products already given in Table XXXIV are not repeated. Many of the reactions have been carried out without isolation of the hydroxide.

Experimental. Trimethyltelluronium Fluoride (E-4): The iodide (2.5 g) was converted into the base by grinding with excess Ag_2O and little water in a

TABLE XXXVI

Reactions of Tris(organo)telluronium Hydroxides

R_3TeOH	KCl	KBr	KI	HF	HCl	HBr	HI	H_2SiF_6	Picric acid	H_2PtCl_6
$(C_6H_5)_3TeOH$		W-3		E-4[a]					L-12	
$(C_6H_5)_2CH_3TeOH$		L-6	L-6		L-6	L-6	L-6			
$(C_6H_5)_2(CH_2CO_2H)TeOH$			L-7		L-7	L-7				
$(C_6H_5)_2(C_6H_5CH_2)TeOH$									L-10	
$p\text{-}CH_3C_6H_4)_3TeOH$									L-12[b]	
$p\text{-}CH_3C_6H_4)_2CH_3TeOH$					L-6					
$o\text{-}CH_3C_6H_4)_3TeOH$									L-12[c]	
$o\text{-}CH_3C_6H_4)_2CH_3TeOH$	L-6									
$(CH_3)_3TeOH$				E-4[d]				E-4[d]		C-1
$(C_2H_5)_3TeOH$					B-6		M-9			C-1

[a] 203° (dec.).
[b] m.p. 194–195°.
[c] m.p. 182°.
[d] m.p. 128° (dec.).
[e] Decrepitated at 210° and dec. at 320–336°.

mortar. The filtered solution was neutralized to phenolphthalein with dilute HF in a platinum dish and then evaporated in vacuum over H_2SO_4. The solid was recrystallized from anhydrous acetone containing the minimum amount of absolute ethanol to dissolve the salt. The anhydrous salt melted at 128° (dec.).

VIII. REACTIONS OF TETRAORGANOTELLURIUM COMPOUNDS

Only a small number of these compounds are known. The preparation of tetraorganotellurium compounds is discussed in Sections VI-3 and I-C-7-c. It is interesting to note that tetraphenyltellurium—Te^{125m} was formed from $Sb^{125}(C_6H_5)_5$ and $(C_6H_5)_4Sb^{125}Cl$ (N-3). The only stable tetraorganotellurium derivatives are tetraphenyltellurium (W-3), tetra(pentafluorophenyl)tellurium (C-25), and bis(2,2′-biphenylylene)tellurium (H-3). The known tetraaryltellurium compounds are thermally rather stable in contrast to their aliphatic analogs.

Tetraphenyltellurium was found to yield benzene, biphenyl, diphenyl telluride, and other nonidentifiable decomposition products upon heating to 115°. Tetrakis(perfluorophenyl)tellurium kept in a sealed tube at 200–220°

$$(C_6H_5)_4Te \longrightarrow C_6H_6 + (C_6H_5)_2 + (C_6H_5)_2Te + residue$$

formed bis(pentafluorophenyl) and bis(pentafluorophenyl) telluride (C-25). Bis(2,2′-biphenylylene)tellurium thermally decomposed into 2,2′-biphenylylene telluride (96% yield), biphenylene (54% yield) and higher polybiphenylene compounds (H-3).

The tetraalkyltellurium compounds have been shown to exist only in solution and are thermally unstable. Hellwinkel and Fahrbach (H-4) obtained, upon distillation of a solution containing tetrabutyltellurium, dibutyl telluride and butane, as well as octane. Decomposition even occurred at room temperature.

Tetraorganotellurium compounds are capable of exchanging organic groups with certain organolithium compounds. The exchange between tetraphenyltellurium and butyllithium was studied by Franzen and Mertz (F-9). The exchange was found to proceed very slowly.

Tetrabutyltellurium was obtained—in solution only—by an alkyl–aryl group exchange between bis(2,2′-biphenylylene)tellurium and butyllithium mixed in a 1:5 molar ratio in ether solution. In reactions with equimolar amounts of reagents, 61% of the starting material was recovered unchanged, while a 1:2 molar ratio produced 32% unchanged bis(2,2′-biphenylylene)tellurium, biphenyl, and 2,2′-biphenylylene telluride. CH_3Li, C_6H_5Li and $4\text{-}(CH_3)_2NC_6H_5Li$ did not give exchange reactions (H-4).

Electrophilic reagents attack the tetra(organo)tellurium compound giving rise to a telluronium salt.

Tributyltelluronium iodide was formed by the reaction of tetrabutyltellurium with bis(2,2'-biphenylylene)arsonium or phosphonium iodide (H-4).

(Bu_4Te) + [bis(2,2'-biphenylylene)arsonium] I^- $\longrightarrow$ $Bu_3Te^+I^-$ + [arsonium]$-C_4H_9$

With triphenylboron tetraphenyltellurium forms, from ether solution, the derivative triphenyltelluronium tetraphenylborate which melts at 217–219° (W-3). Tetrabutyl- and tetramethyltellurium, when treated with aqueous

$$(C_6H_5)_4Te + B(C_6H_5)_3 \longrightarrow [(C_6H_5)_3Te][B(C_6H_5)_4]$$

solutions of potassium iodide, produced the respective telluronium iodides (H-4).

Tetraphenyltellurium is hydrolyzed by water to the telluronium hydroxide. It was obtained as the bromide after addition of potassium bromide to the hydrolysis mixture (W-3).

Bis(2,2'-biphenylylene)tellurium, with HCl, produced the telluronium chloride, while elemental bromine gave 2,2'-biphenylylene 2-bromobiphenylyl-2'-telluronium bromide (H-4).

X = Cl, Y = H, m.p. 269–270°
X = Br, Y = Br, m.p. 264°

When tetraphenyltellurium was dissolved in methylene chloride, or chloroform, triphenyltelluronium chloride was formed (W-3). In chloroform, phenyldichloromethane was formed in accordance with the equation

$$(C_6H_5)_4Te + CHCl_3 \longrightarrow (C_6H_5)_3TeCl + C_6H_5CHCl_2$$

Tetraphenyltellurium was added to a solution of benzaldehyde in ether under a nitrogen atmosphere. After two hours the mixture was decomposed with water. The excess aldehyde was removed as the bisulfite addition compound. The remaining oil, following treatment with HCl, yielded diphenylmethane and triphenyltelluronium chloride (H-3).

$$(C_6H_5)_4Te + C_6H_5CHO \xrightarrow{H_2O} (C_6H_5)_3TeOH + (C_6H_5)_2\overset{\displaystyle H}{\underset{\displaystyle |}{C}}\!\!-OH$$
(not isolated)

$$\downarrow HCl$$

$$(C_6H_5)_3TeCl$$

Tetraphenyltellurium in its reaction with benzaldehyde behaves like a Grignard reagent as is shown by the following equation

$$R_3\overset{..}{Te}\ R + \overset{O}{\underset{}{|O}}\!\!-CH\!\!-R \longrightarrow R_3Te\!\!-\overline{\underline{O}}\!\!-CHR_2 \xrightarrow{H_2O} R_3Te(OH) + R_2CH(OH)$$

$$(R = C_6H_5)$$

The following scheme outlines the reactions of tetraorganotellurium compounds.

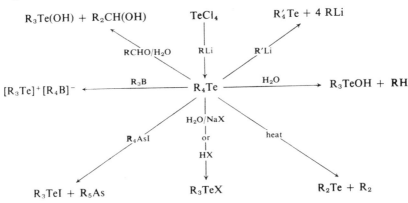

IX. REACTIONS OF ORGANOTELLURIUM TRIHALIDES

The organotellurium trichlorides are the most easily prepared compounds of this series. The condensation of $TeCl_4$ with aromatic or aliphatic compounds containing an activated hydrogen atom yields the trichlorides (see Section I-C). The bromides and iodides are obtained in reactions between ditellurides and elemental halogen (see Section V-1). The ditellurides are also obtained by the reduction of organotellurium trichlorides.

These trihalides are generally stable crystalline substances. Only a few compounds are reported to be unstable. On standing methyltellurium trifluoride disproportionates into TeF_4 and R_2TeF_2 (E-4). 2-Chlorostyryltellurium trichloride gave R_2TeCl_2 on heating in ethanol or glacial acetic acid (M-46) and bis(benzoyl)methyltellurium trichloride easily decomposed (M-17). Butyltellurium triiodide liberated iodine upon heating (B-2). The only known trifluoride, methyltellurium trifluoride, was synthesized by halogen exchange with AgF. The following reactions are reported for organotellurium trihalides:

1. Reduction to bis(organo)tellurium ditellurides, R_2Te_2, by $K_2S_2O_5$, $Na_2S \cdot 9H_2O$, $NaHSO_3$, hydrazine sulfate and zinc in $CHCl_3$.

2. Halogen exchange.

3. Hydrolysis reactions with H_2O, NaOH, Na_2CO_3, or K_2CO_3.

4. Condensation reactions with compounds containing an activated hydrogen atom.

5. Addition to the carbon–carbon double bond.

6. Reactions with organometallic compounds.

7. Reactions with ditellurides.

8. Reactions in the organic part of the molecule.

9. Complex formation with other organotellurium compounds.

Figure 7 summarizes the important reactions of organotellurium trichlorides.

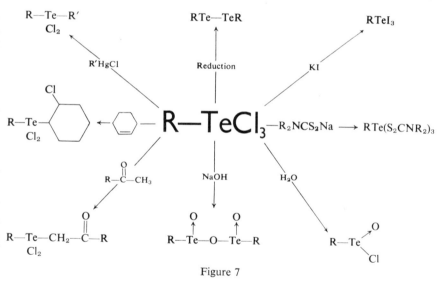

Figure 7

1. Reduction of Organotellurium Trihalides

Organotellurium trichlorides are easily and almost always quantitatively reduced to the strongly colored bis(organo)tellurium ditellurides by potassium metabisulfite or $Na_2S \cdot 9H_2O$.

$$2\ RTeCl_3 \xrightarrow{\text{Red.}} R_2Te_2$$

TABLE XXXVII

Reduction of Organotellurium Trichlorides by $K_2S_2O_5$ and $Na_2S \cdot 9H_2O$ to R_2Te_2

RTeCl₃	$K_2S_2O_5$			$Na_2S \cdot 9H_2O$		
	m.p.	Yield, %	Ref.	m.p.	Yield, %	Ref.
Phenyl	67	76	F-2	60	100	P-5,R-7
4-Methylphenyl	52.5		F-2			
4-Methoxyphenyl	60	91	M-32,R-6			
4-Methoxy-3-hydroxyphenyl[a]	117 (dec.)		M-28			
2-Methoxy-5-bromophenyl[a]			M-32			
4-Methoxy-2-methylphenyl[a]			M-32			
4-Methoxy-3-methylphenyl	78		M-32			
2-Methoxy-5-methylphenyl			M-32			
2,4-Dimethoxyphenyl	135	100	M-28			
4-Ethoxyphenyl	109	~100	M-28,R-6	107	100	P-5,R-7
4-Phenoxyphenyl	88	100	D-11	88	100	P-5,R-7
2-Phenoxyphenyl (?)			D-11			
4-Thiophenoxyphenyl				90	100	P-5
2-Thiophenoxyphenyl				131	100	P-5
1-Naphthyl				124.5	100	P-5
2-Naphthyl				123	100	V-5
—CH₂COOH	140–144 (dec.)		M-26			
—CH₂CO—O—COCH₂—TeCl₃		100	M-26			
—CH₂—TeCl₃			M-26			
—CH(CH₃)COOH	75		M-32			

[a] Position of substituents uncertain.

The reductions with $K_2S_2O_5$ are carried out by the slow addition of a three-fold molar excess of aqueous $K_2S_2O_5$ into the trichloride at 0°. With $Na_2S \cdot 9H_2O$ as the reducing agent, the trichloride is heated with a 15-fold molar excess of the sulfide to 95–100°. The trihalides reduced by these two reagents are listed in Table XXXVII. However, not all trichlorides can be reduced to the ditellurides. 4-Hydroxyphenyl- (P-5), 2-chlorocyclohexyl- (M-45), 2-chlorostyryl- (M-46), O-ethylacetylacetonyl- (M-16) the enol form of heptoylacetonyltellurium trichloride (M-21) and 4-(4'-carboxyquinol-2-yl)-phenyltellurium trichloride (R-8) decompose to elemental tellurium when treated with Na_2S or $K_2S_2O_5$. Among the other trihalides known, only the 4-methoxyphenyltellurium tribromide was investigated with respect to its behavior toward the reducing agent $K_2S_2O_5$ and was found to give the corresponding ditelluride (M-32).

Since the reactions with $K_2S_2O_5$ are carried out in aqueous medium and the trichlorides are hydrolyzed under these conditions (see Sections X and IX-3), the species actually reduced probably are organotellurium oxychlorides.

Organotellurium trichlorides are also reduced by sodium methanethiosulfonate (F-3) and thiourea (F-5). Thiourea has been reported to reduce

$$CH_3O-\!\!\left\langle\bigcirc\right\rangle\!\!-TeCl_3 + 3\ CH_3\overset{\overset{O}{\uparrow}}{\underset{\underset{O}{\downarrow}}{S}}-SNa \longrightarrow$$

$$CH_3O-\!\!\left\langle\bigcirc\right\rangle\!\!-Te-S-\overset{\overset{O}{\uparrow}}{\underset{\underset{O}{\downarrow}}{S}}-CH_3 + S_2(SO_2CH_3)_2 + 3\ NaCl$$

phenyltellurium trichloride to phenyltellurenyl chloride. When four moles of thiourea per mole of tellurium compound were present in the reaction mixture, the complex $C_6H_5Te[SC(NH_2)_2]_2Cl$ was formed, while with a 3:1 ratio only one molecule of thiourea was coordinated to the tellurium atom. Foss and Hauge (F-5) wrote the following equations for these reactions.

$$C_6H_5TeCl_3 + 3\ SC(NH_2)_2 \longrightarrow C_6H_5Te[SC(NH_2)_2]Cl + [SC(NH_2)_2]_2^{2+} + 2\ Cl^-$$
$$C_6H_5TeCl_3 + 4\ SC(NH_2)_2 \longrightarrow C_6H_5Te[SC(NH_2)_2]_2Cl + [SC(NH_2)_2]_2^{2+} + 2\ Cl^-$$

When aqueous solutions of thiourea containing KBr, HNO_3, $HClO_4$, or KCNS were combined with phenyltellurium trichloride, the respective halides, as shown in Table XXXVIII were obtained.

Experimental. Bis(4-ethoxyphenyl) Ditelluride (R-6): 4-Ethoxyphenyl-tellurium trichloride (3 g, 0.0084 moles) were suspended in 30 ml of H_2O. A solution of $K_2S_2O_5$ (5.6 g, 0.025 moles) in 8 ml of H_2O was added dropwise at

TABLE XXXVIII

Tellurenyl Compounds from Organotellurium Trichlorides: $RTeCl_3 \longrightarrow RTeX$

R	X	Ligand(s)	Yield, %	m.p.	Ref.
p-Methoxyphenyl	$S_2O_2CH_3$		68	107	F-3
Phenyl	Cl	1 (tu)[a]	79	166	F-5
	Cl	2 (tu)	85	168	F-5
	Br	1 (tu)	99	178	F-5
	ClO_4	2 (tu)		136	F-5
	NO_3	2 (tu)		135	F-5
	CNS	2 (tu)	58	109	F-5

[a] tu = thiourea.

0°. After stirring for one hour, the product was filtered. Recrystallized from petroleum ether, the ditelluride melted at 109° (Yield: 100%).

Bis(4-thiophenoxyphenyl) Ditelluride (P-5): 4-Thiophenoxyphenyltellurium trichloride (0.42 g, 0.001 mole) was added to $Na_2S \cdot 9H_2O$ (3.6 g, 0.015 moles). The mixture was heated to 95–100° and this temperature was maintained for 15 min. After treatment with water the ditelluride separated quantitatively. Following recrystallization from benzene–petroleum ether, the product melted at 180° with decomposition.

Thiourea(phenyl)tellurenyl Chloride (F-5): Thiourea (2.28 g, 0.03 moles) in 20 ml of warm water was added to phenyltellurium trichloride (3.11 g, 0.01 mole) in 20 ml of methanol. From the clear orange-red solution the compound crystallized in 80% yield on standing. After recrystallization from methanol the substance melted at 166°.

2. Halogen Exchange Reactions

The halogen exchange reactions of organotellurium trihalides have received little attention. An exchange of iodide for chloride with KI in aqueous–ethanolic medium was employed to obtain 4-hydroxyphenyl- (m.p. 125–150° dec.) (P-5) and 2-chlorocyclohexyltellurium triiodide (M-45). The latter compounds could not be prepared in the usual manner, for example, action of iodine upon the ditelluride, because the reducing agents used for the synthesis of these compounds decomposed the trichlorides.

Emeleus and Heal (E-4) obtained an impure sample of methyltellurium trifluoride from the reaction of the triiodide with AgF. Potassium bromide did not react with 4-hydroxyphenyltellurium trichloride. The starting material was recovered unchanged (P-5). Foss (F-4) obtained 4-methoxyphenyltellurium tris(dithiocarbamates) from the trichloride and sodium dithiocarbamates in dioxane solution.

$$CH_3O-\langle\bigcirc\rangle-TeCl_3 + 3\ R_2NC\underset{SNa}{\overset{S}{\diagdown}} \longrightarrow$$

$$CH_3O-\langle\bigcirc\rangle-Te(S-\overset{S}{\overset{\|}{C}}-NR_2)_3 + 3\ NaCl$$

R = CH₃, m.p. 184° (dec.); C₂H₅, m.p. 156° (dec.); piperidyl-, m.p. 178° (dec.)

These compounds dissolved in $CHCl_3$, $C_2H_2Cl_4$ or toluene possess thermo-chromic properties. Upon heating the color changes slowly from greenish-yellow to red. Addition of $(S_2CNR_2)_2$ to the hot red solution caused the color to recede more rapidly on cooling. A possible explanation for these color changes assumes a reversible dissociation into the ditelluride and disulfide or formation of the corresponding radicals.

Experimental. 2-Chlorocyclohexyltellurium Triiodide (M-45): A solution of 2-chlorocyclohexyltellurium trichloride (0.70 g, 0.002 mole) in a small volume of ethanol was treated with excess aqueous KI. Upon the addition of small amounts of benzene the triiodide crystallized (88% yield). Recrystallized from benzene–petroleum ether the reddish crystals melted at 175–190° (dec.).

3. Hydrolysis of Organotellurium Trihalides

Organotellurium trichlorides and tribromides are hydrolyzed to tellurinic acid chlorides or bromides by water and to anhydrides or the free acid by sodium hydroxide.

$$RTeCl_3 + H_2O \longrightarrow RTe\underset{Cl}{\overset{O}{\diagup}} + 2\ HCl$$

$$2\ RTeCl_3 + 6\ NaOH \longrightarrow RTe\overset{O}{\underset{\uparrow}{}}-O-\overset{O}{\underset{\uparrow}{Te}}-R + 6\ NaCl + 3\ H_2O$$

$$RTeCl_3 + 3\ NaOH \longrightarrow RTe\underset{OH}{\overset{O}{\diagup}} + 3\ NaCl + H_2O$$

Morgan and Kellet (M-32) claimed to have obtained 4-methoxyphenyl-tellurium trihydroxide by hydrolyzing the trichloride with 2N NaOH and acidifying the mixture with acetic acid. The same authors (M-32) observed that 2-methyl-4-methoxyphenyltellurium trichloride hydrolyzed upon expo-sure to moisture and that 4-methoxyphenyltellurium tribromide produced

TABLE XXXIX
Hydrolytic Reactions of Organotellurium Trihalides

RTeX$_3$			Product			
R	X	Reagent	Formula	Yield, %	m.p.	Ref.
CH$_3$—	F	H$_2$O	(RTeO)$_2$O[a]			E-4
	Br	H$_2$O	RTeOBr[b]			L-36
	I	H$_2$O	RTeOI[b]			L-36
C$_2$H$_5$—	I	H$_2$O	RTeOI			L-36
C$_6$H$_5$—	Cl	Cold H$_2$O	RTeOCl	88	250	P-4
	Br	Cold H$_2$O	RTeOBr	86	247–249	P-4
		10% Na$_2$CO$_3$ or NaOH	(RTeO)$_2$O	90–95	220–225	P-4
4-Hydroxyphenyl	Cl	Cold H$_2$O	RTeOCl	100	dec.	R-6
		10% Na$_2$CO$_3$ or NaOH	(RTeO)$_2$O	75–80	200 (dec.)	P-4
		2N NaOH	RTeOOH		dec.	R-6
4-Methoxyphenyl	Cl	2N NaOH	RTe(OH)$_3$			M-32
		Cold H$_2$O	RTeOCl	90	225–235 (dec.)[c]	P-4,R-6
		10% Na$_2$CO$_3$ or NaOH	(RTeO)$_2$O	100	200–205	P-4
		2N NaOH	RTeOOH		—[d]	R-6
	Br	Cold H$_2$O	RTeOBr	96	231–234	P-4
		10% Na$_2$CO$_3$ or NaOH	(RTeO)$_2$O			P-4
	I	Hot H$_2$O	RTeOI	67.2	190–195 (dec.)	P-4

				Yield (%)	M.p.	Ref.
4-Ethoxyphenyl	Cl	Cold H_2O	RTeOCl	88	224–226[c]	P-4,R-6
		10% Na_2CO_3 or NaOH	$(RTeO)_2O$	90–95	206–210	P-4
		70–80% NaOH	RTeOOH	100	—[d]	R-6
	Br	Cold H_2O	RTeOBr	91.5	233–236	P-4
4-Phenoxyphenyl	Cl	Cold H_2O	RTeOCl	98.5	185–186 (dec.)	P-4
		10% Na_2CO_3 or NaOH	$(RTeO)_2O$	100	276 (dec.)	P-4
	Br	Cold H_2O	RTeOBr	100	189–190 (dec.)	P-4
2-Naphthyl	Cl	Cold H_2O	RTeOCl	71.5	257 (dec.)	P-4
		10% Na_2CO_3 or NaOH	$(RTeO)_2O$	100	230	P-4,V-5
	Br	Cold H_2O	RTeOBr	85	216	P-4
		10% Na_2CO_3 or NaOH	$(R_2TeO)_2O$			V-5
4-(4'-Carboxyquinol-2'-yl)-phenyl	Cl	H_2O	RTeOCl	100	179	R-8
4-(5',6'-Benzo-4'-carboxyquinol-2'-yl)-phenyl	Cl	H_2O	RTeOCl	100	302	R-8
4-(Acridin-9'-yl)phenyl	Cl	H_2O	RTeOCl	80	276 (dec.)	R-8

[a] Identity uncertain.
[b] Only in solution; conductivity measured.
[c] R-6 reported slow decomposition between 400–500°.
[d] No definite melting point.

281

a white hydroxide when treated with water. 3-Methyl-4-methoxyphenyl-tellurium trichloride was insoluble in cold water, but gave a white hydroxide on warming (M-32).

Moist air hydrolyzed 3-hydroxy-4-methoxyphenyltellurium trichloride when dissolved in an organic solvent (M-28).

Methylenebis(tellurium trichloride) was slowly changed by moist air with the evolution of hydrogen chloride and was hydrolyzed immediately by cold water to a white amorphous oxide or oxychloride (M-26).

Reichel and Kirschbaum (R-6) found that 4-ethoxy- and 4-hydroxyphenyl-tellurium trichlorides dissolved in NaOH were hydrolyzed to tellurinic acids, which were isolated after acidifying the solution with H_2SO_4.

Petragnani and Vicentini (P-4) systematically studied the hydrolysis of aryltellurium trihalides. Their investigation showed that the trichlorides and tribromides are easily hydrolyzed to the acid chlorides and bromides, respectively. The triiodides were stable toward cold water. 4-Methoxyphenyl-tellurium triiodide yielded the tellurinic acid iodide when boiled with water. All the other triiodides investigated gave unidentified products.

With a 10% solution of Na_2CO_3 or NaOH the trihalides formed tellurinic acid anhydrides. The compounds investigated with respect to their hydrolytic behavior are listed in Table XXXIX.

Organotellurium trichlorides obtained by condensation of $TeCl_4$ with 1,3-diketones in ethanol containing $CHCl_3$ (see Section I-C-1) decomposed upon treatment with alikali hydroxide solutions.

$$R-\underset{\underset{O}{\|}}{C}-CH{=}\underset{\underset{OC_2H_5}{|}}{C}-CH_2-TeCl_3 \xrightarrow{KOH} R\cdot\underset{\underset{O}{\|}}{C}-CH{=}\underset{\underset{OC_2H_5}{|}}{C}-CH_3 + KHTeO_3$$

$$R = CH_3 \text{ (M-16)}; \quad R = C_6H_5 \text{ (M-16)}; \quad R = (CH_3)_3C \text{ (M-17)}$$

4-Ethoxyphenyltellurium trichloride developed the odor of ethoxybenzene on warming with an aqueous potassium carbonate solution (M-28).

Experimental. 4-Hydroxybenzenetellurinic Acid (R-6): 4-Hydroxyphenyl-tellurium trichloride was dissolved in 500 ml of 2N NaOH solution. Upon the addition of 2N H_2SO_4 a white solid precipitated which redissolved upon further addition of H_2SO_4. The clear, slightly acidic solution deposited the tellurinic acid upon standing for 24–48 hr. The compound recrystallized from glacial acetic acid decomposed on heating.

Aromatic Tellurinic Acid Chlorides (P-4): The trichlorides either treated directly with cold water or first dissolved in methanol and then added to water form white precipitates, which are separated, washed with water, and dried in a vacuum desiccator over calcium chloride.

Aromatic Tellurinic Acid Anhydrides (P-4): The trihalides were dissolved in a 10% solution of sodium carbonate or sodium hydroxide. The anhydrides separated upon acidification of these solutions with 10% acetic acid. The filtered products were washed with water and dried in a vacuum desiccator over calcium chloride.

4. Condensation Reactions with Compounds Containing an Activated Hydrogen Atom

Organotellurium trichlorides and tribromides condense with organic compounds containing an activated hydrogen atom with the elimination of the hydrogen halide and formation of bis(organo)tellurium dihalides. The compounds prepared are listed in Table XL. The organic compounds tested are: acetone, acetophenone, 4-phenoxy-, 4-methoxy-, 4-ethoxy, 4-dimethyl-amino- and 2,4-dihydroxybenzene. The reactivity of the organotellurium trihalides decreases in the sequence $Cl > Br > I$. This is probably due to the decreasing electrophilic character of the ion $RTe^{\oplus}X_2$ with increasing atomic mass of X (P-7). Methylketones generally do not react with bromides under normal

TABLE XL
Condensation Reactions of Organotellurium Trihalides

$RTeX_3$		$R'—H$	$RR'TeX_2$		
R	X	R'	m.p.	Yield, %	Ref.
$—CH_2—TeCl_3$	Cl	$CH_3COCH_2—$	181	100	M-26
		$C_6H_5COCH_2—$			M-26
p-Methoxyphenyl	Cl	$CH_3COCH_2—$	138	63	P-7
		$C_6H_5COCH_2—$	137	60	P-7
		$4-(CH_3)_2NC_6H_4—$	172 (dec.)	75	P-7
		$2,4-(OH)_2C_6H_3—$	183 (dec.)	70	P-7
	Br	$C_6H_5COCH_2—$	152 (dec.)	29.2	P-7
		$4-(CH_3)_2NC_6H_4—$	184 (dec.)	50.6	P-7
		$2,4-(OH)_2C_6H_3—$	180 (dec.)	6	P-7
p-Ethoxyphenyl	Cl	$4-CH_3O—C_6H_4—$	166		M-38
		$4-C_2H_5O—C_6H_4—$	108		M-28
		$CH_3COCH_2—$	135	75	P-7
		$C_6H_5COCH_2—$	142	57	P-7
		$4-(CH_3)_2NC_6H_4—$	154(dec.)	59.1	P-7
		$2,4-(OH)_2C_6H_3—$	190 (dec.)	70	P-7
	Br	$4-(CH_3)_2NC_6H_4—$	123 (dec.)	49	P-7
		$2,4-(OH)_2C_6H_3—$	191 (dec.)	28	P-7
p-Phenoxyphenyl	Cl	$C_6H_5COCH_2—$	147	45.5	P-7
		$4-(CH_3)_2NC_6H_4—$	195 (dec.)	41	P-7
		$2,4-(OH)_2C_6H_3—$	181 (dec.)	61	P-7
		$4-C_6H_5OC_6H_4—$	158	52	D-11
	Br	$4-(CH_3)_2NC_6H_4—$	189 (dec.)	Small	P-7

conditions, while drastic conditions yield bromoketones and tarry products. Triiodides do not condense with any of the mentioned compounds. From the very limited data available the reactivity of alkoxyphenyltellurium trichlorides seems to be higher than that of the corresponding phenoxy derivative, the latter giving lower yields than the former. With acetone the phenoxyphenyl compound formed a yellow-orange oil. While the condensations with ketones and aromatic compounds activated by amino- or hydroxy- groups proceeded at room temperature with, or without solvent, reactions with aliphatic-aromatic ethers were run by keeping the mixture of the components at approximately 150° for six hours.

Only one halogen atom of the trihalides can be replaced by an organic group in these reactions, the excess organic component is recovered unchanged.

Organotellurium trichlorides, intermediates in the formation of heterocyclic tellurium dichlorides from 1,3-diketones and $TeCl_4$ (see Section I-C-1), react intramolecularly to yield these cyclic tellurium derivatives. However,

$$\underset{\displaystyle \text{C}_6\text{H}_{13}\text{C}}{\overset{\displaystyle \text{O}}{\underset{\displaystyle \|}{}}}\text{—CH}=\underset{\displaystyle \text{C}}{\overset{\displaystyle \text{OH}}{\underset{\displaystyle |}{}}}\text{—CH}_2\text{—TeCl}_3$$ did not form the cyclic compound, even

when heated with $AlCl_3$ (M-21). Drew (D-11) prepared 10,10-dichlorophenoxtellurine from diphenyl ether as well as from 4-phenoxyphenyltellurium trichloride. Condensations with other phenyl ethers gave *p*-substituted products. The ring formation, however, requires a 2-phenoxyphenyltellurium trichloride. Drew formulated the following reaction scheme to explain the formation of the heterocyclic ring.

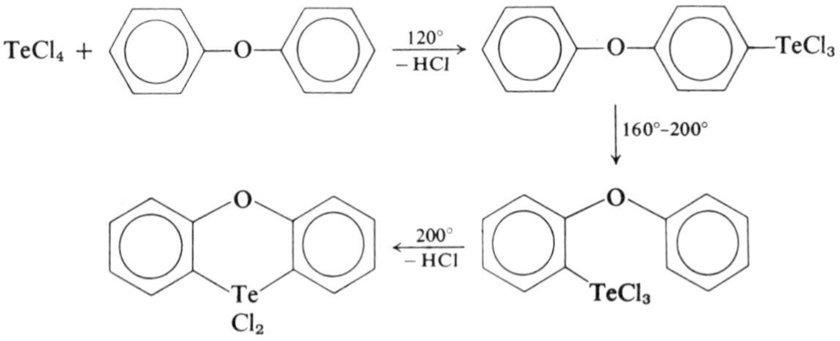

In addition to this cyclic product, bis(4-phenoxyphenyl)tellurium dichloride is also formed. A 52% yield of this linear condensation product was obtained from a 1:1 molar mixture of the trichloride and diphenyl ether at 140–165°.

The reaction scheme is supported by the following experimental results (D-11): The trichloride when heated for 4¾ hr from 150 to 210° gave a 33% yield of 10,10-dichlorophenoxtellurine, melting at 265° without decomposition.

When the trichloride was kept for two hours at 150–160°, a small amount

($\sim$10%) of a tellurium trichloride, possibly the 2-substituted compound, decomposing near 125°, was isolated. $K_2S_2O_5$ reduced it to a ditelluride.

2-(4-Carboxyphenoxy)phenyl- (C-4), 2-(4-methylphenoxy)phenyl- (C-4) and 2-thiophenoxyphenyltellurium trichlorides (P-5) heated for half an hour to 200–240° yielded the cyclic products melting at 319°, 275°, and 265–270°, respectively. The yields obtained were 62, 80, and 42%.

Experimental. 2,4-*Dihydroxyphenyl p-Methoxyphenyltellurium Dichloride* (P-7): A solution of 4-methoxyphenyltellurium trichloride (0.34 g, 0.001 mole) and resorcinol (0.22 g, 0.002 moles) in a small volume of methanol was kept overnight at room temperature. Concentration of the solution to almost dryness gave the crystalline dichloride (70% yield), which recrystallized from ether/petroleum ether melted at 182–183° (dec.).

Bis(4-phenoxyphenyl)tellurium Dichloride (D-11): 4-Phenoxyphenyltellurium trichloride (3.0 g, 0.0097 mole) and diphenyl ether (1.7 g, 0.01 mole) were heated together at 140–165° for six hours in a nitrogen atmosphere. The product was extracted with ether, the ether evaporated and the residue triturated with light petroleum ether and ethanol gave 52% of the crude dichloride. Recrystallized from CCl_4/ether the substance melted at 157–158° without decomposition.

10,10-Dichlorothiophenoxtellurine (P-5): 2-Thiophenoxyphenyltellurium trichloride (2.1 g, 0.005 mole) was heated in a glass tube for 30 min with stirring at 240–250°. On cooling a dark yellow solid was obtained, which recrystallized from acetone yielded 42% cyclic compound melting at 265–270°.

5. The Addition of Organotellurium Trichlorides to Carbon–Carbon Double Bonds

The addition of aryltellurium trichlorides to unsaturated carbon–carbon bonds has been studied by de Moura Campos and Petragnani (M-41,M-42, M-45,M-46). The reactivity of the trichlorides is less than that of tellurium tetrachloride. While $TeCl_4$ easily combined with cyclohexene (M-41) and phenyl- as well as diphenylacetylene (M-46) (see Section I-C-5), 4-ethoxyphenyltellurium trichloride reacted with cyclohexene (M-41) only at reflux temperature. It did not combine at all with the acetylenic compounds (M-41). The addition probably proceeds via a cyclic telluronium ion intermediate, which then undergoes nucleophilic attack by the chloride ion to give the racemic trans-compounds

Cyclohexene and aryltellurium trichlorides in boiling cyclohexene gave the dichlorides listed in Table XLI.

TABLE XLI

Addition of $RTeCl_3$ to Unsaturated Compounds to Yield

$$\begin{array}{c} R \quad\quad Cl \\ \diagdown\;\diagup \\ Te \\ \diagup\;\diagdown \\ R' \quad\quad Cl \end{array}$$

R'	R	m.p.	Yield, %	Ref.
—CH₂—CH——CH₂ (see structure)	Phenyl	188–189	90	M-45
	p-Methoxyphenyl	178–181		M-41
	p-Ethoxyphenyl	193–196		M-41
	p-Phenoxyphenyl	122–125	80	M-45
	1-Naphthyl	209–211	75	M-45
	2-Naphthyl	163–165	73	M-42
	2-Chlorocyclohexyl	95–120[a]	57	M-45
2-Chlorocyclohexyl	Phenyl	129–131 (dec.)	63	M-45
	p-Ethoxyphenyl	97		M-41
	p-Phenoxyphenyl	121–124 (dec.)	72	M-45
	1-Naphthyl	158–160 (dec.)	87	M-45
	2-Naphthyl	136–138 (dec.)	96	M-45
(fluorenone-type structure)	p-Ethoxyphenyl	206–209 (dec.)	58	M-45

[a] Amorphous, white solid.

Styrene, di-isobutylene, and 1,4-diphenyl-1,3-butadiene were unreactive. While 2-chlorocyclohexyltellurium trichloride is stable in excess of cyclohexene, it was found to react with 2,2-diphenyl-4-pentenoic acid (M-45). This γ,δ-unsaturated acid with a double bond activated by the carboxylic acid group also combined with a variety of other organotellurium trichlorides (see Table XLI) with the formation of 2,2-diphenyl-5-(organodichlorotelluro)-4-pentanolactones. The mechanism of the lactone ring closure involves a cyclic oxonium ion intermediate, which then loses HCl.

$$\text{ArTeCl}_3 + \underset{\underset{O}{\overset{\|}{C}}\diagdown_{C_6H_5}}{\overset{CH_2=CH-CH_2}{\underset{|}{HO\diagup}C-C_6H_5}} \longrightarrow \text{ArTe}-CH_2-CH\underset{\underset{O}{\overset{\|}{C}}\diagdown_{C_6H_5}}{-CH_2} + \text{HCl}$$

A more detailed reaction scheme is presented in Section III. 9-Allyl-9-fluorenecarboxylic acid and 4-ethoxyphenyltellurium trichloride reacted similarly with lactone formation (M-45).

Experimental. 4-Phenoxyphenyl 2-Chlorocyclohexyltellurium Dichloride (M-45): 4-Phenoxyphenyltellurium trichloride (0.80 g, 0.002 mole) was refluxed in 10 ml of cyclohexene for one hour and then heated in an open vessel with ethanol to remove excess cyclohexene and precipitate the dichloride, melting at 121–124° (dec.), in 72% yield.

2,2-Diphenyl-5-(4-phenoxyphenyldichlorotelluro)-4-pentanolactone (M-45): 4-Phenoxyphenyltellurium trichloride (0.80 g, 0.002 mole) was treated with a solution of 2,2-diphenyl-4-pentenoic acid (0.50 g, 0.002 mole) in 40 ml of CHCl₃. The yellow solution lost HCl and became colorless. After four hours the chloroform was removed by heating with ethanol. The crystalline dichloride separated on cooling (m.p. 122–125°; yield: 80%).

6. Reactions with Organometallic Compounds

The reaction of organotellurium trichlorides with organomercuric chlorides is the only general method known for the preparation of symmetric, but especially unsymmetric, bis(organo)tellurium dichlorides. The organotellurium trichlorides are obtained from $TeCl_4$ and certain aromatic compounds (see Section I-C-4) or from $TeCl_4$ and organomercuric chlorides (see Section I-C-7). The condensation between organotellurium trichlorides and the organomercuric chloride (1:1 molar ratio) takes place in boiling dioxane. The mercuric chloride precipitates as the dioxane adduct

$$\text{R}-\text{TeCl}_3 + \text{R'HgCl} \longrightarrow \underset{R'}{\overset{R}{\diagdown\diagup}}\text{TeCl}_2 + \text{HgCl}_2$$

and is easily filtered off. The bis(organo)tellurium dichlorides separate upon the addition of dilute hydrochloric acid as yellow oils, which solidify on trituration with the mother liquor. The crude products are reduced to the tellurides with $Na_2S \cdot 9H_2O$. The treatment with Na_2S removes any unreacted mercuric compound as HgS and reduces organotellurium trichlorides to the ditellurides. If the telluride contains the ditelluride as an impurity as indicated by the red color of the product, purification is accomplished by treatment with SO_2Cl_2 in benzene solution which converts the tellurides into the dichlorides and the ditellurides into the trichloride. The tellurium trichloride is removed by washing with alcohol. Reduction of the dichlorides then produces the pure tellurides (R-11). Liquid tellurides are extracted with petroleum ether and purified by distillation, while solid tellurides are isolated by filtration and recrystallized from proper solvents. The unsymmetric tellurides, with the exception of the benzyl derivatives (see Section IV-B-4), are stable and do not convert into the symmetric telluride (V-4). The compounds prepared are given in Table XLII.

Experimental. 4-Methoxyphenyl Ethyl Telluride (V-4): Ethylmercuric chloride (8 g, 0.03 mole) and 4-methoxyphenyltellurium trichloride (10.2 g, 0.03 mole) in 45 ml of dioxane were refluxed for 24 hr. Upon cooling the $HgCl_2 \cdot$ dioxane adduct separated. The filtrate was poured into 150 ml of 1% hydrochloric acid. The oil which separated solidified. The solid after extraction with petroleum ether (50–70°) was added in small portions to 70 g of $Na_2S \cdot 9H_2O$ kept at 95–100°. After stirring for 10 min 100 ml of water was added, and the mixture was cooled with ice water. The telluride was extracted with petroleum ether and purified by distillation (b.p. 120°/1 mm yield: 30%).

7. Reactions with Ditellurides

Petragnani and de Moura Campos (P-6) studied the reactions between diaryl ditellurides and aryltellurium trichlorides. The equimolar mixture of the reagents was refluxed in toluene for two hours. The elemental tellurium which separated was filtered and the bis(organo)tellurium dihalide was precipitated with petroleum ether.

$$\left(R-O-\underset{}{\bigcirc}- \right)_2 Te_2 + RO-\bigcirc-TeCl_3 \longrightarrow \left(RO-\underset{}{\bigcirc}- \right)_2 TeCl_2 + Te$$

R = CH_3 m.p. 182–184°; 95.1% yield,
R = C_2H_5 m.p. 110–111°; 96.6% yield,

For a possible reaction mechanism consult Section V-3.

8. Reactions in the Organic Part of the Molecule

Reactions of organotellurium trichlorides, which affect only the organic part of the molecule are virtually unexplored. Morgan and Drew (M-26) reported the hydrolysis of $(Cl_3TeCH_2CO)_2O$ to carboxymethyltellurium trichloride. The addition of bromine to the unsaturated tellurium trichloride,

$$\bigcirc-\underset{\underset{Cl}{|}}{C}=\underset{\underset{R}{|}}{C}-TeCl_3,$$

TABLE XLII
Reactions of Organotellurium Trichlorides with Organomercuric Chlorides

R—TeCl₃	R′—HgX[a]		RR′TeCl_n		
R	R′	n^b	m.p. or b.p.^{c} °C	Yield, %	Ref.
Phenyl	1-Naphthyl	0	148 (0.01)	80.3	R-11
p-Methoxyphenyl	2-Naphthyl	0	49–50	75.5	R-11
	C₂H₅—	0	120 (1)	30	V-4
	Phenyl	0	60.5	87	R-11
	C₆H₅CH₂—	2	126.0–127.5	100	V-4
	1-Naphthyl	0	72.5	95	R-11
	2-Naphthyl	0	88	83.5	R-11
	—CH₂—CH—CH₂—C(C₆H₅)₂—C=O (ring, O)	2		100	M-45
p-Phenoxyphenyl	—CH₂—CH—CH₂—C(C₆H₅)₂—C=O (ring, O)	2	122.5	100	M-45
	C₆H₅—	0	177 (0.03)	70	R-11
	1-Naphthyl	0	71–72	70	R-11
	2-Naphthyl	0	69–70	87.2	R-11
	Cyclohexyl^d	0	190 (0.024)	59	V-4
	C₆H₅CH₂—	2	128.5–130	80	V-4
2-Naphthyl	1-Naphthyl	0	93–94	91.5	R-11
	2-Naphthyl	0	144–145	92	R-11

[a] X = Cl unless otherwise stated.
[b] The tellurides ($n = 0$) were isolated by treatment of the crude reaction product with Na₂S·9H₂O.
[c] Numbers with pressures in mm Hg in parentheses are boiling points.
[d] Cyclohexylmercuric bromide was employed.

in benzene resulted in the formation of $TeBr_4$ at room temperature (R = H). When R represented the phenyl group no reaction occurred at room temperature, $TeBr_4$ was precipitated only after several hours of refluxing (M-46). No attempt was made to isolate intermediate brominated products.

9. Complex Formation with Other Organotellurium Compounds

For the discussion of complexes formed between methyltellurium trihalides and trimethyltelluronium halides (D-15) refer to Section VII-5. In addition, one should mention that CH_3TeI_3 did not combine with a mixture of TeI_4 and $(CH_3)_2TeI_2$ (D-15).

According to Drew's report (D-15) CH_3TeI_3 adds one mole of KI with the formation of almost black crystals of $K[CH_3TeI_4]$.

X. REACTIONS OF TELLURINIC ACIDS AND THEIR DERIVATIVES

Tellurinic acids, RTeOOH, are products formed by hydrolysis of organotellurium trihalides. The following scheme shows the possible intermediates formed during hydrolysis. The unknown compounds are parenthesized and the general reactions of these substances are shown.

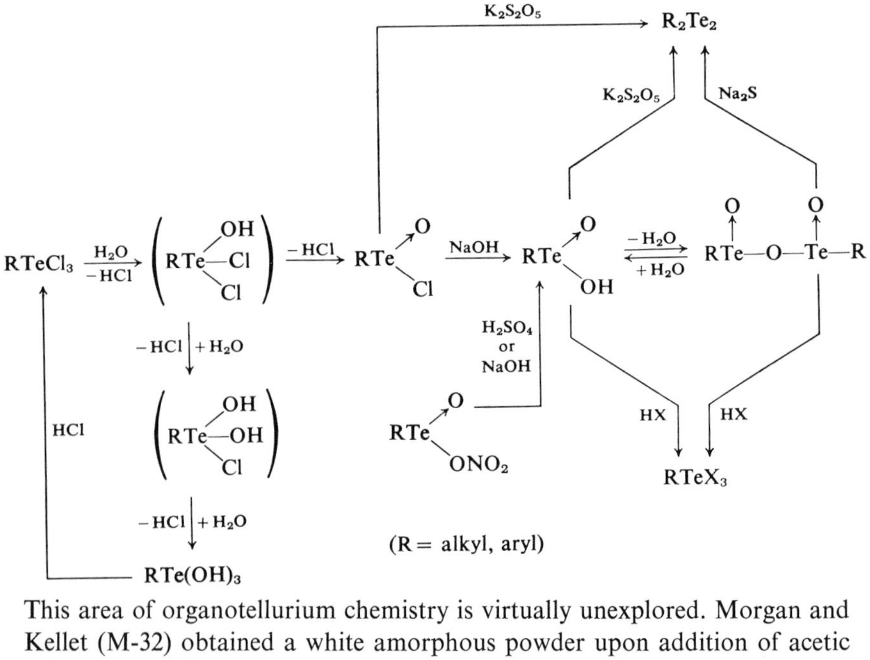

$$
\begin{array}{c}
\xrightarrow{K_2S_2O_5} R_2Te_2 \\[2mm]
K_2S_2O_5 \quad \Big\downarrow Na_2S
\end{array}
$$

$$
RTeCl_3 \xrightarrow[-HCl]{H_2O} \left(RTe\!\!\begin{array}{c} OH \\ \diagdown Cl \\ \diagup \\ Cl \end{array} \right) \xrightarrow{-HCl} RTe\!\!\begin{array}{c} O \\ \diagup \\ \diagdown Cl \end{array} \xrightarrow{NaOH} RTe\!\!\begin{array}{c} O \\ \diagup \\ \diagdown OH \end{array} \underset{+H_2O}{\overset{-H_2O}{\rightleftharpoons}} RTe\!-\!O\!-\!Te\!-\!R
$$

$$
-HCl \Big\downarrow +H_2O \qquad\qquad \Big\uparrow {}^{H_2SO_4}_{\ or\ NaOH}
$$

$$
HCl \qquad \left(RTe\!\!\begin{array}{c} OH \\ \diagdown OH \\ \diagup \\ Cl \end{array} \right) \qquad RTe\!\!\begin{array}{c} O \\ \diagup \\ \diagdown ONO_2 \end{array} \qquad\qquad HX\Big\downarrow \quad \Big\downarrow HX
$$

$$
-HCl \Big\downarrow +H_2O \qquad\qquad (R = alkyl, aryl) \qquad\qquad RTeX_3
$$

$$
RTe(OH)_3
$$

This area of organotellurium chemistry is virtually unexplored. Morgan and Kellet (M-32) obtained a white amorphous powder upon addition of acetic acid to a solution of p-methoxyphenyltellurium trichloride dissolved in 2N NaOH. According to the analytical data the formula of a trihydroxide,

$RTe(OH)_3$, was proposed, although the values were 1% low. When this substance was treated with concentrated hydrochloric acid, $RTeCl_3$ was isolated after evaporation of the solvent. When dissolved in warm dilute nitric acid 2,4-dinitrophenol was produced.

The complex substances $(C_4H_9)(C_2H_5O_2CCH_2)TeO \cdot 2C_4H_9TeOOH$, $(C_4H_9)_2TeO \cdot 3C_4H_9TeOOH$ and $2(C_4H_9)_2TeO \cdot C_4H_9TeOOH$, when suspended in acetone and treated with HI gave $(C_4H_9)_2TeI_2$ and $C_4H_9TeI_3$. (C_4H_9)-(*l*-menthyl $O_2CCH_2)TeO \cdot C_4H_9TeOOH$ reacted with nitric acid and yielded a substance decomposing at 130°, to which the formula $C_4H_9TeO \cdot O \cdot Te$-$(C_4H_9)(NO_3)(CH_2CO_2$-*l*-menthyl) was assigned (B-2).

$$ (CH_3)_3Te\overset{\overset{\textstyle O}{\uparrow}}{—}O—Te—CH_3 $$

The ester $(CH_3)_3Te—O—Te—CH_3$ produced, with hydriodic acid, trimethyltelluronium iodide and methanetellurinic acid, or its anhydride. The latter, with excess HI, was converted to methyltellurium triiodide (D-15). With HBr the corresponding bromides were isolated. Methanetellurinic acid kept in a desiccator for several weeks disproportionated according to the following equation:

$$ 2\ CH_3TeOOH \longrightarrow (CH_3)_2Te\begin{smallmatrix} \diagup OH \\ \\ \diagdown OTe(O)OH \end{smallmatrix} \xrightarrow{H_2O} (CH_3)_2Te(OH)_2 + TeO(OH)_2 $$

Aqueous solutions of CH_3TeOOH were neutral toward litmus. $K_2S_2O_5$ reduced the acid to the ditelluride.

Reichel and Kirschbaum (R-6) investigated a number of aromatic tellurinic acid derivatives. They found that the acid chlorides, $RTe(O)Cl$, are reduced by $K_2S_2O_5$ in acidic medium to the ditellurides ($R = 4$-$CH_3OC_6H_4$, 4-$C_2H_5OC_6H_4$). Treatment with Ag_2O resulted in a tellurium–carbon bond cleavage and formation of the corresponding phenols. 4-Hydroxybenzenetellurinic acid chloride decomposed when reduced with $K_2S_2O_5$.

4-Hydroxybenzenetellurinic acid undergoes the following reactions (yields and melting points of products given). The nitrobenzenetellurinic acids cannot be reduced to the amines. Cleavage of the Te–carbon bond or ditelluride formation was observed to take place upon reduction.

The mixed anhydrides of benzenetellurinic acid and 3,5-dinitro-4-hydroxy-benzenetellurinic acid with nitric acid, $RTe\overset{\overset{\textstyle O}{\nearrow}}{—}ONO_2$, were saponified to the parent acids, the first with $1N$ NaOH (L-10), the latter with 20% H_2SO_4 (R-6). The 3,5-dinitro-4-hydroxybenzenetellurinic acid could not be converted to the ditelluride (R-6). 2-Naphthalenetellurinic acid anhydride was reduced to the ditelluride with Na_2S (V-5).

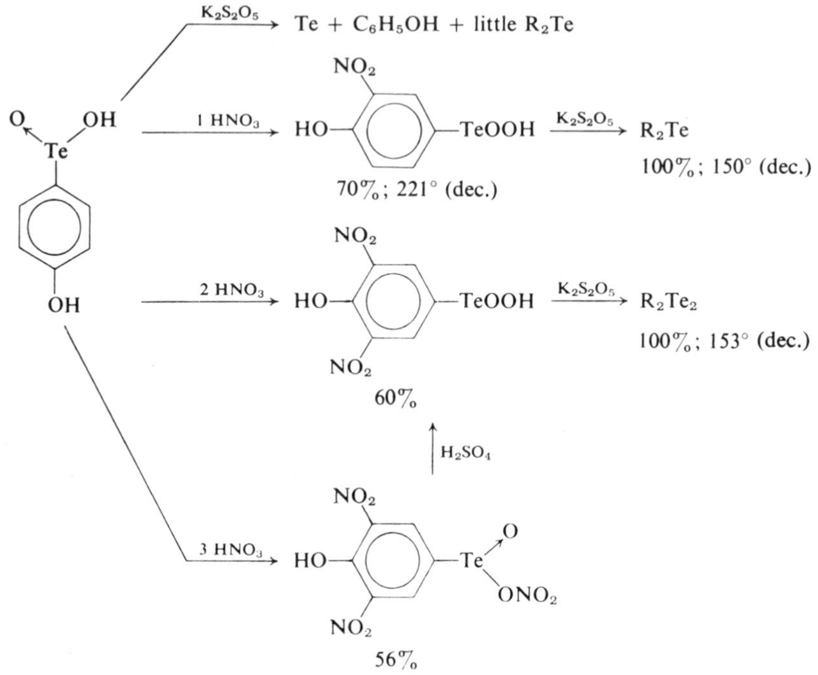

$\xrightarrow{K_2S_2O_5}$ Te + C_6H_5OH + little R_2Te

$\xrightarrow{1\ HNO_3}$ HO—⟨NO_2⟩—TeOOH $\xrightarrow{K_2S_2O_5}$ R_2Te

70%; 221° (dec.)   100%; 150° (dec.)

$\xrightarrow{2\ HNO_3}$ HO—⟨(NO_2)_2⟩—TeOOH $\xrightarrow{K_2S_2O_5}$ R_2Te_2

60%   100%; 153° (dec.)

↑ H_2SO_4

$\xrightarrow{3\ HNO_3}$ HO—⟨(NO_2)_2⟩—Te(=O)ONO_2

56%

According to electrometric titrations (G-2,L-32) methanetellurinic acid is amphoteric.

Experimental. 4-Ethoxybenzenetellurinic Acid (R-6): 4-Ethoxybenzene-tellurinic acid chloride (3 g) was dissolved in 70–80% hot NaOH (10 ml). On cooling, crystals separated which, after drying, were treated with dilute H_2SO_4. The precipitate was collected. The substance, recrystallized from glacial acetic acid, lacked a definite melting point (yield: ~100%).

3-Nitro-4-hydroxybenzenetellurinic Acid (R-6): 4-Hydroxybenzenetellurinic acid (25.2 g, 0.1 mole) was added in small portions with cooling to 100 ml concentrated H_2SO_4. While stirring, a mixture of 6.2 ml of concentrated HNO_3 and 6.2 ml of concentrated H_2SO_4 was dropped into the solution, the temperature of which should not exceed 0°. After completion of the reaction the temperature was allowed to rise to 10°. The mixture was poured in 180 ml of water. After 24 hr the precipitated nitro-compound was filtered and purified by boiling with water (70% yield; m.p. 221° dec.).

Bis(3-nitro-4-hydroxyphenyl) Ditelluride (R-6): 3-Nitro-4-hydroxyben-zenetellurinic acid (2.5 g) was dissolved in 35 ml of 2N NaOH. After addition of $K_2S_2O_5$ (7 g in 15 ml of H_2O) 40 ml of 2N HCl are slowly dropped into the stirred solution. The precipitate was filtered and dried (yield: 100%; m.p. 150° dec.).

XI. TELLUROALDEHYDES AND TELLUROKETONES

Telluroformaldehyde is the only known compound in the telluroaldehyde series. This compound was prepared by the reaction of methylene radicals with tellurium mirrors (see Section I-A-6). The aldehyde is reported to exist in two modifications. The first modification, a deep-red solid, is insoluble in H_2O, quite insoluble in ethanol, ether, and benzene and can be distilled at 100° in high vacuum with only slight decomposition (R-13,P-2). It has an unpleasant odor. The other modification is colorless at $-70°$ and turns gray-green at room temperature. It was found to be readily soluble in acetone, $CHCl_3$, and ether. The compound darkens in air.

Telluroformaldehyde reacts with bromine and iodine with formation of dihalomethanes (P-2).

Williams and Dunbar (W-2) prepared a sample of telluroformaldehyde from a tellurium mirror and methylene radicals generated from diazomethane at 500°. The compound was analyzed in the mass spectrograph. The fragmentation pattern and IR analysis in the range $2.5-50\mu$ indicated a trimeric molecule $(CH_2Te)_3$.

The telluroketones (see Sections I-B-2) are yellow to brown oils, insoluble in water and soluble in ethanol and ether. Dipropyl telluroketone decomposed during distillation at 4 torr.

XII. REACTIONS OF TELLURONES

Tellurones, R_2TeO_2, are the oxidation products of tellurides and telluroxides in which the tellurium atom possesses the same oxidation number as sulfur in sulfones. These are ill-defined substances. Vernon (V-2) claimed that treatment of dimethyl telluride or dimethyltellurium dihydroxide with hydrogen peroxide yielded dimethyl tellurone, a white amorphous powder insoluble in all solvents. It oxidized hydrohalic acids to the free halogens, decolorized permanganate, and had explosive properties. Diethyl tellurone with the same properties as the methyl derivative was formed from diethyl telluroxide upon exposure to air (G-4). It is very doubtful, however, that the tellurones have the simple formula R_2TeO_2.

Vernon (V-2) observed that tellurones formed complex tellurium halides $R_3TeX \cdot RTeX_3$ with hydrohalic acids (see Section VI-A for more information on these complexes). Lowry and Gilbert confirmed these results and proposed for the tellurones the formula $[R_3Te]^+[RTeO_4]^-$.

Balfe and co-workers (B-2) isolated complexes between tellurinic acids and telluroxides which were obtained by the oxidation of tellurides (see Section IV-3). They obtained a white solid melting at 90° from the oxidation of bis-(carbomenthoxymethyl) telluride with 30% hydrogen peroxide. This solid liberated iodine from potassium iodide, bromine from hydrobromic acid, and decolorized potassium permanganate. It was formulated as a hydrogen peroxide addition compound of the telluroxide, $R_2Te(OH)(OOH)$.

Pentamethylene tellurone (M-35), obtained from the telluride and hydrogen peroxide in methanol solution, crystallized with one molecule of water. It showed the same properties as Vernon's dimethyl tellurone. Concentrated nitric and sulfuric acid violently decomposed this heterocyclic tellurone.

Phenoxtellurine, when treated with an excess of 30% hydrogen peroxide in acetone or glacial acetic acid, gave a product with properties like those of the other tellurones (D-12). This substance heated for one hour *in vacuo* is supposed to have formed a compound of the following structure.

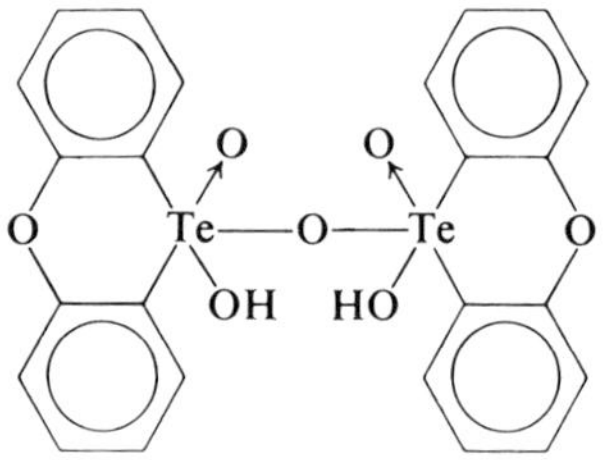

XIII. REACTIONS OF HETEROCYCLIC COMPOUNDS CONTAINING A TELLURIUM ATOM

Investigations of heterocyclic compounds containing a tellurium atom were carried out almost exclusively by Morgan and his co-workers during the decade 1920–1930. Most of the work performed dealt with the telluracyclohexane and the phenoxtellurine ring systems. Five- and six-membered rings containing one or two tellurium atoms, or one tellurium atom and another hetero atom have been successfully synthesized. Attempts to prepare telluracyclobutane failed (see Section IA-2). The reactions of tellurium heterocycles will be discussed in the following sections.

A. Telluracyclopentane and Its Derivatives

The following names have been used for telluracyclopentane:

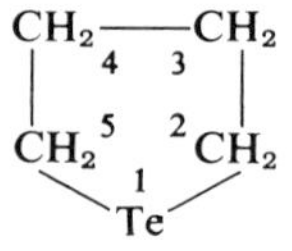

telluracyclopentane
tetrahydrotellurophene
tellurolane
cyclotellurobutane

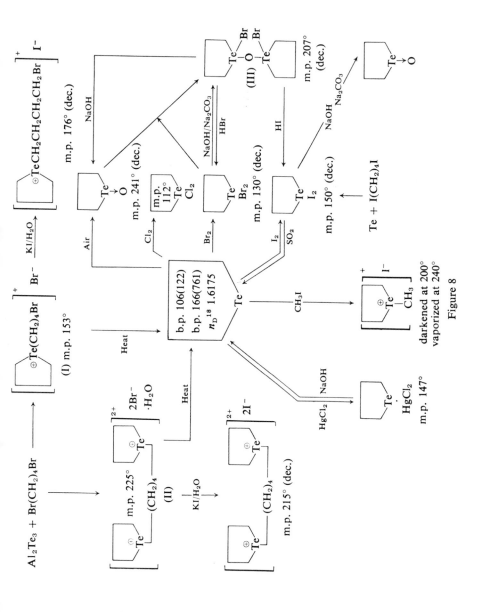

Figure 8

295

TABLE XLIII

Reduction of 1-Tellura-3,5-cyclohexanedione 1,1-Dihalides to Cyclic Tellurides

$$R^3 \quad R^2$$
$$O = \overset{\displaystyle}{\underset{\displaystyle O}{\bigcirc}} \overset{R^1}{\underset{R^4}{\text{Te}}}$$
$$X_2$$

1-Tellura-3,5-cyclohexanedione 1,1-Dihalide						1-Tellura-3,5-cyclohexanedione		
R^1	R^2	R^3	R^4	X	Reagent	m.p.	Yield	Ref.
H	H	H	H	Cl	$KHSO_3$	182	90	M-15–17
				Br	$KHSO_3$			M-16
				I	$KHSO_3$			M-16
H	Cl—	H	H	Cl	$K_2S_2O_5$	154 (dec.)	51	M-17
H	CH_3—	H	H	Cl	$K_2S_2O_5$	170 (dec.)	76	M-17
H	C_2H_5—	H	H	Cl	$K_2S_2O_5$	142	88,96	M-17,M-31
H	C_3H_7—	H	H	Cl	$K_2S_2O_5$	107	—	M-22
H	$i\text{-}C_3H_7$—	H	H	Cl	$K_2S_2O_5$	153	—	M-22
H	C_4H_9—	H	H	Cl	$KHSO_3$	129	—	M-23
H	$i\text{-}C_4H_9$—	H	H	Cl	$K_2S_2O_5$	150	—	M-29
H	$sec\text{-}C_4H_9$—	H	H	Cl	$KHSO_3$	145	—	M-29
H	$(2\text{-}CH_3)C_4H_9$—	H	H	Cl	$NaHSO_3$	139	—	M-29
H	$C_6H_5CH_2$—	H	H	Cl	$K_2S_2O_5$	153 (dec.)	—	M-27

R1	R2	R3	R4	X	Reagent	M.p. (°C)	Yield (%)	Ref.
H	CH₃—	H	H	Cl	K₂S₂O₅	125	100	M-24
H	C₂H₅—	H	H	Cl	K₂S₂O₅	86	100	M-21
H	C₆H₅CH₂—	H	H	Cl	K₂S₂O₅	128	—	M-27
CH₃—	H	H	H	Cl	K₂S₂O₅	100	79	M-19
C₂H₅—	H	H	H	Cl	K₂S₂O₅	110–112	100	M-21
C₃H₇—	H	H	H	Cl	K₂S₂O₅	80	70	M-23
C₄H₉—	H	H	H	Cl	K₂S₂O₅	86	—	M-23
C₅H₁₁—	H	H	H	Cl	K₂S₂O₅	86	—	M-21
C₆H₁₃—	H	H	H	Cl	K₂S₂O₅	75	—	M-27
C₇H₁₅—	H	H	H	Cl	KHSO₃	89	—	M-27
C₈H₁₇—	H	H	H	Cl	KHSO₃	64	—	M-21
C₁₀H₂₁—	H	H	H	Cl	KHSO₃	99 (dec.)	—	M-29
C₆H₅CH₂—	C₆H₅CH₂—	H	H	Cl	K₂S₂O₅	159 (dec.)	—	M-29
CH₃—	CH₃—	H	H	Cl	K₂S₂O₅	125–126	—	M-21
CH₃—	C₂H₅—	H	H	Cl	K₂S₂O₅	109	—	M-22
CH₃—	C₃H₇—	H	H	Cl	K₂S₂O₅	102	—	M-29
CH₃—	i-C₃H₇—	H	H	Cl	K₂S₂O₅	127	—	M-29
CH₃—	C₄H₉—	H	H	Cl	—	93	—	M-23
CH₃—	C₆H₅CH₂—	H	H	Cl	K₂S₂O₅	124 (dec.)	—	M-29
C₂H₅—	C₂H₅—	H	H	Cl	—	113 (dec.)	—	M-21
CH₃—	H	H	CH₃	Cl	KHSO₃	151	—	M-21, M-27
CH₃—	H	H	C₂H₅	Cl	K₂S₂O₅	101–102	—	M-21
C₂H₅—	H	H	C₂H₅	Cl	KHSO₃	97	—	M-22
CH₃—	CH₃—	H	CH₃	Cl	K₂S₂O₅	135 (dec.)	100	M-27
CH₃—	C₂H₅—	H	CH₃	Cl	—	138 (dec.)	—	M-21

297

This ring system can be obtained by dissolving amorphous tellurium in α,δ-tetramethylene diiodide at 130° (see Section IA-2). The other method, already mentioned in Section I-B-2, involves heating of aluminum telluride with tetramethylene dibromide to 125° (M-40). It did not yield the tellura-cyclopentane itself. The excess halide reacted with the cyclic telluride to form the telluronium salts I and II (M-40) (see Figure 8). After two hours at 125° the cooled semi-solid mixture was extracted successively with CCl₄, acetone, ethanol, and water. A black insoluble residue was left.

The acetone and ethanol extracts deposited tetramethylene δ-bromobutyl-telluronium bromide (I). The aqueous extract gave tetramethylene bis(tetra-methylenetelluronium) dibromide (II). Both of these bromides were converted to the respective iodides upon treatment with aqueous potassium iodide solution. The dibromide (II) precipitated AgBr upon addition of AgNO₃ and gave a red complex with platinum (IV) chloride, decomposing at 135°.

When heated above their melting points these telluronium bromides regenerate telluracyclopentane with the loss of tetramethylene dibromide.

Telluracyclopentane 1,1-dichloride was reduced to the telluride by reduction with SO₂ in aqueous solution. Telluracyclopentane was separated by steam distillation and fractionated *in vacuo*. The cyclic tellurahydrocarbon readily adds elemental halogen to form stable dihalides. These compounds melt without decomposition and remain unchanged at temperatures well above their melting points. The dichloride and the dibromide, in contrast to the alkyl compounds, dissolve in hot water from which they are recovered unchanged on cooling. Treatment of the dibromide with one equivalent of aqueous alkali gives a well-characterized compound, bis(telluracyclopentane) 1,1-[oxydibromide(III). Mercuric chloride and methyl iodide yield the mercuric chloride adduct and the methiodide. Concentrated nitric acid reacts explosively while sulfuric acid develops a red coloration.

Telluracyclopentane 1-oxide was prepared by air oxidation of the cyclic telluride, or by treatment of the dihalides with NaOH or Na₂CO₃ solutions. The oxide regenerated the dihalides with hydrohalic acids and formed compound III when warmed with an equimolar amount of the dibromide in water. The reaction scheme of Table XLIII also lists the melting and boiling points of the compounds prepared.

Duffield, Budzikiewicz, and Djerassi studied the mass spectrometric fragmentation of telluracyclopentane (D-19).

B. Tellurophene and Its Derivatives

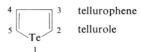

Mack (M-6) successfully prepared tellurophene and a number of its 2,5-

disubstituted derivatives by the reaction of diacetylenic compounds with Na_2Te in methanol (see Section IB-1). The aromatic character of tellurophene is demonstrated by the fact that it undergoes several substitution reactions. Thus, in CH_3OD, with D_2SO_4, 2,5-dideuteriotellurophene is formed while 2,5-bis(acetoxymercuri)tellurophene is obtained from the reaction of mercury(II) acetate with tellurophene in ethanol.

With bromine in methanol tellurophene 1,1-dibromide, which decomposes at 125°, was obtained. An aqueous hydrosulfite solution reduced the dibromide to tellurophene.

Tellurophene, according to cryoscopic molecular mass determinations, is dimeric in benzene, while its 2,5-bis(hydroxymethyl) and 2,5-bis(dimethyl-hydroxymethyl) derivatives were found to be monomeric in the same solvent. Tetrachlorotellurophene (M-5) reacted with chlorine to form quantitatively the dichloride (m.p. 200°), which dissolved in benzene. Tetrachlorotelluro-phene 1,1-dichloride was readily converted to tetrachlorotellurophene by shaking it with sodium bisulfite solution. Tetrachlorotellurophene 1,1-dichloride was rapidly decomposed to 1,2,3,4-tetrachloro-1,3-butadiene and tellurous acid by aqueous base.

$$\underset{\underset{Cl}{\overset{Cl}{\diagup}}\overset{Cl}{\underset{Cl}{\diagdown}}}{\diagup\!\!\!\!\diagdown} \;\; + \; 2\;NaOH \; + \; H_2O \; \longrightarrow \; \overset{Cl\;\;Cl\;\;Cl\;\;Cl}{\underset{}{HC=C-C=CH}} \; + \; H_2TeO_3 \; + \; 2\;NaCl$$

Tetrachlorotellurophene was not cleaved by base.

Tetraphenyltellurophene (see Sections I-B-1 and I-C-7-c) failed to undergo the Diels–Alder reaction with maleic anhydride even under drastic conditions (21 hr at 220°). Attempts to form a telluronium salt by heating methyl iodide with the tellurophene for 19 hr at 150° were also unsuccessful. Tetraphenyl-tellurophene gave a small quantity of a complex of the formula $(C_{28}H_{20})Te-Fe(CO)_3$, when it was refluxed with $Fe_3(CO)_{12}$ in a toluene/benzene mixture. This experiment could not be successfully repeated. $Fe(CO)_5$ and $Fe_2(CO)_9$ did not react. These results suggest that tellurophenes have an aromaticity similar to that of thiophenes and selenophenes. Tetraphenyltellurophene, however, readily added bromine with the formation of tetraphenyltelluro-phene 1,1-dibromide melting at 243–245° (dec.) (B-18).

Tetraphenyltellurophene 1,1-dibromide when heated with maleic anhydride in 30 ml of dry benzene in a sealed tube at 140° for 4.5 hr yielded crystalline tellurium, unchanged dibromide and small amounts of 1,4-diphenyl-2,3,5,6-dibenzopentalene and 1,4-dihydro-1,4-diphenyl-2,3,5,6-dibenzopentalene (B-18). When the reaction was carried out at 180° for 13 hr, 81% tellurium and 41% tetraphenylfuran were obtained.

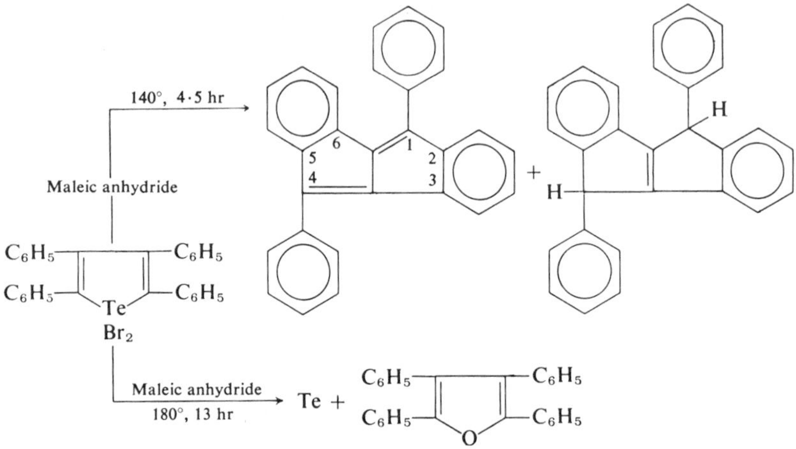

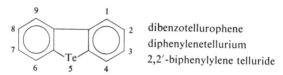

C. Dibenzotellurophene and Its Derivatives

dibenzotellurophene
diphenylenetellurium
2,2'-biphenylylene telluride

Dibenzotellurophene, in contrast to tetraphenyltellurophene, reacted with methyl iodide to form the telluronium iodide, which, however, is not very stable. It regenerates its constituents upon recrystallization from ethanol (H-4). The telluronium iodide, when treated with 2,2'-dilithiobiphenyl, produced bis(2,2'-biphenylylene)tellurium, which, upon hydrolysis gave biphenylylene biphenylyltelluronium hydroxide (H-4). This base was converted to the telluronium iodide with potassium iodide (H-3).

Dibenzotellurophene was converted to the dihalides by treatment with elemental halogen in ether (C-27,H-3). The dihalides are easily reduced back to the telluride (C-27,H-3). The dichloride hydrolyzed to the dibenzotellurophene oxide, which with hydrohalic acids regenerated the dihalides (C-27). The dibromide decomposed on heating to 4,4'-dibromobiphenyl (C-27). The dichloride produced bis(2,2'-biphenylylene)tellurium and biphenylylene biphenylyltelluronium chloride in its reaction with 2,2'-dilithiobiphenyl (H-3).

Dibenzotellurophene and butyllithium when mixed in a 1:2 molar ratio exchanged their metal atoms to yield dilithiobiphenyl and dibutyl telluride (H-4).

Finally, the tellurium atom was eliminated upon heating the telluride with sulfur with the formation of dibenzothiophene (C-27).

Figure 9 summarizes the reactions of dibenzotellurophene.

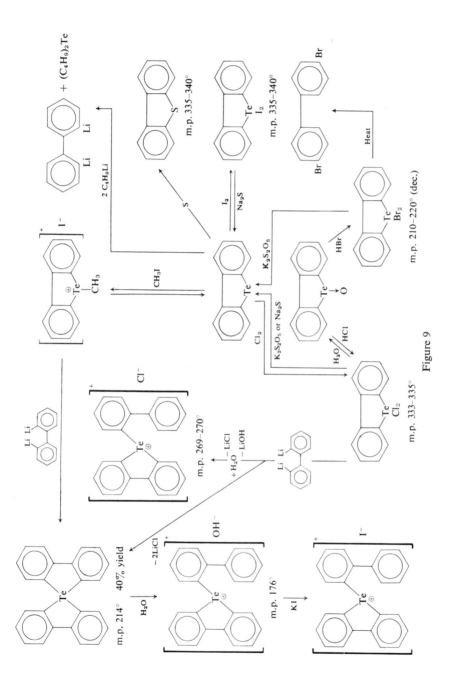

Figure 9

301

D. Benzotellurophene

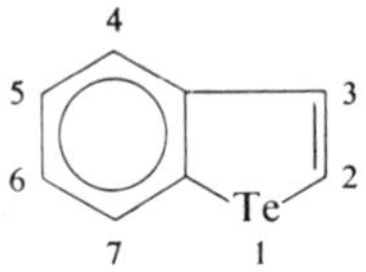

benzotellurophene
telluranaphthene

3-Hydroxybenzotellurophene was synthesized by Mazza and Melchionna (M-11) from bis(2-carboxyphenyl) telluride. Farrar (F-2), however, in attempting to repeat this work obtained different results. The structure of all compounds derived from this telluride including hydroxybenzotellurophene must be considered questionable.

It is reported (M-11), that 3-hydroxybenzotellurophene was converted to the acetate by acetic anhydride.

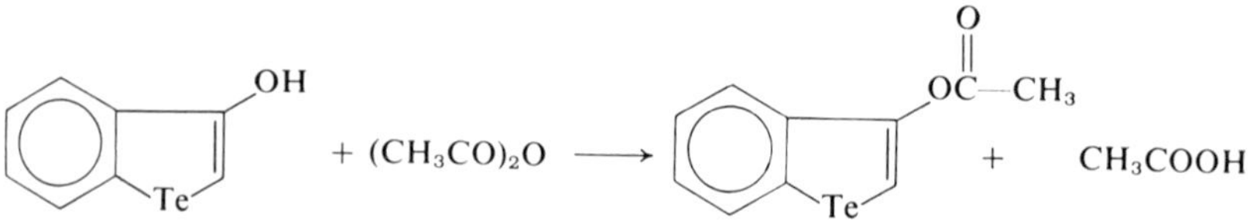

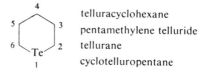

The hydroxy compound could not be oxidized to telluroindigo.

E. Telluracyclohexane and Its Derivatives

telluracyclohexane
pentamethylene telluride
tellurane
cyclotelluropentane

The synthesis of this heterocyclic ring system is discussed in Sections I-A-2, I-B-1, and I-B-2. The reaction of aluminum telluride with pentamethylene dihalides produced small amounts of telluracyclohexane 1,1-dihalides and the telluronium salts I and II (see Figure 10) by combination of the telluride with excess halide (M-34,M-35). The telluronium salts decomposed on heating to telluracyclohexane. The pentamethylene bis(pentamethylenetelluronium) dibromide and diiodide gave with elemental bromine and iodine the compounds III (Figure 10) (G-3,M-35). Telluracyclohexane additively combined with elemental halogens to form the respective telluracyclohexane 1,1-dihalides (F-1,M-35), which are reducible to the parent telluride. With hydrogen peroxide the telluracyclohexane was converted into the tellurone (M-35), while the dibromide and silver oxide gave telluracyclohexane 1-oxide or dihydroxide (G-3). The diiodide formed with iodine the tetraiodide(IV), a black solid (G-3), and gave upon boiling in concentrated K_2CO_3 solution bis(pentamethylene iodo tellur)oxide (G-3).

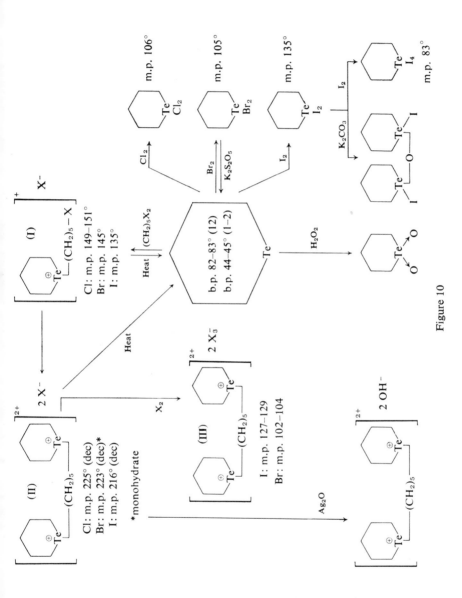

Figure 10

303

UV and visible spectral data and conductivity studies suggest that the telluracyclohexane 1,1-dihalides exist in aqueous solutions as pentamethylene hydroxytelluronium halides (G-3).

The telluronium salts I and II are binary and ternary electrolytes, respectively. While telluracyclohexane 1,1-tetraiodide hydrolyzed to $[C_5H_{10}Te-OH]^+I_3^-$, dissociation of compound III (X = I) was proposed in order to account for the observed conductivity data.

The dibromides and diiodides II when ground with moist silver oxide formed the dihydroxide (G-3).

Telluracyclohexane 1,1-diiodide seems to exist in two crystalline forms—one deep red and the other orange. The red modification changes to the high-temperature orange form at 105°. Both forms melt at 135–136° (M-35).

Figure 10 summarizes the reactions of telluracyclohexane.

F. 1-Tellura-3,5-cyclohexanedione and Its Derivatives

1-tellura-3,5-cyclohexanedione

The preparation of telluracyclohexanediones has been discussed in Section I-C-1. In these reactions the initial product is the 1-tellura-3,5-cyclohexanedione 1,1-dichloride. The dichlorides and the other dihalides, when carefully reduced with sulfurous acid, alkali hydrogen sulfite, or potassium metabisulfite, yield the telluracyclohexanediones listed in Table XLIII. These cyclic tellurides react with a mole of halogen to form 1-tellura-3,5-cyclohexanedione 1,1-dihalide (see Table XLIV).

Treatment of these cyclic diketones with hydroxylamine sulfate in boiling water yielded dioximes. Steric hindrance seems to favor the formation of a monoxime. 2,4-Dimethyl-1-tellura-3,5-cyclohexanedione monoxime was obtained on oximation in dilute acidic acid, while its dioxime formed under more

TABLE XLIV

1-Tellura-3,5-cyclohexanediones 1,1-Dihalides

R^1	R^2	R^3	R^4	Cl m.p.	Cl Yield, %	Br m.p.	Br Yield, %	I m.p.	I Yield, %	Ref.
H	H	H	H			~160 (dec.)	—	141 (dec.)	—	M-15–M-17
H	CH$_3$	H	H			153 (dec.)	100	176 (dec.)	—	M-17
H	C$_2$H$_5$	H	H	185–190(dec.)	—	161–170(dec.)	100	176 (dec.)	100	M-17
CH$_3$	H	H	H			156 (dec.)	—	190 (dec.)	100	M-19

305

TABLE XLV
Oximes of 1-Tellura-3,5-cyclohexanediones[a]

R¹	R²	R³	R⁴	Monoxime m.p.	Dioxime m.p.	Ref.
H	H	H	H		190–207 (dec.)	M-24
H	C₂H₅	H	H		192 (dec.)	M-24
H	C₆H₅CH₂	H	H		168–170 (dec.)	M-27
H	CH₃	CH₃	H		235 (dec.)	M-24
CH₃	H	H	H	184 (dec.)	161.5	M-29
C₂H₅	H	H	H		149 (dec.); 183 (dec.)[b]	M-24
CH₃	CH₃	H	H	164 (dec.)	198 (dec.)	M-24
CH₃	H	H	CH₃		168–170 (dec.)	M-27
CH₃	CH₃	H	CH₃		170 (dec.)	M-27
CH₃	C₂H₅	H	H	157	182 (dec.)	M-29

[a] Yields are not reported.
[b] Two isomers isolated.

drastic condition in alkaline solutions (M-24). The monoximes of 2,4-substituted cyclic tellurides should exist in two isomeric forms.

The dioximes should be capable of existing as one of four isomers, one each with a syn- and anti- configuration and two with amphi- configurations in the case of an unsymmetric substituted cyclic telluride. Morgan and Drew (M-24) isolated two forms of the 2-ethyl-1-tellura-3,5-cyclohexanedione dioxime. The question of the stereochemistry of these compounds remains unsettled at the moment. The known oximes are listed in Table XLV.

Some of the cyclic tellurides are decomposed by alkali or concentrated hydrochloric acid with the elimination of elemental tellurium (M-15–M-17).

The respective cyclic dichlorides were transformed upon treatment with ethyl chloride and hydrogen chloride in $CHCl_3$ solution to a linear tellurium trichloride.

$$H_3C-\underset{\underset{OC_2H_5}{|}}{C}=CH-CO-CH_2-TeCl_3$$

According to Morgan and Drew (M-24), the unstable enolized form becomes fixed by alkylation of the hydroxyl group. The strained ring is then broken by hydrogen chloride at the tellurium–carbon bond.

The bactericidal properties of the telluracyclohexanediones and their derivatives have been investigated. (G-11,L-32,L-40,M-17,M-18,M-20,M-25,M-33).

G. Telluroisochroman

telluroisochroman
3,4-dihydro-1-H-2-benzotellurine

The preparation and study of this heterocyclic compound is due to Mann and Holliman (H-7,M-8). Telluroisochroman is prepared in 50% yield by the reaction between o-(β-bromoethyl)benzyl bromide and Na_2Te (see Section I-B-1).

The reactions of telluroisochroman are outlined in the following reaction scheme.

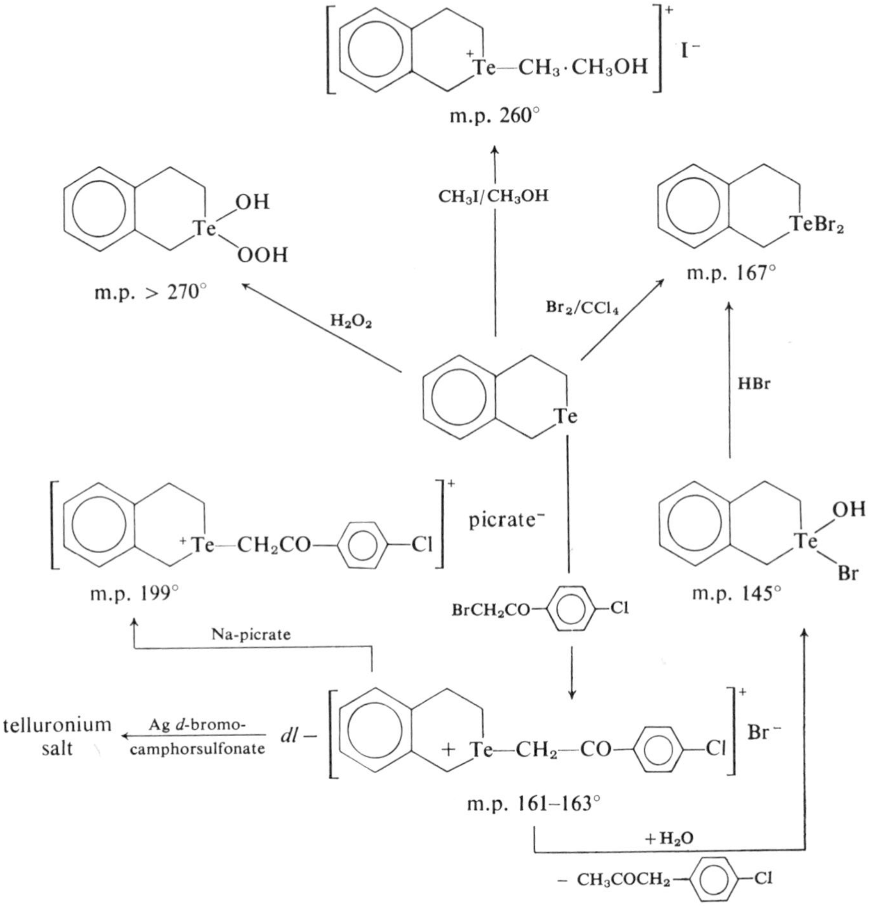

For details on the preparation of an optically active telluronium salt see Section XV.

H. 1-Oxa-4-telluracyclohexane

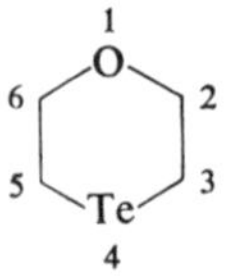

1-oxa-4-telluracyclohexane
1,4-oxatellurane

Farrar and Gulland (F-1) described the preparation of 1,4-oxatellurane and its reactions. For the synthetic procedure consult Section I-B-1.

Reduction of oxatellurane dichloride with potassium metabisulfite ($K_2S_2O_5$) yields the oxatellurane

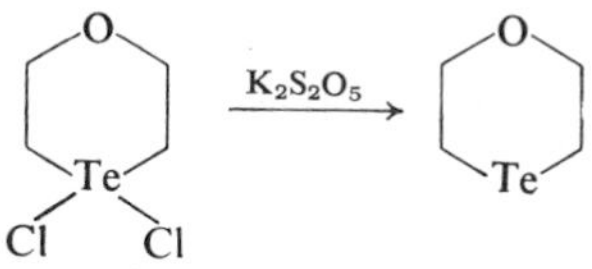

The reduction is carried out in an aqueous solution of the metabisulfite in the presence of carbon tetrachloride. The oxatellurane is insoluble in water and enters the organic solvent. 1,4-Oxatellurane boils at 90° at 21 mm and melts at 6°. With methyl iodide it forms the methyltelluronium salt

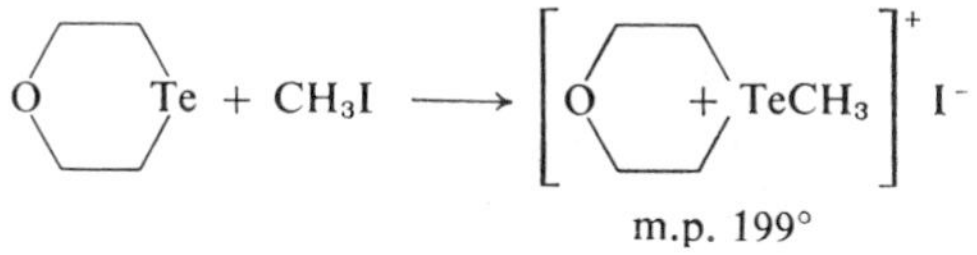

m.p. 199°

Upon the addition of bromine the dibromide (m.p. 158°, dec.) is obtained. In contact with air oxatellurane is converted into the dihydroxide.

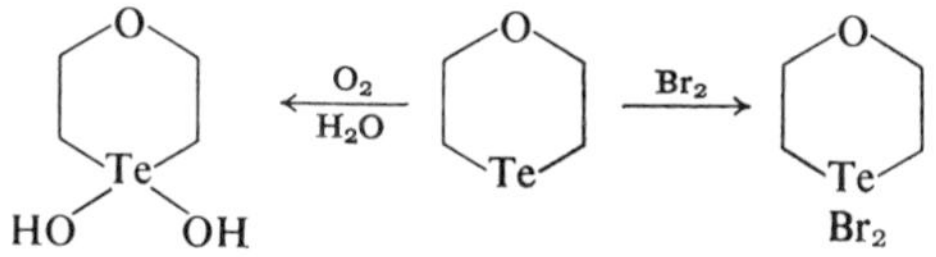

The other reactions of oxatellurane derivatives are outlined below.

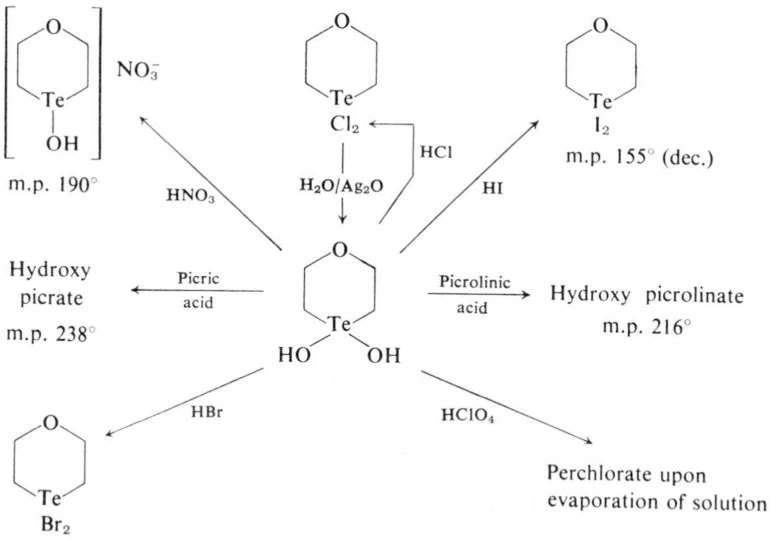

m.p. 190°

m.p. 155° (dec.)

Hydroxy picrate
m.p. 238°

Hydroxy picrolinate
m.p. 216°

Perchlorate upon evaporation of solution

I. 1-Thia-4-telluracyclohexane

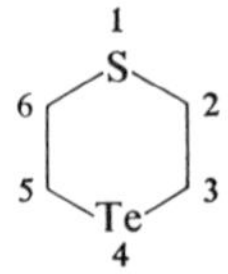

1-thia-4-telluracyclohexane

1,4-thiatellurane

McCullough (M-1) has described the preparation of 1,4-thiatellurane in great detail. The procedure involves the reaction between sodium telluride and bis(β-chloroethyl) sulfide (mustard gas). The yield,

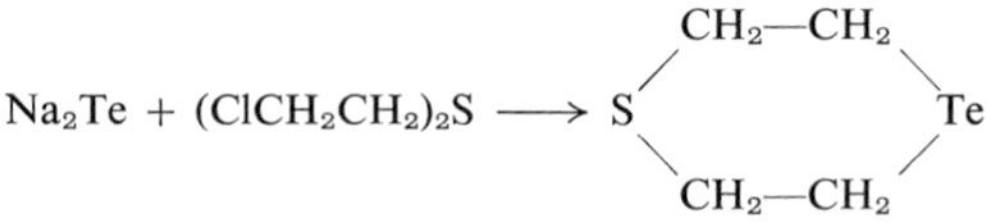

$$Na_2Te + (ClCH_2CH_2)_2S \longrightarrow$$

based on the mustard, is 6.6%. 1,4-Thiatellurane can be recrystallized from methanol in the form of a nearly white solid which melts at 69.5° (see also Section I-B-1). The reactions of this compound are summarized below.

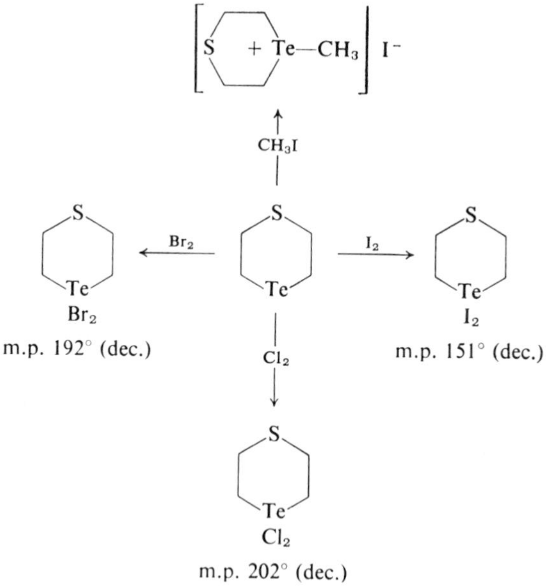

m.p. 192° (dec.)

m.p. 151° (dec.)

m.p. 202° (dec.)

J. Phenoxtellurine

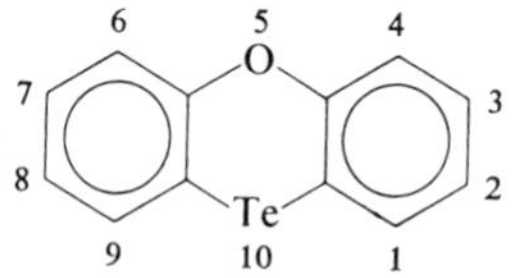

phenox(a)tellurine

Phenoxtellurine 10,10-dichlorides are obtained from $TeCl_4$ and diphenyl ether and its substitution products (D-11,D-13) or by heating the appropriate tellurium trichlorides (C-4) (see Sections I-C-6 and IX-4).

The dichlorides can easily be reduced by $K_2S_2O_5$ or $KHSO_3$ to the parent phenoxtellurine. The cyclic tellurides prepared in this way are listed in Table XLVI.

TABLE XLVI
Phenoxtellurines

by Reduction of Their 10,10-Disubstituted Derivatives

Phenoxtellurine	m.p.	Yield, %	Ref.
2-Nitro[a]	129	20	C-4,D-13
4-Nitro[a]	104		D-13
2,8-Dinitro[a]	228		D-13
4,8-Dinitro[a]	198		D-13
2-Amino[b]	157	65	C-4,D-13
2,8-Diamino[b]	198		D-13
4,8-Diamino[b]	156		D-13
2-Methyl[c]	50–52	93	C-4
2-Carboxy[c]	231–233		C-4
2-Chloro-8-methyl[c]	68		C-4
Phenoxtellurine[c]	79	100	D-11,13,14
Phenoxtellurine[d]	79	100	R-7

[a] From phenoxtellurine by nitration followed by reduction of the nitro-10,10-dinitrate.
[b] By reduction of nitro-compounds with Sn/HCl.
[c] By $K_2S_2O_5$ reduction of 10,10-dichloride.
[d] By $Na_2S \cdot 9H_2O$ reduction of 10,10-dichloride.

Phenoxtellurines with nitro substituents in the ring cannot be prepared by the condensation of $TeCl_4$ with nitrodiphenyl ethers. Direct nitration, however, produced the desired compounds (D-13). Phenoxtellurines upon treatment with warm concentrated HNO_3 (D-11) or upon crystallization from HNO_3 (D-12), yield quantitatively the 10,10-dinitrates (see Table XLVII). 4-Nitro- and 4,8-dinitrophenoxtellurine yielded only the hydroxy nitrate monohydrates under the same conditions, decomposing at 243° (D-12) and 239° (D-12,D-13), respectively. Under suitable conditions this action is then followed by nitration of the nucleus. The nitro derivatives of phenoxtellurine are obtained by alkali bisulfite reduction of the 10,10-dinitrates (see Table XLVI).

TABLE XLVII
10,10-Disubstituted Phenoxtellurine Derivatives

Phenoxtellurine	10,10-Dichloride		10,10-Dibromide		10,10-Diiodide		10,10-Dinitrate	
	m.p.	Ref.	m.p.	Ref.	m.p.	Ref.	m.p.	Ref.
Phenoxtellurine			290 (dec.)	D-11	270 (dec.)	D-11	258 (dec.)	D-11,D-12
2-Nitro							197 (dec.)	C-4,D-12,D-13
2,8-Dinitro							259 (dec.)	D-12,D-13
4-Nitro			302 (dec.)	D-13			243 (dec.)[a]	D-12
4,8-Dinitro							239 (dec.)[a]	D-12
2-Chloro-8-methyl	300 (dec.)	D-14						

[a] Only 10-hydroxide 10-nitrates were isolated, the melting points of which are given.

The nitro derivatives of phenoxtellurine can be reduced to the corresponding amino derivatives (see Table XLVI).

Concentrated nitric acid leads chiefly to formation of mononitro derivatives, the principle product being 2-nitrophenoxtellurine with 4-nitrophenoxtellurine formed as a byproduct.

When fuming hot nitric acid is used, a second nitro group is substituted into the ring system and, following reduction, 2,8-dinitrophenoxtellurine is obtained as the main product and 4,8-dinitrophenoxtellurine as a secondary product.

Further nitration was found to be difficult and tri- or higher nitro derivatives were not isolated.

The nitro- and dinitrophenoxtellurines are well crystallized yellow to orange-red solids of considerable stability. The proof of the orientation of the nitro groups was obtained in a straightforward manner. This was done by decomposing the 10,10-dinitrate, dichloride, or similar salt, with warm alkali. The product formed is a nitrated diphenyl ether derivative which is readily characterized (D-13).

The influence of the Te(NO$_3$)$_2$ group as regards orientation of the incoming nitro groups appears to be negligible in comparison with that of the ether oxygen atom. The tellurium dinitrate group may possibly be *meta*-directing, in which case the separate influences of the two groups would be reinforcing. The entry of a second nitro group into the phenoxtellurine ring occurs almost exclusively in the unnitrated phenylene group, as expected (D-13).

The nitrophenoxtellurines are readily reduced by means of tin and hydrochloric acid to the corresponding amino derivatives. The amines are crystalline, stable, and appreciably water soluble. Their hydrochlorides are readily diazotized or tetrazotized, giving yellow to red diazonium salts which may be coupled with reactive phenols, etc., to give azo dyes and pigments (D-13) (see Table XLVI).

During the course of this same investigation (D-13) interesting stable addition compounds between various phenoxtellurine molecules were discovered. For example, two moles of 2,8-dinitrophenoxtellurine combine with one of the 2-nitro derivative to form a stable, recrystallizable entity. Another example is the 1:1 complex which was found to form between 2-chloro-8-methylphenoxtellurine and phenoxtellurine. The nature of these addition compounds is not understood at the present time. Phenoxtellurines add a mole of halogen with ease to form the 10,10-dihalides (see Table XLVII).

Phenoxtellurine 10,10-dichloride with moist silver oxide yields the colorless oxide (D-16).

Phenoxtellurine 10,10-diacetate melting at 205–207° was obtained in 90% yield from the telluride upon treatment with glacial acetic acid and H$_2$O$_2$. The diacetate is readily reduced by aqueous KHSO$_3$ (D-12,D-16). Phenoxtellurine sulfates crystallize from a reaction mixture containing the diacetate and sulfuric acid (D-12,D-16). The 2-chloro-8-methylphenoxtellurine diacetate melted at 230–232° (D-14). Tellurones were formed by treating phenoxtellurine or its diacetate with hydrogen peroxide (D-12) (for 2-methyl derivative see Section C-4).

Observations of remarkable interest concerning the development of colors by phenoxtellurines were made by Drew (D-12). During the conversion of phenoxtellurine to the 10,10-dinitrate or during the reduction of the latter the formation of a bluish-violet intermediate was noted.

$$\text{(phenoxtellurine)} \quad \underset{\text{red.}}{\overset{\text{HNO}_3}{\rightleftharpoons}} \quad \begin{array}{c}\text{bluish-violet}\\\text{intermediate}\end{array} \quad \underset{\text{red.}}{\rightleftharpoons} \quad \text{(phenoxtellurine dinitrate)}$$

Drew (D-12,D-14) has attributed the formation of the colored complexes to the interaction of phenoxtellurine with a salt of the type formed when phenoxtellurine is dissolved in sulfuric or nitric acid. The colored complex would be of the following type

The complexes containing nitrate groups undergo considerable decomposition in warm glacial acetic acid, the most favorable medium for forming the colored complexes. This led Drew to substitute the less strongly oxidizing bisulfate groups for nitrate groups.

In concentrated sulfuric acid the dissolution of phenoxtellurine is accompanied by the evolution of SO_2. A red solution is formed and the addition of water slowly brings about the separation of red crystals described by Drew as "diphenoxtellurylium dibisulfate disulfuric acid trihydrate." Drew postulates the existence of a Te—Te bond between the heterocyclic molecules. On porous tile this hygroscopic substance loses sulfuric acid and is quantitatively transformed into a violet-blue substance, "diphenoxtellurylium dibisulfate di- or trihydrate."

Various other colored substances were formed when the quantity of water added to the phenoxtellurine sulfuric acid solutions was increased. Drew attributed these differences in color to removal of water of hydration and to replacement of bisulfate groups by hydroxy groups.

When the dry compounds are mixed and rubbed together, an intense violet color is also developed. The original substances can be recovered unchanged. Table XLVIII taken from Drew (D-12) lists the reagents and the colors obtained.

Farrar (F-2) has shown that the stoichiometries suggested by Drew are consistent with the formulation of salts containing a radical cation and has

TABLE XLVIII

Colored Complexes Containing Phenoxtellurine and its Derivatives

Phenoxtellurine	Phenoxtellurine dinitrate or hydroxynitrate	Color
Phenoxtellurine	Phenoxtellurine	Intense violet
2-Nitro	Phenoxtellurine	Intense violet
Phenoxtellurine	2,8-Dinitro	Feeble violet
2-Nitro	2-Nitro	Rather feeble violet
2,8-Dinitro	Phenoxtellurine	None
2,8-Dinitro	2,8-Dinitro	None
Phenoxtellurine	4,8-Dinitro	Intense violet
4,8-Dinitro	4,8-Dinitro	None

suggested that the color is due to the presence of these ions. The question has received very little attention and must be considered unsettled.

Campbell and Turner (C-4) showed that 2-carboxyphenoxtellurine can be decarboxylated in quinoline over Cu-bronze. Barry, Cauquis, and Maurey (B-5) studied the electrochemical oxidation of phenoxtellurine in acetonitrile.

Hieber and Kruck (H-6) obtained $Mn(CO)_3(C_{12}H_8OTe)_2Cl$ from $Mn(CO)_5$-Cl and phenoxtellurine. Drew (D-16) prepared salts of the type.

K. Thiophenoxtellurine

phenothiatellurine
thiophenoxtellurine

Petragnani (P-5) prepared the 10,10-dichloride of this heterocyclic ring system by heating o-thiophenoxyphenyl tellurium trichloride (see Sections

I-C-7-b and IX-4). Reduction of the dichloride with hydrated sodium sulfide gives phenothiatellurine in 96% yield. It is obtained in the form of pale yellow needles which, on recrystallization from ethanol, melt at 122–123.5°. No further reactions have been investigated.

L. 1,3,5-Tritelluracyclohexane

$$
\begin{array}{c}
\text{Te} \\
\text{CH}_2 \qquad \text{CH}_2 \\
| \qquad\qquad | \\
\text{Te} \qquad\quad \text{Te} \\
\text{CH}_2
\end{array}
$$

The trimeric telluroformaldehyde prepared by Williams and Dunbar (W-2) (see Section XI) can be considered as a heterocyclic ring system containing three tellurium atoms. Its chemistry is virtually unexplored.

XIV. REACTIONS OF COMPOUNDS CONTAINING TELLURIUM-METAL BONDS

A variety of compounds are known in which tellurium atoms form one or two bonds to a metal atom. Some of these substances, for example, RTe— MgX, $R_3GeTeLi$, etc., are stable in solution only under an inert atmosphere, while disubstituted compounds, $(R_3M)_2Te$, are rather stable and can be distilled and recrystallized.

The reactions between $(C_2H_5)_3M$—Te—C_2H_5 (M = Si or Ge) and tri-ethyltin hydride to yield $(C_2H_5)_3M$—Te—$Sn(C_2H_5)_3$ have already been described (see Sections IV-A-4).

The formation of phenyltelluromagnesium bromide from elemental tellurium and the Grignard reagent in ether was discussed in Section I-A-3-a. This compound, $C_6H_5TeMgBr$, which precipitated as a paste on addition of petroleum ether to the reaction mixture, is sensitive toward water and oxygen with the formation of benzenetellurol, diphenyl telluride, elemental tellurium, and hydrogen telluride (G-6,P-9,P-10). Treatment of the hydrolyzed products with SO_2Cl_2 resulted in the formation of diphenyltellurium dichloride and phenyltellurium trichloride (see Section I-A-3-a).

Bowden and Braude (B-14) investigated the reaction of phenyltelluro-magnesium bromide with ethyl iodide to form the unsymmetric telluride.

$$C_6H_5TeMgBr + C_2H_5I \longrightarrow C_6H_5\text{—Te—}C_2H_5 + MgBrI$$

Pourcelot and co-workers (P-16,P-17) obtained propargyl phenyl tellurides from $C_6H_5TeMgBr$ and the propargyl halide.

$$C_6H_5TeMgBr + Br\text{—}CH(R)\text{—}C{\equiv}CH \longrightarrow C_6H_5\text{—Te—}CH(R)\text{—}C{\equiv}CH + MgBr_2$$
$$(R = H, CH_3)$$

The propargyl tellurides isomerize to allenic tellurides (see Section IV-C).

Tetrahydrofuran solutions containing $(C_6H_5)_3MTeLi$ (M = Ge, Sn, Pb) (see Section I-A-3-c) react with triphenylmetal halides to form the symmetric and unsymmetric tellurides listed in Table XLIX (S-7,S-9,S-10)

$$(C_6H_5)_3MTeLi + (C_6H_5)_3M'X \longrightarrow (C_6H_5)_3M—Te—M'(C_6H_5)_3 + LiX$$
$$(M = Ge, Sn, Pb; M' = Ge, Sn, Pb; X = Cl, Br)$$

The stability of these compounds in the presence of moisture and oxygen increases in the order $(R_3Ge)_2Te < (R_3Sn)_2Te \ll (R_3Pb)_2Te$ and $(R_3Ge)_2Te < R_3GeTeSnR_3 < R_3GeTePbR_3$ ($R = C_6H_5$). While the symmetric germanium and tin tellurides decompose readily in moist air, the lead compound was kept for months under these conditions without effect (S-10).

Bis(trimethylsilyl) telluride, prepared from $C_6H_5TeMgBr$ and trimethyl-chlorosilane, gave a 100% yield of $(CH_3)_3SiI$ in a reaction with AgI which proceeds rapidly at room temperature (H-8).

$$[(CH_3)_3Si]_2Te + 2\,AgI \longrightarrow 2\,(CH_3)_3SiI + Ag_2Te$$

Compounds of the general formula R—C≡C—Te—Na, have been pre-pared from sodium acetylide in liquid ammonia (B-15,B-17,R-3) (see Section I-A-3-a). The alkynyl sodium telluride was permitted to react in liquid ammonia without being isolated to yield the alkynyl alkyl telluride in good yields.

$$R—C≡C—Na + Te \longrightarrow R—C≡C—Te—Na$$
$$R—C≡C—Te—Na + R'X \longrightarrow R—C≡C—Te—R' + NaX$$

The compounds prepared by this method are listed in Table XLIX.

Experimental. Phenyl Ethyl Telluride (B-14): A suspension of tellurium (25.6 g, 0.2 moles) in ether (100 ml) was added to a stirred solution of C_6H_5MgBr (0.2 moles) in a nitrogen atmosphere. The mixture was refluxed for three hours and allowed to stand for a day. Ethyl iodide (31 g, 0.2 moles) in ether (50 ml) was then dropped into the mixture, refluxed for one hour cooled and hydrolyzed with ice cold hydrochloric acid. The separated ethereal layer was washed with 30% aqueous NaOH, then with water, dried over Na_2SO_4 and distilled. Phenyl ethyl telluride (20 g, 43%) was collected as a pale yellow liquid at 107–108°/22 mm.

Bis(triphenylplumbyl) Telluride (S-10): An equimolar amount of triphenyl-plumbyl chloride dissolved in 15 ml of dry tetrahydrofuran was dropped with-in five minutes into the well-stirred THF solution of lithium triphenylplumbyl telluride (see Section I-A-3-c). After two hours of stirring the solution was evaporated *in vacuo* at room temperature. The oily residue was dissolved in benzene. The lithium chloride was filtered off under nitrogen in a closed system. The solvent was removed under vacuum and the residue was

TABLE XLIX

Reactions of Compounds Containing a Tellurium–Metal Bond

| R—Te—Y | | | | | m.p., b.p. | |
R	Y	Reagent	Product	Yield	(mm Hg)	Ref.
C_6H_5—	MgBr	C_2H_5I	C_6H_5—Te—C_2H_5	43	107–108(22)	B-14
		HC≡C—CH_2Br	C_6H_5Te—CH_2—C≡CH	—	—	P-16,P-17
		HC≡C—CH(CH_3)Br	C_6H_5Te—CH(CH_3)C≡CH	—	—	P-16,P-17
		$(CH_3)_3$SiCl	$[(CH_3)_3Si]_2$Te	15	40–42(0.25)	H-8
$(C_6H_5)_3$Ge—	Li	$(C_6H_5)_3$GeBr	$[(C_6H_5)_3Ge]_2$Te	11	120 (dec.)	S-9
		$(C_6H_5)_3$SnCl	$(C_6H_5)_3$GeTeSn$(C_6H_5)_3$	48	142–146	S-9
		$(C_6H_5)_3$PbCl	$(C_6H_5)_3$GeTePb$(C_6H_5)_3$	42	115–117	S-9
$(C_6H_5)_3$Sn—	Li	$(C_6H_5)_3$SnCl	$[(C_6H_5)_3Sn]_2$Te	59	149	S-6,S-7
		$(C_6H_5)_3$PbCl	$(C_6H_5)_3$SnTePb$(C_6H_5)_3$	27	136 (dec.)	S-7
$(C_6H_5)_3$Pb—	Li	$(C_6H_5)_3$PbCl	$[(C_6H_5)_3Pb]_2$Te	60	128–129	S-10
CH_3—C≡C—	Na	C_2H_5Br	CH_3C≡C—Te—C_2H_5	65	71 (11)	B-15,B-17
CH_2=CH—C≡C—	Na	C_2H_5Br	CH_2=CH—C≡C—Te—C_2H_5	58	80–81 (10)	P-14
		CH_3I	CH_2=CH—C≡C—Te—CH_3	50	83–84 (20)	P-14
CH_2=C(CH_3)—C≡C—	Na	C_2H_5Br	CH_2=C(CH_3)—C≡C—Te—CH_3		92–95 (20)	R-3

recrystallized several times from benzene/hexane. The purified product melting at 128–129° was obtained in 60% yield.

Bis(trimethylsilyl) Telluride (H-8): An ethereal solution of C_6H_5MgBr (0.2 moles) was filtered through a sintered glass disk onto a stirred suspension of finely powdered tellurium (25.4 g, 0.2 moles) in 50 ml of ether. After refluxing for three hours trimethylchlorosilane (22.0 g, 0.2 moles) was added. Refluxing for two hours led to the production of a white precipitate and a considerable amount of tellurium. All volatile material was removed under vacuum with cautious warming and condensed at −78°. The condensate was fractionated to yield bis(trimethylsilyl) telluride (15%) boiling at 40–42°/0.25 mm.

Ethyl Propynyl Telluride (B-15): To 0.5 mole of NaC≡CH in 0.5 liter of liquid ammonia was added 0.5 mole of ethyl bromide at the b.p. of ammonia. When the reaction was complete a solution containing 0.45 mole of $NaNH_2$ in liquid ammonia at −80° was added to the propyne solution. This was followed by the addition of 0.45 mole of powdered Te. The addition was carried out quickly, but in small portions since the reaction was quite vigorous. Immediately following the addition of the Te, 0.6 moles of ethyl bromide were added. The ammonia was permitted to evaporate and the residue was extracted with ether and washed with water. The ether layer was dried. Following evaporation of the ether, the liquid residue was vacuum distilled. The telluride, $b_{11}71°$, n_D^{20} 1.5890, was obtained in 65% yield.

XV. OPTICAL ACTITITY IN ORGANOTELLURIUM COMPOUNDS

The existence of optically active tris(organo)telluronium compounds is to be expected because of the tetrahedral arrangement of groups bonded to the central tellurium atom with the lone electron pair occupying the fourth position. Lederer (L-20) made an unsuccessful attempt to prepare optically active 4-methylphenyl phenyl methyltelluronium iodide. Lowry and Gilbert (L-37) treated this racemic iodide with silver *d*-α-bromocamphor-π-sulfonate and *d*-10-camphorsulfonate and obtained the salts of the *l*- and *d*-telluronium cations, respectively. These optically active salts showed mutarotation in solution. The final rotation observed after the solutions were allowed to stand for over an hour corresponded to the rotation of the active sulfonate ions alone. The telluronium iodides which precipitated upon treatment of the optically active telluronium sulfonates also showed mutarotation. The molecular rotation of the cation $[M]_{5461}^{18}$ was estimated to be higher than 85°. Reichel and Kirschbaum (R-6) prepared *l*-4-methoxyphenyl ethyl methyltelluronium *d*-α-bromocamphor-π-sulfonate with a molecular rotation of $[M]_D^{20} = +198.5°$ in $CHCl_3$. The iodide obtained from this salt was inactive.

d-10-Camphorsulfonic acid mixed with 4-methylphenyl phenyl telluroxide gave an active salt with a rotation of $[M]_D^{18} = +450.05°$ in acetone. Lowry and Hüther (L-38) employed the following reaction sequence in their preparation of optically active telluronium salts.

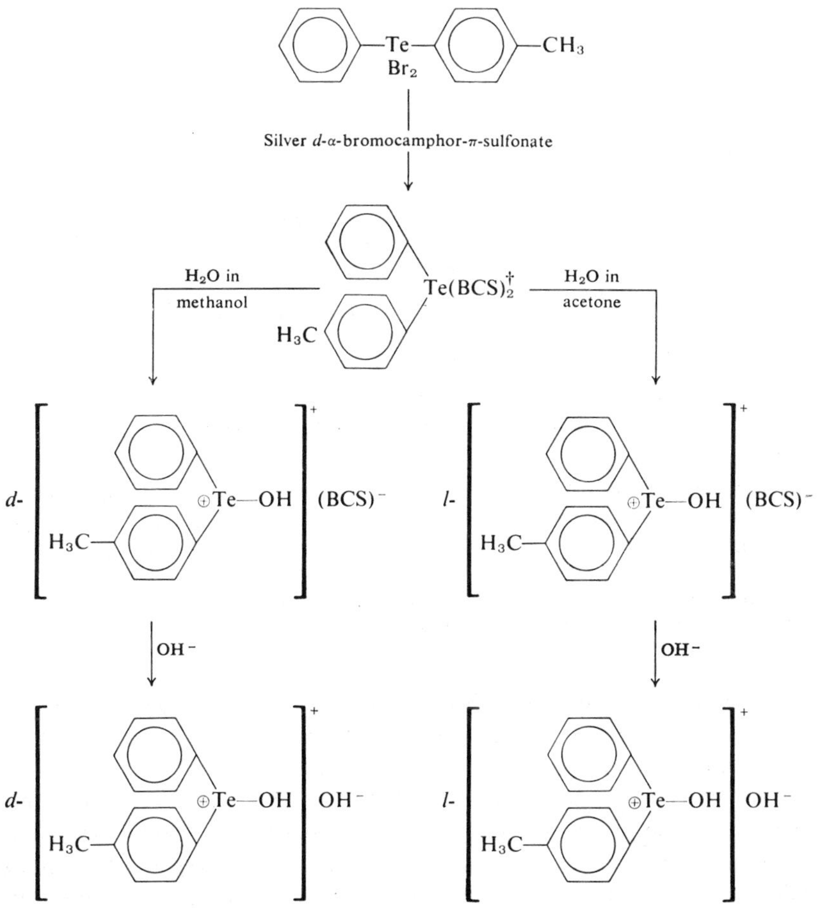

* (BCS) = *d*-α-bromocamphor-π-sulfonate.

All these salts exhibit mutarotation. The hydroxy sulfonate, from acetone, changes to a higher value, while the hydroxy sulfonate, from methanol, changes to a lower, limiting value. The two samples of the base show mutarotation in the same direction as sulfonates from which they were prepared. Ter Horst (H-9) further investigated the optical activity of tellurium compounds of the type $RR'TeX_2$ and $RRTeX_2$. The following results were obtained:

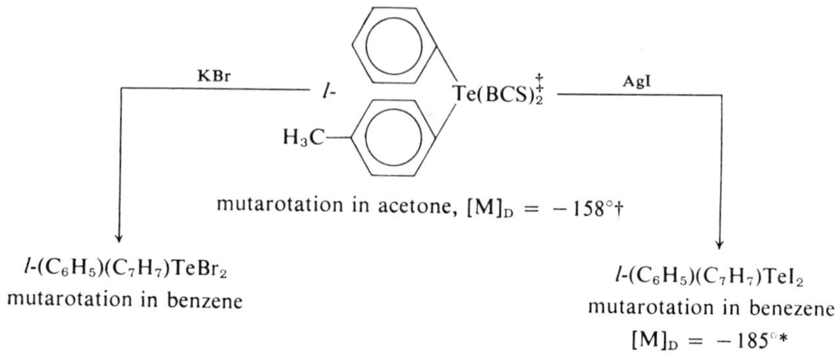
KBr
l-
H₃C
Te(BCS)₂‡
AgI
mutarotation in acetone, [M]D = −158°†
l-(C6H5)(C7H7)TeBr2
mutarotation in benzene
l-(C6H5)(C7H7)TeI2
mutarotation in benezene
[M]D = −185°*

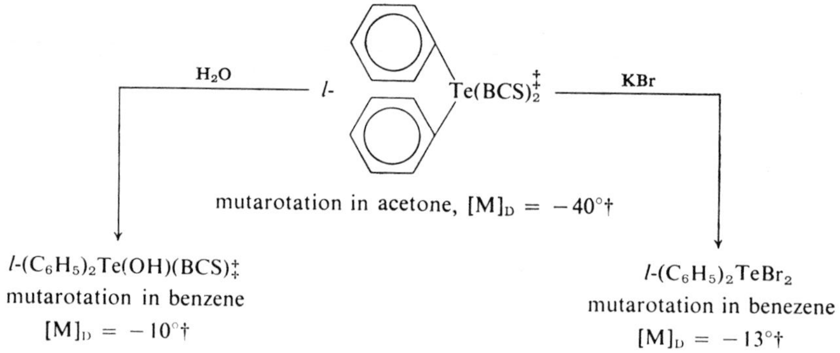
H₂O
l-
Te(BCS)₂‡
KBr
mutarotation in acetone, [M]D = −40°†
l-(C6H5)2Te(OH)(BCS)‡
mutarotation in benzene
[M]D = −10°†
l-(C6H5)2TeBr2
mutarotation in benezene
[M]D = −13°†

† molecular rotation due to the telluronium cation only.
‡ (BCS) = *d*-α-bromocamphor-π-sulfonate.

The effects in the diphenyl series are rather small and should be considered with reservations (L-38). Experiments in the dimethyl series should be valuable to elucidate the role played by the aromatic groups.

Campbell and Turner (C-4) attempted to resolve 2-carboxy- and 2-aminophenoxtellurine into its optically active antipodes, which should exist under the assumption that the molecule is folded along the oxygen–tellurium axis. Salts of these phenoxtellurine derivatives with optically active acids and bases were prepared, but could not be resolved. Balfe, Chaplin, and Phillips (B-2) obtained dibutyl carbo-*l*-menthoxymethyltelluronium bromide, which had a specific rotation of $[\alpha]_{5461}^{17.5} = -23.6°$ in $CHCl_3$, from dibutyl telluride and the *l*-menthyl ester of bromoacetic acid. Thermal decomposition of the telluronium salt gave butylcarbo-*l*-menthoxymethyl telluride, $[\alpha]_{5461}^{20} = -5.2°$, which with phenacyl bromide formed a telluronium salt with $[\alpha]_{5461}^{20} = -23.7°$ in $CHCl_3$. The other decomposition product, bis(carbo-*l*-menthoxymethyl) telluride had $[\alpha]_{5461}^{15} = -27.6°$ in ether and $[\alpha]_{5461}^{17.5} = -34.4$ in $CHCl_3$. The activity in these compounds is probably caused by the *l*-menthyl groups. No attempt was made to resolve butyl carbo-*l*-menthoxymethyl phenacyltelluronium bromide.

Holliman and Mann (H-7) prepared the following heterocyclic bromide.

They partially resolved this salt into *d*-telluronium *d*-α-bromocamphor-π-sulfonate, and *l*-telluronium *d*-α-bromocamphor-π-sulfonate and prepared the respective picrates. Their samples were optically impure. The picrates showed a slow mutarotation.

XVI. SPECTROSCOPIC AND STRUCTURAL INVESTIGATIONS OF ORGANOTELLURIUM COMPOUNDS

Fritz and Keller (F-15) were the first to investigate the infrared spectra of organotellurium compounds. They reported spectra in the region 250–4000 cm^{-1} for $(CH_3)_2Te$, $(CH_3)_2TeI_2$, $(CH_3)_3TeI$, $(C_2H_5)_2Te$, $(C_2H_5)_2TeI_2$, $(C_2H_5)_3TeI$, $(C_2H_5)_2CH_3TeI$, $(C_6H_5)_2Te$, $(C_6H_5)_2TeCl_2$, $(C_6H_5)_3TeI$, $(C_6H_5)_4$-Te, $(CH_3)(C_6H_5)Te$, $(CH_3)(C_6H_5)TeI_2$, $C_6H_5TeCl_3$, and $(C_6H_5)_2Te_2$. The tellurium–carbon (alkyl) frequencies were found in the region from 477 to 529 cm^{-1} with a force constant of about 2.27 millidynes/Å. Bands between 455 and 487 cm^{-1} were tentatively assigned to a tellurium–phenyl vibration. Absorptions at 290 cm^{-1} were attributed to tellurium–chlorine vibrations.

Dimethyl telluride was reexamined by Allkins and Hendra (A-3) and treated theoretically by Freeman and Henshall (F-10). Chen and George (C-21) recorded the spectrum of $(CH_3)_2Te_2$ in the liquid state and assigned a band at 122 cm^{-1} to the tellurium–tellurium stretching vibration.

Mack (M-5) reported infrared frequencies for tetrachlorotellurophene. Cyclic and linear condensation products between $TeCl_4$ and acetylacetone were investigated with infrared techniques to elucidate the structures of these compounds (D-7) (see Section I-C-1). Schumann and Schmidt (S-8), Hooton and Allred (H-8), and Egorochkin and co-workers (E-2,E-3) studied the spectra of compounds of the general formula R_3M—Te—$M'R_3$, where M and M' represent Si, Ge, Sn, and Pb, and R stands for ethyl or phenyl. The frequencies corresponding to the M—C modes of vibration were assigned. Te—M vibrations, which are expected to occur below 250 cm^{-1}, were not detected. Infrared spectra of transition metal complexes containing an organotellurium compound as a ligand have been reported (B-18,C-16,C-18,H-6, H-7) (see Section IV-A-4).

TABLE L
Nuclear Magnetic Resonance Studies of Organotellurium Compounds

Compound	Solvent/standard	Chemical shifts (δ) (position)	Coupling constants	Relative intensities	Ref.
C_6H_5—Te—CH_2—C≡CH	CCl_4/TMS	CH_2 3.32; ≡CH 1.98	J_{H-H} 2.75		S-14
[tellurium-containing diketone ring structure]	$CDCl_3$/TMS	$CH_2(4)$ 3.45; $CH_2(2,6)$ 3.90		1:2	D-7
[Cl_2Te chelate structure]$_2$	$CDCl_3$/TMS	CH_3 2.33; CH_2 4.87; OH 1.46		3:2:1	D-7
[tellurophene diol structure]	$CDCl_3$/TMS	CH(3,4) 7.43; CH_2 4.85; OH 2.16			M-6
[tellurophene di(dimethylcarbinol) structure]	$CDCl_3$/TMS	CH(3,4) 7.29; CH_3 1.62 OH 2.74			M-6

324

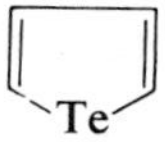

Compound	Solvent/Ref.	Chemical shift	Reference
(tellurophene ring) Te	CDCl₃/TMS	CH(3,4) 7.83; CH(2,5) 8.94	M-6
[(CH₃)₃Si]₂Te	CCl₄/TMS	CH₃ 0.6	H-8
[(C₂H₅)₃Si]₂Te	—/TMS	CH₂ 0.81; CH₃ 1.00	E-1
[(C₂H₅)₃Ge]₂Te	—/TMS	CH₂ 1.05; CH₃ 1.05	E-1
[(C₂H₅)₃Sn]₂Te	—/TMS	CH₂ 1.09; CH₃ 1.20	E-1
(C₂H₅)₃Sn / (C₂H₅)₃Si, Te	—/TMS	CH₂ 1.09; CH₃ 1.31 (Sn); CH₂ 0.80; CH₃ 0.97 (Si)	E-1
(C₂H₅)₃Ge / (C₂H₅)₂Te, Te	—/TMS	CH₂ 1.08; CH₃ 1.20 (Sn); CH₂ 1.05; CH₃ 1.05 (Ge)	E-1
(CH₃)₂Te	—/TMS		B-19

For (CH₃)₂Te:

$J_{C^{13}H}(CH_3)\ 127.1 \pm 0.4$
$J_{C^{13}H}(CH_2)\ 141.0 \pm 0.4$
$J_{HH}\ 7.6 \pm 0.1$
$J_{Te^{125}C-H}\ \mp 24.8 \pm 0.2$
$J_{Te^{125}C-C-H}\ \mp 22.2 \pm 0.3$
$J_{Te^{125}C^{13}}\ + 158.5 \pm 0.5$
$J_{C^{13}H}\ + 140.7 \pm 0.2$
$J_{Te^{125}C-H}\ - 20.8 \pm 0.2$
$J_{C-Te-C}\ \pm 0.2 \pm 0.1$
$J_{C-Te-C-H}\ + 3.1 \pm 0.2$
$J_{H-C-Te-C-H}\ \pm 0.1$

References: D-10, M-3, K-3

The ultraviolet and visible spectra of the following compounds were recorded: tetraphenyltellurophene (B-18), tetrachlorotellurophene (M-5), 1-tellura-3,5-cyclohexanedione and its 1,1-dichloride (D-7), bis(acetylacetonyl)tellurium dichloride (D-7), dibutyl ditelluride (B-8), bis(carboxymethyl) ditelluride (B-8), phenyl ethyl telluride (B-14), diethyl phenyltelluronium iodide (B-14), 2-naphthyl phenyl telluride (V-5), 2-naphthyltellurenyl iodide (V-5), dimethyltellurium diiodide (Vernon's α- and β- forms) (L-35), and trimethyltelluronium iodide (L-35).

A number of organotellurium compounds has been investigated by nuclear magnetic resonance spectroscopy. A review on coupling constants, including those for ^{125}Te $-$ H in Te—C—H and Te—C—C—H systems has been published (F-12). Simonnin (S-14) who compared the NMR spectra of phenyl propargyl telluride with the corresponding sulfur and selenium compounds, found a dependence of the chemical shifts and coupling constants on the electronegativity of the heteroatom. See Section I-C-1 for the NMR spectra of tellurium derivatives of acetylacetone (D-7). Lambert and Keske (L-1) investigated 1-telluracyclohexane 1,1-dibromide using NMR techniques and proposed a conformation in which part of the ring is flattened and part considerably puckered, as shown.

The other NMR investigations are summarized in Table L.

Structural investigations of organotellurium compounds were first undertaken in 1950. Before this time a variety of physical properties of these compounds had been measured in order to elucidate their structures. The magnetic susceptibilities of the dimethyltellurium dihalides were measured (B-9,B-10, D-8), parachors were determined (B-22,S-13), dipole moments calculated (J-2,J-3), and the symmetry of crystals determined (B-12). Jensen proposed a trigonal bipyramidal structure for bis(organo)tellurium dihalides with one lone electron pair occupying a position in the equatorial plane (J-2). Gillespie discussed the stereochemistry of octahedral, trigonal-bipyramidal and related molecules of nontransition elements and included organotellurium compounds (G-5).

Structural investigations have been reported for the following compounds: $(CH_3)_2TeCl_2$ (C-23), $(CH_3)_2TeI_2$ (G-1), $(C_6H_5)_2TeBr_2$ (C-22), (4-ClC$_6$H$_4$)$_2$-TeI$_2$ (C-11), (4-CH$_3$C$_6$H$_4$)$_2$Te (B-13), (4-ClC$_6$H$_4$)$_2$Te$_2$ (K-10), C$_6$H$_5$Te-(thiourea)X [X = Cl, Br] (F-5,F-6,F-8) and C$_6$H$_5$Te(thiourea)$_2$Y [Y = Cl] (F-5–F-7), [Y = NO$_3$, ClO$_4$, SCN] (F-5). The tellurium–carbon bond length varied in these compounds from 2.05 to 2.16 Å, the C—Te—C bond angles were found to lie between 96 and 101°, the X—Te—X bond angles (X = Cl, Br, I) between 172 and 178°, and the Te—Te—C bond angle in bis(4-chlorophenyl) ditelluride was found to be 94°. The bis(organo)tellurium dihalides have the shape of a highly distorted tetrahedron or distorted trigonal bipyramid with the lone electron pair occupying one equatorial position.

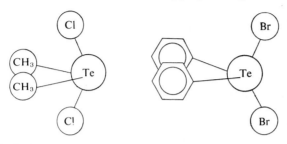

In dimethyltellurium dichloride the Cl—Te—Cl angle is bent toward the methyl groups, while in diphenyltellurium dibromide the bromine atoms point away from the phenyl groups.

REFERENCES

A-1.   S. C. Abrahams, *Quart. Rev.*, **10**, 407 (1956).
A-2.   M. Adloff and J. P. Adloff, *Bull. Soc. Chim. France*, **1966**, 3304.
A-3.   J. R. Allkins and P. J. Hendra, *Spectrochim. Acta*, **22**, 2075 (1966).
A-4.   H. H. Anderson, *J. Chem. Eng. Data*, **9**, 272 (1964).
A-5.   E. E. Aynsley, *J. Chem. Soc.*, **1953**, 3016.
A-6.   E. E. Aynsley and R. H. Watson, *J. Chem. Soc.*, **1955**, 2603.
B-1.   K. W. Bagnall, *The Chemistry of Selenium, Tellurium and Polonium*, Elsevier, Amsterdam, 1966.
B-2.   M. P. Balfe, C. A. Chaplin, and H. Phillips, *J. Chem. Soc.*, **1938**, 341.
B-3.   M. P. Balfe and K. N. Nandi, *J. Chem. Soc.*, **1941**, 70.
B-4.   A. Baroni, *Atti Accad. Lincei Classe Sci. Fis., Mat. Nat.*, **27**, 238 (1938); *Chem. Abstr.*, **33**, 163 C. **1938** II. 2918.
B-5.   C. Barry, G. Cauquis, and M. Maurey, *Bull. Soc. Chim. France*, **1966**, 2510.
B-6.   F. Becker, *Ann.*, **180**, 257 (1875).
B-7.   L. Belchetz and E. Rideal, *J. Am. Chem. Soc.*, **57**, 1168 (1935).
B-8.   G. Bergson, *Acta Chem. Scand.*, **11**, 571 (1957).
B-9.   S. S. Bhatnagar and T. K. Lahiri, *Current Sci.*, **1**, 380 (1933).
B-10. S. S. Bhatnagar and T. K. Lahiri, *Z. Physik*, **84**, 671 (1933).
B-11. E. Billows, *Z. Kryst.*, **40**, 289 (1902).
B-12. M. L. Bird and F. Challenger, *J. Chem. Soc.*, **1939**, 163.
B-13. W. R. Blackmore and S. C. Abrahams, *Acta Cryst.*, **8**, 317 (1955).
B-14. K. Bowden and E. A. Braude, *J. Chem. Soc.*, **1952**, 1068.
B-15. L. Brandsma, H. Wijers, and J. F. Arens, *Rec. Trav. Chim.*, **81**, 583 (1962).
B-16. L. Brandsma and H. Wijers, *Rec. Trav. Chim.*, **82**, 68 (1963).
B-17. L. Brandsma, H. E. Wijers, and C. Jonker, *Rec. Trav. Chim.*, **83**, 208 (1964).

B-18. E. H. Braye, W. Hübel, and I. Caplier, *J. Am. Chem. Soc.*, **83**, 4406 (1961).
B-19. V. Breuninger, H. Dreeskamp, and G. Pfisterer, *Ber. Bunsenges. Physik. Chem.*, **70**, (6), 613 (1966).
B-20. E. Buchta and K. Greiner, *Naturwissenschaften*, **46**, 532 (1959).
B-21. E. Buchta and K. Greiner, *Ber.*, **94**, 1311 (1961).
B-22. F. H. Burstall and S. Sugden, *J. Chem. Soc.*, **1930**, 229.
C-1.   A. Cahours, *Ann.*, **135**, 352 (1865).
C-2.   A. Cahours, *Compt. Rend.*, **60**, 620 (1865).
C-3.   A. Cahours, *Ann. Chim.*, **10**, 50 (1877).
C-4.   I. G. M. Campbell and E. E. Turner, *J. Chem. Soc.*, **1938**, 37.
C-5.   F. Carr and T. G. Pearson, *J. Chem. Soc.*, **1938**, 282.
C-6.   E. D. Cerwenka, Jr., and W. C. Cooper, *Arch. Environ. Health*, **3**, 189 (1961).
C-7.   F. Challenger, *Chem. Rev.*, **36**, 315 (1945).
C-8.   F. Challenger, *J. Proc. Roy. Inst. Chem.* (Gt. Britain and Ireland), **1945**, 105.
C-9.   F. Challenger, *Sci. Progr.*, **35**, 396 (1947).
C-10. Chang-Hsueh Yao and Cheng E. Sun, *J. Chinese Chem. Soc.*, **5**, 22 (1937).
C-11. G. Y. Chao and J. D. McCullough, *Acta Cryst.*, **15**, 887 (1962).
C-12. J. Chatt, L. A. Duncanson, and L. M. Venanzi, *Chem. Ind.*, **1955**, 749.
C-13. J. Chatt and L. M. Venanzi, *J. Chem. Soc.*, **1955**, 2787.
C-14. J. Chatt and L. M. Venanzi, *J. Chem. Soc.*, **1955**, 3858.
C-15. J. Chatt, L. A. Duncanson, and L. M. Venanzi, *J. Chem. Soc.*, **1955**, 4456.
C-16. J. Chatt, L. A. Duncanson, and L. M. Venanzi, *J. Chem. Soc.*, **1955**, 4461.
C-17. J. Chatt and L. M. Venanzi, *J. Chem. Soc.*, **1957**, 2351.
C-18. J. Chatt and L. M. Venanzi, *J. Chem. Soc.*, **1957**, 2445.
C-19. J. Chatt, L. A. Duncanson, B. L. Shaw, and L. M. Venanzi, *Discussions Faraday Soc.*, No. 26, 131 (1958).
C-20. J. Chatt, G. A. Gamlen, and L. E. Orgel, *J. Chem. Soc.*, **1959**, 1047.
C-21. M. T. Chen and J. W. George, *J. Organometal. Chem.*, **12**, 401, (1968).
C-22. G. D. Christofferson and J. D. McCullough, *Acta Cryst.*, **11**, 249 (1958).
C-23. G. D. Christofferson, R. A. Sparks, and J. D. McCullough, *Acta Cryst.*, **11**, 782 (1958).
C-24. G. E. Coates, *J. Chem. Soc.*, **1951**, 2003.
C-25. S. C. Cohen, M. L. N. Reddy, and A. G. Massey, *Chem. Commun*, **1967**, 451.
C-26. S. C. Cohen, M. L. N. Reddy, and A. G. Massey, *J. Organometal. Chem.*, **11**, 563 (1968).
C-27. C. Courtot and M. G. Bastani, *Compt. Rend.*, **203**, 197 (1936).
C-28. D. V. Cow and W. E. Dixon, *J. Physiol.*, **56**, 42 (1922).
C-29. N. M. Cullinane, A. G. Rees, and C. A. J. Plummer, *J. Chem., Soc.*, **1939**, 151.
D-1.   A. Damiens, *Compt. Rend.*, **171**, 1140 (1920).
D-2.   A. Damiens, *Compt. Rend.*, **173**, 300 (1921).
D-3.   A. Damiens, *Compt. Rend.*, **173**, 583 (1921).
D-4.   J. N. E. Day, *Sci. Progr.*, **23**, 211 (1928).
D-5.   E. Demarcay, *Bull. Soc. Chim. France*, **40**, 99 (1883).
D-6.   G. H. Denison, Jr., and P. C. Condit, U.S. Patent 2,398,414 (1946), to Calif. Research Corp.
D-7.   D. H. Dewar, J. E. Fergusson, P. R. Hentschel, C. J. Wilkins, and P. P. Williams, *J. Chem. Soc.*, **1964**, 688.
D-8.   S. S. Dharmatti, *Proc. Indian Acad. Sci.*, **12A**, 212 (1940).
D-9.   W. von E. Doering and A. K. Hoffmann, *J. Am. Chem. Soc.*, **77**, 521 (1955).
D-10. H. Dreeskamp and G. Pfisterer, *Mol. Phys.*, **14**, 295 (1968).
D-11. H. D. K. Drew, *J. Chem. Soc.*, **1926**, 223.

D-12. H. D. K. Drew, *J. Chem. Soc.*, **1926**, 3054.

D-13. H. D. K. Drew and R. W. Thomason, *J. Chem. Soc.*, **1927**, 116.

D-14. H. D. K. Drew, *J. Chem. Soc.*, **1928**, 506.

D-15. H. D. K. Drew, *J. Chem. Soc.*, **1929**, 560.

D-16. H. D. K. Drew, *J. Chem. Soc.*, **1934**, 1790.

D-17. H. D. K. Drew, in *Textbook of Inorganic Chemistry*, Vol. XI, pt. 4, N. Friend, Ed., Griffin & Co., Ltd. London, p. 186, 1937.

D-18. M. Dubien, *Rev. Gen. Sci.*, **37**, 366 (1926).

D-19. A. M. Duffield, H. Budzikiewicz, and C. Djerassi, *J. Am. Chem. Soc.*, **87**, 2920 (1965).

E-1. A. N. Egorochkin, N. S. Vyazankin, G. A. Razuvaev, O. A. Kruglaya, and M. N. Bochkarev, *Dokl. Akad. Nauk. SSSR*, **170**, 333 (1966).

E-2. A. N. Egorochkin, N. S. Vyazankin, M. N. Bochkarev, and S. Ya. Khorshev, *Zh. Obshch. Khim.*, **37**, 1165 (1967).

E-3. A. N. Egorochkin, S. Y. Khorshev, N. S. Vyazankin, M. N. Bochkarev,, O. A. Kruglaya, and G. S. Semchikova, *Zh. Obshch. Khim.*, **37**, 2308 (1967).

E-4. H. J. Emeléus and H. G. Heal, *J. Chem. Soc.*, **1946**, 1126.

F-1. W. V. Farrar and J. M. Gulland, *J. Chem. Soc.*, **1945**, 11.

F-2. W. V. Farrar, *Research* (*London*), **4**, 177 (1951).

F-3. O. Foss, *Acta Chem. Scand.*, **6**, 306 (1952).

F-4. O. Foss, *Acta Chem., Scand.*, **7**, 227 (1953).

F-5. O. Foss and S. Hauge, *Acta Chem. Scand.*, **13**, 2155 (1959).

F-6. O. Foss, S. Husebye, and K. Maroey, *Acta. Chem. Scand.*, **17**, 1806 (1963).

F-7. O. Foss and K. Marøy, *Acta Chem. Scand.*, **20**, 123 (1966).

F-8. O. Foss and K. Husebye, *Acta Chem. Scand.*, **20**, 132 (1966).

F-9. V. Franzen and C. Mertz, *Ann.*, **643**, 24 (1961).

F-10. J. M. Freeman and T. Henshall, *J. Mol. Structure*, **1**, 31 (1967/8).

F-11. J. J. Frewing, C. de Graaf, and R. S. Airs, Brit. Patent 576,740 (1946).

F-12. H. Frischleder, G. Klose, and J. Ranft, *Wiss. Z. Karl-Marx Univ., Leipzig, Math. Naturw. Reihe*, **14** (4), 863 (1965).

F-13. E. Fritsman and V. Krinitskii, *J. Appl. Chem.* (*U.S.S.R.*), **11**, 195 (1938).

F-14. E. Kh. Fritsman and V. V. Krinitskii, *J. Appl. Chem.* (*U.S.S.R*), **11**, 1610 (1938).

F-15. H. P. Fritz and H. Keller, *Ber.*, **94**, 1524 (1961).

F-16. E. Fritzman, *J. Russ. Phys. Chem. Soc.*, **47**, 588 (1915).

F-17. E. Ch. Fritzman, *Z. Anorg. Allgem. Chem.*, **133**, 119 (1924).

F-18. M. Frommes, *Z. Anal. Chem.*, **96**, 447 (1934).

F-19. H. Funk and W. Weiss, *J. Prakt. Chem.*, [4], **1**, 33 (1954).

G-1. E. E. Galloni and J. Pugliese, *Acta Cryst.*, **3**, 319 (1950).

G-2. F. L. Gilbert and T. M. Lowry, *J. Chem. Soc.*, **1928**, 1997.

G-3. F. L. Gilbert and T. M. Lowry, *J. Chem. Soc.*, **1928**, 2658.

G-4. F. L. Gilbert and T. M. Lowry, *J. Chem. Soc.*, **1928**, 3179.

G-5. R. J. Gillespie, *Can. J. Chem.*, **39**, 318 (1961).

G-6. M. Giua and F. Cherchi, *Gazz. Chim. Ital.*, **50**, 362 (1920); *Chem. Abstr.*, **15**, 521.

G-7. H. H. Glazebrook and T. G. Pearson, *J. Chem. Soc.*, **1936.**, 1777.

G-8. H. H. Glazebrook and T. G. Pearson, *J. Chem. Soc.*, **1937**, 567.

G-9. H. H. Glazebrook and T. G. Pearson, *J. Chem. Soc.*, **1939**, 589.

G-10. A. E. Goddard, *A Textbook of Inorganic Chemistry*, Vol. XI, Pt. IV, Griffin, London, 1937, p. 166.

G-11. J. M. Gulland and W. V. Farrar, *Anales Farm. Bioquim.* (*Buenos Aires*), **15**, 73 (1944).

H-1. H. G. Heal, *J. Chem. Ed.*, **35**, 192 (1958).

H-2. M. Heeren, *Inauguraldissertation, Göttingen*, 1861; *Chem. Zentr.*, **1861**, 916.
H-3. D. Hellwinkel and G. Fahrbach, *Tetrahedron Letters*, No. 23, **1965**, 1823.
H-4. D. Hellwinkel and G. Fahrbach, *Ber.*, **101**, 574 (1968).
H-5. W. Hieber and K. Wollmann, *Ber.*, **95**, 1552 (1962).
H-6. W. Hieber and T. Kruck, *Ber.*, **95**, 2027 (1962).
H-7. F. G. Holliman and F. G. Mann, *J. Chem. Soc.*, **1945**, 37.
H-8. K. A. Hooton and A. L. Allred, *Inorg. Chem.*, **4**, 671 (1965).
H-9. M. G. ter Horst, *Rec. Trav. Chim.*, **55**, 697 (1936).
H-10. K. W. Hübel and E. H. Braye, U.S. 3,149,101 (1964), to Union Carbide Corp.
H-11. K. W. Hübel, E. H. Braye, and I. H. Caplier, U.S. 3,151,140 (1964), to Union Carbide Corp.
J-1. K. A. Jensen, *Z. Anorg. Allgem. Chem.*, **231**, 365 (1937).
J-2. K. A. Jensen, *Z. Anorg. Allgem. Chem.*, **250**, 245 (1943).
J-3. K. A. Jensen, *Z. Anorg. Allgem. Chem.*, **250**, 268 (1943).
K-1. L. V. Kaabak, A. P. Tomilov, and S. L. Varshavskii, *Zh. Vses. Khim. Obshchestva D. I. Mendeleeva*, **9**, 700 (1964).
K-2. T. Kaiwa. *Tôkohu. J. Exptl. Med.*, **20**, 163 (1932).
K-3. G. Klose, *Z. Naturforsch.*, **16a**, 528 (1961).
K-4. F. Krafft and W. Vorster, *Ber.*, **26**, 2813 (1893).
K-5. F. Krafft and R. E. Lyons, *Ber.*, **27**, 1768 (1894).
K-6. F. Krafft and O. Steiner, *Ber.*, **34**, 560 (1901).
K-7. E. Krause and G. Renwanz, *Ber.*, **62B**, 1710 (1929).
K-8. E. Krause and G. Renwanz, *Ber.*, **65B**, 777 (1932).
K-9. T. Kruck and M. Höfler, *Ber.*, **96**, 3035 (1963).
K-10. F. H. Kruse, R. E. Marsh, and J. D. McCullough, *Acta Cryst.*, **10**, 201 (1957).
L-1. J. B. Lambert and R. G. Keske, *Tetrahedron Letters*, No. 47, 4755 (1967).
L-2. J. O. M. van Langen and Th. v. d. Plas, *Research Correspondence*, Suppl. to *Research (London)*, **7**, 512 (1954).
L-3. K. Lederer, *Compt. Rend.*, **151**, 611 (1910).
L-4. K. Lederer, *Ber.*, **44**, 2287 (1911).
L-5. K. Lederer, *Ann.*, **391**, 326 (1912).
L-6. K. Lederer, *Ann.*, **399**, 260 (1913).
L-7. K. Lederer, *Ber.*, **46**, 1358 (1913).
L-8. K. Lederer, *Ber.*, **46**, 1810 (1913).
L-9. K. Lederer, *Ber.*, **47**, 277 (1914).
L-10. K. Lederer, *Ber.*, **48**, 1345 (1915).
L-11. K. Lederer, *Ber.*, **48**, 1422 (1915).
L-12. K. Lederer, *Ber.*, **48**, 1944 (1915).
L-13. K. Lederer, *Ber.*, **48**, 2049 (1915).
L-14. K. Lederer, *Ber.*, **49**, 334 (1916).
L-15. K. Lederer, *Ber.*, **49**, 345 (1916).
L-16. K. Lederer, *Ber.*, **49**, 1071 (1916).
L-17. K. Lederer, *Ber.*, **49**, 1076 (1916).
L-18. K. Lederer, *Ber.*, **49**, 1082 (1916).
L-19. K. Lederer, *Ber.*, **49**, 1385 (1916).
L-20. K. Lederer, *Ber.*, **49**, 1615 (1916).
L-21. K. Lederer, *Ber.*, **49**, 2529 (1916).
L-22. K. Lederer, *Ber.*, **49**, 2002 (1916).
L-23. K. Lederer, *Ber.*, **49**, 2532 (1916).
L-24. K. Lederer, *Ber.*, **49**, 2663 (1916).
L-25. K. Lederer, *Ber.*, **50**, 238 (1917).

L-26. K. Lederer, *Ber.*, **52B**, 1989 (1919).

L-27. K. Lederer, *Ber.*, **53B**, 712 (1920).

L-28. K. Lederer, *Ber.*, **53B**, 1430 (1920).

L-29. K. Lederer, *Ber.*, **53B**, 1674 (1920).

L-30. K. Lederer, *Ber.*, **53B**, 2342 (1920).

L-31. V. Lenher, *J. Am. Chem. Soc.*, **22**, 136 (1900).

L-32. C. Levaditi, *Ann. Inst. Pasteur*, **41**, 369 (1927).

L-33. D. T. Lewis, *J. Chem. Soc.*, **1940**, 831.

L-34. Y. Llabador and J. P. Adloff, *Radiochim. Acta*, **6**, 49 (1966).

L-35. T. M. Lowry, R. R. Goldstein, and F. L. Gilbert, *J. Chem. Soc.*, **1928**, 307.

L-36. T. M. Lowry and F. L. Gilbert, *J. Chem. Soc.*, **1929**, 2076.

L-37. T. M. Lowry and F. L. Gilbert, *J. Chem. Soc.*, **1929**, 2867.

L-38. T. M. Lowry and F. Hüther, *Rec. Trav. Chim.*, **55**, 688 (1936).

L-39. A. Lowy and R. F. Dunbrook, *J. Am. Chem. Soc.*, **44**, 614 (1922).

L-40. H. Lüers and F. Weinfurtner, *Wochschr. Brau.*, **43**, 25, 35, 45 (1926).

L-41. R. E. Lyons and G. S. Bush, *J. Am. Chem. Soc.*, **30**, 831 (1908).

L-42. R. E. Lyons and E. D. Scudder, *Ber.*, **64B**, 530 (1931).

M-1. J. D. McCullough, *Inorg. Chem.*, **4**, 862 (1965).

M-2. D. H. McDaniel, *Science*, **125**, 545 (1957).

M-3. W. McFarlane, *Molec. Phys.*, **12**, 243 (1967).

M-4. F. A. McMahon, T. G. Pearson, and P. L. Robinson, *J. Chem. Soc.*, **1933**, 1644.

M-5. W. Mack, *Angew. Chem.*, **77**, 260 (1965).

M-6. W. Mack, *Angew. Chem.*, **78**, 940 (1966).

M-7. J. W. Mallet, *Ann.*, **79**, 223 (1851).

M-8. F. G. Mann and F. G. Holliman, *Nature*, **152**, 749 (1943).

M-9. A. Marquardt and A. Michaelis, *Ber.*, **21**, 2042 (1888).

M-10. H. Matsuo, *J. Sci. Hiroshima Univer.*, *Ser. A*, **22**, 51 (1958).

M-11. F. P. Mazza and E. Melchionna, *Rend. Accad. Sci. Napoli*, **34**, 54 (1928).

M-12. H. Miller, U.S. Patent, 2,980,500 (1961), to Monsanto Chemical Company.

M-13. E. Montignie, *Z. Anorg. Allgem. Chem.*, **307**, 109 (1960).

M-14. E. Montignie, *Z. Anorg. Allgem. Chem.*, **315**, 102 (1962).

M-15. G. T. Morgan and H. D. K. Drew, *J. Chem. Soc.*, **117**, 1456 (1920).

M-16. G. T. Morgan and H. D. K. Drew, *J. Chem. Soc.*, **119**, 610 (1921).

M-17. G. T. Morgan and H. D. K. Drew, *J. Chem. Soc.*, **121**, 922 (1922).

M-18. G. T. Morgan, E. A. Cooper, and A. W. Burtt, *Biochem. J.*, **17**, 30 (1923).

M-19. G. T. Morgan and H. G. Reeves, *J. Chem. Soc.*, **123**, 444 (1923).

M-20. G. T. Morgan, E. A. Cooper, and A. W. Burtt, *Biochem. J.*, **18**, 190 (1924).

M-21. G. T. Morgan and H. D. K. Drew, *J. Chem. Soc.*, **125**, 731 (1924).

M-22. G. T. Morgan and R. W. Thomason, *J. Chem. Soc.*, **125**, 754 (1924).

M-23. G. T. Morgan and E. Holmes, *J. Chem. Soc.*, **125**, 760 (1924).

M-24. G. T. Morgan and H. D. K. Drew, *J. Chem. Soc.*, **125**, 1601 (1924).

M-25. G. T. Morgan, E. A. Cooper, and F. J. Corby, *J. Soc. Chem. Ind.*, **43**, 304 (1924).

M-26. G. T. Morgan and H. D. K. Drew, *J. Chem. Soc.*, **127**, 531 (1925).

M-27. G. T. Morgan and C. J. A. Taylor, *J. Chem. Soc.*, **127**, 797 (1925).

M-28. G. T. Morgan and H. D. K. Drew, *J. Chem. Soc.*, **127**, 2307 (1925).

M-29. G. T. Morgan, F. J. Corby, O. C. Elvins, E. Jones, R. E. Kellett, and C. J. A. Taylor, *J. Chem. Soc.*, **127**, 2611 (1925).

M-30. G. T. Morgan and O. C. Elvins, *J. Chem. Soc.*, **127**, 2625 (1925).

M-31. G. T. Morgan and A. E. Rawson, *J. Soc. Chem. Ind.*, **44**, 462T (1925).

M-32. G. T. Morgan and R. E. Kellett, *J. Chem. Soc.*, **1926**, 1080.

M-33. G. T. Morgan, E. A. Cooper, and A. E. Rawson, *J. Soc. Chem. Ind.*, **45**, 106 (1926).

M-34. G. T. Morgan and H. Burgess, Brit. Patent, 292,222 (1927).

M-35. G. T. Morgan and H. Burgess, *J. Chem. Soc.*, **1928**, 321.

M-36. G. T. Morgan and H. Burgess, *J. Chem. Soc.*, **1929**, 1103.

M-37. G. T. Morgan and H. Burgess, *J. Chem. Soc.*, **1929**, 2214.

M-38. G. T. Morgan and F. H. Burstall, *J. Chem. Soc.*, **1930**, 2599.

M-39. G. T. Morgan, *J. Chem. Soc.*, **1935**, 554.

M-40. G. T. Morgan and F. H. Burstall, *J. Chem. Soc.*, **1931**, 180.

M-41. M. de Moura Campos and N. Petragnani, *Tetrahedron Letters*, No. 6, **1959**, 11.

M-42. M. de Moura Campos and N. Petragnani, *Ber.*, **93**, 317 (1960).

M-43. M. de Moura Campos, N. Petragnani, and C. Thome, *Tetrahedron Letters*, No. 15, **1960**, 5.

M-44. M. de Moura Campos, *Selecta Chim.*, No. 19, **1960**, 55.

M-45. M. de Moura Campos and N. Petragnani, *Tetrahedron*, **18**, 521 (1962).

M-46. M. de Moura Campos and N. Petragnani, *Tetrahedron*, **18**, 527 (1962).

M-47. M. de Moura Campos, E. L. Suranyi, H. de Andrade, Jr., and N. Petragnani, *Tetrahedron*, **20**, 2797 (1964).

M-48. A. N. Murin. V. D. Nefedov, and O. V. Larionov, *Radiokhimiya*, **3**, 90 (1961).

N-1. G. Natta, *Giorn. Chim. Ind. Applicata*, **8**, 367 (1926); *Chem. Abstr.*, **20**, 3273, C. 1927, I, 415.

N-2. V. D. Nefedov, I. S. Kirin, and V. M. Zaitsev, *Radiokhimiya*, **4**, 351 (1962).

N-3. V. D. Nefedov, I. S. Kirin, and V. M. Zaitsev, *Radiokhimiya*, **6**, 78 (1964).

N-4. V. D. Nefedov, V. E. Zhuravlev, and M. A. Toropova, *Zh. Obshch. Khim.*, **34**, 3719 (1964).

N-5. V. D. Nefedov, M. A. Toropova, V. E. Zhuravlev, and A. V. Levchenko, *Radiokhimiya*, **7**, 203 (1965).

N-6. V. D. Nefedov, V. E. Zhuravlev, M. A. Toropova, L. N. Gracheva, and A. V. Levchenko, *Radiokhimiya*, **7**, 245 (1965).

N-7. V. D. Nefedov, V. E. Zhuravlev, M. A. Toropova, S. A. Grachev, and A. V. Levchenko, *Zh. Obshch. Khim.*, **35**, 1436 (1965).

N-8. A. N. Nesmeyanov, *Uspekhi Khim.*, **14**, 261 (1945).

N-9. P. Nylen, *Z. Anorg. Allgem. Chem.*, **246**, 227 (1941).

P-1. T. G. Pearson and R. H. Purcell, *J. Chem. Soc.*, **1936**, 253.

P-2. T. G. Pearson, R. H. Purcell, and G. S. Saigh, *J. Chem. Soc.*, **1938**, 409.

P-3. M. A. Perkins, U.S. Patent 2,030,035 (1936) to E. I. du Pont de Nemours and Company.

P-4. N. Petragnani and G. Vicentini, *Univ. Sao Paulo, Fac. Filosof. Cienc. Letras Bol. Quim.*, No. 5, **1959**, 75.

P-5. N. Petragnani, *Tetrahedron*, **11**, 15 (1960).

P-6. N. Petragnani, and M. de Moura Campos, *Ber.*, **94**, 1759 (1961).

P-7. N. Petragnani, *Tetrahedron*, **12**, 219 (1961).

P-8. N. Petragnani and M. de Moura Campos, *Anais da Associacao Brasileira de Quimica*, *XXII*, 3 (1963).

P-9. N. Petragnani *Ber.*, **96**, 247 (1963).

P-10. N. Petragnani and M. de Moura Campos, *Ber.*, **96**, 249 (1963).

P-11. N. Petragnani and M. de Moura Campos, *Chem. Ind. (London)*, **1964**, 1461.

P-12. N. Petragnani and M. de Moura Campos, *Tetrahedron*, **21**, 13 (1965).

P-13. N. Petragnani and M. de Moura Campos. *Organometal. Chem. Rev.*, **2**, 61 (1967).

P-14. A. A. Petrov, S. I. Radchenko, K. S. Mingaleva, I. G. Savich, and V. B. Lebedev, *Zh. Obshch. Khim.*, **34**, 1899 (1964).

P-15. H. A. Potratz and J. M. Rosen, *Anal. Chem.*, **21**, 1276 (1949).
P-16. G. Pourcelot, M. Lequan, M. P. Simonnin, and P. Cadiot, *Bull. Soc. Chim. France*, **1962**, 1278.
P-17. G. Pourcelot, *Compt. Rend.*, **260**, 2847 (1965).
R-1.   S. I. Radchenko, V. N. Chistokletov, and A. A. Petrov, *Zh. Obshch. Khim.*, **35**, 1735 (1965).
R-2.   S. I. Radchenko, V. N. Chistokletov, and A. A. Petrov, *Zh. Organ. Khim.*, **1**, 51 (1965).
R-3.   S. I. Radchenko and A. A. Petrov, *Zh. Organ. Khim.*, **1**, 2115 (1965).
R-4.   G. L. Ratcliffe, U.S. Patent 2,622,987 (1952), to National Lead Co.
R-5.   A. I. Razumov, B. G. Liorber, M. B. Gazizov, and Z. M. Khammatova, *Zh. Obshch. Khim.*, **34**, 1851 (1964).
R-6.   L. Reichel and E. Kirschbaum, *Ann.*, **523**, 211 (1936).
R-7.   L. Reichel and E. Kirschbaum, *Ber.*, **76B**, 1105 (1943).
R-8.   L. Reichel and K. Ilberg, *Ber.*, **76B**, 1108 (1943).
R-9.   H. Rheinboldt and E. Giesbrecht, *J. Am. Chem. Soc.*, **69**, 2310 (1947).
R-10. H. Rheinboldt, in *Methoden der organischen Chemie*, Houben-Weyl, Ed., Vol. IX, Thieme, Stuttgart, 1955, p. 917.
R-11. H. Rheinboldt and G. Vicentini, *Ber.*, **89**, 624 (1956).
R-12. H. Rheinboldt and N. Petragnani, *Ber.*, **89**, 1270 (1956).
R-13. F. O. Rice and A. L. Glasebrook, *J. Am. Chem. Soc.*, **56**, 2381 (1934).
R-14. F. O. Rice and A. L. Glasebrook, *J. Am. Chem. Soc.*, **56**, 2472 (1934).
R-15. F. O. Rice and M. D. Dooley, *J. Am. Chem. Soc.*, **56**, 2747 (1934).
R-16. F. Rogoz, *Dissertationes Pharm.*, **16**, 157 (1964); *Chem. Abstr.*, **62**, 11722.
R-17. E. Rohrbaech, *Ann.*, **315**, 9 (1901).
R-18. E. Rust, *Ber.*, **30**, 2828 (1897).
S-1.   R. B. Sandin, F. T. McClure, and F. Irwin, *J. Am. Chem. Soc.*, **61**, 2944 (1939).
S-2.   R. B. Sandin, R. G. Christiansen, R. K. Brown, and S. Kirkwood, *J. Am. Chem. Soc.*, **69**, 1550 (1947).
S-3.   M. Schmidt and H. Schumann, *Z. Naturforsch.*, **19b**, 74 (1964).
S-4.   O. Schmitz-DuMont and B. Ross, *Angew. Chem.*, **6**, 1071 (1967).
S-5.   A. Schoenberg, E. Rupp, and W. Gumlich, *Ber.*, **66**, 1932 (1933).
S-6.   H. Schumann, K. F. Thom, and M. Schmidt, *Angew. Chem.*, **75**, 138 (1963).
S-7.   H. Schumann, K. F. Thom, and M. Schmidt, *J. Organometal. Chem.*, **2**, 361 (1964).
S-8.   H. Schumann and M. Schmidt, *J. Organometal. Chem.*, **3**, 485 (1965).
S-9.   H. Schumann, K. F. Thom, and M. Schmidt, *J. Organometal. Chem.*, **4**, 22 (1965).
S-10. H. Schumann, K. F. Thom, and M. Schmidt, *J. Organometal. Chem.*, **4**, 28 (1965).
S-11. H. Schumann and M. Schmidt, *Angew. Chem.*, **77**, 1049 (1965).
S-12. A. Scott, *Proc. Chem. Soc.*, **20**, 156 (1904).
S-13. M. Shinagawa, H. Matsuo, and H. Sunahara, *Japan. Analyst*, **3**, 204 (1954).
S-14. M. P. Simonnin, *Compt. Rend.*, **257**, 1075 (1963).
S-15. B. Singh and R. Krishen, *J. Indian Chem. Soc.*, **12**, 711 (1935).
S-16. R. Sochacka and A. Szuchnik, *Polska Akad. Nauk. Inst. Badán Jadrowych*, No. 149, **XIII**, 1 (1960).
S-17. E. A. L. Suranyi and H. de Andrade, Jr., *Bol. Dept. Eng. Quim. EPUSP*, No. 19, **1964**, 31.
S-18. O. Steiner, *Ber.*, **34**, 570 (1901).
T-1.   H. Taniyama, F Miyoshi, E. Sakakibara, and H. Uchida, *Yakugaku Zasshi*, **77**, 191 (1957).

T-2. L. Tschugaeff and W. Chlopin, *Ber.*, **47**, 1269 (1914).

V-1. R. H. Vernon, *J. Chem. Soc.*, **117**, 86 (1920).

V-2. R. H. Vernon, *J. Chem. Soc.*, **117**, 889 (1920).

V-3. R. H. Vernon, *J. Chem. Soc.*, **119**, 687 (1921).

V-4. G. Vicentini, *Ber.*, **91**, 801 (1958).

V-5. G. Vicentini, E. Giesbrecht, and L. R. M. Pitombo, *Ber.*, **92**, 40 (1959).

V-6. M. Vobetsky, V. D. Nefedov, and E. N. Sinomova, *Zh. Obshch. Khim.*, **35**, 1684 (1965).

V-7. N. S. Vyazankin, M. N. Bochkarev, and L. P. Sanina, *Zh. Obshch. Khim.*, **36**, 166 (1966).

V-8. N. S. Vyazankin, M. N. Bochkarev, and L. P. Sanina, *Zh. Obshch. Khim.*, **36**, 1154 (1966).

V-9. N. S. Vyazankin, M. N. Bochkarev, and L. P. Sanina, *Zh. Obshch, Khim.*, **37**, 1037 (1967).

W-1. W. A. Waters, *J. Chem. Soc.*, **1938**, 1077.

W-2. F. D. Williams and F. X. Dunbar, *Chem. Comm.*, **1968**, 459.

W-3. G. Wittig and H. Fritz, *Ann.*, **577**, 39 (1952).

W-4. F. Wöhler, *Ann.*, **35**, 111 (1840).

W-5. F. Wöhler, *Ann.*, **84**, 69 (1852).

W-6. F. Wöhler and J. Dean, *Ann.*, **93**, 233 (1855).

W-7. F. Wohler and J. Dean, *Ann.*, **97**, 1 (1856).

Z-1. F. Zeiser, *Ber.*, **28**, 1670 (1895).

Z-2. R. A. Zingaro, *J. Organometal. Chem.*, **1**, 200 (1963).

Z-3. R. A. Zingaro, B. H. Steeves, and K. Irgolic, *J. Organometal. Chem.*, **4**, 320 (1965).

Reactions of Organomercury Compounds, Part 2

L. G. MAKAROVA

Institute of Organoelemento Compounds, Academy of Sciences of U.S.S.R.,
Moscow, U.S.S.R.

Part 2, Chapters 12–17

Chapter 12. Radical Exchange Reactions between Organomercury Compounds and Organic Compounds of Other Elements

Chapter 13. Reactions with Nitrogen-Containing Compounds: Nitric Oxides, Nitric Oxide Derivatives, Diazo Compounds, and Carbodiimides

Chapter 14. Reactions with Ketenes and Their Derivatives

Chapter 15. Reactions of RHgX → R₂Hg and Reverse Reactions

Chapter 16. Decomposition of Organomercury Compounds by Oxygen and Other Oxidants, and Light and Thermal Treatment

Chapter 17. Reactions with Alkalis

Chapter 12

Radical Exchange Reactions between Organomercury Compounds and Organic Compounds of Other Elements

I. INTRODUCTION

Organomercury compounds may participate in radical exchange reactions with organic compounds of other elements according to the following reactions:

$$n R_2 Hg + R'_n Me^n \longrightarrow n R'_2 Hg + R_n Me^n \tag{1}$$
$$n R_2 Hg + R'_{n-m} Me^n X_m \longrightarrow n R'_2 Hg + R_{n-m} Me^n X_m \tag{2}$$
$$n RHgX + R'_n Me^n \longrightarrow n R' HgX + R_n Me^n \tag{3}$$
$$n RHgX + R'_{n-m} Me^n X_m \longrightarrow n R' HgX + R_{n-m} Me^n X_m \tag{4}$$

The rate and degree of exchange depend on such factors as the nature of the element of exchanging organo compounds, the nature of radicals connected with both this element and with mercury, and the conditions of the reaction.

II. REACTION MECHANISM

The exchange of radicals between organomercury compounds is typical of reactions listed above (see also Section III-B). The mechanism of such an exchange was studied by Reutov and co-workers and then by Ingold and co-workers with the use of the following isotopic exchange reactions of organomercury compounds:

$$R_2 Hg + R'_2 Hg^* \rightleftharpoons R_2 Hg^* + R'_2 Hg \tag{5}$$
$$R_2 Hg + R' Hg^* X \rightleftharpoons RHg^* R' + RHgX \tag{6}$$
$$RHgX + R' Hg^* X \rightleftharpoons RHg^* X + R' HgX \tag{7}$$
$$(\text{where } Hg^* = {}^{203}Hg)$$

These reactions often proceed at room temperature and therefore the possibility of a free radical mechanism is eliminated, since free radicals are usually not formed under such mild conditions. Reaction (5) proceeds more rapidly under mild conditions when one of the radicals is aryl and containing electron donor substituents in the para position; this reaction proceeds with difficulty if electron acceptor substituents[1] are present and indicates an electrophilic mechanism for the reaction. Radicals might pass from one metal atom to another simultaneously in a single elementary act, or one by one (through a four-center transition state of type **1**) with intermediate formation of asymmetric organomercury compounds which then disproportionate:

$$R-Hg-R \quad\quad R-Hg-R \quad\quad\quad R-Hg \quad R$$
$$+ R'-Hg^*R' \;\rightleftharpoons\; R-Hg^*-R' \;\rightleftharpoons\; \underset{R'}{|} + \underset{Hg^*-R'}{|}$$

$$\text{(1)}$$

$$R\,HgR' + R\,Hg^*R' \;\rightleftharpoons\; \text{(1)} \;\rightleftharpoons\; RHg^*R + R'HgR'$$

By using mass spectroscopy on the systems containing $(CD_3)_2Hg$ and $(CH_3)_2Hg$ (in tetrahydroduran, 65°C, 78 hr) and for $(CH_3)_2Hg$ or CH_3HgBr (shorter period) Dessy et al.[2] have found the presence of $CD_3-Hg-CH_3$ molecules, which indicates the possibility of a type **1** transitory state for the exchange of aliphatic compounds of mercury (reaction symbol, S_F2).[3]

An $ArHgAr'$ compound was not detected, however, in the reaction between Ar_2Hg and $Ar'_2{}^{203}Hg$ and therefore it is possible that in case of aromatic compounds all four groups migrate simultaneously.[4]

However, reactions such as:

$$R_2Hg + R'_2Hg \;\rightleftharpoons\; 2\,RHgR'$$

(Hg is nonlabeled mercury) analogous to reaction (5) are catalyzed by peroxides and this indicates the possibility of a homolytic mechanism for such reactions[3] (see below, Section III-B).

It was found that trialkyl exchange reactions such as (6) ($R=CH_3CH-(CH_2)_2-CH(CH_3)_2$ and sec-C_4H_9R[7], R' for each case is, respectively, the same optically active radical in alcohol, 60° C) are bimolecular ones. Such reactions have an overall second-order[6,7] rate constant (first order with respect to each component) and proceed with retention of configuration.[5–7] For R = sec-C_4H_9, the reaction rate increases considerably with the increase of anion polarity: $X = Br < CH_3COO < NO_3$.[7] Addition of LiX' salts increases the reaction rate: $X' = CH_3COO < NO_3 < Br < ClO_4$. It is evident that the transition state is more polar than the initial compound. According to Charman, Ingold, and others,[7] the presence of a transition state of open type (S_E2 mechanism) in this reaction is indicated by such factors as the dependency of the reaction rate on the anion polarity, the high polarity of the transition state, and the retention of configuration. Reutov and others[6] consider that nonpolar and slightly polar solvents should facilitate transition state **2** (S_Ei mechanism) while the highly polar solvents are favorable for transitory state **3**

(S_E2 mechanism) (as the result of the decomposition of this state, $\overset{\diagdown}{\underset{\diagup}{C}}-Hg^+$

and X^- ions should form and these ions should then associate into the $\overset{\diagdown}{\underset{\diagup}{C}}-HgX$ molecule):

$$(2) \qquad (3)$$

Reaction (7) (R $= C_6H_5CHCO_2C_2H_5$, R$' = C_6H_5$, X$' =$ Br in pyridine at 100°C[8] and for R $= CH_2{=}CH{-}CH_2$, R$' = p\text{-}(CH_3)_2NC_6H_4$, X $=$ Br[9]) has also overall second order (first order with respect to each component) and probably also is an electrophilic substitution reaction (S_E2 mechanism). The following conclusion may be derived on the basis of the data mentioned above: the radical exchange reactions between organometallic compounds in the majority of cases are probably electrophilic substitution reactions.

However, the regularities derived for fast reactions of isotopic exchange of organomercury compounds (reactions (5)–(7)) cannot be mechanically transferred to the radical exchange reactions between organomercury and organometallic compounds of other metals if the nature of the latter and kinetic character of exchange are not taken into account.

III. APPLICATION OF THE REACTION

A. Exchange with Compounds of Other Elements

The lower the bond strength between C—element and C—Hg, which is determined by both the nature of the element and the nature of the radicals being exchanged, the more easily the exchange occurs. For instance, $(C_6H_5)_4Si$ with its inert C—Si bond practically does not participate in the exchange reaction with $(^{14}C_6H_5)_2Hg$ at 100°C during 20 hr[10] while under the same conditions $(C_6H_5)_4Pb$ exchanges (88%). $(^{14}C_2H_5)_2Hg$ does not exchange with $(C_2H_5)_2Pb$ under these conditions and $(^{14}C_2H_5)_2Hg$ exchanges with C_2H_5Na only to a small degree with prolonged interaction.[11] There are no exchanges between $(CH_3)_2Hg$[12] or $(CD_3)_2Hg$[2] and $(CH_3)_2Zn$ or $(CH_3)_2Cd$. However, $(CD_3)_2Hg$ exchanges with $(CH_3)_2Mg$ or CH_3MgBr.[2] The exchange rate in both cases is approximately the same.

Addition of CH_3Li facilitates CH_3-group exchange between molecules of $(CH_3)_2Hg$ in ether and THF.[39] R_2Hg readily undergoes transmetallation with RLi.[40–42] For example: $(C_6H_5)_2Hg$ with $n\text{-}C_4H_9Li$ undergoes Li—Hg exchange readily even at -78°C. A mixture of $(C_6F_5)_2Hg$ and tetrameric mercurial o,o'-$(C_6F_4—C_6F_4Hg)_4$ was obtained by the reaction of o,o'-$(LiC_6F_4)_2$ with $C_6F_5H_8Cl$.[43]

A kinetic study of the electrophilic substitution of R_2Zn with C_6H_5HgCl has been carried out.[44]

The mercury mobility in organomercury compounds facilitating the exchange may be caused, for example, by σ,π-conjugation of the mercury with carbonyl group. As was shown by Nesmeyanov and Pecherskaya,[13] who corrected data obtained by Grignard and Abelmann,[14,15] the effect of a Grignard reagent upon a halo-mercuryacetophenone is to leave the carbonyl group unchanged and effect only a radical exchange:

$$C_6H_5\underset{\underset{O}{\|}}{C}—CH_2HgX + RMgX \longrightarrow RHgX + C_6H_5\underset{\underset{O}{\|}}{C}—CH_2MgX$$

[tautomeric with $C_6H_5C(OMgX){=}CH_2$]

Similarly, o-halomercuryphenol reacts with the Grignard reagent:[16]

$+ 2 C_2H_5MgBr \longrightarrow$ (OMgBr, MgBr) $+ C_2H_6 + C_2H_5HgCl$

Less mobile mercury atoms do not take part in the exchange reaction:[15]

$+ C_2H_5MgBr \longrightarrow$

In some cases the exchange is facilitated by the partial ionic character of the C—Hg bond. For instance, the interaction between the bis(bromoacetylenyl)mercury with phenylmagnesium bromide results in radical exchange:[17]

$$(BrC{\equiv}C)_2Hg + 2 C_6H_5MgBr \longrightarrow 2 BrC{\equiv}CMgBr + (C_6H_5)_2Hg$$

This reaction demonstrates the absence of quasicomplex properties in the bis-(bromocetylenyl)mercury:[17] the effect of C_6H_5MgBr upon such a quasicomplex mercury compound as $trans$-β-chlorovinylmercury chloride, has resulted in the formation of diphenylmercury with the separation of acetylene:[18]

$$Cl—CH{=}CH—HgCl + 2 C_6H_5MgBr \longrightarrow (C_6H_5)_2Hg + HC{\equiv}CH + MgBr_2 + MgCl_2$$

Other mercury acetylenides, such as bis(phenylacetylenyl)mercury, under mild conditions also fully exchange radicals with phenylmagnesium bromide giving $(C_6H_5)_2Hg$.[17] The greater the difference between the electronegativities of radicals being exchanged, the more easily the radical exchange occurs.[19]

For instance, a considerable exchange takes place between perfluoroalkyl and perfluorovinylmercury compounds and ethyl- or phenylmagnesium bromide (in ether or tetrahydrofuran).[20]

The stability of carbanions is determined according to their ability to make bonds with Mg:

$$R_2Hg + R_2'Mg \rightleftharpoons R_2'Hg + R_2Mg$$

Because of decreasing stability, some carbanions, $R:^-$, are arranged in the following series: $C_6H_5C{\equiv}C > C_6H_5 > C_2H_5 > CH(CH_3)_2$; (steric factors are also important in this series). ($tert$-$C_4H_9)_2Mg$ does not exchange with (i-$C_3H_7)_2Hg$ and $(C_6H_5)_2Hg$ during 5 hr at 20°C. The greater the difference in stability, the faster the reaction; however, different dialkylmagnesium compounds react with $(C_6H_5)_2Hg$ approximately at the same rate. Taking into account the steric factors, authors[21] proposed that the reaction proceeds through a multi-centered transition state,

$$R{-}Hg \overset{\displaystyle R'}{\underset{\displaystyle R}{\diamond}} Mg{-}R'$$

as it is difficult to explain a sharp difference in reactivities of i-C_3H_7 and $tert$-C_4H_9 if an ionic exchange is proposed:

$$R_2'Mg \rightleftharpoons R':^- + R'Mg^+; \; R':^- + R_2Hg \longrightarrow \text{reaction product.}$$

Transmetallation by alkyllithium proceeds under mild conditions,[22-25] and

$$R_2Hg + 2\,AlkLi \longrightarrow 2\,RLi + Alk_2Hg$$
$$RHgX + AlkLi \longrightarrow RLi + AlkHgX$$

is widely used for obtaining organolithium compounds. Methyllithium may be prepared by the exchange between soluble alkyllithium and dimethylmercury:[22]

$$(CH_3)_2Hg + 2\,AlkLi \longrightarrow 2\,CH_3Li + Alk_2Hg$$

The reaction of $(CH_3)_2Hg$ with metallic lithium is not suitable as a preparation. It is easier experimentally to obtain phenyllithium by a transmetallation reaction, i.e., by the reaction of alkyllithium with diphenylmercury.[22]

Transmetallation reaction by use of various AlkLi and chloromercuryferrocene, diferrocenylmercury, and 1,1'-dichloromercuryferrocene (in ether) were used for the preparation of mono- and 1,1'-dilithiumferrocene.[26,27]

Exchange occurs between RLi and (p-$BrC_6H_4)_2Hg$, and bromine is not substituted for Li:[28]

$$(p\text{-}BrC_6H_4)_2Hg + 2\,RLi \longrightarrow R_2Hg + 2p\text{-}BrC_6H_4Li \xrightarrow{CO_2} 2\,BrC_6H_4COOH$$

Besides this reaction, RLi (R = CH_3) interacts with p-BrC_6H_4Li with the formation of dilithiumbenzene; carboxylation resulted in terephthalic acid.[28]

The reaction of R_2Hg with $R'I$ in the presence of a small quantity of $R'Li$ is shown by the following reactions:

$$R_2Hg + 2\,R'Li \rightleftharpoons R_2'Hg + 2\,RLi \tag{8}$$

$$2\,R'I + 2\,RLi \rightleftharpoons 2\,RI + 2\,R'Li \tag{9}$$

$$R_2Hg + 2\,R'I \rightleftharpoons R_2'Hg + 2\,RI \tag{10}$$

For instance, in the presence of such halogens as o-bromo- and even o-iodoanisole which can be easily substituted by Li, the exchange reaction of radicals combined with Hg and Li takes place initially, (this was supported by the formation of $(n\text{-}C_4H_9)_2Hg$ and the absence of n-valeric and benzoic acids after carboxylation in the reaction $(C_6H_5)_2Hg + 2n\text{-}C_4H_9Li \rightarrow (n\text{-}C_4H_9)_2Hg + 2C_6H_5Li$ which was conducted in the presence of o-bromo- or o-iodoanisole); this reaction is followed by the reaction of RHal with C_6H_5Li:[28]

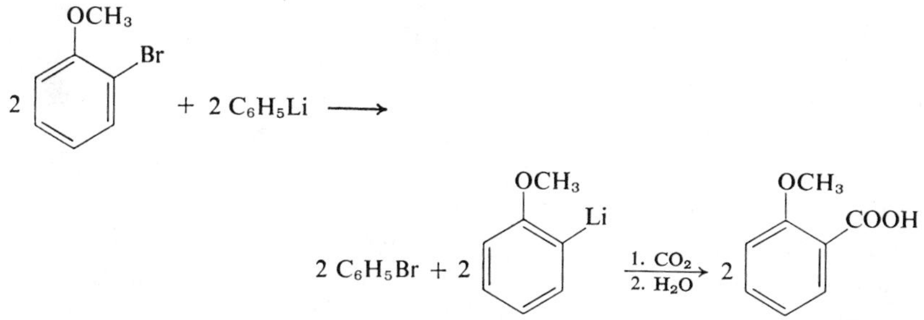

Upon the reaction between R_2Hg and RLi and alkali metal (Na, Cs) in ether, the complexes $(R\text{—}Li \leftarrow R)Me$ (R = Alk and Ar) are formed through the intermediate formation of an organic compound of the alkali metal.[24,29]

Stirring with KNa alloy is required for obtaining such complexes with potassium; stirring with metallic potassium is not feasible as the latter replaces lithium from its organic compounds.

In the majority of cases radical exchange reactions between R_2Hg and organic compounds of other metals are reversible with the exception of those cases when the reaction products are insoluble, for example when insoluble methyllithium is obtained as a result of the reactions of dimethylmercury with alkyllithium (see above). For this reason, radical exchange reactions result in various equilibrium mixtures of products. As the separation of radicals from polyalkylated organometallic compounds (separation of radicals from R_2Hg) is stepwise, the more alkylated organometallic compounds participate in the reaction, the more complex is the equilibrium mixture. In such redistribution reactions, the composition of the obtained mixture corresponds usually to the distribution on the basis of equal probability (random equilibrium mixtures).[30-33] However, when radicals pass from one metal to another, a relative affinity of the given radical with the given metal does not depend on the nature of other radicals connected with the metal. For instance, in the reaction,

$$CH_3PbR_3 + C_2H_5HgR \rightleftharpoons CH_3HgR + C_2H_5PbR_3$$

radicals HgR and PbR$_3$ do not affect the equilibrium state of this reaction. The elucidation of the composition of mixtures obtained shows that mercury has a higher affinity for methyl radicals than with ethyl when compared with lead. Therefore, strictly speaking, the random redistribution is only an extreme case in such reactions. The composition of the equilibrium mixture depends on a relative value of the energy of the bond of the given radical with metal and mercury. On the basis of the mass interaction law, the equilibrium displacement may be expressed by the "relative affinity constant," K, of metal and radical, which, for the given reaction, is as follows:[32]

$$K = \frac{[CH_3\text{—}Hg][C_2H_5\text{—}Pb]}{[C_2H_5\text{—}Hg][CH_3\text{—}Pb]}$$

where the symbols in brackets show relative concentration of these four linkages in the final product. For this example, K is approximately equal to 4. On the basis of experimental determination of this constant value for a single case, it is possible to predict the composition of an equilibrium mixture composed of the same two metals and radicals used in different ratios. The radical exchange reactions of R$_2$Hg with R$_4'$Pb (as well as of R$_2$Hg with R$_2'$Hg, see part B below) usually are conducted in the presence of AlCl$_3$ catalyst without solvent at a temperature from 60 to 80°C. The exchange does not take place in the absence of catalyst. For deviation from randomness see.[45,46]

B. Radical Redistribution Reactions between Organomercury Compounds

The following reactions belong in the category of radical redistribution reactions between organomercury compounds:

R$_2$Hg + R$_2'$Hg $\rightleftharpoons$ 2 RHgR'	analogous to reaction (5)	(11)
R$_2$Hg + R'HgX $\rightleftharpoons$ RHgX + RHgR'	analogous to reaction (6)	(12)

Reaction (11) is a reversible reaction of the formation of a fully substituted asymmetric mercury compound from two symmetric compounds and reaction (12) is the alkylation of RHgX to RHgR' (see above). The reaction, as written from left to right, is catalyzed by peroxides[3] (this may indicate the homolytic nature of the process), while in the presence of trace amounts of RHgX or previously used catalysts (such as, for example, AlCl$_3$)[30–33] this process is catalyzed, as was recently shown,[34] by admixtures of Hg$_2$X$_2$ and Hg. Exchange (11) (R = C$_6$F$_5$, R^1 = CH$_3$) is promoted by traces of halide.[47]

The interaction between R$_2$Hg and R$_2'$Hg usually results in the equilibrium mixture of R_2Hg, $R_2'Hg$, and $RHgR'^2$. In some cases, vigorous reaction conditions (100–150°C, sealed tubes)[35] are required for obtaining the equilibrium. This is especially true for perfluoroalkyl and perfluoroarylmercury compounds for a number of which exchange is highly nonrandom. In other cases the reaction proceeds under sufficiently mild conditions but in the presence of the catalysts mentioned above. Various R and R' give various compositions of the equilibrium mixture.[45]

Belonging to reaction class (12) is a phenyl exchange in the system $^{14}C_6H_5HgBr + (C_6H_5)_2Hg$ which proceeds rapidly in pyridine.[36]

An attempt to obtain nonsymmetrical mercury compounds, RHgR', by the alkylation of RHgX with R'MgX or other active organometallic compound often results in a radical redistribution reaction only. For instance, Frankland[37] reacted $(C_2H_5)_2Zn$ with CH_3HgI and obtained only $(C_2H_5)_2Hg$. He also obtained $(CH_3)_2Hg$ and $(C_2H_5)_2Hg$ from $(CH_3)_2Zn$ and C_2H_5HgCl, respectively. $(C_2H_5)_2Hg$ was also obtained from C_2H_5MgBr and C_6H_5HgBr or $C_{10}H_7HgBr$.[38] Frankland supposed that this reaction partially resulted in the formation of a nonsymmetrical mercury compound, RHgR', which decomposed into the symmetrical compounds R_2Hg and $R_2'Hg$. (See explanation of the reaction by Calingaert.[31])

REFERENCES

1. O. A. Reutov, T. A. Smolina and K. V. Khu, *Izv. Akad. Nauk SSSR, Otd. Khim. Nauk*, **1959**, 559.
2. R. E. Dessy, F. Kaplan, C. R. Coe, and R. M. Salinger, *J. Am. Chem. Soc.*, **85**, 1191 (1963).
3. R. E. Dessy, W. L. Budde, and C. Woodruff, *J. Am. Chem. Soc.*, **84**, 1172 (1962).
4. O. A. Reutov, *Rec. Chem. Progr.*, **22**, 1 (1961).
5. O. A. Reutov, T. P. Karpov, E. V. Uglova, and V. A. Malyanov, *Dokl. Akad. Nauk SSSR*, **134**, 360 (1960); *Izv. Akad. Nauk SSSR, Otd. Khim. Nauk*, **1960**, 1311; *Tetrahedron Letters*, **1960**, N19, 6.
6. T. P. Karpov, V. A. Malyanov, E. V. Uglova, and O. A. Reutov, *Izv. Akad. Nauk SSSR, Otd. Khim. Nauk*, **1964**, 1580.
7. H. B. Charman, E. D. Hughes, C. K. Ingold, and F. G. Thorpe, *J. Chem. Soc.*, **1961**, 1121.
8. O. A. Reutov, K. V. Khu, I. P. Beletskaya, and T. A. Smolina, *Zh. Fiz. Khim.*, **35**, 2424 (1961).
9. T. A. Smolina, V. A. Kalyavin, and O. A. Reutov, *Izv. Akad. Nauk SSSR, Otd. Khim. Nauk.*, **1963**, 2235.
10. L. M. Nazarova, *Zh. Obshch. Khim.*, **31**, 1119 (1961).
11. L. M. Nazarova, *Zh. Obshch. Khim.*, **29**, 2671 (1959).
12. C. R. McCoy and A. L. Allred, *J. Am. Chem. Soc.*, **84**, 912 (1962).
13. A. M. Nesmeyanov and K. A. Pecherskaya, *Izv. Akad. Nauk SSSR, Otd. Khim. Nauk*, **1941**, 67.
14. V. Grignard and A. Abelmann, *Bull. Soc. Chim. France*, [4], **19**, 18 (1916).
15. A. Abelmann, *Ber.*, **47**, 2931 (1914).

16. A. N. Nesmeyanov and K. A. Pecherskaya, *Izv. Akad. Nauk SSSR, Otd. Khim. Nauk*, **1943**, 317.
17. A. N. Nesmeyanov and N. K. Kochetkov, *Izv. Akad. Nauk SSSR, Otd. Khim. Nauk*, **1949**, 587; *Uchen. Zap. Moskov. Gos. Univer.*, **132**, 51 (1950).
18. A. N. Nesmeyanov, R. Kh. Freidlina, and A. E. Borisov, Yubil. Sb. posvjashch. 30-let. Vel. Okt. Soz. Rev., part I, Moskau, *Izdat. Akad. Nauk SSSR*, **1947**, p. 658.
19. R. N. Sterlin, L. P. Pinkina, L. F. Nezgovorov, and I. L. Knunyants, *Khim. Nauka i prom.*, **6**, 809 (1959).
20. R. N. Sterlin, V. G. Li, and I. L. Knunyants, *Zh. Vsesojuzn. Khim. Obshch.*, **6**, 108 (1961).
21. R. M. Salinger and R. E. Dessy, *Tetrahedron Letters*, **11**, 729 (1963).
22. W. Schlenk and J. Holtz., *Ber.*, **50**, 262 (1917).
23. F. Hein, E. Petzchner, K. Wagner, and F. A. Segitz, *Z. Anorg. Allgem. Chem.*, **141**, 161 (1924).
24. G. Wittig and F. Bickelhaupt, *Chem. Ber.*, **91**, 883 (1958).
25. G. Wittig and E. Benz, *Chem. Ber.*, **91**, 873 (1958).
26. D. Seyferth, H. P. Hofmann, R. Burton, and Y. F. Helling, *Inorg. Chem.*, **1**, 227 (1962).
27. M. D. Rausch, *Inorg. Chem.*, **1**, 414 (1962).
28. H. Gilman and R. G. Jones, *J. Am. Chem. Soc.*, **63**, 1443 (1941).
29. G. Wittig and F. Bickelhaupt, *Chem. Ber.*, **91**, 865 (1958).
30. G. Calingaert, H. A. Beatty, and H. R. Neal, *J. Am. Chem. Soc.*, **61**, 2755 (1939).
31. G. Calingaert, H. Soroos, and V. Hnizda, *J. Am. Chem. Soc.*, **62**, 1107 (1940).
32. G. Calingaert, H. Soroos, and W. Thomson, *J. Am. Chem. Soc.*, **62**, 1542 (1940).
33. G. Calingaert, H. Soroos, and H. Shapiro, *J. Am. Chem. Soc.*, **63**, 947 (1941).
34. W. E. French, N. Inamoto, and G. F. Wright, *Can. J. Chem.*, **42**, 2228 (1964).
35. M. Rausch and J. R. Van Wazer, *Inorg. Chem.*, **3**, 761 (1964).
36. V. D. Nefedov and E. N. Sinotova, *Sb. Rabot po Radiokhim. Izd. Leningr. Gos. Univer*, **1955**, p. 110.
37. E. Frankland, *Ann Chem.*, **111**, 57 (1859).
38. S. Hilpert and G. Grütner, *Ber.*, **48**, 906 (1915).
39. L. M. Seitz and S. D. Hall, *J. Organometal. Chem.*, **15**, 7 (1968).
40. H. Daniel and J. Paetsch, *Ber.*, **98**, 1915 (1965).
41. N. E. Alexandrou, *J. Organometal. Chem.*, **5**, 30 (1966).
42. S. Toppet, G. Slinekx, and G. Smets, *J. Organometal. Chem.*, **9**, 205 (1967).
43. S. C. Cohen and A. G. Massey, *Chem. and Ind.*, 252 (1968).
44. M. H. Abraham and P. H. Rolfa, *J. Organometal. Chem.*, **8**, 395 (1967).
45. R. E. Dessy, W. Kitching, T. Psarras, R. Salinger, A. Chen, and T. Chivers, *J. Am. Chem. Soc.*, **88**, 460 (1966).
46. G. F. Reynolds and S. R. Daniel, *Inorg. Chem.*, **6**, 480 (1967).
47. R. D. Chambers, G. E. Coates, J. G. Livingstone, and W. K. R. Musgrave, *J. Chem. Soc.* **1962**, 4367.

Chapter 13

Reactions with Nitrogen-Containing Compounds: Nitric Oxides, Nitric Oxide Derivatives, Diazo Compounds and Carbodiimides

I. REACTIONS WITH NITRIC OXIDES

Organomercury compounds do not react with nitric oxide.

Reactions of some R_2Hg and of organomercury salts with nitrosyl halides result in the formation of nitroso compounds, for example:

$$R_2Hg + 2\,NOBr \longrightarrow 2\,RNO + HgBr_2 \qquad (R = aryl\ or\ polyfluoroalkyl)$$

Bayer[1] was the first to obtain nitrosoaryl compounds by this method. Under the action of NOCl the atom of mercury sometimes is substituted by chlorine in polyfluoroalkyl compounds of mercury, while there are some cases when the hydrogen is substituted by chlorine in partially fluorinated compounds.[2]

Nitroso compounds are formed by the effect of nitrogen dioxide on both R_2Hg (R = Alk or Ar[3-5]) and $ArHgX$[5] in chloroform:

$$R_2Hg + N_2O_4 \longrightarrow RNO + RHgNO_3$$
$$ArHgX + N_2O_4 \longrightarrow ArNO + HgXNO_3$$

Dinitrogen trioxide extracts one aryl radical from diarylmercury producing a 50% yield of aryldiazonium salt: [3,4]

$$Ar_2Hg + 2\,N_2O_3 \longrightarrow ArHgNO_3 + ArN_2NO_3$$

Nitrosobenzene is produced simultaneously.

A mixture of nitric oxide and dinitrogen trioxide extracts both phenyls from diphenylmercury producing an 85% yield of phenyldiazonium salt.[6] Under the combined effect of both these oxides on the monoarylated mercury compound, phenylmercury acetate, the latter is completely dearylated with quantitative formation of phenyldiazonium salt.

Unlike aromatic mercury compounds, aliphatic dimethylmercury does not react with dinitrogen trioxide to form aliphatic diazocompounds but, evidently, forms a compound which may be imidodihydroxamic acid.[7]

$$HON{=}C{-}NH{-}C{=}NOH$$
$$\qquad\ \ | \qquad\qquad |$$
$$\qquad\ \ HO \qquad\quad OH$$

Formation of nitroso-, nitro-, and diazo compounds when organomercury compounds react with nitric acid containing nitric oxides is discussed in Chapter 9.

A. Formation of Diazo Compounds by the Reaction of Nitric Oxides with Organomercury Compounds

Dinitrogen trioxide, evidently, reacts as products of its dissociation:

$$2 N_2O_3 \longrightarrow N_2O_4 + 2 NO$$

Nitroso aryl compounds are formed first from nitrogen dioxide, which is the dissociation product of dinitrogen trioxide:

$$Ar_2Hg + N_2O_4 \longrightarrow ArNO + ArHgNO_3$$

The reaction of nitrosobenzene with nitric oxide then produces the diazo compound:

$$ArNO + 2 NO \longrightarrow ArN_2NO_3$$

This mechanism is confirmed by the fact that during combined action of dinitrogen trioxide and nitric oxide on the organomercury compound Ar_2Hg, the yield of the diazo compound is higher than from the reaction of dinitrogen trioxide only.[6] It may be also due to the fact that under the effect of both nitric oxides, both aryls are utilized.

II. REACTIONS WITH DIAZO COMPOUNDS

The organomercury compounds, where mercury is activated by conjugation with a carbonyl group such as in ethyl α-bromomercuryphenylacetate or bromomercurymethyl *tert*-butyl ketone, react with aryldiazonium fluoborate in media of high dielectric constant—the first reacts in dimethylsulfoxide, nitromethane, dimethylformamide-methylacetamide mixture (3:1) but not in dimethylformamide; the second reacts in dimethylformamide substituting the arylazo group for the mercury and producing the arylhydrazones of carbonyl compounds:[8]

$$C_6H_5\underset{\underset{HgBr}{|}}{CH}COOC_2H_5 + ArN_2BF_4 \longrightarrow Ar\underset{\underset{NNHAr}{\|}}{C}COOC_2H_5 + HgBrBF_4$$

$$(Ar = p\text{-}NO_2C_6H_4 \text{ and } 2,4\text{-}(NO_2)_2C_6H_3)$$

Depending on the nature of the solvent the yield varies between 40–90%.

Bromomercurymethyl *tert*-butyl ketone reacts with $2,4\text{-}(NO_2)_2C_6H_3N_2BF_4$ in the presence of catalytic amounts of KCN producing *cis*- and *trans*-2,4-dinitrophenylhydrazone of *tert*-butylglyoxal and a small amount of 1,5-*bis*-(2,4-dinitrophenyl)-3-pivalylformazane:

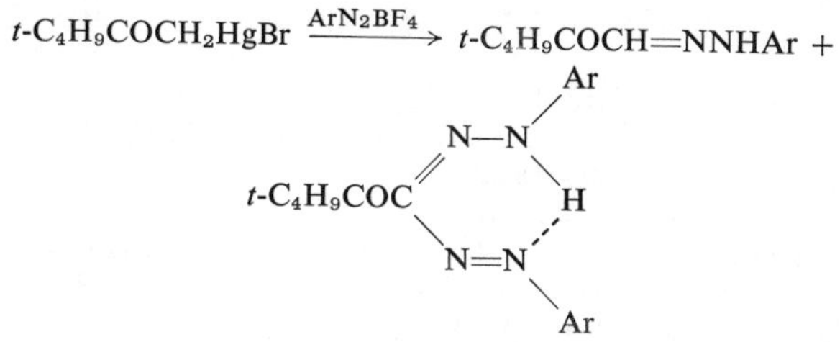

$$t\text{-}C_4H_9COCH_2HgBr \xrightarrow{ArN_2BF_4} t\text{-}C_4H_9COCH{=}NNHAr +$$

α-Benzyl-α-bromomercurycamphor produces an 80% yield of the respective azo compound (conditions are not specified).[8]

The reaction of diethylmercury and C_2H_5HgCl with ArN_2BF_4 is discussed in Ref. 9.

III. REACTIONS OF PHENYL(BROMODICHLOROMETHYL)MERCURY WITH CARBODIIMIDES

Phenyl(bromodichloromethyl)mercury reacts with carbodiimides according to the following equation:[10]

$$C_6H_5HgCCl_2Br + RN{=}C{=}NR \longrightarrow C_6H_5HgBr + \underset{(A)}{RN{=}CCl_2} + \underset{(B)}{RN{\equiv}C}$$

$$(R = i\text{-}C_3H_7,\ cyclo\text{-}C_6H_{11})$$

The reaction is conducted in chlorobenzene at 80°C.

As greater amounts of the organomercury compounds are used as compared with carbodiimide, product (A) is produced in greater amounts as compared with product (B). The reaction of mercurial with such strong nucleophiles is considered as bimolecular and the following scheme was proposed:

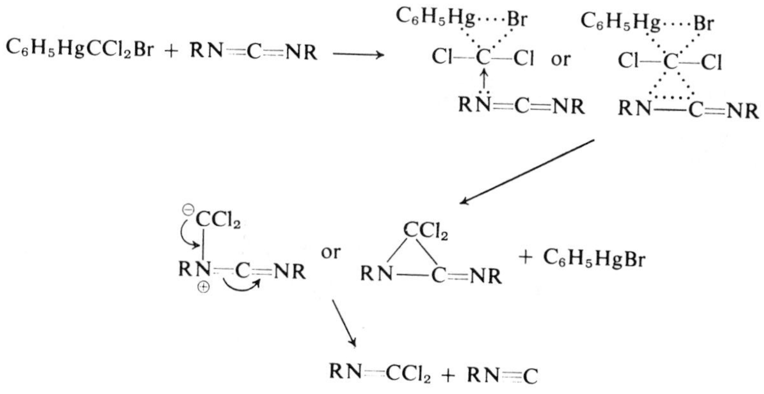

REFERENCES

1. A. Baeyer, *Ber.*, **7**, 1638 (1874).
2. P. Tarrant and D. E. O'Connor, *J. Org. Chem.*, **29**, 2012 (1964).
3. E. Bamberger, *Ber.*, **30**, 506 (1898).
4. I. Kunz, *Ber.*, **31**, 1528 (1898).
5. L. I. Smith and F. L. Taylor, *J. Am. Chem. Soc.*, **57**, 2460 (1935).
6. L. G. Makarova and A. N. Nesmeyanov, *Zh. Obshch. Khim.*, **9**, 771 (1939).
7. E. Bamberger, *Ber.*, **32**, 3546 (1899).
8. I. P. Beletskaya, K. P. Butin, and O. A. Reutov, *Izv. Akad. Nauk. Otd. Khim. Nauk*, **1964**, 1711.
9. P. F. Hu, *Acta Cimica Sinica*, **21**, 272 (1955).
10. D. Seyferth and R. Damrauer, *Tetrahedron Letters*, **1966**, 189.

Chapter 14

Reactions with Ketenes and Their Derivatives

I. INTRODUCTION

Depending on the nature of radical R and sometimes on the nature of radical X, organomercury compounds R_2Hg and $RHgX$ react with ketenes and their derivatives in different ways. R_2Hg and $RHgX$, according to a reaction which has not been explored to any great depth, react with ketenes and produce ketones. $ArHgOH$ is symmetrized by reaction with ketene diacetal.

Ketenes react differently with organomercury compounds where the mercury is affected by σ,σ- or σ,π-conjugation in the molecule: β-mercurated ethers constitute addition products of mercury salts to olefin double bonds (products of alkoxymercuration) obtained by use of Schoeller and Schrauth's method[1,2] which uses alcohol as the solvent:

$$CH_2{=}CH_2 + HgX_2 + AlkOH \longrightarrow AlkOCH_2CH_2HgX + HX$$

Schoeller and Schrauth's method was developed from Hofmann and Sand's methods[3-6] of addition of mercury salts to olefins in a water medium:

$$CH_2{=}CH_2 + HgX_2 + H_2O \longrightarrow HOCH_2CH_2HgX + HX$$

α-Mercurated oxocompounds, obtained in accordance with the method developed by Nesmeyanov and Lutsenko,[7] are produced by the addition of mercury salts to vinyl ethers:

$$CH_2{=}CROAlk + HgX_2 + H_2O \longrightarrow \left[XHgCH_2C\underset{\textstyle OH}{\overset{\textstyle OAlk}{R}} \right] \longrightarrow XHg{-}CH_2CRO$$

$$(R = H \text{ or } Alk)$$

Syntheses of these compounds are reviewed in Ref. 8.

Reactions of ketene with addition products of mercury salts to double bonds (β-mercurated ethers and α-mercurated oxocompounds) occur rather specifically. In these reactions the ketene, a compound of high-activity multiple bond, displaces ethylene and vinyl alkyl ether, compounds of low-activity multiple bonds from the organomercury compounds, thus producing derivatives of mercurated acetic acid. In a similar way, mercurybisacetalde-hyde, a mercurated oxocompound, reacts with the dimer of ketene giving ester of mercurybisacetoacetic acid.

II. FORMATION OF KETONES

A. Introduction and Application of the Reaction

Some organomercury compounds, R_2Hg or $RHgX$, $(R =$ phenyl and derivatives of furyl) react with ketene producing ketones, for example:[9-11]

$$RHgX + CH_2=C=O \longrightarrow R-\underset{\underset{O}{\|}}{C}-CH_3$$

B. Reaction Mechanism

As the authors believe, the reaction forming ketones from ketenes and organomercury compounds goes through preliminary addition of the organomercury compound to the carbonyl that is similar to the reaction of the Grignard reagent with ketenes:

$$RHgX + CH_2=C=O \longrightarrow \underset{\underset{OHgX}{|}}{CH_2=C-R} \xrightarrow{HX} \underset{\underset{OH}{|}}{CH_2=C-R} \longrightarrow \underset{\underset{O}{\|}}{CH_3CR}$$

C. Reaction Conditions

The reaction takes place under mild conditions and can be carried out in a number of solvents. With use of solvents such as acetone, acetone + hydroquinone, chloroform, diethyl sulfide, glycol dimethyl ether, and carbon tetrachloride, for the reaction producing furyl methyl ketone, the purest product was obtained in $CHCl_3$ or CCl_4. Solvents taken for the purpose should be absolutely dry. Ketones have not been produced in pyridine.

III. FORMATION OF UNSYMMETRICAL ORGANOMERCURY COMPOUNDS

Arylmercury hydroxide reacts with unsaturated compounds, as it has been shown by Lutsenko et al.,[12] and with diethylacetal of ketene producing unstable unsymmetrical mercury compounds,

$$ArHgOH + CH_2=C(OC_2H_5)_2 \longrightarrow ArHgCH_2C(OR)(OC_2H_5)_2$$

which are immediately decomposed into a mixture of two symmetrical compounds:

$$2\,ArHgCH_2C(OH)(OC_2H_5)_2 \longrightarrow Ar_2Hg + Hg(CH_2COOC_2H_5)_2 + 2C_2H_5OH$$

These reactions constitute a symmetrization method for aromatic mercury compounds (see Chapter 15).

IV. REACTIONS OF KETENES WITH β-MERCURATED ETHERS

A. Application and Conditions of the Reaction

β-Mercurated ethers react with ketene in a different way. During the reactions stated above, the unsaturated compounds of a high-activity multiple bond (ketene) displaces the unsaturated compound of a low-activity multiple bond (olefin) from the organomercury compound.[13] The reaction was conducted in various media: water, alcohols, benzene, and ethyl bromide. By the reactions of ketenes on water solutions of mercurated ethers, ethylene was separated and halomercury acetic acid was produced. In water the reaction occurs slowly; therefore, 10–12-fold excess amounts of ketene have to be passed through the reaction mixture.

The reaction in alcohol medium is accompanied as well by isolation of ethylene and reaction times are faster producing the esters of mercurated acetic acid. In this case the ester group is always formed only due to the presence of alcohol used as a solvent:

$$XHgCH_2CH_2OCH_3 + CH_2{=}C{=}O + ROH \longrightarrow$$
$$XHgCH_2COOR + CH_2{=}CH_2 + CH_3OH$$
$$(X = Cl, Br; R = H, CH_3, i\text{-}C_4H_9)$$

Exchange of alkoxy groups between original and terminal organomercury compounds and the solvent does not occur.

The reaction fails to occur in benzene even in the presence of a 25-fold excess of ketene. However, excess ketene in ethyl bromide reacts with bromomercuryethyl methyl ether and has produced an 82% yield of the methyl ester of bromomercury acetic acid. Thus, the reaction depends on polarity of the solvent indicating that the reaction requires additional polarization of the molecule of mercurated ether.

B. Reaction Mechanism

Two possible mechanisms of the reactions are described below:[14]

1. If the reactions with ketene in hydroxylic media and in ethyl bromide occur by different mechanisms, the following reaction scheme is plausible for water and alcohol medium:

$$CH_3OCH_2CH_2HgX + ROH \rightleftharpoons CH_3OCH_2CH_2HgOR + HX$$
$$CH_3OCH_2CH_2HgOR + CH_2{=}C{=}O \longrightarrow CH_3OCH_2CH_2HgCH_2COOR$$
$$HX + CH_3OCH_2CH_2HgCH_2COOR \longrightarrow XHgCH_2COOR + CH_2{=}CH_2 + CH_3OH$$
$$\text{where } R = H \text{ or Alk}; X = Cl, Br.$$

It is possible that the reaction in ethyl bromide goes through a six-centered intermediate complex:

$$BrHg\underset{CH_2=C=O}{\overset{CH_2-CH_2}{\diagdown}}OCH_3 \longrightarrow BrHgCH_2CO_2CH_3 + CH_2=CH_2$$

2. If the mechanism of reactions with ketenes is similar, it is possible that the reaction scheme in which the polarity of the solvent is important may be expressed as follows:

$$\begin{matrix}CH_2\\\|\\C\\\|\\O\end{matrix} + \underset{CH_2CH_2OCH_3}{\overset{X}{Hg}} \longrightarrow \begin{matrix}CH_2\\\|\\C\cdots Hg\\\|\\O\end{matrix}\overset{X}{\underset{CH_2CH_2OCH_3}{\diagdown}} \xrightarrow[\text{by solvent}]{\text{dissociation}} \left[\begin{matrix}CH_2\\\|\\C\cdots HgX\\\|\\O\end{matrix}\right]^+ + CH_2=CH_2 \\ + OCH_3^-$$

ROH *(in hydroxyl-containing medium)*

(in medium not containing hydroxyl) CH_3O^-

$XHgCH_2COOR + H^+$ $XHgCH_2COOCH_3$

V. REACTIONS WITH α-MERCURATED OXOCOMPOUNDS

A. Application and Conditions of the Reaction

α-Mercurated aldehydes and ketones having the more reactive conjugate bond system (σ-π), react with ketenes much more easily than mercurated ethers (having conjugate bond system σ-σ), thus the reaction occurs in hydroxyl-containing solvents (water, alcohol) and in benzene.[13]

During the reaction of ketene with α-mercurated oxocompounds in water or alcohol medium, the structure of the organomercury compounds obtained in almost quantitative yield does not depend on the organomercury compounds but rather is solvent determined: the reaction in water medium results in production of mercurated acetic acid, while the reaction in alcohols results in the production of its esters:

$$Hg(CH_2COR)_2 + 2 R'OH \xrightarrow{CH_2=C=O} Hg(CH_2COOR')_2$$
$$XHgCH_2COR + R'OH \xrightarrow{CH_2=C=O} XHgCH_2COOR'$$
$$R' = H, Alk; R = H, Alk; X = Cl, Br$$

At least four moles of ketene per mole of mercury bisaldehyde or ketone is required. At lower amounts of ketene, noncrystallizable oils are formed.

During the reaction of α-mercurated oxocompounds with ketenes in benzene, vinyl and substituted vinyl esters of mercurated acetic acid are formed in approximately quantitative yields:

$$Hg(CH_2COR)_2 + 2CH_2{=}C{=}O \longrightarrow Hg(CH_2C(O)OCR{=}CH_2)_2$$
$$ClHgCH_2COR + CH_2{=}C{=}O \longrightarrow ClHgCH_2C(O)OCR{=}CH_2$$
$$R = H, Alk$$

$ClHgC(CH_3)_2COCH_3$ and

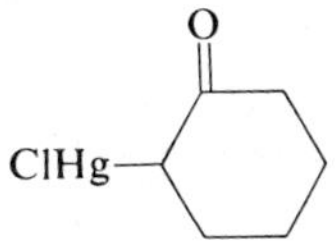

may also react.[13] In these cases, yields are also nearly quantitative.

B. Reaction Mechanism

It has been proposed that the intermediate product of the reaction of ketene with mercurybisacetaldehyde in ethanol is the acetal of mercurated acetaldehyde (acetals of mercurated aldehydes constitute, in most cases, oil-like products) whose formation can be explained by the following reaction scheme:[13] mercurybisacetaldehyde reversibly reacts with ethyl alcohol and produces semiacetal; ketene reacts with semiacetal and produces semiacetal-semiacylal of mercurated acetaldehyde and thus displaces the equilibrium of the first reaction to the right:

$$Hg(CH_2CHO)_2 + 2\,C_2H_5OH \rightleftharpoons Hg\left(CH_2{-}CH\begin{matrix}OH\\OC_2H_5\end{matrix}\right)_2 \xrightarrow{CH_2=C=O} Hg\left(CH_2{-}CH\begin{matrix}OCOCH_3\\OC_2H_5\end{matrix}\right)_2$$

It is known that similar compounds in excess alcohol can be easily transformed into the mercurated acetal:

$$Hg\left(CH_2{-}CH\begin{matrix}OCOCH_3\\OC_2H_5\end{matrix}\right)_2 \xrightarrow[\text{excess}]{C_2H_5OH} Hg\left(CH_2{-}CH\begin{matrix}OC_2H_5\\OC_2H_5\end{matrix}\right)_2$$

The unsaturated compound with highly active multiple bonds (ketene) then displaces the unsaturated compound with less active multiple bonds (vinyl ether) from the acetal of mercurybisacetaldehyde:

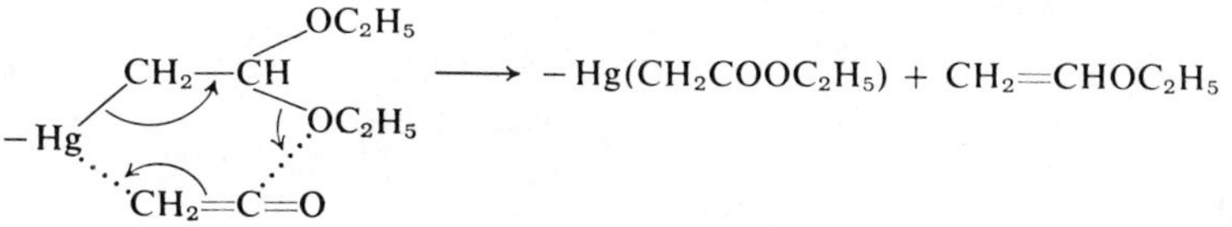

It is possible that the reaction of halogen salts of organomercury oxocompounds with ketenes occurs in a similar way. The reaction of ketene with α-mercurated oxocompounds in benzene probably goes through the intermediate formation of a six-centered complex:

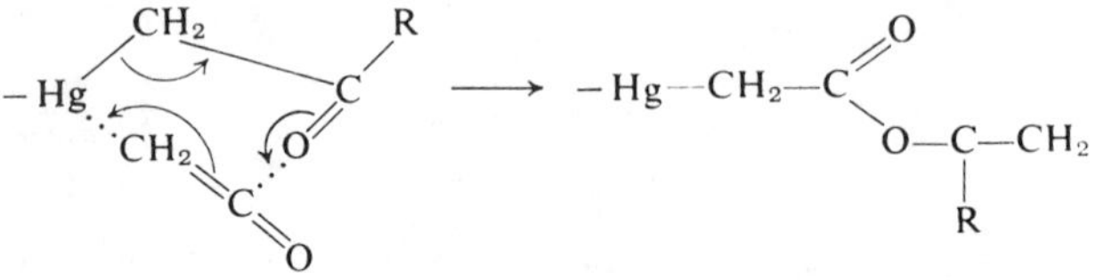

VI. REACTIONS OF α-MERCURATED CARBONYL COMPOUNDS WITH KETENE DIMER

A. Application and Conditions of the Reaction

Organomercury salts of aldehydes and ketones as well as mercurybisketones do not react with the ketene dimer. The reaction takes place with mercurybisacetaldehyde only because it is the compound of highest reactivity.[14]

During the reaction of mercurybisacetaldehyde with the ketene dimer in ethanol, the ethyl ester of mercurybisacetoacetic acid, vinyl ethyl ether, acetone and carbon dioxide are formed.

B. Reaction Mechanism

Formation of these compounds can be explained by the following series of successive reactions:[14]

$$Hg(CH_2CHO)_2 + 2\,C_2H_5OH \rightleftharpoons Hg\left(CH_2CH\begin{matrix}OC_2H_5\\OH\end{matrix}\right)_2$$

$$Hg\left(CH_2CH\begin{matrix}OC_2H_5\\OH\end{matrix}\right)_2 + 2\,CH_3COCH{=}C{=}O \longrightarrow Hg\left(CH_2CH\begin{matrix}OC_2H_5\\OCOCH_2COCH_3\end{matrix}\right)_2$$

$$Hg\left(CH_2CH\begin{matrix}OC_2H_5\\OCOCH_2COCH_3\end{matrix}\right)_2 + 2\,C_2H_5OH \longrightarrow Hg\left(CH_2CH\begin{matrix}OC_2H_5\\OC_2H_5\end{matrix}\right)_2 + CH_3COCH_2COOH \longrightarrow CH_3COCH_3 + CO_2$$

TABLE VIII

R in R_2Hg or $RHgX$	Reaction conditions	Products	Yield, %	Ref.
	Reactions with $CH_2=C=O$			
CH_2CHO	H_2O, 5–7°	$Hg(CH_2COOH)_2$	78	13
CH_2CHO	CH_3OH, 5–6°C	$Hg(CH_2COOCH_3)_2$	80	13
CH_2CHO	C_2H_5OH, cool	$Hg(CH_2COOC_2H_5)_2$	97	13
CH_2CHO	$i\text{-}C_4H_9OH$, cool	$Hg(CH_2COOC_4H_9\text{-}i)_2$	Quant.	13
CH_2CHO	$t\text{-}C_4H_9OH$, 5–7°C	$Hg(CH_2COOC_4H_9\text{-}t)_2$	Quant.	13
$ClHgCH_2CHO$	C_6H_6	$ClHgCH_2COOCH=CH_2$	Quant.	13
$ClHgC(CH_3)_2CHO$	C_6H_6	$ClHgCH_2COOCH=C(CH_3)_2$	85	13
CH_2COCH_3	C_6H_6, cool	$Hg(CH_2COOC(CH_3)=CH_2)_2$	90	13
$CH_2COC_2H_5$	CH_3OH	$Hg(CH_2COOCH_3)_2$	Quant.	13
$CH_2COC_2H_5$	C_6H_6, cool	$Hg(CH_2COOC(C_2H_5)=CH_2)_2$	83	13
$CH_2COC_3H_7$	CH_3OH, 5–7°C	$Hg(CH_2COOCH_3)_2$	80	13
$CH_2COC_3H_7$	C_6H_6, cool	$Hg(CH_2COOC(C_3H_7)=CH_2)_2$	95	13
$CH_2COC_4H_9$	C_6H_6, cool	$Hg(CH_2COOC(C_4H_9)=CH_2)_2$	Quant.	13
$HC{<}^{COCH_3}_{COOCH_3}$	CH_3OH, 0–5°C, 30 min	$Hg(CH_2COOCH_3)_2$	79	14

(continued)

359

TABLE VIII (*Continued*)

R in R₂Hg or RHgX	Reaction conditions	Products	Yield, %	Ref.
$HC\begin{smallmatrix}COCH_3\\COOCH_3\end{smallmatrix}$	C_6H_6, 0–5°C, 1 hr	$Hg(CH_2{-}COOC{=}CHCOOCH_3)_2$ with CH_3	Quant.	14
$ClHgCH_2COCH_3$	H_2O, heat	$ClHgCH_2COOH$	Quant.	13
$ClHgCH_2COCH_3$	CH_3OH	$ClHgCH_2COOCH_3$	Quant.	13
$ClHgCH_2COCH_3$	$t\text{-}C_4H_9OH$, cool	$ClHgCH_2COOC_4H_9\text{-}t$	Quant.	13
$ClHgCH_2COCH_3$	C_6H_6, cool	$ClHgCH_2COOC{=}CH_2$ with CH_3	Quant.	13
$BrHgCH_2COCH_3$	H_2O, heat	$BrHgCH_2COOH$	75	13
$BrHgCH_2COCH_3$	CH_3OH	$BrHgCH_2COOCH_3$	Quant.	13
$BrHgCH_2COCH_3$	$t\text{-}C_2H_5OH$, cool	$BrHgCH_2COOC_2H_5$	Quant.	13
$ClHgC(CH_3)_2COCH_3$	C_6H_6, cool	$ClHgCH_2COOC{=}C(CH_3)_2$ with CH_3	95	13
cyclohexanone with $ClHg$	C_6H_6, cool	$ClHgCH_2COO{-}$cyclohexenyl	Quant.	13
$ClHgCH_2CH_2OCH_3$	Hot H_2O	$ClHgCH_2COOH$	90	13
$ClHgCH_2CH_2OCH_3$	CH_3OH	$ClHgCH_2COOCH_3$	90	13
$ClHgCH_2CH_2OCH_3$	$i\text{-}C_4H_9OH$	$ClHgCH_2COOC_4H_9\text{-}i$	Quant.	13
$BrHgCH_2CH_2OCH_3$	Hot H_2O	$BrHgCH_2COOH$	93	13
$BrHgCH_2CH_2OCH_3$	$t\text{-}CH_3OH$,	$BrHgCH_2COOCH_3$	Quant.	13
$BrHgCH_2CH_2OCH_3$	C_2H_5Br	$BrHgCH_2COOCH_3$	82	13
$CH_3COOHgCH_2CH(OC_2H_5)OCOCH_3$	CH_3OH	$Hg(CH_2COOCH_3)_2$	97	13
C_6H_5HgCl	—	$CH_3COC_6H_5$	20–50	9

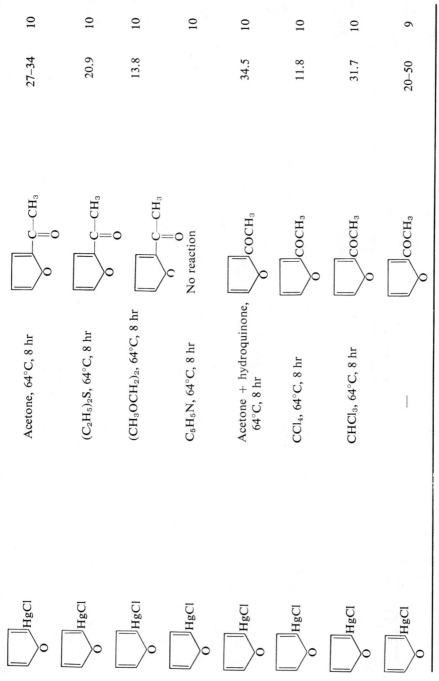

Reactant	Conditions	Product	Yield (%)	Ref.
2-furyl-HgCl	Acetone, 64°C, 8 hr		27–34	10
2-furyl-HgCl	(C₂H₅)₂S, 64°C, 8 hr		20.9	10
2-furyl-HgCl	(CH₃OCH₂)₂, 64°C, 8 hr		13.8	10
2-furyl-HgCl	C₅H₅N, 64°C, 8 hr	No reaction		10
2-furyl-HgCl	Acetone + hydroquinone, 64°C, 8 hr		34.5	10
2-furyl-HgCl	CCl₄, 64°C, 8 hr		11.8	10
2-furyl-HgCl	CHCl₃, 64°C, 8 hr		31.7	10
2-furyl-HgCl	—		20–50	9

(continued)

TABLE VIII (Continued)

R in R_2Hg or RHgX	Reaction conditions	Products	Yield %	Ref.
(furan ring)—HgCl	—	(furan ring)—$COCH_3$	20–50	9
ClHg—(furan ring with CH_2OCOCH_3)	$CHCl_3$, 64°C, 12.5 hr	CH_3CO—(furan ring)—CH_2OCOCH_3	14	10
ClHg—(furan ring with HgCl)	—	CH_3CO—(furan ring)—$COCH_3$	20–50	9
Reaction with $CH_2{=}C(OC_2H_5)_2$				
C_6H_5HgOH	20°C, overnight	$(C_6H_5)_2Hg$	8	12
		$ClHgCH_2COOC_2H_5$	88	
Reactions with $CH_3COCH{=}C{=}O$				
CH_2CHO	CH_3OH, 60°C	$Hg\left(CH{<}^{COCH_3}_{COOCH_3} \right)_2$	82	14
CH_2CHO	C_2H_5OH, 20°C	$CH_2{=}CH{-}OC_2H_5$	65	14

$$Hg\left(CH_2CH \overset{OC_2H_5}{\underset{OC_2H_5}{}} \right)_2 + 2\ CH_3COCH{=}C{=}O \longrightarrow$$

$$Hg\left(CH \overset{COCH_3}{\underset{COOC_2H_5}{}} \right)_2 + 2\ C_2H_5OCH{=}CH_2$$

The intermediate formation of acetoacetic acid (identified in the decomposition products) in the medium of absolute alcohol is direct evidence of the reaction mechanism. The last stage is probably effected through intermediate formation of a reactive six-centered complex:

REFERENCES

1. W. Schoeller, W. Schrauth, and R. Struensee, *Ber.*, **43**, 695 (1910).
2. W. Schoeller, W. Schrauth, and R. Struensee, *Ber.*, **44**, 1048 (1911).
3. K. A. Hofmann and J. Sand, *Ber.*, **33**, 1340 (1900).
4. K. A. Hofmann and J. Sand, *Ber.*, **33**, 1353 (1906).
5. J. Sand, *Ber.*, **34**, 1385 (1901).
6. J. Sand and F. Singer, *Ber.*, **35**, 3170 (1902).
7. A. N. Nesmeyanov, I. F. Lutsenko, and N. I. Wereshchagina, *Izv. Akad. Nauk SSSR, Otd. Khim. Nauk*, **1947**, 63.
8. L. G. Makarova and A. N. Nesmeyanov, *Methods of Organoelements Chemistry*, Mercury, Ed. "Nauka", Moskau 1965, Chapt. 6.
9. H. Gilman, B. L. Wooley, and G. F. Wright, *J. Am. Chem. Soc.*, **55**, 2609 (1933).
10. W. J. Chute, W. M. Orchard, and G. F. Wright, *J. Org. Chem.*, **6**, 157 (1941).
11. H. Gilman and J. Nelson, *Rec. Trav. Chim.*, **55**, 518 (1936).
12. I. F. Lutsenko and E. I. Yurkova, *Izv. Akad. Nauk SSSR, Otd. Khim. Nauk*, **1956**, 27.
13. V. L. Foss, M. A. Zhadina, I. F. Lutsenko, and A. N. Nesmeyanov, *Zh. Obshch. Khim.*, **33**, 1927 (1963).
14. M. A. Kazankova, I. F. Lutsenko, and A. N. Nesmeyanov, *Zh. Obshch. Khim.*, **35**, 1447 (1965).

Chapter 15

Reactions of RHgX $\longrightarrow$ R₂Hg and Reverse Reactions

I. INTRODUCTION

The reaction of symmetrization is considered as one of the most important reactions of organomercury compounds, i.e., transformation of organomercury compounds into compounds of mercury which are totally organically substituted:

$$2\,RHgX \underset{b}{\overset{a}{\rightleftharpoons}} R_2Hg + HgX_2$$

This reaction is reversible. The reverse reaction of symmetrization, i.e., preparation of the organomercury compounds from completely substituted compounds of mercury and mercury salts, is of paramount preparative significance.

In order to force the equilibrium of the above reaction to the right, it is necessary to isolate the mercury salt produced by addition of "symmetrization agents" which either bind the mercury salt into a complex, reduce it to metallic mercury, or cause precipitation of the mercury in the form of an insoluble mercury salt. The direction b is thermodynamically favored and occurs readily unless other agents are added.

Significance of the symmetrization reaction for organomercury compounds comes from the fact that use of most synthetic compounds for the preparation of organomercury compounds results in the formation of RHgX.

The methods in question are as follows: Mercuration:[1-3]

$$RH + HgX_2 \longrightarrow RHgX + HX$$

is the most important and widely used synthetic method for aliphatic and, particularly, various aromatic and heterocyclic mercury compounds.

The Grignard reaction occurs readily and the first product is the organomercury salt 4,5:

$$RMgX + HgX_2 \longrightarrow RHgX + MgX_2$$

which then is additionally alkylated to form the completely substituted mercury compound:

$$RMgX + RHgX \longrightarrow R_2Hg + MgX_2$$

Usually the simplest alkyl compounds of mercury are produced by this method.

In the aromatic series, this method is restricted since it is unsuited for synthesis of compounds having active substituents. A method discovered by

Nesmeyanov[6,7] and referred to as the all-purpose diazo method for the synthesis of aromatic mercury salts is free of the above restrictions:

$$ArN_2X \cdot HgX_2 + 2\,Cu \longrightarrow ArHgX + N_2 + 2\,CuX$$

which produce aromatic mercury compounds with various substituents in an individual state free of isomers. The variation of the methods:

$$2\,ArN_2X \cdot HgX_2 + 6\,Cu + NH_3 \longrightarrow Ar_2Hg + 2\,N_2 + 6\,CuX \cdot NH_3$$

The original RHgX produced initially is further alkylated to R_2Hg in the presence of copper and ammonia.

The compounds RHgX are also produced as a result of most reactions displacing acid groups such as $COOH$,[8-10] BO_2H,[11-13] $-SO_2H$,[14] and $-IO_2$[15] in organic compounds by mercury or by the reaction of alkyl halides with metallic mercury:[16-18]

$$RHal + Hg \longrightarrow RHgHal$$

The widely known cases of addition of mercury salts to multiple bonds (well characterized by Hofmann, Sand,[19] Schoeller, Schrauth,[20] Jenkins,[21] Nesmeyanov,[22-26] Freidlina,[22-23] Borisov,[24,25] Lutsenko,[26] and others) may be presented by the following equations:

$$\overset{\diagdown}{\underset{\diagup}{C}}{=}\overset{\diagup}{\underset{\diagdown}{C}} + HgX_2 + ROH \longrightarrow RO-\overset{|}{\underset{|}{C}}-\overset{|}{\underset{|}{C}}-HgX$$

$$-C{\equiv}C- + HgX_2 \longrightarrow X-\overset{|}{C}{=}\overset{|}{C}-HgX$$

and produce specific compounds of the type RHgX as well.

The reaction of alkyl halides with sodium amalgam[27] produces a symmetrical compound of mercury

$$2\,RX + Hg \cdot nNa \longrightarrow R_2Hg + nNaX$$

However, symmetrizing agents such as sodium amalgam or metallic sodium are present which probably symmetrize the originally produced RHgX.

II. MECHANISM OF SYMMETRIZATION AND REVERSE REACTIONS

Depending on the nature of the symmetrizing agent, the symmetrization of the organomercury salt can occur in an electrophilic or homolytic way that has been proved kinetically. The mechanism of the equilibrium reaction of symmetrization defines the stereochemistry of its course. The reaction of desymmetrization is an electrophilic substitution reaction (S_E2 mechanism) and proceeds with retention of configuration.

The kinetics, mechanism, and stereochemistry of the symmetrization reaction have been relatively comprehensively studied for the examples of symmetrization of compounds with labile mercury: esters of α-bromomercuryarylacetic acids and other compounds by ammonia.

Reutov and others have proved that symmetrization reactions of esters of α-bromomercuryphenylacetic acid *l*-menthyl[28-30] and ethyl[30] diastereomers proceed by reaction with ammonia with retention of configuration and are considered to be bimolecular (they are second order; first order for each component). On this basis and on the basis of the investigation of other symmetrization reactions, Nesmeyanov and Reutov[28,29] stated a rule delineating conditions for retention of configuration during electrophilic substitution reactions and loss of radical configuration during reactions of radical substitution on a carbon-saturated atom. This rule was then experimentally proved simultaneously by Reutov and Ingold.

On the basis of the aforementioned kinetic and stereochemical data, Reutov[31,32] has proposed that these as well as other reactions, referred to as bimolecular electrophilic substitution reactions, proceed by the formation of a cyclic four-centered transition state, type (A):

$$\underset{/}{\overset{\backslash|}{C}}-Hg\overset{X}{\underset{Z}{\diagdown}} + Hg\overset{Y}{\underset{Z}{\diagup}} \longrightarrow \underset{/}{\overset{\backslash|}{C}}-Hg\underset{\cdot}{\overset{X}{\diagdown}}\ \overset{\cdot}{\underset{Z}{\diagup}}Hg-Y \longrightarrow \underset{/}{\overset{\backslash|}{C}}-Hg-Z + \overset{X}{\underset{|}{Hg}}-Y$$

(A)

(for symmetrization reactions X = Y = Hal; Z = Alk; for desymmetrization reactions X = Alk; Z = Y = Hal). This four-centered system was confirmed by the fact that reactions proceed in nonpolar and low-polarity media, i.e., under conditions when intermediate ion formation is hardly possible.

It is possible that the symmetrization reaction by ammonia goes through an intermediate formation of a complex of the organomercury salt, for example:

$$RHgX + NH_3 \underset{\longleftarrow}{\overset{fast}{\longrightarrow}} RHgX \cdot NH_3$$

$$2\,RHgX \cdot NH_3 \overset{slow}{\longrightarrow} R_2Hg + HgX_2(NH_3)_2$$

Intermediate formation of a complex is supported by preparation of complexes of organomercury salts with ammonia[33] and by an acceleration effect of added bases. The four-centered transition state with consideration of complex formation can be expressed as (A'), (A") or (A‴):[34]

(A')   (A")   (A‴)

That (A), (A'), (A") or (A‴) as a transition state will not affect the final conclusions relative to the mechanism of the reaction (S_E2 mechanism) and of the similarity of the symmetrization reaction to isotope exchange reactions and with other reactions of organomercury compounds which have a four-centered transition state are accepted as well.

However, the order assumed by substituents R and specified by their influence on the rate of symmetrization of ethyl aryl-α-bromomercury-acetates by ammonia

was found to be the reverse as compared with the usual influences exerted by substituents in S_E2 reactions. For example, the maximum rate constant of the second-order reaction is observed for R = p-NO$_2$, the minimum one for R = p-(CH$_3$)$_3$C, etc.[35] This is explained by the fact that for organomercury compounds with a labile C—Hg bond, the S_E2 reaction can be very close to an S_E1 reaction.

The presence of transition state A (or A', etc.) is confirmed by the fact that the symmetrization rate of bromomercury-α-carbethoxybenzylmercury is higher if various compounds are introduced into the symmetrization reaction; compounds **1** and **2** containing opposite substituent effects (electron-donor X and electron-acceptor Y) in the para position of the benzene ring:[32,36]

Moreover, use of radioactive mercury has proved that the mercury atom is transferred from the original molecule (**1**) chiefly into the complex HgBr$_2$·· 2NH$_3$, while the atom of mercury from molecule (**2**) is transferred into the final compound R$_2$Hg as would be if transition state (B) were inferred.

Formation of transition state (B) facilitates the symmetrization reaction, as can be seen from the fact that it is possible to introduce into this "co-symmetrization" reaction compounds containing substituents such as p-Br, as well as benzylmercury bromide which are not separately symmetrized by ammonia.[37]

Jensen and co-workers[38,39] have found that symmetrization of bromo-mercury carbethoxybenzyl esters by ammonia in chloroform depends on a fourth-order equation (second-order in RHgX and NH_3). The rate equation includes the equilibrium constant K for the formation of complex RHgBr with ammonia:

$$\text{rate} = Kk[\text{RHgBr}]^2[\text{NH}_3]^2$$

The symmetrization reaction proceeds according to the scheme:

$$\text{RHgBr} + 2\,\text{NH}_3 \underset{\text{fast}}{\overset{K}{\rightleftharpoons}} \text{RHg} \begin{matrix} \overset{+}{N}H_3 \\ \\ \underset{+}{N}H_3 \end{matrix} \quad 2\,^-\text{Br}$$

$$\text{RHgBr(NH}_3)_2 + \text{RHgBr} \xrightarrow{\underset{k}{\text{slow}}} R_2\text{Hg} + \text{HgBr}_2\cdot 2\,\text{NH}_3$$

The authors explain the irregular effect of the substituents in this reaction by an increase of the equilibrium constant of complex K caused by electron-accepting substituents.

The symmetrization of 3-bromomercury-l- and -d-camphor by sodium thiosulfate proceeds with retention of configuration; consequently this reaction involving this symmetrizing agent occurs in an electrophilic manner.[40]

Symmetrization reactions accompanied by inversion of configuration by sodium stannite as was illustrated by the examples of symmetrization of 2-methoxycyclohexylmercury iodide[41] and sec-butylmercury bromide,[42] by hydrazine-hydrate (symmetrization of 3-bromomercurycamphor,[43] some reaction products of addition to cyclic alkenes and their derivatives[44]) and symmetrization of sec-butylmercury bromide by electrolysis[45] are referred to as reactions proceeding homolytically.

Symmetrization of some diastereoisomers characterized by wide-range stability by sodium stannite or hydrazine can proceed with retention of configuration.[45a]

According to Jensen, symmetrization which results in reducing the mercury to metallic mercury by magnesium L-($-$)sec-butylmercury bromide and cis- and $trans$-4-methylcyclohexylmercury bromides are also stereospecific:[46]

$$2\,\text{RHgBr} + \text{Mg} \longrightarrow R_2\text{Hg} + \text{MgBr}_2 + \text{Hg}$$

The following scheme is suggested by Jensen for the symmetrization:

$$RHgBr + Mg \longrightarrow RHg^- + \overset{+}{Mg}Br$$
$$RHg^- + RHgBr \longrightarrow RHgHgR + Br^-$$
$$RHgHgR \longrightarrow RHgHg$$
$$\overset{|}{R}$$
$$RHgHg \longrightarrow RHgR + Hg$$
$$\overset{|}{R}$$

The symmetrization reaction which proceeds with diphenylmercury occurs in two stages:[47–49]

$a.$ $RHgX + (C_6H_5)_2Hg \longrightarrow C_6H_5HgX + C_6H_5HgR$
$b.$ $RHgC_6H_5 + RHgX \longrightarrow R_2Hg + C_6H_5HgX$

and was confirmed experimentally by use of radioactive mercury. Suggested for step a is a transition state, type (C):

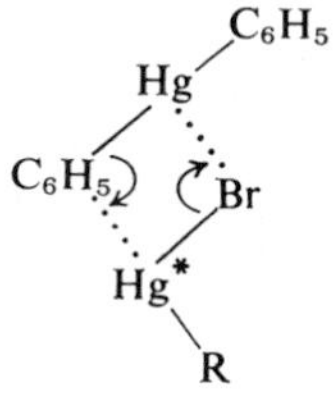

(C)

Due to the fact that for $R = ArCH(COOR')$ the effect on reaction rate exerted by substituents in the p-position of Ar in the attacked molecule ArHgC(H)COOR'Ar: $tert$-$C_4H_9 > H > Cl > I$ is opposite to the effect of substituents in the attacking molecule $ArCH(HgBr)COOR'$: $NO_2 > Hal > H > Alk$, the transition state for reaction b was described by Reutov as D (or D'):

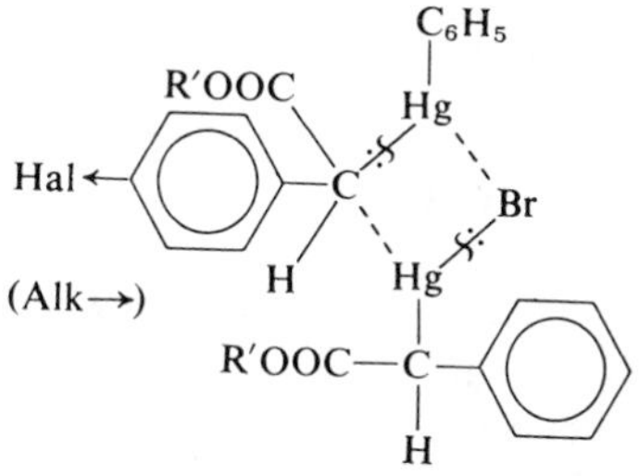

(D) 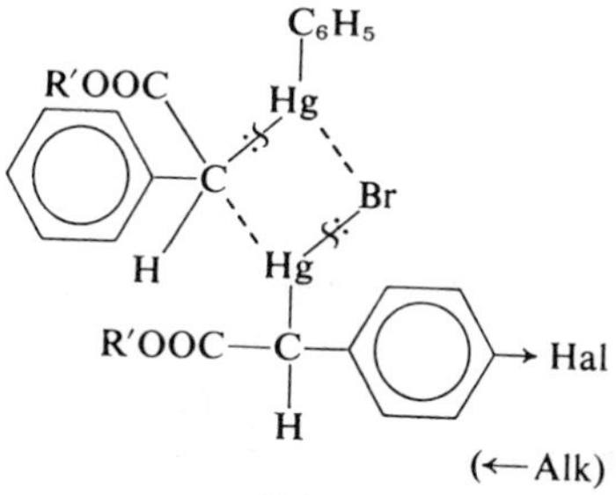   (D')

Reaction b is typical only for compounds with a fairly labile C—Hg bond. Thus, with respect to this reaction, the reactivity of RHgX is as follows

$$R = CH_3COCH_2 > ClCH{=}CH > C_6H_5\overset{|}{C}(H)\, CO_2C_2H_5 > CH_3OOC—$$

As stated above, the desymmetrization reaction proceeds with retention of configuration[42,50-55] and is considered to be second order (first order for each component, S_E2 mechanism). The reaction mechanism may be represented as a four-centered transition state, type A (see above) similar to the symmetrization reaction.[51-53,56-59]

The S_E2 mechanism of the desymmetrization reaction

$$R_2Hg + HgX_2 \longrightarrow 2\ RHgX$$

was confirmed by the fact that the reaction occurs more easily if the HgX bond in HgX_2 is more ionic. Dessy, therefore, proposed that ion pairs are involved in the transition state:[55]

$$\overset{\displaystyle R\text{---}Hg\text{---}R}{RHgR + XHg^\oplus X^\ominus \longrightarrow XHg^\oplus X^\ominus \longrightarrow 2\ RHgX}$$

During desymmetrization of ArHgAlk, as was shown[60,61] by Nic. Nesmeyanov and Reutov by use of radioactive mercury, contrary to the data presented by Brodersen[62] and Dessy,[63] the aryl group is labile:

$$ArHgAlk + Hg^*Hal_2 \longrightarrow ArHg^*Hal + AlkHgHal$$

even in the case when Ar = mesityl in which steric hindrance is a factor.

Statistical[64] distribution of radioactive mercury in R'HgX and R"HgX produced during the interaction of cis-2-methoxycyclohexylneophylmercury with $^{203}HgCl_2$:

$$R'HgR'' + Hg^*Cl_2 \longrightarrow R'Hg^*Cl + R''HgCl$$

may be, evidently, explained by the reaction mechanism found later[65] (see Chapter 1)

$$R_2Hg + R_2'Hg^* \rightleftharpoons R_2Hg^* + R_2'Hg$$

III. SYMMETRIZATION AGENTS

In many cases each symmetrization agent has its own use, some of them are widely applicable, while others are very specific. It is possible that many symmetrization agents now in use for symmetrizing various classes of organo-mercury compounds may be used on a wider scale and may be found to be suitable for the symmetrization of other groups of organomercury compounds. For example, ammonia was used specifically as a symmetrization agent for β-chlorovinylmercury compounds and then widely used for symmetrization of other quasicomplex, α-mercurated oxo- and carboxy-compounds and compounds of other classes.

A. Reducing Symmetrizators

1. Sodium Stannite

$$2 RHgX + Na_2SnO_2 + H_2O \longrightarrow R_2Hg + Hg + 2 NaX + H_2SnO_3$$

An aqueous alkali solution of sodium stannite is one of the most convenient and all-purpose symmetrizators. Symmetrization with this solution is usually carried out under cold conditions with organomercury salts during stirring from one to several hours. If the organomercury salt fails to be dissolved in water it can be dissolved in organic solvents such as alcohol, acetone, or pyridine.

The alkali solution of sodium stannite is used for the symmetrization of almost all types of organomercury compounds including those having mercury on a carbon at a double bond,[41,66,67] of γ-alkanolmercury salts,[68,69] other addition products of mercury salts at a double bond[70] (see also Section IV-A-2), but not quasicomplex compounds, which are completely decomposed into the unsaturated compound and mercury. The alkali solution of sodium stannite is used in a larger scale for the symmetrization of various groups of aromatic organomercury compounds,[71-73] including those containing a rather strong C—Hg bond. It also symmetrizes organomercury compounds of aliphatic,[42] alicyclic,[74] aralkyl,[75] heterocyclic compounds[76,77] and mercury derivatives of metallocene compounds.[78] As a rule, yields of symmetrical organomercury compounds are good.

The alkali solution of sodium stannite is unfit for symmetrization of quasicomplex compounds of mercury: β-alkanolmercury salts, β-chlorovinyl compounds of mercury, and α-mercurated oxocompounds were decomposed.

Naturally, it is advisable to avoid use of, or to use carefully, the alkali solution of sodium stannite for symmetrization of substances containing carbalkoxyl and similar groups. In some particular cases, use of sodium stannite results not in symmetrization of the organomercury compound but in reducing the latter. The compound, $CF_3CClFHgF$, is not symmetrized by sodium stannite but mercury is replaced by hydrogen.[79] See also[80,81].

When stereoisomeric organomercury compounds are symmetrized by sodium stannite or by other reducing symmetrizators, inversion of configuration (R = 2-CH_3O-cyclo-$C_6H_{10}HgI$)[41] or racemization (R = sec-C_4H_9)[42] was observed.

2. Metals and Metal–Base Alloys

a. Copper. Copper is an excellent symmetrizator having a wide range of application, particularly in the presence of organic (pyridine and its homologous compounds, (see F. Hein[82,83]) or inorganic (ammonia[7]) bases:

$$2 RHgX + 2 Cu + B \longrightarrow R_2Hg + 2 CuXB + Hg$$

Symmetrization by copper in the absence of a base (ammonia or pyridine) is recommended for those cases when the compound is reactive toward the base. Ethyl *p*-chloromercurybenzoates is saponified by ammonia and therefore produces better yields if the symmetrization is conducted without ammonia.[84] Use of copper in the absence of ammonia requires rather long heating in an organic solvent.

During symmetrization by copper in the presence of pyridine, the latter is used as the reaction medium; during symmetrization by copper in the presence of ammonia (usually aqueous solution), acetone[7] mixed with water is usually used as the solvent.

In the presence of a base, copper was used for symmetrization of mercury derivatives of fatty hydrocarbons,[82] chloromercuryacetic acid,[85] derivatives of alicyclic compounds,[86] and various types of aromatic compounds of mercury.[7,82,83,87-94]

b. Sodium and Its Alloys

1. Sodium.

$$2\,RHgX + 2\,Na \longrightarrow R_2Hg + 2\,NaX + Hg$$

Use of metallic sodium as a symmetrization agent is not a very widespread method. The metallic sodium is used in some cases. Only metallic sodium can symmetrize β-thienylmercury chloride.[95] In some cases it was used for symmetrizing a thiophene mercurated in the α-position[96] which can be readily symmetrized by other more universal symmetrization agents. Ferrocenylmercury chloride was symmetrized by sodium.[78] Symmetrization by sodium is usually carried out in an unreactive solvent (fatty or aromatic hydrocarbon).

2. Sodium Amalgam.

$$2\,RHgX + Na_2Hg \longrightarrow R_2Hg + 2\,NaX + 2\,Hg$$

In comparison with metallic sodium, sodium amalgam is more widely used as a symmetrizing agent. It is used mainly for symmetrization of mercury derivatives of hydrocarbons of both fatty[97,98] and aromatic series.[99] Adams and co-workers[100] were the first to use sodium amalgam to symmetrize the mercury double bond addition product (2-chloromercurymethyl-2,3-dihydrobenzofuran—the product of oxymercuration of *o*-allylphenol) which was one of the principle foundations for proving the true organometallic structure of these compounds. However, other products of oxy(alkoxy) mercuration of alkenes by sodium amalgam are not symmetrized (see below). Symmetrization reactions by sodium amalgam are conducted in alcohol (absolute) or in an organic solvent, sometimes in water but mostly under boiling conditions.

Sodium amalgam, metallic sodium, and other sodium alloys are naturally not fit for symmetrization of mercury compounds containing an organic radical if the substituents are reactive to sodium.

In some cases the effect of sodium amalgam on organomercury compounds results not in symmetrization but in replacement of their mercury by hydrogen. This reduction is particularly typical of the addition products of mercury salts to a double bond; β-mercurated alcohols,[101] their ethers,[102] or addition products of mercury salts in a secondary amine medium[103] ("quasicomplex" compounds of mercury), for example:

$$2\ HOCH_2CH_2HgX \xrightarrow[ROH]{Na(Hg)} 2\ HOCH_2CH_3 + 2\ Hg + 2\ NaX$$

$$C_5H_{11}NCH_2CH_2HgX \xrightarrow[ROH]{Na(Hg)} C_5H_{11}NCH_2CH_3 + 2\ Hg + NaX$$

Organomercury salts of other types, under the action of sodium amalgam, particularly in water, can be subjected not to symmetrization but to protodemercuration.

3. Tin-Sodium Alloy.

$$2\ RHgX + Na_2Sn \longrightarrow R_2Hg + 2\ NaX + HgSn$$

Tin–sodium alloy (15% sodium) was used for the symmetrization of isomeric chloromercurystilbenes[104] in aromatic hydrocarbons.

c. Cadmium, Copper, and Silver Amalgams.

$$2\ C_nF_{(2n+1)}HgI + Cd(Hg) \longrightarrow (C_nF_{2n+1})_2Hg + Hg + CdI_2$$

Amalgams of silver, copper, and, better, cadmium are the sole symmetrization agents of lower perfluoroalkylmercury compounds.[105] Symmetrization is conducted without solvent during heating in a sealed tube (due to volatility of R_f Hg compounds). The amalgam of cadmium and silver produces an 80–90% yield, while the amalgam of copper, without solvent, produces a 20% yield. The yield obtained in acetone is 50–60%.

d. Zinc.

$$2\ RHgX + Zn \longrightarrow R_2Hg + ZnX_2 + Hg$$

Zinc may be used as a symmetrization agent. Its pure symmetrizing effect was revealed in only one case: ethyl p-chloromercurybenzoate was symmetrized producing a good yield during prolonged boiling with zinc dust in toluene.[84] In some other cases where it was used, symmetrization was a side reaction. Bis-γ-acetopropylmercury was produced during ketone decomposition of ethyl α-[(2-chloromercuryethyl]acetoacetate by zinc dust in alkali solution.[106] Intermediate formation of symmetrical organomercury compounds proceeds when γ-mercurated secondary and tertiary alcohols are heated with zinc dust.[107]

e. Magnesium.

$$2 \, RHgBr + Mg \longrightarrow R_2Hg + MgBr_2 + Hg$$

The use of magnesium as a symmetrization agent has already been described for three cases. See Section II. Reaction conditions were not specified.

3. Hydrazine, Its Derivatives, Hydroxylamine

$$4 \, RHgX + NH_2NH_2 \longrightarrow 2 \, R_2Hg + 4 \, HX + N_2 + Hg$$

Hydrazine is widely used as a symmetrization agent in reactions. It was used in some cases for symmetrization of mercury derivatives of aliphatic hydrocarbons,[108] a wide range of mercury derivatives of aromatic compounds,[108-110] and some superaromatic heterocyclic compounds.[109,111] Hydrazine is widely used for symmetrizing the addition products of mercury salts to cyclic alkenes,[112-114] (including the products of the latter's ammoniomercuration)[115] and to other unsaturated compounds (but not to other quasicomplex compounds). Symmetrization of stereoisomeric organomercury compounds by hydrazine results mostly in inversion of configuration of the radical bound with mercury. In some cases symmetrization by hydrazine proceeds without inversion of configuration. The effect of hydrazine hydrate on the products of oxy-(alkoxy)-mercuration of cyclic unsaturated compounds may result not only in their symmetrization but in replacement of mercury by hydrogen,[116] sometimes in the breaking of RO group (but not OH group)[117] and in regeneration of an unsaturated compound. Symmetrization by hydrazine is conducted in alkali-alcohol aqueous or dioxane solution, usually without heating.

The derivatives of hydrazine, phenylhydrazine[118] and semicarbazide,[119] as well as hydroxylamine can act as symmetrizing agents for mercury derivatives of aromatic hydrocarbons.

4. Electrolytic Symmetrization

Electrolysis of various salts of AlkHgX—derivatives of saturated aliphatic hydrocarbons which may be presented by the scheme:

$$2 \, RHgX \rightleftarrows 2 \, RHg\cdot + 2 \, X\cdot \longrightarrow R_2Hg + Hg + X_2$$

is one of the best methods of symmetrization of organomercury compounds of this group.[120-122] The reaction proceeds in aqueous solution. Chiefly introduced into the reaction are readily soluble mercury salts, such as sulfate or succinate. For the purpose of increasing the salts' solubility with other anions and increasing electrical conductivity of solutions, pyridine is added which creates better conditions for electrolysis of alkylmercury halides. Good yields of Alk_2Hg are obtained during symmetrization by electrolysis.

If electrolysis of alkylmercury salts is conducted in liquid ammonia or, as in the case of methylmercury halides, in alcohol or water, unstable compounds of the type RHg which undergo transformations at room temperature are produced:[123]

$$2\,RHg \longrightarrow R_2Hg + Hg$$

Electrolysis is not considered as a preparative method for the symmetrization of the respective compounds.

It is probable that RHgX symmetrization reactions by electrolysis go through an intermediate RHg radical which then undergoes disproportionation.

Electrolysis of alkylmercury salts is conducted in aqueous solutions at temperatures of 20–80°C and with a current density of 0.01–0.05 A/cm². Under these conditions, the temperature does not affect yield; however, as temperature increases a certain reduction of the final product yield[121] is expected to occur. When attempts were made to electrolyze $ClCH{=}CHHgCl$ in liquid ammonia with the purpose of producing RHg, symmetrization of the compound occurred.[122]

Symmetrization of organomercury compounds of other classes by electrolysis has not been performed.

5. Alcoholates of Alkali Metals

Symmetrization by sodium alcoholate

$$4\,RHgX + R'CH_2ONa + 4\,NaOH \longrightarrow$$
$$2\,R_2Hg + R'COONa + 4\,NaX + 2\,Hg + 3\,H_2O$$

is not convenient and thus not widely used. It was used only in four cases; however, the results were different. Thus, for example, an 80% yield of *cis-o*-nitrophenylmercury[124] was obtained, while attempts to produce di-β-chloro-naphthyl, di-phenyl, -*di-p*-tolyl-mercury gave poorer results.[125]

6. Ferrous Oxide Hydrate and Its Salts

$$2\,RHg^+ + 2\,Fe^{2+} \longrightarrow R_2Hg + Hg + 2\,Fe^{3+}$$

Ferrous oxide hydrate and its salts can be used as symmetrization agents. Symmetrization of mercurated nitrobenzoic acids by ferrous oxide hydrate is accompanied by reduction of the nitro group.[126]

7. Sodium Hydrosulfite

Symmetrization by sodium hydrosulfite was used only for the salt of nitrophenylmercury.[124]

8. Salts of Divalent Chromium

$$2\,RHg^+ + 2\,Cr^{2+} \longrightarrow R_2Hg + Hg + 2\,Cr^{3+}$$
$$R = \text{benzyl, cyclohexyl, phenyl, 4-phenylbutyl; } X = \text{perchlorate}$$

$CrSO_4$ and $Cr(OH)_2$ reduce RHgX, symmetrizing it to R_2Hg. The reaction proceeds in dimethylformamide. A right order of addition is needed. Yields of symmetrical compounds are close to quantitative. For X = Hal, the reaction is heterogeneous and yields are reduced to a certain extent.[127]

B. Symmetrizators Entraining Hg^{2+} in the Form of Un-Ionized Compounds

1. Alkali Metal Iodides

Alkali metal iodides, KI and NaI, are widely used as symmetrization agents and were thoroughly studied by Whitmore.[128] However, they have the following limitations: the symmetrization reaction by potassium (sodium) iodide is reversible:

$$2\,RHgX + 2\,KI \rightleftharpoons R_2Hg + K_2HgX_2I_2$$

Therefore, in order to force the reaction to completion, it is necessary to use an excessive amount (up to 10 times) of K(Na)I (for exceptions see below). In addition, in those cases in which the symmetrical product is more soluble in the reaction medium than the original organomercury salt, for example, in the case with R = C_6H_5, the reaction in alcohol medium proceeds from right to left even when excess NaI is involved.

The effect of K(Na)I (which usually occurs in a medium containing hydroxyl) on some organomercury compounds results in replacement of mercury by hydrogen. The organomercury derivatives of malonic ester[129] and malonic[130] and cyanoacetic acids are decomposed by potassium iodide according to the equation (see also Chapter 7, Section II-f):

$$RHgX + KI + H_2O \longrightarrow RH + HgIX + KOH$$

Dimercurated derivatives of ethyl acetoacetate, $CH_3COC(HgOCOR)_2$-$COOC_2H_5$, are decomposed in a similar manner; their stability to the solution reduces in the series: R = H > CH_3 > C_2H_5 > C_3H_7 > CH_2Cl > CCl_3.[131]

Mercury derivatives of naphthylamines[132] and, evidently, other aromatic amines are decomposed by potassium iodide.

Quasicomplex compounds could not be symmetrized by potassium (sodium) iodide; addition products of mercury salts to olefins[133] and acetylenes[134] eliminate the unsaturated compound, for example:

$$ROCH_2CH_2HgCl + 4\,KI + H_2O \longrightarrow C_2H_4 + K_2HgI_4 + ROH + KOH + KCl$$
$$ClCH{=}CHHgCl + 4\,KI \longrightarrow CH{\equiv}CH + K_2HgI_4 + 2\,KCl$$

α-Mercurated aldehydes and ketones are decomposed by potassium iodide in protonic solvents, for example:[135]

$$ClHgCH_2CHO + 4\,KI + H_2O \longrightarrow K_2HgI_4 + CH_3CHO + KCl + KOH$$

Mercurated phenols, not symmetrized by KI, have unprotected hydroxyl groups and are decomposed according to the following equation:[136]

$$HOC_6H_4HgX + 3\ KI + H_2O \longrightarrow C_6H_5OH + K_2HgI_3X + KOH$$

However, the alkali metal iodides do not affect the mercurated nitrophenols, i.e., they neither decompose nor symmetrize them.[137]

Taking these limitations into account, potassium (sodium) iodide is considered a symmetrization agent having a wide use. It is an excellent symmetrizator for perhaloaliphatic mercury compounds.[138]

Alkenyl compounds of mercury, having mercury at the double bond and which are not considered to be quasicomplexes, are symmetrized by potassium iodide.[104] Potassium iodide is widely used as a symmetrizator for aromatic mercury compounds,[125,128,136–142] mercurated ethers of phenols,[178] acylated phenols,[178] substituted aromatic amines,[142] heterocyclics, and ferrocene derivatives.[143]

Symmetrization by alkali metal iodides is carried out in aqueous alcohol, acetone, and other solutions, chiefly at cold temperatures but occasionally with heating. Heating may be rather prolonged if nonlabile mercury compounds are involved in the reaction.

2. Potassium (Sodium) Cyanide

$$2\ RHgX + 4\ KCN \longrightarrow R_2Hg + 2\ KX + K_2Hg(CN)_4$$

Potassium cyanide was used for the first time for symmetrization of lower aliphatic mercury compounds. It is considered the best agent of symmetrization for allylmercury iodide.[144] Aromatic compounds of mercury are symmetrized by potassium cyanide in good yield, though it was seldom used in this manner.[145] Like potassium iodide and other complexing agents it is unsuited for symmetrization of quasicomplex mercury compounds, α-mercurated oxocompounds, and, probably, phenols having unprotected hydroxyl group.

Hydrogen usually replaces mercury in the reaction of potassium cyanide with polymercury products produced during mercuration of acetaldehyde and acetic acid in aqueous solution[146,147] (see also Chapter 7).

3. Potassium (Sodium) Thiocyanate

$$2\ RHgX + 4\ KCNS \longrightarrow R_2Hg + 2\ KX + K_2Hg(CNS)_4$$

Potassium thiocyanate may be used as a symmetrization agent in the same areas as potassium cyanide. It is considered to be the best symmetrization agent for perhalovinylmercury salts which require only one equivalent of KCNS for symmetrization.[138]

It is not suited for symmetrization of quasicomplex compounds. It is probable that in the same cases in which KCN and possibly KI react irregularly, for example, replace HgX by H, KCNS reacts in a similar manner but its reactions have not been studied in great detail.

4. Sodium Thiosulfate

$$2\ RHgX + 2Na_2S_2O_3 \longrightarrow R_2Hg + 2\ NaX + Na_2[Hg(S_2O_3)_2]$$

Sodium thiosulfate, in aqueous solution, is very often used, particularly for symmetrization of such organomercury compounds where the atom of mercury is labile, as in mercurated amines[71,148-151] and phenols.[71,89] However, it is widely used as well for symmetrization of organomercury compounds of other groups: mercury derivatives of terpenes,[40,43] some heterocyclic compounds,[94,152-154] and metallocene compounds.[143,155-157]

During symmetrization by sodium thiosulfate, use was made of a method[43] to prevent interaction of the R_2Hg produced with the mercury salt. The reaction proceeds in a heterogeneous medium, benzene–water; R_2Hg is readily soluble in benzene, and is thus removed from the reaction medium.

Like other symmetrizators (complexing agents), sodium thiosulfate does not symmetrize quasicomplex compounds and α-halomercuryoxo- and -carboxy compounds, but decomposes them, releasing the unsaturated compound in the first case and aldehyde or ketone in the second case (in this case the reaction of protodemercuration occurs; refer to Chapter 9).

Symmetrization of mercury derivatives of stereoisomeric terpenes by sodium thiosulfate occurs with retention of configuration.[40]

5. Ammonia

$$2\ RHgHal + 2\ NH_3 \longrightarrow R_2Hg + HgHal_2 \cdot (NH_3)_2$$

One of the mildest and, probably, one of the most widely used symmetrization agents is ammonia in an unreactive solvent such as chloroform, benzene, or carbon tetrachloride.

Saturated aliphatic organomercury salts, AlkHgX, and aromatic organomercury compounds, ArHgX (probably with various Ar) can be symmetrized by ammonia. Ammonia is the only agent which symmetrizes alkenyl quasicomplex compounds: addition products of mercury salts to triple bonds, β-chlorovinyl compounds of mercury,[21,24] and α-halomercuryoxocompounds[157a] (but not halomercuryacetaldehyde, which has not been symmetrized by any means, even by ammonia) which are decomposed by most other symmetrization agents in aqueous solutions as well as in alcohols. It has been used as well for symmetrization of esters of α-halomercuryarylacetic acids.

Symmetrization by ammonia has been thoroughly investigated from the point of view of reaction mechanism and stereochemistry. Retention of configuration is observed in the reaction with ammonia.

During symmetrization of halomercuryketones, in order to avoid the harmful effect of excess ammonia which reduced the yield of mercurybisketone, the solution of ammonia in dichloroethane or chloroform is added to the solution of halomercuryketone until formation of the precipitate stops.[157a] The reaction is conducted at room temperature since an increase in temperature reduces the yield.

Symmetrization by ammonia should be used in those cases when a mild symmetrization agent is required. Esters of chloromercuryacetic acid do not react with ammonia, but do form complex compounds. Extremely reactive compounds such as addition products of acetylenic alcohols,[158] ketones,[159] and acids[160] under the effect of ammonia as well as under the effect of other symmetrization agents are not symmetrized but undergo decomposition. Products of the oxymercuration of acyclic alkenes are not symmetrized by ammonia. Benzylmercury halide is not symmetrized, but in the presence of other RHgX compounds halides of carbalkoxybenzylmercury undergoes a "co-symmetrization" reaction (preparation of RHgR')[37] with ammonia.

6. Pyridine

$$2 \, RHgX + C_5H_5N \longrightarrow R_2Hg + C_5H_5N \cdot HgX_2$$

Only one case was described when pyridine was used as a symmetrization agent for mercury derivatives of indandione.[161]

7. Tertiary Phosphines

Tertiary phosphines react in a different way with organomercury salts and one of the possible paths of the reaction is through the formation of a symmetrical compound. The effect of tertiary phosphines, R_3P, on $R'HgX$ depends not only on the nature of R in R_3P and R' and X in $R'HgX$ but also on the reaction medium. Thus $R'HgX$ with the same R_3P placed in various solvents can produce a phosphonium salt, $(R'Hg \cdot R_3P)Hal$[162] or complex $R_3P \cdot HgCl_2$,[162] or a symmetrical compound, apart from the complex, according to the following equation:[138]

$$2 \, R'HgX + R_3P \longrightarrow R_2'Hg + R_3P \cdot HgX_2$$

$ClCH{=}CHHgCl$ reacts with Ph_3P as free $HgCl_2$ giving $(C_6H_5)_3P \cdot HgCl_2$.[163]

The reaction forming R_2Hg probably proceeds through intermediate formation of the complex salt, $(R'HgR_3P)Hal$. The decomposition rate of the salt $CH_3Hg(C_6H_3)_3PHal$ depends on the nature of halide: $Cl < Br < I$ and increases with an increase in the dielectric constant of the reaction medium: benzene $<$ tetrahydrofuran $<$ acetone $<$ methanol.[164] Halogen phosphonium salts of this type are decomposed faster than the salts of other anions which were investigated. Addition of NaBr or NaI to an acetone solution of $[C_6H_5HgP(C_2H_5)_3NO_3]$ results in the formation of $(C_6H_5)_2Hg$.[164]

On the other hand C_6H_5HgCl reacts with $(C_6H_5)_3P$ to produce $[C_6H_5Hg-(C_6H_5)_3P]Cl$,[165] which may be explained by the low solubility of C_6H_5HgCl. The reaction of the more soluble m-tolylmercury chloride with $(C_6H_5)_3P$ or $(C_2H_5)_3P$ proceeds faster in acetone and produces $R_2'Hg$ and $(R_3P)_2HgCl_2$ in 90% yield.[164]

The symmetrization reaction by tertiary phosphines is usually conducted at low temperatures; however, the reaction of CH_3HgHal with $(C_6H_5)_3P$ was conducted in boiling benzene.[164] Benzene, ethyl alcohol, and acetone are used as solvents.

CH_3HgBr is symmetrized in benzene by $(C_2H_5)_3P$.[162] If the reaction is conducted in ether, $CH_3Hg(C_2H_5)_3PBr$[162] is produced which does not disproportionate. On the other hand $(C_6H_5)_3P$ reacts in ether with CH_3HgCl to produce the double salt,[162] $(C_6H_5)_3P \cdot HgCl_2$, but $CCl_2{=}CClHgCl$ reacting with $(C_6H_5)_3P$ in ether or acetone, along with formation of such double salt, produces a symmetrical compound according to the above equation.[138]

With triphenylphosphine in benzene a symmetrical compound

$$[CH_3OC(O)]_2Hg$$

was obtained, from quasicomplex compounds of the addition product of the mercury salt to carbon monoxide in alcohol medium, $CH_3OC(O)HgCl$.[166]

Symmetrization of phenylmercury acetate during heating with tributylphosphine in benzene proceeds according to the following equation:[167]

$$2\ C_6H_5HgOCOCH_3 + (n\text{-}C_4H_9)_3P \longrightarrow (C_6H_5)_2Hg + (CH_3CO)_2O + (n\text{-}C_4H_9)_3PO$$

8. Triethanolamine

$$2\ C_6H_5HgO_2CCH_3 + 2\ N(CH_2CH_2OH)_3 \longrightarrow$$
$$2\ C_6H_5HgOOCCH_3 \cdot N(CH_2CH_2OH)_3\ (1) \longrightarrow$$
$$(C_6H_5)_2Hg + Hg(O_2CCH_3)_2 \cdot N(CH_2CH_2OH)_3$$

The given reactions according to patent data[168] are representative of the symmetrization of phenylmercury acetate. The originally produced double compound, (1) is decomposed during heating and produces a symmetrical compound and double salt of mercury acetate with triethanolamine. The reaction is carried out in the presence of lactic or citric acids, the participation of which is not clear.

9. Iron Pentacarbonyl and Its Derivatives

$$2\ RHgOH + Fe(CO)_5 \longrightarrow (RHg)_2Fe(CO)_4 \longrightarrow R_2Hg + HgFe(CO)_4$$

These reactions are irreversible.

When $RHgOH$ reacts with $Fe(CO)_5$,[169] $Fe(CO)_4H_2$,[169] or $CaFe(CO)_4$,[170] symmetrization occurs, thus producing R_2Hg and $HgFe(CO)_4$. It is probable that the reaction goes through the intermediate formation of $(RHg)_2Fe(CO)_4$.

The reaction is conducted in alcohol where R = Alk or Ar. The reaction has no preparative significance due to the existence of various agents of symmetrization which are more convenient to handle.

10. Alkali Sulfides

Alkali sulfides react with organomercury salts to produce sulfides which, when heated to a rather high temperature, are sometimes decomposed producing mercury(II) sulfide:

$$2\ RHgX + Na_2S \longrightarrow (RHg)_2S + 2\ NaX \longrightarrow R_2Hg + HgS$$

This is the basis for the symmetrization reaction caused by alkali sulfides. It is a well-known method used for symmetrization of organomercury compounds of aliphatic compounds,[171] various aromatic compounds,[172–177] and heterocyclic organomercury compounds, but is very seldom used now since it has been replaced by more suitable preparative methods. However, it has been recently investigated in detail for the purpose of symmetrization and produced a good yield of phenylmercury acetate.[179] Heating of α-camphenonylmercury sulfide to 220°C made it possible to produce a good yield of the appropriate symmetrical compound.[180]

It is probable that intermediate formation of arylmercury sulfide occurs as exemplified by the following equations:[181]

$$2\ C_6H_5HgOH + 2\ CS_2 \longrightarrow (C_6H_5Hg)_2S + 2\ COS + H_2S$$
$$(C_6H_5Hg)_2S \longrightarrow HgS + (C_6H_5)_2Hg$$

11. Potassium (Sodium) Hydroxide

Aqueous solutions of caustic alkalis were tested for reactivity as symmetrizators.[182] According to the equation:

$$2\ RHgX + 2\ KOH \rightleftharpoons R_2Hg + 2\ KX + H_2O + HgO$$

mercury is released in the form of slightly soluble mercury(II) oxide which reacts with R_2Hg producing RHgOH. Thus, symmetrization by alkalis proceeds incompletely as compared to the action of other symmetrization agents and this method is of no great practical significance. Mercurated pyrazoles were symmetrized by sodium hydroxide.[183]

12. Calcium Chloride

A surprising symmetrization by calcium chloride[184] was found in two cases:

$$2\ RHgX + CaCl_2 \longrightarrow R_2Hg + Ca[HgCl_2X_2]$$

Usually $CaCl_2$ does not show a symmetrizing effect on RHgX and it is widely used as a source of chloride ions with the purpose of converting RHg-acetates into RHgCl.

C. Symmetrizators Producing Completely Substituted Unsymmetrical Organomercury Compounds which then Disproportionate

1. Unsaturated Compounds

The unsaturated compounds, vinyl ethers and diacetal of ketene, react with arylmercury hydroxide to produce an unsymmetrical compound:[185]

$$2 \, ArHgOH + CH_2{=}CHOR \xrightarrow{ROH} ArHgCH_2CH(OR)_2$$

which immediately becomes a mixture of two symmetrical compounds:

$$2 \, ArHgCH_2CH(OR)_2 \longrightarrow Ar_2Hg + Hg[CH_2CH(OR)_2]_2$$

It is not necessary to isolate the original arylmercury hydroxide from the reaction mixture during its preparation by the effect of alkali on the aryl-mercury salt. The produced α-mercurybisoxo or carboxy compound (in case of the diacetal of ketene) after separation of Ar_2Hg is isolated in the form of the RHgX compounds by addition of mercury salts.

Symmetrization of unsaturated compounds is accomplished under mild conditions, at room temperature in alcohol and is used mainly for the preparation of aromatic compounds of mercury.

2. n-Butyllithium

Through formation of unsymmetrical compounds which are then decomposed into a mixture of two symmetrical compounds, RHgX is symmetrized by n-butyllithium.

$$RHgX + n\text{-}C_4H_9Li \longrightarrow n\text{-}C_4H_9HgR + LiX$$
$$2 \, n\text{-}C_4H_9HgR \longrightarrow R_2Hg + (n\text{-}C_4H_9)_2Hg$$

Symmetrization by n-butyllithium does not represent a widely used method and was accomplished for two cases—production of mercurybis-stilbenes (cis- and trans-)[186] and $[C_5H_4Mn(CO)_3]_2Hg$.[187]

3. Diphenylmercury

Symmetrization by diphenylmercury is based on formation of an unsymmetrical compound of mercury which is likely to be attacked by another molecule of the symmetrized compound:[47-49]

$$RHgCl + (C_6H_5)_2Hg \longrightarrow RHgC_6H_5 + C_6H_5HgCl$$
$$RHgC_6H_5 + RHgCl \longrightarrow R_2Hg + C_6H_5HgCl$$

In this case, a low-soluble phenylmercury halide was isolated. A symmetrical compound was found in the filtrate. Diphenylmercury was used for symmetrization for organomercury compounds containing a "labile" mercury bond. This method has not been used widely in practice. The reaction is usually conducted in chloroform.

4. Disproportionation of Unsymmetrical Compounds

The bulk of unsymmetrical, completely substituted mercury compounds, RHgR′, have a tendency to disproportionate into two symmetrical compounds:

$$2\,RHgR' \longrightarrow RHgR + R'HgR'$$

This tendency may be considered to be a method of symmetrization. Each RHgR′ is labile to this disproportionation in its specific way. Compounds of RHgR′ such as RHgCN, RHgC≡CR′,[165] and other compounds having, evidently, a partially ionic nature of one of the C—Hg bonds, are stable and not labile to disproportionation. As a rule, common RHgR′ compounds— AlkHgAlk′, ArHgAlk, ArHgAr′—disproprtionate more or less readily upon standing or, particularly, at increased temperatures.[188,189] Nesmeyanov and Kravtsov[190] showed that it was impossible to isolate the unsymmetrical compounds $C_6H_5HgC_6H_4COOH$-p [immediately decomposed to $(C_6H_5)_2Hg$ and $(p\text{-}HOOCC_6H_4)_2Hg$]. On the other hand, while the compound 9-phenyl-mercurifluorene is unstable, compounds $C_6H_5HgCH_2SO_2C_6H_5$, CH_3C-$(HgC_6H_5)(SO_2C_6H_5)_2$, and 9-phenylmercurifluoren-9-yl phenyl sulfone exist.[190,191]

In order to make conclusions on regularities which control the stability of unsymmetrical compounds (RHgR′), further systematic investigations are needed.

D. Symmetrizator Acting as an Adsorbent

1. Aluminum Oxide

The symmetrizing effect of aluminum oxide shown during chromatography in the mixture of hydrocarbons of o-iodophenylmercury iodide[192] is based, evidently, on different adsorption of all or part of the components contained in an equilibrium mixture:

$$2\,RHgX \rightleftarrows R_2Hg + HgX_2$$

E. Thermal Symmetrization

$$2\,RHgX \xrightarrow{\Delta} R_2Hg + HgX_2$$

A number of different organomercury[88,192–195] salts are symmetrized upon heating to rather high temperatures with or without solvents, or in a vacuum.

Organomercury compounds, symmetrized in this way, are related to various classes and there is no possibility as yet to delineate relationships between the structures of the organomercury salts and their ability to be thermally symmetrized.

IV. SYMMETRIZATION REACTION, APPLICATION OF THE REACTION

A. Aliphatic Mercury Compounds

1. Mercury at a Saturated Carbon Atom

a. Alkylmercury Salts. Electrolysis is the best and the most developed method known for symmetrization of alkymercury salts. However, a number of other agents may be used for the symmetrization of mercury derivatives of aliphatic hydrocarbons.

The saturated aliphatic compounds of mercury can be successfully symmetrized by the action of all-purpose agents of symmetrization such as copper (in the presence of pyridine) or sodium stannite. They are symmetrized also by potassium cyanide,[176] while *n*-butylmercury salt is symmetrized by hydrazine.[108] The latter may be used for symmetrization of other AlkHgX. Also used for symmetrization are $H_2Fe(CO)_4$ and $CaFe(CO)_4$. CH_3HgHal was symmetrized by tertiary phosphines.

ω,ω'-Dimercurated aliphatic compounds[98,99,113] are symmetrized by sodium stannite with production of cyclic compounds in which mercury is a member of the cycle, for example:[133]

$$\begin{array}{c}
\text{CH}_2\text{CH}_2\text{HgCl} \\
| \\
\text{CH}_2\text{CH}_2\text{HgCl}
\end{array} \xrightarrow{\text{Na}_2\text{SnO}_2} \begin{array}{c}
\text{CH}_2\text{—CH}_2 \\
| \qquad\qquad \diagdown \\
\qquad\qquad\qquad \text{Hg} \\
| \qquad\qquad \diagup \\
\text{CH}_2\text{—CH}_2
\end{array}$$

Low perfluoroalkyl compounds of mercury are symmetrized only by copper, silver, or best, cadmium amalgams.[105] The perfluoroalkyl compounds of mercury are also symmetrized by other common reagents such as alkali solutions of sodium stannite or ferrous oxide. Sodium iodide or cyanide in aqueous or acetone solution will not produce symmetrical compounds. In all these cases, the replacement of mercury by hydrogen and formation of $CF_{2n+1}H$ occur.

Allylmercury salts[144] are symmetrized best of all by a concentrated aqueous solution of KCN. Quantitative symmetrization occurs in the cold.

b. Mercury Derivatives of γ-Alcohols and Ethers. 3-Alkanol mercury salts and their ethers were symmetrized by sodium stannite.[68,69]

2. Olefinic Derivatives with Mercury at an Unsaturated Carbon Atom which are not Addition Products of Mercury Salts to Multiple Bonds

Vinylmercury salts[196] and isopropenylmercury salts[197] are symmetrized by sodium stannite. Mercury derivatives of stilbene[104] were symmetrized by potassium iodide, sodium stannite, tin–sodium alloy, and ammonia. Perchlorovinylmercury salts are readily symmetrized in good yield by triphenylphosphine in alcohol or ether. They are also symmetrized by equimolar amounts of potassium iodide or thiocyanate.[138] The latter is considered to be the best symmetrization agent for these compounds.

Perfluorovinylmercury chloride was symmetrized by an equimolar amount of KI in acetone.[138]

3. Addition Products of Mercury Salts to Unsaturated Compounds

a. Quasicomplex Compounds of Mercury

1. Addition Products to Olefins. Quasicomplex compounds—addition products of mercury salts to noncyclic double bonds, β-mercurated alcohols and ethers, cannot be symmetrized—they are decomposed by all agents of symmetrization and formation of the unsaturated compound results, see Section III-B-1. This, according to Nesmeyanov,[198,199] specifies the presence of a σ,σ-conjugate system of bonds existing in the compound:

$$R\overset{\frown}{O}\text{---}CH_2\text{---}\overset{\frown}{C}H_2\text{---}HgCl$$
$$1 \quad\quad 2 \quad\quad 3 \quad\quad 4$$

and depending on the reaction, occurs with breakage of 1–2 or 3–4 bonds as is typical of conjugate systems.

Olefin addition products of mercury salts in secondary amines as solvent are not symmetrized.[103] The product of the ammonia mercuration of cyclohexene was symmetrized by hydrazine or $NaBH_4$ producing mixtures of isomers.[115] Products of oxy(alkoxy) mercurated cyclic alkenes, cyclohexene, its homologs and derivatives, are symmetrized by hydrazine,[114] sodium stannite, or potassium iodide.

Apparently other products of oxy(alkoxy) mercuration, which do not show striking quasicomplex properties can be symmetrized by such means.

2. Addition Products to Acetylene. Ammonia is the only symmetrizator for the addition products of mercury salts to acetylene (β-chlorovinyl compounds of mercury) in nonreactive solvents, chloroform, benzene, or carbon tetrachloride.[21,24] Symmetrization is carried out at room temperature or lower and usually produces good yields. Ammonia symmetrization results in retention of configuration.

3. Addition Products to Methylphenylacetylenylketone[160] and to Acetylenic Acids.[159] These compounds have higher lability due to the presence of the conjugate bond of XHg with C=C and C=O bonds (see Chapter 9, Section IV-A) and are not symmetrized by symmetrization agents but are decomposed producing the unsaturated compound. Addition products of $HgCl_2$ to acetylenic alcohols[158] are decomposed as well by KI, KCN, and $Na_2S_2O_3$.

4. Addition Products to Carbon Monoxide. Carbomethoxymercury chloride, the addition product of mercury salt to carbon monoxide in alcohol, was symmetrized only by triphenylphosphine:[166]

$$2\ ClHgCOOCH_3 + (C_6H_5)_3P \longrightarrow Hg(COOCH_3)_2 + (C_6H_5)_3P \cdot HgCl_2$$

The reaction is carried out in cold benzene. Other symmetrizators decompose the compound releasing CO.

Addition products of mercury salts to carbon monoxide in the medium of secondary amines were not symmetrized.

b. Addition Products of Mercury Salts to Unsaturated Compounds which are not Considered to be Quasicomplex Compounds

1. Addition Products of Mercury Salts to Substituted Acetylenes. Alkenyl compounds of mercury similar to salts of isomers of dimethylethenylmercury were symmetrized by ammonia in benzene or KI in water or acetone.[200]

2. α-Mercurated Oxocompounds. α-Mercurated oxocompounds (addition products of mercury salt to vinyl ether) as quasicomplex compounds of mercury-addition products of $HgCl_2$ to acetylene are symmetrized solely by ammonia. α-Halomercuryketones[157a] were symmetrized by ammonia in an inert solvent. However, the salt of α-mercuryacetaldehyde is not symmetrized by ammonia, nor by other symmetrization agents.

3. α-Mercurated Carboxycompounds. The ester of α-chloromercuryacetic acid was symmetrized by copper or pyridine producing a small yield of the symmetrical product.[201] $C_6H_5CH(OCH_3)CH$—CO, which, in some respects,

$$\underset{Hg-O}{\overset{\mid\qquad\quad\mid}{}}$$

shows quasicomplex properties, was symmetrized by KI.

4. Anhydrization Products of Ethanolmercury Salts and Their Derivatives. Mercury derivatives of ethyl ether, $(XHgCH_2CH_2)_2O$ are produced during addition of mercury salts to ethylene, along with β-mercurated alcohols; the anhydrization products of ethanolmercury salts, as distinct from the latter can be symmetrized by sodium stannite.[202] A cyclic compound containing mercury and having, according to X-ray investigations,[203] a dimeric structure was produced from ethylene in this case:

$$\begin{array}{c} CH_2CH_2OCH_2CH_2 \\ Hg \diagup \qquad \diagdown Hg \\ \diagdown CH_2CH_2OCH_2CH_2 \diagup \end{array}$$

2-Chloromercurymethyl-2,3-dihydrobenzofuran, the product of oxymercuration of o-allylphenol produced as a result of further cyclization, was symmetrized by sodium amalgam.[100]

B. Mercury Derivatives of Aralkyls

Evidently, for benzylmercury salts, the best method of symmetrization is the recently suggested method of reduction involving salts of divalent chromium.[127] It is advisable to use benzylmercury perchlorate which produces a quantitative yield of dibenzylmercury. However, benzylmercury chloride, whose reaction proceeds in a heterogeneous way, nevertheless produces rather high yields of the desired product. Benzylmercury salts were also successfully symmetrized by sodium stannite.[75]

Esters of α-anionmercuryarylacetic acids were symmetrized by ammonia (see Section II and Section III-B-5).

C. Mercury Derivatives of Alicyclic Compounds

Sodium stannite was successfully used for the symmetrization of mercury derivatives of alicyclic compounds which are not considered to be quasicomplex organomercury compounds i.e., hydrocarbons,[45a,205] alcohols,[101] and ketones such as camphor[74] including those with mercury at a bridgehead.[204]

Symmetrization of cyclohexylmercury perchlorate produced a very high yield when reacted with the salts of chromium(II) oxide.[127] Mercury derivatives of camphane[45a] and camphor[43] were symmetrized by hydrazine hydrate. Camphor derivatives were symmetrized by diphenylmercury and by $Na_2S_2O_3$.[40,43,47]

The only method of symmetrization of α-mercurycamphenilone was by heating with sulfide and resulted in production of di-α-camphenilonyl mercury in 80% yield.[180]

D. Mercury Derivatives of Aromatic Compounds

1. Mercurated Hydrocarbons, Their Halogen and Nitro Derivatives

Monomercurated aromatic hydrocarbons are readily symmetrized by most symmetrization agents: sodium stannite,[71-74] potassium iodide,[125,139-142] (but never diphenylmercury; refer to limitations typical of KI as a symmetrizator[128]) potassium (sodium) cyanide and thiocyanate,[176] sodium amalgam,[172,206] copper in the presence of a nitrogen base,[7,82,83] chromium(II) oxide,[127] alkali sulfides,[172,179] hydrazine,[108] phenylhydrazine, triethanolamine,[168] ammonia and tri-n-butylphosphine in benzene,[167] (the last three reagents were used for symmetrization of phenylmercury acetates) unsaturated compounds, and iron pentacarbonyl and its derivatives. The fluoride[194] and triphenylsilyloxyl[195] of phenylmercury are symmetrized upon heating.

Halogen derivatives of aromatic hydrocarbons were symmetrized by copper in the presence of a base,[7,82,83] by hydrazine[110] (in particular, this symmetrizator was used for symmetrization of pentafluorophenylmercury salt[110]) and sodium iodide, used for symmetrization of mercurated chloronaphthalene,[125] for which symmetrization by sodium alcoholates produced poor yields.[125] o-Iodophenylmercury iodide was symmetrized during chromatography on aluminum oxide (solvent benzene-cyclohexane) with a 90% yield[192] and when heated to 100°C in vacuum for 20 hr.[192] The acetate of 2,5-dichlorophenyl-mercury[88] was symmetrized upon heating to 145°C in p-xylene or mesitylene.

Many reagents symmetrize mercury salts of aromatic nitro derivatives: copper in the presence of a nitrogen base,[7,82,83] sodium alcoholates,[124] alkali sulfides,[173] and sodium hydrosulfite.[124]

2. Mercurated Phenols

The best symmetrization agent for mercurated phenols is sodium thio-sulfate.[178,89] They were also symmetrized by copper in the presence of ammonia[7] and by sodium sulfide[177] and are not symmetrized by potassium iodide but rather decomposed.[136]

Phenolic ethers are readily symmetrized by sodium stannite,[72] sodium thiocyanate,[178] or copper in the presence of ammonia.[7,83,84] They are not decomposed by potassium (sodium) iodide and therefore can be symmetrized by these reagents.[178]

3. Mercurated Aromatic Amines

Sodium thiosulfate[71,148–151] is the best agent of symmetrization of mercurated amines. Mercurated amines and alkylamines were symmetrized also by copper in pyridine[82,83] and alkali sulfides;[175] haloalkylamines were symmetrized by potassium iodide.[142] Potassium iodide is unsuited for symmetrization of mercurated amines with unprotected amine groups since it attacks them.

4. Mercurated Aromatic Ketones

Mercurated benzanthrone is symmetrized upon heating.[193] Mercury derivatives of indandione, constituting, according to Hantzsch, C—Hg compounds, are symmetrized by pyridine in the cold.[161] Derivatives of benzophenone are symmetrized[93] by Cu.

5. Mercury Derivatives of Aromatic Acids, Their Esters, and Halogen- and Nitro-Substituted Compounds

Mercurated benzoic acids and their esters are symmetrized by the action of sodium stannite,[73] potassium iodide, alkali sulfides[174] and copper in the presence of ammonia,[92] but for the esters of acids, it is better to use copper in the absence of ammonia[84] whose saponifying effect reduces yields.

Potassium iodide[207] was used for symmetrizing the mercurated esters of nitrobenzoic acids. Symmetrization of mercurated nitrobenzoic acids by iron(II) hydroxide is accompanied by the reduction of the nitro group to an amino group [126]

Symmetrization of mercury derivatives of halogen, nitro-carboxy- and carbalkoxyderivatives of aromatic compounds also occurs with those symmetrizators that do not react with the aforementioned groups.

E. Mercury Derivatives of Heterocyclic Compounds

1. Mercury Derivatives of Superaromatic Heterocyclic Compounds

Thiophene, its alkyl-, halo-, nitro-, carbalkoxy-, and other substituted mercurated and bimercurated compounds in the α-position, are symmetrized by alkali metal iodides,[76,95,96,208-210] thiocyanates,[76,211] and sodium stannite.[76,77] For symmetrization of salts of nonsubstituted α-thienylmercury and halogenated α-thienylmercury, use was made of hydrazine.[145]

Thiophene mercurated in the α-position was also symmetrized by metallic sodium,[96] however, milder symmetrizing agents are more often used. Thiophene mercurated in the β-position and its derivatives were symmetrized by metallic sodium in xylene.[95]

Selenophene, mercurated in the α-position, was symmetrized by sodium iodide.[212]

Furan, mercurated in the α- and β-position,[152] and 5-mercury-2-furyl alcohol were symmetrized by sodium thiosulfate[153] and furan, mercurated in α-position, by hydrazine as well.[152]

2. Mercury Derivatives of Other Heterocyclic Compounds

A mild agent, sodium thiosulfate, was used for the symmetrization of pyridine mercurated in the β-position[94] and quinoline mercurated in the 8-position.[154]

Mercurated, substituted pyrazoles were symmetrized by a concentrated water solution of caustic sodium (simultaneous isolation of the mercury salt complex), for example:[183]

F. Mercury Derivatives of Metallocenes

For symmetrization of mercurated ferrocene, used with equal success, were such symmetrizators of usually so different a range of characteristics as metallic sodium,[78] potassium iodide,[78] sodium thiosulfate,[156,157] and sodium stannite.[78] Ferrocene dimercurated, in different rings, was symmetrized by potassium iodide and $Na_2S_2O_3$ with formation of a polymeric compound, $(C_{10}H_8FeHg)_n$.[143]

Sodium thiosulfate[155] as well as n-butyllithium[187] symmetrized chloromercurycyclopentadienyltricarbonylmanganese.

G. Cases when Symmetrization Reactions Fail to Occur

As was already described, symmetrization does not occur with mild reagents such as ammonia in a nonpolar organic solvent, or other symmetrization agents with such labile organomercury compounds as α-chloromercuryacetaldehyde and addition products of mercury salts to acetylenic alcohols,[158] ketones,[160] and acids.[159] The latter are not symmetrized but are decomposed producing the unsaturated compound with all symmetrization agents. Also, the products of oxy(alkoxy) mercuration of noncyclic alkenes, β-anionmercuryderivatives of alcohols and ethers, are not symmetrized.

V. REVERSE SYMMETRIZATION REACTION

The reverse symmetrization reaction—the interaction of completely substituted compounds of mercury with the mercury salt—is considered of preparative importance[206]:

$$R_2Hg + HgX_2 \longrightarrow 2\,RHgX$$

This reaction constitutes a convenient method for producing an organomercury salt of the desired anion in a pure form without admixture of salts containing other anions. Being thermodynamically more favorable[50] than the symmetrization reaction of organomercury compounds, this reaction almost always proceeds readily when R_2Hg and HgX_2 are heated in an equimolecular relationship. It occurs sometimes in the cold, and in solvents such as acetone, alcohol, tetrahydrofuran or even without a solvent, particularly when one of the compounds is a liquid. $(C_6Cl_5)_2Hg$ was desymmetrized by $HgCl_2$ in nitrobenzene with two hours of boiling.[214]

Reactions with mercury salts of organic acids are carried out in a medium of the same acid[196] or without solvents; the reaction with mercury salts of tribasic acids is also carried out in this manner in the presence of the same amount of water.[213]

The reverse symmetrization reaction always proceeds stereospecifically (see Section II).

Cyclic organomercury compounds, containing mercury as a member of the ring, react with mercury salts, thus opening the ring.[215]

The desymmetrization reaction was not successful for R_2Hg having a C—Hg bond with a partially ionic nature, i.e., RHgX is not obtained by the following reactions: $(C_5H_6C\equiv C—)_2Hg$ and probably other acetylenides

$(RC\equiv C-)_2Hg$ with HgI_2, $(CCl_2\equiv CCl-)_2Hg$ with HgI_2, $[(CF_3)_2CF]_2Hg$ with $HgCl_2$ and other low perfluoroalkyl compounds of mercury, $(C_nF_{2n+1})_2$-Hg with HgX_2. Formation of $C_6H_5C\equiv CCl$ see.[165]

VI. PREPARATIVE SYNTHESES

A. Symmetrization of Alkylmercury Salts by Electrolysis[121]

Electrolysis was conducted in a cylindrical flask, 18 mm in diameter, on the bottom of which was placed a mercury cathode. A platinum wire, 0.5 mm in size, was used as an anode, with a mercury cathode surface area equal to 2.5 cm^2. A 6.2-V lead storage battery served as a power supply source. To maintain the desired electrolysis temperature the electrolysis flask was kept in a water bath heated to the required temperature.

1. Electrolysis of Propylmercury Sulfate [121]

The electrolyte was prepared by dissolving 1.5 g of propylmercury sulfate and 0.1 g of sodium sulfate in 15 ml of water. Dissolution was carried out with heating due to poor solubility of propylmercury sulfate, and upon cooling the solution to 20°C the electrolysis was accomplished. The electrolysis was finished after 2–3 hr, which was evident by the appearance of hydrogen bubbles on the cathode. Current density was maintained at 0.02–0.03 A/cm^2. With electrolysis over, the dipropylmercury produced was separated from the mercury in a microseparatory funnel, washed with a dilute solution of sulfuric acid, and dried over calcium chloride to give 0.7 g, boiling point 185–186°C.

2. Electrolysis of Ethylmercury Succinate [121]

One gram of ethylmercury succinate and 0.1 g of sodium sulfate were dissolved with heating in 15 ml of water. Electrolysis of the prepared solution was carried out at a temperature of 80–90°C due to poor solubility of the ethylmercury succinate in cold water. Current density was 0.04 A/cm^2. At the end of the electrolysis, the electrolyte was cooled and the prepared diethylmercury was separated from mercury and water and dried over anhydrous calcium chloride. The yield was approximately 70%, b.p. 156°C.

B. Preparation of 1,6-Dimercuracyclodecane [113]

A sample of 0.05 g (10^{-4} mole) of 1,4-dichloromercury butane was treated with 40 ml of 10% aqueous sodium hydroxide and then with a solution of 1 g of stannous chloride dihydrate in 20 ml of 20% aqueous sodium hydroxide. The black suspension was filtered off when no further color change occurred. The black product was extracted with boiling xylene which was evaporated at low temperature leaving 0.17 g (66%), m.p. 37–40°C. Crystallization from ether–benzene raised this melting point to 43–45°C.

C. Preparation of Bistrifluoromethylmercury[105]

Trifluoromethylmercury iodide (2.0 g) was heated at 140°C with silver amalgam (silver, 10 g; mercury, 25 g) for 20 hr in a rocking furnace, the excess amalgam was removed, and the residual solid extracted with ether. The ethereal extract upon evaporation yielded a white solid which was purified by sublimation at 70°C (1 atm). The yield of pure product, m.p. 163°C (sealed tube), was 80%. With cadmium amalgam at 120–130°C and reaction times of 10 hr, dimercurial yields of 80–90% were readily obtained.

D. Preparation of Diallylmercury[144]

On treating allylmercury iodide with an excess of a concentrated solution of KCN the characteristic smell of diallylmercury was immediately noticed. Gradually all the allylmercury iodide disappeared and a colorless heavy liquid settled to the bottom leaving the aqueous layer quite clear. The diallylmercury was extracted with ether, b.p. 58–58.5°C (1.5 mm), n_D^{20} 1.6309, d_4^{20} 2.3180.

E. Symmetrization of Acetic Acid Salts of 3-Oxy-, 3-Methoxy-, and 3-Ethoxy-2,2,3-trimethylbutylmercury[68]

A solution of 25 ml of 40% caustic sodium solution and 10 g of $SnCl_2 \cdot 2H_2O$ and 5 ml of water is added to a solution of acetic acid salts of 3-oxy-2,2,3-trimethylbutylmercury (6 g) (or to the emulsion of the same salt) or methoxy-alkylmercury (6.1 g), or ethoxyalkylmercury (6.4 g) in water (60 ml) during continuous, vigorous stirring. The reaction mixture was stirred for an hour; the black suspension was filtered off, washed with water, and extracted by acetone. Water was added to the acetone solution until a suspension appeared. The crystals obtained during cooling were dried over calcium chloride and were recrystallized from ethyl alcohol.

F. Preparation of Bis(perchlorovinyl)mercury[138]

To a solution of perchlorovinylmercury chloride (5.0 g) in anhydrous acetone, a solution containing a molar equivalent of NaI in acetone was added dropwise with rapid stirring. The NaCl precipitate was filtered and R_2Hg isolated from the filtrate by evaporation of solvent. The R_2Hg was separated from the organomercury salt by extraction with hot pentane. Yield 2.8 g (88%), m.p. 73°C.

G. Preparation of cis,cis-Di-β-chlorovinylmercury[24]

Into a solution of 2.6 g of cis-β-chlorovinylmercury chloride (m.p. 73°C) in 35 ml of dry benzene, ammonia is bubbled for 6 min at a rate of 80–100 bubbles per minute. The precipitate is separated from the filtrate, and the solvent is slowly evaporated at a temperature of 30–35°C. The weight of the colorless liquid obtained is 1 g (70.9% yield). cis,cis-Di-β-chlorovinylmercury does not solidify upon cooling to $-70°C$, d_4^{20} 2.8090; n_D^{20} 1.6308.

H. Preparation of Bis-carbomethoxymercury[166]

A solution of carbomethoxymercury chloride (23 g, 0.08 mole) in benzene (250 ml) was added to triphenylphosphine (20.8 g, 0.08 mole) dissolved in benzene (250 ml) and immediate precipitation of $(C_6H_5)_3P \cdot HgCl_2$ occurred and was filtered off. The filtrate was evaporated to dryness under vacuum below 50°C. Recrystallization from ether by cooling yielded a white solid, m.p. 84–85°C.

I. Preparation of Diacetonylmercury[157a]

Added to a solution of 9 g (0.03 mole) of chloromercuryacetone in 250 ml dichloride was a solution of ammonia in dichloroethane until solid stopped precipitating. The residue is filtered, washed with dichloroethane. The solutions are combined, and the dichloroethane is evaporated in the cold affording a quantitative yield of diacetonylmercury, 4.5 g. Recrystallization from a mixture of benzene and heptane gave a melting point of 69°C.

J. Preparation of Dibenzylmercury[127]

The organomercury perchlorates were prepared by addition of 1 equiv. of 70% perchloric acid to 1 equiv. of organomercury acetate dissolved in 100 ml of degassed dimethylformamide. The resultant solution was kept under nitrogen. The chromium(II) sulfate was injected into the system through a serum cap. Finely divided mercury was immediately produced and a green solution resulted. The reaction mixture was centrifuged and the liquid layer was removed. The solid residue was washed consecutively with dimethylformamide, water, acetone, and ether. Metallic mercury remains. The disubstituted organomercury compound was obtained by dilution of the dimethylformamide solution with water and isolation of the resulting solid by filtration. Recrystallization from either benzene or ethanol yielded 99% dibenzylmercury, m.p. 111°C.

K. Symmetrization of Diastereomer of 2-Chloromercurycamphane by Hydrazine Hydrate[45a]

Diastereomer (1.1 g) with $[d]_D^{18} + 15.4 \pm 1.2°C$ is dissolved in 25 ml of hot ethyl alcohol. Hydrazine hydrate (0.4 g) and 0.3 g of potash are added to the solution; the reaction mixture is boiled for 3 hr. The solution is filtered until all the residue is collected. The filtrate is concentrated to dryness. The residue is washed by water and a small amount of ethyl alcohol to give 0.35 g (63%) of 2,2'-mercury-biscamphane. Upon recrystallization from ethyl alcohol the compound melted at 143–150°C and had $[d]_D^{18}$ 7.3 ± 1.2°C.

L. Preparation of Di-p-bromophenylmercury [82]

A solution of 6.1 g of p-bromophenylmercury chloride in 40 ml of pyridine and 7.2 g of copper is shaken in a Schlenck flask filled with N_2 or CO_2. After the pyridine is driven off *in vacuo* at a temperature of the bath not exceeding 40°C, 3.7 g of R_2Hg (92%) remained. Upon recrystallization from acetone, the product melted at 244–245°C.

M. Preparation of Di-p-aminophenylmercury [149,150]

A solution of 16 g of sodium thiosulfate in 200 ml of water was heated, while stirred, to 45–50°C and added to a 20 g suspension of p-aminophenyl-mercury acetate in 75 ml of water which was also heated to 45–50°C. With this, the phenylmercury acetate is dissolved and a small precipitate of mercury sulfide is formed. The mixture is heated to a temperature not exceeding 65–70°C, filtered off, and the solution of 90 g of sodium thiosulfate in 70 ml of water is added to the filtrate having a temperature of not more than 55°C. The precipitate of di-p-aminophenylmercury is filtered, washed with water, and dried after two-hour aging. The yield is 8 g (73%). Upon recrystallization from dioxane the m.p. is 174°C (capillary is taken into a bath previously heated to 165°C). In a similar way, other compounds containing a dialkyl-amino group were symmetrized in 70–75% yield.

N. Preparation of Di-m-oxyphenylmercury [89]

Eight grams of m-oxyphenylmercury chloride are dissolved in a solution of crystalline sodium thiosulfate in 100 ml of water. Soon after dissolution, the solution becomes cloudy and a precipitate forms. The precipitate is filtered and washed after standing one day; 4.6 g is obtained (quantitative yield), which was recrystallized from alcohol. The substance does not melt and decomposes above 265°C.

O. Preparation of Mercury-bis-p-benzoic Acid Ethyl Ester by Copper without Ammonia [84]

p-Chloromercurybenzoic acid ethyl ester (3.84 g, 0.01 mole) and 3 g of copper in 150 ml of benzene are boiled for 10 hr; 2.4 g (96%) of the compound were produced. Recrystallization from alcohol gave a solid with m.p. = 192°C.

P. Preparation of Mercury-bis-*p*-benzoic Acid Methyl Ester by Copper and Ammonia[92]

Methyl *p*-chloromercurybenzoate (6.6 g) is mixed with 6 g of copper powder; 60 ml of acetone are added and the mixture thoroughly stirred. During shaking, 30 ml of 25% aqueous ammonia solution are poured in and the mixture is shaken energetically for 10 min. The solution is poured into excess water (300 ml) and the precipitate is filtered. After drying at 100°C the residue is extracted by hot ethyl acetate to give 2.85 g (68%) of product. In recrystallization from chloroform or acetone, the m.p. is 264–265°C. The yield of the hardly saponifiable ethyl ester under similar conditions is 88.5%.

Q. Preparation of Di-*o*-phenetolylmercury[178]

A solution of 2 g of *o*-iodomercuryphenetole in 50 cc of alcohol together with 2 g of potassium iodide is refluxed for 1 hr. When this product is cooled and diluted with water, 1.5 g of di-*o*-phenetolylmercury separates. After recrystallization from alcohol m.p. = 81°C.

R. Preparation of Bispentafluorophenylmercury[110]

Hydrazine hydrate (0.5 g) was added dropwise to pentafluorophenylmercury chloride (0.2 g) in ethanol (20 ml). An immediate reaction occurred and after 5 min, a grey precipitate appeared. After 2 hr at room temperature the ethanol solution was decanted from the mercury mirror (*ca.* 0.05 g) into water (50 ml), yielding 0.07 g. The precipitate has a m.p. of 117–118°C. Upon standing the m.p. = 136–137°C.[217] According to Ref. 218, the m.p. after sublimation is 142°C.

S. Preparation of 2,2′-Difurylmercury[152]

To a solution of 50 g of sodium thiosulfate in 200 cc of water was added 0.1 mole of the mercurial; the mixture was shaken vigorously for a few minutes, and after 8 hr the precipitate was filtered to give a 95% yield of crude difurylmercury, which may be purified by recrystallization from hot acetone–water or by a careful distillation under reduced pressure, the material sometimes decomposing violently. The compound distills at 156°C (7 mm) and melts at 114°C.

T. Preparation of Bis(2,4-diphenylselenophen-5-yl)mercury[212]

The mixture of 1.5 g of the 5-chloromercury derivative and 1 g of sodium iodide in acetone solution was stirred for 6 hr, the precipitate removed, washed with water, dried, and recrystallized from benzene, m.p. 236.7°C (corr.).

U. Preparation of Di-3-pyridylmercury[94]

To a solution of 50 g of sodium thiosulfate in 50 ml of water was added 10 g of 3-pyridylmercury acetate. The solid dissolved, but the solution became cloudy in 3 min, and after 30 min 4.5 g (85%) of white solid separated, m.p. 225–226°C. After recrystallization from alcohol the m.p. was 239°C.

V. Preparation of Diferrocenylmercury[78]

A mixture of 600 ml of 95% ethanol, 10.5 g of NaI, and 2.1 g (0.005 mole) of chloromercuryferrocene was refluxed for 2 hr. After filtering the hot mixture a yellow-orange solid separated, m.p. 215–225°C (corr.). Two recrystallizations from xylene produced 0.9 g (64% yield) of diferrocenylmercury, m.p. 235–236°C with decomposition. Upon cooling and concentrating the ethanolic filtrate 0.5 g (35% yield) of orange, crystalline solid was obtained, m.p. 245–248°C. Recrystallization from xylene produced a second form of diferrocenylmercury, m.p. 248–249°C, with decomposition.

W. Preparation of Ethylmercury Phosphate[213]

A mixture of 25 g of diethylmercury, 25.5 g of mercury phosphate and 5 ml of water was heated to 115°C on an oil bath in a flask fitted with a reflux condenser. The reaction goes to completion in 15 min and the contents of the flask are transformed into a liquid state. Cooling gave a solid which, upon recrystallization from water and drying at a temperature of 85°C, gave a yield of 50 g (approximately 98%) of a white, crystalline solid, m.p. 179–180°C. It is recrystallized from water as the crystalline hydrate $(C_2H_5Hg)_3$-$PO_4 \cdot H_2O$ which has a sharp m.p. at approximately 110°C (due to loss of part of the water during heating). During continuous drying it completely loses the water and becomes an anhydrous salt which melts at 179–180°C. When kept in the air the dry ethylmercury phosphate slowly absorbs water, thus developing into the crystalline hydrate.

X. Preparation of Ethylmercury Nitrate[213]

Diethylmercury (8.5 g) gradually added to 11.5 g of mercury nitrate during continuous stirring. During the reaction the mixture is heated and becomes liquid. To complete the reaction the mixture was heated for 15 min on a boiling water bath in a flask fitted with a reflux condenser. Upon cooling, the mixture crystallized. The crude ethylmercury nitrate was recrystallized from water, affording 16–17 g (80–85%) of solid, m.p. 86–86.5°. It crystallizes from water with one molecule of H_2O.

REFERENCES

1. O. Dimroth, *Ber.*, **32**, 758, 798 (1899); **35**, 2032, 2853 (1902).
2. L. Pesci, *Gazz. Chim. Ital.*, **23**, II, 52 (1893); *Ber.*, **27**, 127 (1894); *Gazz.*, **28**, II, 101 (1898); *Z. Anorg. Allgem. Chem.*, **15**, 208 (1897); **17**, 276 (1898).
3. W. J. Klapproth and F. H. Westheimer, *J. Am. Chem. Soc.*, **72**, 4461 (1950).
4. P. Pfeiffer and P. Truskier, *Ber.*, **37**, 1125 (1904).
5. H. Gilman and R. E. Brown, *J. Am. Chem. Soc.*, **51**, 928 (1929); **52**, 3314 (1930).
6. A. N. Nesmeyanov, *Zh. Russk. Phyz.-Khim. Obshch.*, **61**, 1393 (1929); *Ber.*, **62**, 1010 (1929); A. N. Nesmeyanov, N. F. Gluschnev, P. F. Epifanski, and A. M. Flegontov, *Zh. Obshch. Khim.*, **4**, 713 (1934); A. N. Nesmeyanov and L. G. Makarova, *Zh. Obshch. Khim.*, **1**, 598 (1931); A. N. Nesmeyanov and E. M. Toropova, *Zh. Obshch. Khim.*, **4**, 664 (1934); A. N. Nesmeyanov, L. G. Makarova, and I. V. Polovjanjuk, *Zh, Obshch. Khim.*, **35**, 681 (1965).
7. A. N. Nesmeyanov and E. I. Kahn, *Zh. Russk. Fiz.-Khim. Obshch.*, **61**, 1407 (1929); *Ber.*, **62**, 1018 (1929).
8. L. Pesci, *Atti Accad. naz. Lincei, Rend. Classe Sci. fiz. mat. nat.*, (5) **10**, 1, 362 (1901)
9. M. S. Kharasch and F. Stavely, *J. Am. Chem. Soc.*, **45**, 2961 (1923); M. S. Kharasch, *J. Am. Chem. Soc.*, **43**, 2238 (1921).
10. A. N. Nesmeyanov, I. F. Lutsenko, and S. N. Anachenko, *Uchen. Zap. Moskov. Gos. Univer.*, **132**, N7, 136 (1950).
11. A. Michaelis and P. Becker, *Ber.*, **15**, 180 (1882); **27**, 244 (1894).
12. E. Khotinsky and M. Melamed, *Ber.*, **42**, 3090 (1909).
13. W. König and W. Scharrnbeck, *J. Prakt. Chem.*, **128**, 153 (1930).
14. W. Peters, *Ber.*, **38**, 2567 (1905).
15. A. N. Nesmeyanov and L. G. Makarova, *Zh. Obshch. Khim.*, **3**, 257 (1933); *Ber.*, **66**, 199 (1933).
16. E. Frankland, *Ann. Chem.*, **85**, 365 (1853).
17. N. N. Zinin, *Ann. Chem.*, **96**, 363 (1855).
18. A. N. Nesmeyanov and O. A. Reutov, *Izv. Akad. Nauk SSSR, Otd. Khim. Nauk.*, **1953**, 655.
19. K. A. Hofmann and J. Sand, *Ber.*, **33**, 1340, 1353, 1358, 2692 (1900); J. Sand, *Ber.* **34**, 1385, 2906 (1901).
20. W. Schoeller, W. Schrauth, and W. Essers, *Ber.*, **46**, 2864 (1913).
21. E. T. Chapman and W. J. Jenkins, *J. Chem. Soc.*, **115**, 847 (1919); W. J. Jenkins, *J. Chem. Soc.*, **119**, 747 (1921).
22. A. N. Nesmeyanov and R. Kh. Freidlina, *Zh. Obshch. Khim.*, **7**, 43 (1937); *Ber.*, **69**, 1631 (1936).
23. A. N. Nesmeyanov and R. Kh. Freidlina, *Dokl. Akad. Nauk SSSR*, **26**, 59 (1940).
24. A. N. Nesmeyanov, A. E. Borisov, and A. N. Gus'kova, *Izv. Akad. Nauk SSSR, Otd. Khim. Nauk*, **1945**, 639.
25. A. N. Nesmeyanov and A. E. Borisov, *Tetrahedron*, **1**, 158 (1957).
26. A. N. Nesmeyanov, I. F. Lutsenko, and N. I. Vereshchagina, *Izv. Akad. Nauk SSSR, Otd. Khim. Nauk*, **1947**, 63.
27. E. Frankland and B. F. Duppa, *Ann. Chem.*, **130**, 104 (1864).
28. A. N. Nesmeyanov, O. A. Reutov, and S. S. Poddubnaya, *Izv. Akad. Nauk SSSR*, **1953**, 850; *Dokl. Akad. Nauk SSSR*, **88**, 479 (1953).
29. A. N. Nesmeyanov and O. A. Reutov, *Izv. Akad. Nauk SSSR, Otd. Khim. Nauk*, **1955**, 739.
30. A. N. Nesmeyanov, O. A. Reutov, I. P. Beletskaya, and R. E. Mardaleishvili, *Dokl. Akad. Nauk SSSR*, **116**, 617 (1957).

31. O. A. Reutov, *Angew. Chem.*, **72**, 198 (1960).
32. O. A. Reutov and I. P. Beletskaya, *Dokl. Akad. Nauk SSSR*, **131**, 853 (1960).
33. O. A. Reutov, I. P. Beletskaya, and G. A. Artamkina, *Zh. Obshch. Khim.*, **34**, 2817 (1964).
34. O. A. Reutov, *Dokl. Akad. Nauk SSSR*, **163**, 909 (1965).
35. O. A. Reutov, I. P. Beletskaya, and G. A. Artamkina, *Sb.* "Kinetika i Kataliz," in *Izdat. Akad. Nauk SSSR, Moskau*, **1960**, p. 55.
36. I. P. Beletskaya, G. A. Artamkina, and O. A. Reutov, *Dokl. Akad. Nauk SSSR*, **149**, 90 (1963).
37. I. P. Beletskaya, G. A. Artamkina, and O. A. Reutov, *Izv. Akad. Nauk SSSR, Otd. Khim. Nauk*, **1963**, 765; **1964**, 1737.
38. F. R. Jensen and B. Rickborn, *J. Am. Chem. Soc.*, **86**, 3784 (1964).
39. F. R. Jensen, B. Rickborn, and J. I Miller, *J. Am. Chem. Soc.*, **88**, 340 (1966).
40. O. A. Reutov and T. T. Lu, *Zh. Obshch. Khim.*, **29**, 182 (1959).
41. T. G. Traylor and S. Winstein, *J. Org. Chem.*, **23**, 1796 (1958).
42. F. R. Jensen, L. D. Whipple, D. K. Wedegaertner, and J. A. Landgrebe, *J. Am. Chem. Soc.*, **81**, 1262 (1962).
43. O. A. Reutov and T. T. Lu, *Dokl. Akad. Nauk SSSR*, 110, 575 (1956); *Zh. Obshch. Khim.*, **29**, 1207 (1959).
44. A. G. Brook, R. Donovan, and G. F. Wright, *Can. J. Chem.*, **31**, 536 (1953).
45. B. H. M. Billinge and B. G. Gowenlock, *J. Chem. Soc.*, **1962**, 1201.
45a. O. A. Reutov and T. T. Lu, *Zh. Obshch. Khim.*, **29**, 1617 (1959).
46. F. R. Jensen and J. A. Landgrebe, *J. Am. Chem. Soc.*, **82**, 1004, (1960).
47. I. P. Beletskaya, G. A. Artamkina, and O. A. Reutov, *Dokl. Akad. Nauk SSSR*, **166**, 1347 (1966).
48. O. A. Reutov, I. P. Beletskaya, and L. P. Filippenko, *Nauchn. Dokl. Vissch. Shkoly. Khim. i. Khim. Tekhnolog.*, **1958**, 4, 754.
49. F. Jensen and J. Miller, *J. Am. Chem. Soc.*, **86**, 4735 (1964).
50. G. F. Wright, *Can. J. Chem.*, **30**, 268 (1952).
51. A. N. Nesmeyanov, O. A. Reutov, J. T. U, and T. C. Lu, *Izv. Akad. Nauk SSSR, Otd. Khim. Nauk*, **1958**, 1327.
52. O. A. Reutov and E. V. Uglova, *Izv. Akad. Nauk SSSR, Otd. Khim. Nauk*, **1959**, 1691.
53. H. B. Charman, E. D. Hughes, and C. K. Ingold, *J. Chem. Soc.*, **1959**, 2523, 2530.
54. F. R. Jensen, *J. Am. Chem. Soc.*, **82**, 2469 (1960).
55. R. E. Dessy and I. K. Lee, *J. Am. Chem. Soc.*, **82**, 689 (1960).
56. H. B. Charman, E. D. Hughes, C. K. Ingold, and F. G. Thorpe, *J. Chem. Soc.*, **1961**, 1121.
57. E. D. Hughes, C. K. Ingold, F. G. Thorpe, and H. C. Volger, *J. Chem. Soc.*, **1961**, 1133.
58. H. B. Charman, E. D. Hughes, C. K. Ingold, and H. C. Volger, *J. Chem. Soc.*, **1961**, 1142.
59. E. D. Hughes and H. C. Volger, *J. Chem. Soc.*, **1961**, 2359.
60. N. A. Nesmeyanov and O. A. Reutov, *Dokl. Akad. Nauk SSSR*, **144**, 126 (1962).
61. N. A. Nesmeyanov and O. A. Reutov, *Tetrahedron*, **20**, No. 12, 2803 (1964).
62. K. Brodersen and U. Schlenker, *Chem. Ber.*, **94**, 3304 (1961).
63. R. E. Dessy, I. K. Lee, and J. Y. Kim, *J. Am. Chem. Soc.*, **83**, 1163 (1961).
64. S. Winstein, T. G. Traylor, and C. S. Garner, *J. Am. Chem. Soc.*, **77**, 3741 (1955).
65. O. A. Reutov, T. A. Smolina, and K. V. Khu, *Izv. Akad. Nauk. SSSR, Otd. Khim. Nauk*, **1959**, 559.

66. A. E. Borisov and N. V. Novikova, *Izv. Nauk SSSR, Otd. Khim. Nauk* **1957**, 1258.

67. A. N. Nesmeyanov, A. E. Borisov, and N. V. Novikova, *Izv. Akad. Nauk SSSR, Otd. Khim. Nauk*, **1959**, 1216.

68. R. Ja. Levina and V. N. Kostin, *Zh. Obshch. Khim.*, **23**, 1054 (1953).

69. R. Ja. Levina, V. N. Kostin, and V. A. Tartakovskii, *Zh. Obshch. Khim.*, **27**, 881 (1957).

70. G. Spengler and A. Weber, *Brennst.-Chem.*, **43**, 234 (1962).

71. O. Dimroth, *Ber.*, **35**, 2853 (1902).

72. A. N. Nesmeyanov and R. Kh. Shatskaya, *Zh. Obshch. Khim.*, **5**, 1268, (1935).

73. F. C. Whitmore and G. L. Woodward, *J. Am. Chem. Soc.*, **48**, 533 (1926).

74. J. D. Loudon, *J. Chem. Soc.*, **1933**, 825.

75. I. L. Maynard, *J. Am. Chem. Soc.*, **54**, 2118 (1933).

76. W. Steinkopf, W. Bielenberg, and H. Augenstad-Jensen, *Ann. Chem.*, **430**, 41 (1923).

77. E. Profft and K. H. Otto, *J. Prakt. Chem.*, **36**, 8, 156 (1959).

78. M. Rausch, M. Vogel, and H. Rosenberg, *J. Org. Chem.*, **22**, 900 (1957).

79. H. Goldwhite, R. N. Haszeldine, and R. N. Mukherjee, *J. Chem. Soc.*, **1961**, 3825.

80. H. J. Emeléus and R. N. Haszeldine, *J. Chem. Soc.*, **1949**, 2948, 2953.

81. P. E. Aldrich, E. G. Howard, W. J. Linn, W. J. Middleton, and W. H. Sharkey, *J. Org. Chem.*, **28**, 184 (1963).

82. F. Hein and K. Wagler, *Ber.*, **58**, 1499 (1925).

83. F. Hein and K. Wagler, DRP 444666; *Chem. Zentr.*, **1927**, II, 740.

84. I. T. Eskin, *Izv. Akad. Nauk SSSR, Otd. Chim. Nauk*, **1942**, 302.

85. V. L. Foss, I. F. Lutsenko, and A. N. Nesmeyanov, *Zh. Obshch. Khim.*, **35**, (1965).

86. J. D. Loudon, *J. Chem. Soc.*, **1935**, 535.

87. G. Drefahl and G. Stange, *J. Prakt. Chem.*, **9**, 311 (1959).

88. P. I. Petrovich, *Zh. Vses. Khim. Obshchestva D. I. Mendeleeva*, **5**, 106 (1960).

89. A. N. Nesmeyanov and E. M. Toropova, *Zh. Obshch. Khim.*, **4**, 667 (1943).

90. I. T. Eskin, *Izv. Akad. Nauk SSSR, Otd. Khim. Nauk*, **1947**, 405.

91. G. Rodighiero, *Ann. Chim. Applikata*, **39**, 621 (1949).

92. A. N. Nesmeyanov and L. G. Makarova, *Zh. Obshch. Khim.*, **1**, 598 (1931).

93. I. T. Eskin, *Izv. Akad. Nauk SSSR, Otd. Khim. Nauk*, **1942**, 297.

94. C. D. Hurd and C. J. Morissey, *J. Am. Chem. Soc.*, **77**, 4658 (1955).

95. W. Steinkopf, *Ann. Chem.*, **413**, 310 (1917).

96. W. Steinkopf and M. Bauermeister, *Ann Chem.*, **403**, 50 (1914).

97. S. Hilpert and G. Grüttner, *Ber.*, **47**, 186 (1914).

98. J. Braun, *Ber.*, **47**, 490 (1914).

99. E. Dreher and R. Otto, *Ann Chem.*, **154**, 93 (1870).

100. R. Adams, F. L. Roman, and W. N. Sperry, *J. Am. Chem. Soc.*, **44**, 1789 (1922).

101. J. Sand and F. Singer, *Ber.*, **35**, 3170 (1902).

102. M. J. Abercrombie, A. Rodgman, K. R. Bharucha, and G. F. Wright, *Can. J. Chem.*, **37**, 1328 (1959).

103. R. Kh. Freidlina and N. S. Kochetkova, *Izv. Akad. Nauk SSSR, Otd. Khim. Nauk*, **1945**, 198.

104. A. N. Nesmeyanov, A. E. Borisov, and N. A. Vol'kenau, *Izv. Akad. Nauk SSSR, Otd. Khim. Nauk*, **1956**, 162.

105. H. J. Emeléus and R. N. Haszeldine, *J. Chem. Soc.*, **1949**, 2953.

106. K. Ichikawa, H. Ouchi and S. Fukushima, *J. Org. Chem.*, **24**, 1129 (1959).

107. R. J. Levina and V. N. Kostin, *Zh. Obshch. Khim.*, **28**, 3307 (1958).

108. H. Gilman and M. M. Barnett, *Rec. Trav. Chim.*, **55**, 563 (1936).
109. H. Gilman and H. L. Yale, *J. Am. Chem. Soc.*, **72**, 8 (1950).
110. J. Burdon, P. L. Coe, M. Fulton, and J. C. Tatlow, *J. Chem. Soc.*, **1964**, 2673.
111. A. Brook and G. F. Wright, *J. Org. Chem.*, **22**, 1314 (1957).
112. G. F. Wright, *Ann. N.Y. Acad. Sci.*, **66**, 436 (1957).
113. H. Sawatzky and G. F. Wright, *Can. J. Chem.*, **36**, 1555 (1958).
114. A. G. Brook, R. Donovan, and G. F. Wright, *Can. J. Chem.*, **31**, 536 (1953).
115. G. F. Wright, D. Chow, and J. H. Robson, *Can. J. Chem.*, **43**, 312 (1965).
116. G. F. Wright, *Can. J. Chem.*, **30**, 268 (1952).
117. H. B. Henbest and B. Nicholls, *J. Chem. Soc.*, **1959**, 227.
118. O. A. Seide, S. M. Sherling, and G. I. Braz, *J. Prakt. Chem.*, **138**, (1), 55 (1933).
119. A. Rodgman and G. F. Wright, *J. Am. Chem. Soc.*, **76**, 1382 (1954).
120. I. L. Maynard and H. C. Howard, *J. Chem. Soc.*, **123**, 960 (1923).
121. N. N. Mel'nikov and M. S. Rokitskaya, *Zh. Obshch. Khim.*, **7**, 2596 (1937).
122. B. G. Gowenlock and J. Trotman, *J. Chem. Soc.*, **1957**, 2114.
123. C. A. Kraus, *J. Am. Chem. Soc.*, **35**, 1732 (1913).
124. G. Sachs and K. Fürst, *Monatsh. Chem.*, **53/54**, 550 (1929).
125. R. W. Beattie and F. C. Whitmore, *J. Am. Chem. Soc.*, **55**, 1567 (1922).
126. DRP 249729; *Chem. Zentr.*, **1912**, II, 777.
127. R. J. Ouellette and B. G. Van Leuwen, *J. Org. Chem.*, **30**, 3966 (1966).
128. F. C. Whitmore and R. J. Sobatzki, *J. Am. Chem. Soc.*, **55**, 1128 (1923).
129. E. Biilmann, *Ber.*, **35**, 2581 (1902).
130. E. Biilmann and J. Witt, *Ber.*, **43**, 1070 (1909).
131. K. S. Patel and B. N. Mankad, *Current. Sci.*, **30**, 335 (1961).
132. R. Brieger and W. Schulemann, *J. Prakt. Chem.* (2), **89**, 105 (1914).
133. H. Laubie, *Bull. Soc. Pharm. Bordeaux*, **96**, 65 (1957); *Chem. Abstr.*, **51**, 15889 (1957).
134. A. N. Nesmeyanov and R. Kh. Freidlina, *Izv. Akad. Nauk SSSR, Otd. Khim. Nauk*, **1945**, 150.
135. A. N. Nesmeyanov and I. F. Lutsenko, *Dokl. Akad. Nauk SSSR*, **59**, 707 (1948).
136. F. C. Whitmore and E. Middleton, *J. Am. Chem. Soc.*, **43**, 622 (1921).
137. A. Hantzsch and S. M. Auld, *Ber.*, **39**, 1110 (1906).
138. D. Seyferth and R. H. Towe, *Inorg. Chem.*, **1**, 185 (1962).
139. R. Adams, et al., *Organic Syntheses*, under red., III, 65 (1923).
140. S. Lamdan, *Rev. Assoc. Bioquim. Argentina*, **114**, 295 (1947); *Chem. Abstr.*, **43**, 4236 (1949).
141. M. D. Rausch, *Inorg. Chem.*, **1**, 414 (1962).
142. F. C. Whitmore, *J. Am. Chem. Soc.*, **41**, 1841 (1915).
143. M. D. Rausch, *J. Org. Chem.*, **28**, 3337 (1963).
144. K. V. Vijanaraghavan, *J. Indian. Chem. Soc.*, **20**, 318 (1943).
145. D. Spinelli and A. Salvemini, *Ann. Chim. (Roma)*, **50**, 1423 (1960).
146. K. A. Hofmann and H. Kirmreuther, *Ber.*, **42**, 4232 (1909); **41**, 314 (1908).
147. K. A. Hofmann, *Ber.*, **31**, 1898; 2213 (1898); **32**, 875 (1899); **37**, 4460 (1904).
148. W. Schoeller and W. Schrauth, *Ber.*, **53**, 637 (1920).
149. V. P. Chalov, *Zh. Obshch. Khim.*, **18**, 608 (1958).
150. A. E. Borisov, private communication.
151. L. Vecchiotti, *Gazz. Chim. Ital.*, **58**, 243 (1928).
152. H. Gilman and G. F. Wright, *J. Am. Chem. Soc.*, **55**, 3302 (1933).
153. W. J. Chute, W. M. Orchard, and G. F. Wright, *J. Org. Chem.*, **6**, 157 (1941).
154. T. Ukal, *J. Pharm. Soc. Japan*, **48**, 171 (1927); *Chem. Zentr.*, 1929, I, 1108.

155. A. N. Nesmeyanov, K. N. Anisimov, and Z. P. Valueva, *Izv. Akad. Nauk, Otd. Khim. Nauk*, **1962**, 1683.
156. A. N. Nesmeyanov, V. A. Sazonova, V. N. Drozd, and L. A. Nikonova, *Dokl. Akad. Nauk SSSR*, **131**, 1088 (1960).
157. A. N. Nesmeyanov, E. G. Perevalova, R. V. Golovnja, and O. A. Nesmeyanova, *Dokl. Akad. Nauk SSSR*, **97**, 459 (1954).
157a. A. N. Nesmeyanov, I. F. Lutsenko, and R. M. Khomutov, *Dokl. Akad. Nauk SSSR*, **88**, 837 (1953).
158. A. N. Nesmeyanov and N. K. Kochetkov, *Izv. Akad. Nauk SSSR, Otd. Khim. Nauk*, **1949**, 76.
159. A. N. Nesmeyanov and N. K. Kochetkov, *Izv. Akad. Nauk SSSR, Otd. Khim. Nauk*, **1949**, 305.
160. A. N. Nesmeyanov, N. K. Kochetkov, and V. M. Dashunin, *Izv. Akad. Nauk SSSR, Otd. Khim. Nauk*, **1950**, 77.
161. A. Hantzsch and F. Gajewski, *Ann. Chem.*, **392**, 302 (1912).
162. R. J. Cross, A. Lauder, and G. E. Coates, *Chem. Ind.*, **47**, 2013 (1962).
163. A. N. Nesmeyanov and A. E. Borisov, *Izv. Akad. Nauk SSSR, Otd. Khim. Nauk*, **1945**, 146.
164. G. E. Coates and A. Lauder, *J. Chem. Soc.*, **1965**, 1857.
165. R. E. Dessy, W. I. Budde and C. Woodruff, *J. Am. Chem. Soc.*, **84**, 1172 (1962).
166. F. E. Paulik and R. E. Dessy, *Chem. Ind.*, **37**, 1650 (1962).
167. T. Mukajama, J. Kuwajima, and Z. Suzuki, *J. Org. Chem.*, **28**, 2024 (1963).
168. F. J. Sowa, U.S. Patent 2423261; *Chem. Abstr.*, **42**, 1376 (1948).
169. F. Hein and H. Pobloth, *Z. Anorg. Allgem. Chem.*, **248**, 84 (1911); *Chem. Abstr.*, **37**, 2676 (1943).
170. F. Hein and E. C. Heuser, *Z. Anorg. Allgem. Chem.*, **249**, 293 (1942).
171. S. Hilpert and P. Ditmar, *Ber.*, **46**, 3740 (1913).
172. L. Pesci, *Gazz. Chim. Ital.*, **28**, II, 446 (1898).
173. S. Coffey, *J. Chem. Soc.*, **128**, 639 (1926).
174. W. Schoeller, W. Schrauth, and R. Hueter, *Ber.*, **53**, 637 (1920).
175. L. Pesci, *Gazz. Chim. Ital.*, **23**, II, 529 (1893).
176. G. B. Buckton, *Ann. Chem.*, **108**, 105 (1858).
177. E. Fourneau and A. Vila, *J. Pharm. Chim.*, **6**, VII, 443 (1912); *Zentralbl.*, **1913**, I, 20.
178. F. C. Whitmore and E. B. Middleton, *J. Am. Chem. Soc.*, **45**, 1753 (1923).
179. R. T. McCutchan and K. A. Kobe, *Ind. Eng. Chem.*, **46**, 675 (1954).
180. A. N. Nesmeyanov and I. I. Kritskaya, *Dokl. Akad. Nauk SSSR*, **121**, 477 (1958).
181. W. T. Reichle, *Inorg. Chem.*, **1**, 650 (1962).
182. F. C. Whitmore, E. R. Hanson, and F. L. Carnahan, *J. Am. Chem. Soc.*, **51**, 894 (1929).
183. I. N. Grandberg, A. N. Kost, and N. I. Shabel'skaya, *Zh. Obshch. Khim.*, **30**, 2831 (1960).
184. E. D. Venus-Danilova, *Tr. Leningrad. Technol. Instit. im. Lensoveta*, **60**, 32 (1960).
185. I. F. Lutsenko and E. I. Yurkova, *Izv. Akad. Nauk SSSR, Otd. Khim. Nauk*, **1956**, 27.
186. A. N. Nesmeyanov, A. E. Borisov, and N. A. Vol'kenau, *Izv. Akad. Nauk SSSR, Otd. Khim. Nauk*, **1954**, 992.
187. M. Cais and J. Kozikowsky, *J. Am. Chem. Soc.*, **82**, 5667 (1960).
188. S. Hilpert and G. Grüttner, *Ber.*, **48**, 906 (1915).
189. M. S. Kharasch and R. Marker, *J. Am. Chem. Soc.*, **48**, 3130 (1926).

190. A. N. Nesmeyanov and D. N. Kravtsov, *Izv. Akad. Nauk SSSR, Otd. Khim. Nauk*, **1962**, 431.

191. A. N. Nesmeyanov, D. N. Kravtsov, B. A. Faingor, and L. I. Petrovskaya, *Izv. Akad. Nauk SSSR, Ser. Khim*, 534 (1968).

192. G. Wittig and H. F. Ebel, *Ann. Chem.*, **650**, 20 (1961).

193. A. Bernardi, *Gazz. Chim. Ital.*, **67**, 380 (1937).

194. G. F. Wright, *J. Am. Chem. Soc.*, **58**, 2653 (1936).

195. A. K. Ghosh, C. E. Hansing, A. I. Stutz, and A. G. MacDiarmid, *J. Chem. Soc.*, **1962**, 403.

196. A. N. Nesmeyanov, A. E. Borisov, I. S. Savel'eva, and E. I. Golubeva, *Izv. Akad. Nauk SSSR, Otd. Khim. Nauk*, **1958**, 1490.

197. A. N. Nesmeyanov, A. E. Borisov, and N. V. Novikova, *Dokl. Akad. Nauk SSSR*, **96**, 289 (1954).

198. A. N. Nesmeyanov, *Uchen. Zap. Moskov. Gos. Univer.*, **132**, 5 (1950).

199. A. N. Nesmeyanov and M. I. Kabachnik, *Zh. Obshch. Khim.*, **25**, 41 (1955).

200. A. N. Nesmeyanov, A. E. Borisov, and V. D. Vil'chevskaya, *Dokl. Akad. Nauk SSSR*, **90**, 383 (1953).

201. V. L. Foss, Dissertation, Moskau, 1963.

202. J. Sand, *Ber.*, **34**, 2909 (1901).

203. D. Grdenić, *Acta Cryst.*, **5**, 367 (1952).

204. S. Winstein and T. G. Traylor, *J. Am. Chem. Soc.*, **78**, 2597 (1956).

205. O. A. Reutov, T. C. Lu, and Yu. G. Bundel, *Vestn. Mosk. Univ.*, **13**, No. 5, 111 (1958).

206. R. Otto, *Ann. Chem.*, **154**, 190 (1870).

207. F. C. Whitmore and E. Middleton, *J. Am. Chem Soc.*, **44**, 1546 (1922).

208. W. Steinkopf, *Ann. Chem.*, **424**, 40 (1921).

209. W. Steinkopf and P. Leonhardt, *Ann. Chem.*, **495**, 166 (1932).

210. F. Challenger and S. A. Miller, *J. Chem. Soc.*, **1939**, 1005.

211. W. Steinkopf, *Ann. Chem.*, **428**, 123 (1922).

212. M. T. Bogert and P. P. Herrera, *J. Am. Chem. Soc.*, **45**, 328 (1923).

213. N. N. Mel'nikov and M. S. Rokitskaya, *Zh. Obshch. Khim.*, **11**, 592 (1941).

214. F. E. Paulik, S. Y. E. Green, and R. E. Dessy, *J. Organometall. Chem.*, **3**, 229 (1965).

215. G. Wittig and F. Bickelhaupt, *Ber.*, **91**, 883 (1958).

Chapter 16

Decomposition of Organomercury Compounds by Oxygen and Other Oxidants, and Light and Thermal Treatment

I. INTRODUCTION

The reactions of organomercury compounds mentioned in the title are included in one chapter due to the fact that all these reactions (except for some reactions of autooxidation described below) proceed by a free-radical mechanism. The nature, mechanism, and behavior of radicals formed by oxidants, light, and higher temperatures on organomercury compounds may be very similar, in some cases, while in other cases may differ considerably.

II. DECOMPOSITION BY OXYGEN AND OTHER OXIDANTS

Organomercury compounds, (in particular with a mercury at a secondary or tertiary atom of carbon) during continuous contact with atmospheric oxygen, and compounds with the mercury at a primary carbon atom (using oxygen in solution) are oxidized, isolating metallic mercury. The organic residue produces alcohols, ketones, products of interactions with the solvent, etc. Metallic mercury is always isolated, and many products are formed during oxidation of the organomercury compounds by peroxides, the salts of divalent mercury, and by some other oxidants. These reactions have no preparative significance.

A. Oxidation by Oxygen

1. Introduction

Aliphatic organomercury compounds, R_2Hg with mercury at a secondary or tertiary carbon atom during long-time aging, are gradually oxidized by atmospheric oxygen, which is evident by the darkening of these compounds due to isolation of metallic mercury. The products of oxidation of the organic part of the molecule are ketones and alcohols.

Oxidation of these and other organomercury compounds by oxygen in solutions has been thoroughly investigated. The products of oxidation produced in solutions (in benzene, aliphatic fatty hydrocarbons, alcohols, chloroform, carbon tetrachloride) are as follows: Hg, RHgOR, RHgOH, products of oxidation of R of original organomercury compounds (respective alcohols and ketones), unsaturated and saturated hydrocarbons formed by the reaction of radicals with organomercury compounds as well as products of oxidation of the solvent, and products of interaction of the solvent with the original organomercury compound, provided aliphatic halide-containing compounds are used as a solvent.

The primary organomercury compound, dibenzylmercury, in a solvent (benzene, $CHCl_3$, CCl_4), is oxidized by oxygen, thus producing benzaldehyde (31% in benzene). Oxidizability is evidently a common property of alkylmercury compounds in contact with oxygen under certain conditions.

2. Reaction Mechanism

Oxidation of R_2Hg by oxygen is effected through formation of free radicals and constitutes a combination of a number of parallel and successive reactions.

Razuvaev, et al., who have thoroughly studied the oxidation of organomercury compounds, have suggested the following oxidation scheme of R_2Hg which explains the formation of the obtained products:[1-4]

$$R_2Hg + O_2 \longrightarrow [R_2Hg \cdot O_2] \longrightarrow \left[\begin{matrix} ROOHgR \\ (A) \end{matrix}\right] \xrightarrow{\begin{matrix} R_2Hg \\ \end{matrix}} \begin{matrix} \xrightarrow{} 2\ ROHgR \quad a \\ (B) \\ \searrow RO_2 \cdot + RHg \cdot \quad b \\ (C) \end{matrix} \qquad (1)$$

Oxygen attack on the metal is accompanied by subsequent migration of the alkyl group from metal to oxygen, thus producing an oxide (A). The latter reacts with an unoxidized organomercury compound a, producing an alkoxycompound or is partially decomposed b into peroxyalkyl and alkylmercury radicals.

The peroxyalkyl radical evidently reacts with the original compound as an oxidizer producing alcohols, ketones, and an RHg· radical:

$$RO_2 \cdot + R_2Hg \longrightarrow ROH + R(-H)=O + RHg \cdot \qquad (2)$$

Formation of the RHg· radical is proved by formation of substantial amounts of RHgCl when the reaction proceeds in halide-containing solvents.

Besides, the RHg· radical can be further oxidized c with formation of an alkylmercuryperoxide radical (D) or is decomposed into mercury and an alkyl radical d:

$$RHg \cdot \begin{cases} \xrightarrow{O_2} RHgO_2 \cdot \quad c \\ \qquad\quad (D) \\ \longrightarrow R \cdot + Hg \quad d \end{cases} \qquad (3)$$

Alkylradicals R· can be produced by thermal decomposition of R_2Hg:

$$R_2Hg \longrightarrow RHg \cdot + R \cdot \qquad (4)$$

Thermal decomposition of RHgOR occurs as follows:

$$RHgOR \longrightarrow RHg \cdot + RO \cdot$$

Radicals R· react with the solvent producing common products of similar reactions: RHal, provided halide-containing solvents are involved, and RH, in case hydrogen-containing aliphatic solvents are involved. Radicals R·, RHgO·, and RO· react with R_2Hg as well, for example:

$$R· + R_2Hg \longrightarrow C_nH_{2n+2} + C_nH_{2n} + RHg·$$
$$RHgO·(RO·) + R_2Hg \longrightarrow RHgOH(ROH) + C_nH_{2n} + RHg·$$

As di-tertiaryalkylmercury is oxidized (di-*tert*-amylmercury was investigated) a tertiary alkylperoxide radical $RO_2·$ is produced in accordance with *b* of Eq. (1) and is decomposed according to the following equation:

$$C_2H_5(CH_3)_2CO_2· \longrightarrow (CH_3)_2CO + C_2H_5O·$$

3. Application of the Reaction

When dicyclohexylmercury is continuously in contact with atmospheric oxygen without a solvent, it is oxidized releasing metallic mercury and producing cyclohexanol and cyclohexanone.[5] The reaction probably proceeds according to the following equation:

$$(C_6H_{11})_2Hg \xrightarrow{O_2} [(C_6H_{11})_2Hg·O_2] \longrightarrow C_6H_{11}OH + C_6H_{10}O + Hg$$

As R_2Hg is oxidized by oxygen in a solvent, formation of other products was observed as well as active participation of the solvent in the reaction.

If dicyclohexylmercury is oxidized by free oxygen in benzene[6] at 60°C, the main products are cyclohexanol (57%), cyclohexanone (45%), and metallic mercury (100%). In addition to the oxidation products, disproportionation produces cyclohexane, cyclohexene, and a small amount of biphenyl. (Radicals RO· are the source of the dimerizable radicals $C_6H_5·$).

When diisopropylmercury is oxidized by oxygen in benzene,[1] the latter does not participate in the reaction to any great extent (only 5% yield of the sum of its derivatives, such as diphenylmercury, diphenyl, cumene, etc., is produced). The main products of the reaction are isopropylmercury isopropylate and hydroxide, isopropyl alcohol, acetone, mercury, propylene, and traces of propane. Similar products are formed when the oxidation is conducted in cyclohexane;[1] in this case the yield of acetone is higher than that of isopropyl alcohol. However, if the reaction is conducted in cyclohexene,[1] the yield of isopropyl alcohol is higher than that of acetone; propane is produced instead of propylene, and isopropylmercury hydroxide is produced instead of isopropylmercury isopropylate.

The reaction products, similar to those produced by diisopropylmercury, are formed during oxidation of di-*tert*-amylmercury in benzene and cyclohexane: *tert*-amyl alcohol, mercury, 2-methylbutene, and 2-methylbutane.[1] In addition, the organic residue (the *tert*–amyl radical) is decomposed and acetone is formed.

However, an organomercury compound with the mercury at the primary atom of carbon, such as dibenzylmercury, is oxidized when oxygen is passed into its solution (in C_6H_6, $CHCl_3$, CCl_4); benzaldehyde is produced in about 31% yield (in benzene at 50°C for 120 hr).[1]

The same main reaction products such as ketones, alcohols, and metallic mercury are produced during oxidation of R_2Hg $(R = i$-C_3H_7,[7] $cyclo$-C_6H_{11}[5]).

During oxidation of R_2Hg, diisopropyl-,[1,7] dicyclohexyl-,[5] and di-$tert$-amyl-mercury[1] in halide-containing solvents such as CCl_4 and $CHCl_3$, $RHgCl$ was always produced [apart from oxidation of $(tert$-$C_5H_{11})_2Hg]$, as were CO and CO_2, the latters due to oxidation of the solvent. Diisopropylmercury in CCl_4 produces RCl as well. The other products of oxidation are mainly similar to those produced during oxidation in hydrocarbons and alcohols: ketones, alcohols, Hg, and saturated and unsaturated hydrocarbons (in some cases); Hg and i-C_3H_7OH are not produced during oxidation of $(i$-$C_3H_7)_2Hg$ in CCl_4 at a temperature slightly higher than room temperature (44°C for 18 hr). In $CHCl_3$, the quantity of alcohol increases and RCl was not formed.[7]

B. Oxidation by Acyl Peroxides

1. Application of the Reaction

Some R_2Hg compounds were oxidized by benzoyl peroxide. With $(C_6H_5)_2Hg$,[8] $(C_2H_5)_2Hg$,[8,9] and $(i$-$C_3H_7)_2Hg$,[10] the major reaction products are RH-benzoates.

Upon oxidation of $(C_2H_5)_2Hg$ (without a solvent), by benzoyl peroxide, $C_2H_5HgOCOC_6H_4C_2H_5$, benzoic and ethylbenzoic acids, C_2H_6, C_2H_4, $C_2H_5C_6H_5$, CO_2, Hg, and n-C_4H_{10} are produced.[8,9] The products obtained from the reaction of $(i$-$C_3H_7)_2Hg$ with the peroxide (without a solvent) are i-C_3H_7Hg-benzoate, benzoic acid, propane, propylene, 2,3-dimethylbutane, isopropylbenzene, isopropyl benzoate, CO_2, Hg, and a small amount of other products.[10] As distinct from the reaction with diethylmercury, the isopropyl group does not enter into the benzene nucleus of the peroxide.[10]

Use of deuterobenzene as a solvent for the oxidation of $(i$-$C_3H_7)_2Hg$ by benzoyl peroxide showed that the solvent participates in the formation of the reaction products CH_3CHDCH_3 and C_6H_5COOD. Isopropylmercury benzoate (154%) and the usual products that obtained during decomposition[10] were also produced.

2. Reaction Mechanism

Razuvaev and co-workers[10] believe that the reaction of R_2Hg $(R = C_2H_5$ and i-$C_3H_7)$ with benzoyl peroxide proceeds through formation of an intermediate complex, for example, in the case of $(i$-$C_3H_7)_2Hg$:

$$(i\text{-}C_3H_7)_2Hg + (C_6H_5CO_2)_2 \longrightarrow [(i\text{-}C_3H_7)_2Hg \cdot (C_6H_5COO)_2] \text{ (E)} \longrightarrow$$
$$i\text{-}C_3H_7HgO_2CC_6H_5 + C_6H_5COOH + C_3H_6 \quad (4)$$

The C_6H_5COOH formed partially reacts with the original $(i\text{-}C_3H_7)_2Hg$:

$$(i\text{-}C_3H_7)_2Hg + C_6H_5COOH \longrightarrow i\text{-}C_3H_7HgOCOC_6H_5 + C_3H_8 \qquad (5)$$

They also believe that along with reaction (4), free-radical processes play a substantial part in the reactions of the solvent since it is necessary to expect the interaction of the benzoate radicals formed as a result of dissociation of a benzoyl peroxide with diisopropyl mercury to explain a high yield of i-propylmercury benzoate. A small part of these radicals is decomposed into CO_2 and C_6H_5 radicals. The presence of the isopropylbenzene and 2,3-dimethylbutane proves that $i\text{-}C_3H_7$ radicals are involved in the reaction and are consumed in a substantial amount during polymerization of the propylene. Isolation of mercury was caused by decomposition of $(i\text{-}C_3H_7)_2Hg$ initiated by the free radicals.

During interaction of $(C_2H_5)_2Hg$[8,9] with benzoyl or acetylbenzoyl peroxide (without a solvent at 70–95°C for 10–12 hr), by probable formation of an intermediate complex, type (E), $C_2H_5HgOCOR$ (R = C_6H_5— and CH_3—) and $C_2H_5C_6H_4$—, which is of particular interest, are produced, which proves that C_2H_5 enters into the benzene nucleus. The respective free acids including $C_2H_5C_6H_4COOH$, ethane, ethylene, butane, and a small amount of metallic mercury are produced as well.

When $(C_2H_5)_2Hg$ and $(C_6H_5)_2Hg$ are oxidized by benzoyl peroxide in benzene, the main reaction products are ethyl- and phenylmercury benzoates.[8,9] In the former case the ethyl radical enters into the benzene nucleus of benzoyl peroxide (formation of ethylbenzoic acid and ethyl ethylmercury-benzoate) and into the solvent, causing the formation of ethylbenzene.

C. Oxidation by tert-Butyl Hydroperoxide

In the sole reaction investigated, i.e., the reaction of diisopropylmercury with tert-butyl hydroperoxide[10] (in benzene at 70°C), the latter reacts as an organic acid, releasing a hydrocarbon:

$$R_2Hg + t\text{-}C_4H_9OOH \longrightarrow RH + RHgOOC_4H_9\text{-}tert$$
$$(F)$$

The tert-butyrate of isopropylmercury, one of the main reaction products (43.8% yield), oxidizes the original R_2Hg:

$$RHgO_2C_4H_9\text{-}t + R_2Hg \longrightarrow RHgOR + RHgOC_4H_9\text{-}t$$

This reaction course is proved by the formation of propane and isopropyl-mercury isopropylate and the peroxide compound (F) in equimolar amounts.

The observed small deviation from the expected reaction course is explained by low thermal stability of $RHgOC_4H_9\text{-}tert$, which readily dissociates into radicals:

$$RHgOC_4H_9\text{-}tert \longrightarrow RHg\cdot + tert\text{-}C_4H_9O\cdot$$
$$\downarrow$$
$$R\cdot + Hg$$

As a result, hydrocarbon RH, alcohols $i\text{-}C_3H_7OH$ and $tert\text{-}C_4H_9OH$ and metallic mercury are produced. Thus, the reaction under consideration proceeds through formation of an organomercury peroxide. This confirms the assumption that similar peroxides participate in the oxidation by oxygen of organomercury compounds and are then reduced by the original R_2Hg to the respective $RHgOR'$.

D. Oxidation by Divalent Mercury Salts

1. Application of the Reaction

Some organomercury compounds which have the mercury at the secondary carbon atom are advantageously oxidized by divalent mercury salts.[11,11a]

Depending on the reaction medium, the following products are produced: alcohols (and products of their further oxidation) in water, ethers, and metallic mercury in alcohols.

The oxidant, Hg(II) salt, reduces to Hg(I) or to metallic mercury. If $RHgNO_3$ is oxidized by the action of $Hg(NO_3)_2$, the esters of nitric acid are produced,[11] the formation of which is unusual under the experimental conditions.

$RHgNO_3$ is oxidized by $Hg(NO_3)_2$ faster than $RHgO_2CCH_3$ is by $Hg(O_2CCH_3)_2$.[11] Oxidation of $(CH_3)_2C(OCH_3)CH(C_6H_5)HgO_2CCH_3$ by $Hg(O_2CCH_3)_2$ in CH_3OH was carried out in the presence of $BF_3 \cdot O(C_2H_5)_2$.[11]

The oxidation reaction by HgX_2 is second order and stereospecific and proceeds with Walden inversion. The reaction is of a homopolar nature.

2. Reaction Mechanism

The authors suggested[11] the following reaction scheme for the radical process in the system $R'HgX + HgX_2 + ROH$:

$$HgX_2 \rightleftharpoons X\cdot + \cdot HgX \tag{1}$$
$$HgX_2 + ROH \rightleftharpoons ROHgX + HX \tag{2}$$
$$ROHgX \rightleftharpoons RO\cdot + \cdot HgX \tag{3}$$

(The processes proceed so slowly that the chains are broken by oxygen.)

$$R'HgX \rightleftharpoons R'\cdot + \cdot HgX \tag{4}$$

(This reaction proceeds faster and is able to neutralize the action of O_2.)

$$\cdot HgX + ROHgX \longrightarrow RO\cdot + Hg_2X_2 \tag{5}$$
$$RO\cdot + R'HgX \longrightarrow R'OR + \cdot HgX \tag{6}$$

or

$$\cdot HgX + HgX_2 \longrightarrow X\cdot + Hg_2X_2 \tag{7}$$
$$X\cdot + R'HgX \longrightarrow R'X + \cdot HgX \tag{8}$$

(This reaction is the source of the nitrate type products.)

The reactions (9)–(13) terminate the radical chain:

$$RO\cdot \longrightarrow H\cdot + (R-H)\,O \tag{9}$$
$$H\cdot + ROHgX \longrightarrow ROH + \cdot HgX \tag{10}$$
$$RO\cdot + H\cdot \longrightarrow ROH \tag{11}$$
$$2\cdot HgX \longrightarrow Hg_2X_2 \tag{12}$$
$$2\cdot HgX + O_2 \longrightarrow XHgOOHgX \tag{13}$$

Equations (5) and (6) explain why the uncatalyzed reaction, (6), has second order (under reaction conditions the reaction rate depends on the concentration of ROHgX and RHgX), while the reaction catalyzed by BF_3 has first order. BF_3 labilizes the C—Hg bond and the reaction rate depends only on the concentration of ROHgX. The acid facilitates breakage of the C—Hg bond in R'HgX during reaction (6), creating an excess of $\cdot$HgX radicals, and thus the reaction depends only on the concentration of ROHgX.

E. Oxidation by Other Oxidants

Oxidation of cyclohexylmercury acetate (in CH_3OH) by peracetic acid yields the same products as oxidation by $Hg(NO_3)_2$ in CH_3OH—HgX_2: Hg_2X_2 and cyclohexanone (due to oxidation of the originally produced cyclohexanol).[11]

The homopolar oxidation of 4-hydroxycyclohexylmercury acetate by Fenton's reagent occurs as follows:[11]

F. Autooxidation of Organomercury Salts

1. Application of the Reaction

Organomercury compounds, having a mercury at the secondary carbon atom, $RHgO_2CCH_3$ ($R = cyclo$-C_6H_{11}, i-C_3H_7, and $(C_2H_5)_2CH$) in the presence of $BF_3\cdot O(C_2H_5)_2$, cyclohexylmercury trifluoromethylacetate and nitrate, without a catalyst, undergo monomolecular autoooxidation according to a heteropolar mechanism.[12] In this case, mercury and olefins are produced. The reaction takes place at $50°C$ in CH_3OH in the presence of HgX_2 which combines with the olefinic double bond (methoxymercuration reaction). It is proposed that catalysts such as Lewis acids increase the polarity of a C—Hg bond. $C_6H_5HgO_2CCH_3$ and the primary RHgX (n-$C_4H_9HgO_2CCH_3$) do not undergo this type of oxidation.

2. Reaction Mechanism

A possible scheme for the heteropolar process (R = *cyclo*-C_6H_{11}) is as follows:[12]

$$RHgOCOCH_3 + BF_3 \rightleftarrows RHg^+ + [BF_3OCOCH_3]^-$$
$$RHg^+ \longrightarrow Hg^\circ + R^+$$
$$R^+ \longrightarrow H^+ + C_6H_{10}$$

However, a reaction which goes through a cyclic, six-membered transition complex is also possible:

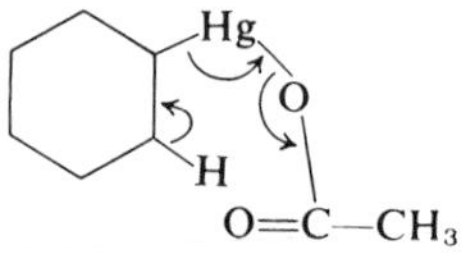

The autooxidation (in cases with R = *cyclo*-C_6H_{11}) is accompanied by certain homopolar oxidations. This is more evident in the case with *c*-$C_6H_{11}HgONO_2$ of higher reactivity, which is mainly oxidized in a homopolar way in CH_3OH in the presence of $Hg(NO_3)_2$ producing $C_6H_{11}OH$ and C_6H_{11}-ONO_2, and partially produces a cyclohexene which is transformed into a 2-methoxycyclohexylmercury nitrate, further oxidized in a homopolar way until formylcyclopentane is obtained. Hetero- and homopolar processes are typical for the system 1-methyl-1-cyclohexylmercuryacetate: $Hg(O_2CCH_3)_2$: CH_3OH as well. The reaction proceeds without a catalyst and shows the electron donor effect of the CH_3 group. The isolated products were 1-methoxy-1-methylcyclohexane, the product of the homopolar oxidation of 1-methyl-1-cyclohexylmercury acetate, and 2-methoxy-2-methylcyclohexylmercury chloride, which was formed from 1-methylcyclohexene, the product of the heteropolar oxidation.

G. *trans,cis* Isomerization of Organomercury Compounds of the Ethylene Series by Peroxides

The isomerization of trans isomers of organomercury compounds having the mercury at the double bond into their cis isomers by peroxides is of preparative significance. Transformation of *trans*-α-chlorovinylmercury chloride[12a] into the cis isomers occurs in xylene containing peroxides as well as in the presence of benzoyl peroxide, acetyl peroxide, or sodium peroxide in toluene, xylene, dioxane, and carbon tetrachloride upon heating to 95–100°C (in CCl_4, sealed flasks):

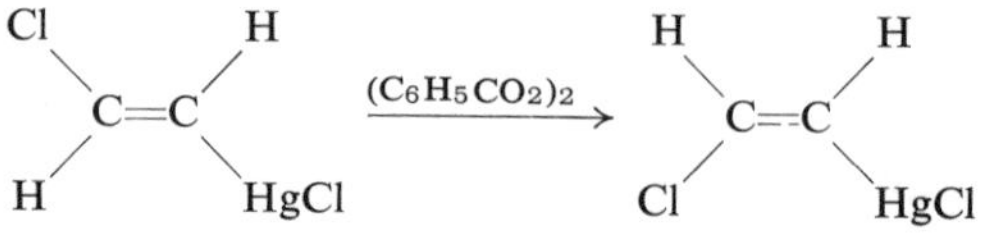

Addition of hydroquinone hinders the isomerization. In a similar way *trans-ω-*styrylmercury bromide was isomerized into the cis isomers[12b] by benzoyl peroxide in toluene.

III. LIGHT-INDUCED REACTIONS OF ORGANOMERCURY COMPOUNDS

A. Introduction

Various organomercury compounds, irradiated with ultraviolet light, decompose into radicals. In many cases, a large range of wavelengths was used for the decomposition; however, the C—Hg bond is broken by UV of shorter wavelength than the C—I bond.[13] Completely substituted mercury compounds are decomposed according to the following equations:

$$R_2Hg \xrightarrow{h\nu} RHg\cdot + R\cdot \tag{1}$$

$$RHg\cdot \xrightarrow{h\nu} R\cdot + Hg \tag{2}$$

Photodecomposition is also a strong possibility in one step:

$$R_2Hg \xrightarrow{h\nu} 2R\cdot + Hg \tag{3}$$

Organomercury salts and hydroxides are decomposed according to the following scheme:

$$RHgX \xrightarrow{h\nu} RHg\cdot + X\cdot \tag{4}$$

The RHg· radical can then be decomposed according to Eq. (2), but if RHgX is decomposed, the mercury will be isolated in the form of a mercury(I) salt as well. Depending on the solvent or, in some cases, on the nature of R during the decomposition of the organomercury compounds in solution, the decomposition either stops at steps (1) or (4) or continues by step (2). The radicals R·, RHg·, and X· which are produced react with the solvent; however, the radical R· can be partially or completely dimerized, disproportionated, etc.

The effect of γ-radiation on organomercury compounds in solution is similar to that of ultraviolet irradiation.[18]

Decomposition without a solvent in the gas phase (studied for R_2Hg) proceeds according to Eq. (1) and (2), or probably, in one stage, according to equation (3).

When some organomercury compounds having the mercury at the olefinic carbon atom are treated by UV light, *cis–trans* isomerization of these compounds occurs.

B. Photolysis in Solution

1. Aliphatic Solvents Free of Halides

Decomposition of R_2Hg and $RHgOH$ in solvents free of halides (in aliphatic hydrocarbons, alcohols, ethers, and esters) proceeds according to Eq. (1) and (2) or (3), and for $RHgOH$ according to Eq. (4) and (2), respectively, i.e. with isolation of metallic mercury. In this case, the radical R usually extracts hydrogen from the solvent (refer to Chapter 7). In some cases, dimerization of R· may occur. $ClHgCH_2CH_2OH$ irradiated by ultraviolet light in CH_3OH, in water, or in a mixture of CH_3OH–benzene (1:7) produces 1,4-butanediol (in CH_3OH the yield is 40%) and Hg_2Cl_2. As suggested by Weinmayr[10a] in this example, the reaction proceeds by a different mechanism, not with the formation of free radicals but rather with separation of the C—Hg bond, thus producing pseudoradicals which do not diffuse into the reaction medium. This reaction fails in pyridine and dimethylformamide. As a rule, photo-decomposition of $ClHgCH_2CH_2OH$ in i-C_3H_7OH produces ethanol and acetone.

2. Aromatic Compounds Used as Solvents

The radical R·, produced during photolysis of completely substituted mercury compounds[14,15] and organomercury salts[16] or hydroxides,[17] placed in a solvent (aromatic compound) usually reacts with the aromatic nucleus. However, some cases are known in which the reaction of aromatic substitution is not observed. For example, if the benzene solution of G is irradiated with ultraviolet light, a 25% yield of G' and metallic mercury (quantitative) is produced:[10a]

$$O \underset{CH_2CH_2HgCH_2CH_2}{\overset{CH_2CH_2HgCH_2CH_2}{<}} O \xrightarrow[\text{benzene}]{h\nu} O \underset{CH_2CH_2CH_2CH_2}{\overset{CH_2CH_2CH_2CH_2}{<}} O + 2\,Hg$$

$$\text{(G)} \qquad\qquad\qquad\qquad\qquad \text{(G')}$$

If the aromatic compound contains a substituent such as isopropyl, the radical R· produced during photodecomposition of the organomercury compound extracts hydrogen from the substituent[13,16] and does not enter into the aromatic nucleus, for example:[16]

$$CH_3HgI + C_6H_5CH(CH_3)_2 \longrightarrow CH_4 + C_6H_5\overset{\cdot}{C}(CH_3)_2 + \cdot HgI$$
$$2\,C_6H_5\overset{\cdot}{C}(CH_3)_2 \longrightarrow C_6H_5C(CH_3)_2C(CH_3)_2C_6H_5$$
$$2\cdot HgI \longrightarrow Hg_2I_2$$

The effect of ^{60}Co γ-radiation on diphenylmercury in benzene produces biphenyl and polymeric products.[18]

3. Aliphatic Halogen-Containing Solvents

Photochemical decomposition of organomercury compounds in halogen-containing solvents proceeds in a different manner compared with halogen-free solvents. The reaction is confined to Eqs. (1) and (4) and mainly does not go up to steps (2) and (3), the isolation of metallic mercury. The RHg· radical extracts halide from the solvent, producing RHgHal. During decomposition in chloroform the photolysis may partially proceed according to reaction (2) with isolation of metallic mercury and formation of Hg_2Cl_2. (See Chapter 10.)

Radiolysis of diphenylmercury in $CHCl_3$ by ^{60}Co γ-rays produced $HgCl_2$ and phenylmercury chloride.[18]

C. Photolysis in the Gas Phase

The properties and reactions of the radicals produced during photolysis of different R_2Hg compounds have been thoroughly studied. These reactions also have no preparative significance, like the aforementioned reactions of photodecomposition of organomercury compounds in solution. We present herein only the respective references relating to some mercury compounds.

Photolysis: $(CH_3)_2Hg$[19–47]

Photolysis of the mixture of $(CH_3)_2Hg$ and $(CD)_3Hg$ and formation of CH_3HgCD_3 are discussed in Refs. 48 and 49. The following is a list of compounds photolyzed and the references: $(C_2H_5)_2Hg$,[50–55] $(C_2F_5)_2Hg$,[56] $(n\text{-}C_3H_7)_2Hg$,[57] $(n\text{-}C_4H_9)_2Hg$,[58,59] $(CH_2{=}CH)_2Hg$,[60] $(C_6H_5)_2Hg$,[61] and polyfluoroarylmercury compounds.[61a] This list is not exhaustive.

D. cis,trans Isomerization of Organomercury Compounds of the Ethylene Series by Ultraviolet Light

During treatment for many hours by ultraviolet light, the *trans* isomers of β-chlorovinyl compounds of mercury, R_2Hg and RHgCl, the former without a solvent, the latter in solution (absolute alcohol, toluene or, better, benzene) are isomerized to the respective *cis* isomers:[62]

$$
\begin{array}{c}
\underset{H}{\overset{Cl}{>}}C{=}C\underset{Hg}{\overset{H}{<}} \quad \underset{H}{\overset{H}{>}}C{=}C\underset{Hg}{\overset{Cl}{<}} \quad \xrightarrow{h\nu} \quad \underset{H}{\overset{H}{>}}C{=}C\underset{Hg}{\overset{Cl}{<}} \quad \underset{Cl}{\overset{H}{>}}C{=}C\underset{Hg}{\overset{H}{<}}
\end{array}
$$

The amount of isomerization is proportional to the length of irradiation, i.e. a 76 hr treatment of *trans,trans*-di-β-chlorovinylmercury produces about a 65% yield of the respective *cis*-, *cis* isomer. The yields for RHgCl compounds are substantially lower. The mercury derivatives of stilbene, R_2Hg and

RHgCl, irradiated by ultraviolet light, cause isomerizations of *cis* isomers to *trans* isomers but not the reverse:[63]

$$\left(\begin{array}{c}C_6H_5 \\ \diagdown \\ \\ H \diagup\end{array} C=C \begin{array}{c}C_6H_5 \\ \diagup \\ \\ \diagdown\end{array}\right)_2 Hg \xrightarrow{h\nu} \left(\begin{array}{c}C_6H_5 \\ \diagdown \\ \\ H \diagup\end{array} C=C \begin{array}{c} \diagup \\ \\ \diagdown C_6H_5\end{array}\right)_2 Hg$$

IV. THERMAL DECOMPOSITION OF ORGANOMERCURY COMPOUNDS

A. Introduction

Organomercury compounds R_2Hg and $RHgX$, heated to 150–200°C and sometimes to 300°C and, in some cases, at lower temperatures, are decomposed into radicals $R\cdot$; the cases with R_2Hg are decomposed into metallic mercury; the cases with $RHgX$, apart from formation of radicals $R\cdot$ and metallic mercury, are decomposed into HgX_2 and Hg_2X_2.

The products of the pyrolysis reaction of the same organomercury compound depend on the reaction temperature.

B. Pyrolysis in the Absence of Solvent

The thermal decomposition scheme presented below is most probable for R_2Hg:[64]

$$R_2Hg \xrightarrow{\Delta} RHg\cdot + R\cdot \tag{5}$$

$$RHg\cdot \xrightarrow{\Delta} R\cdot + Hg \tag{6}$$

The radicals, $R\cdot$, which are produced, may initiate decomposition of R_2Hg:

$$R\cdot + R_2Hg \longrightarrow RH + RHgR\cdot(-H) \longrightarrow R\cdot + Hg + R\cdot(-H) \tag{7}$$

It may be assumed that the decomposition mechanism of R_2Hg in one step is most probable.[65,66]

$$R_2Hg \longrightarrow Hg + 2\,R\cdot$$

Disproportionation and dimerization of the radicals produced then occurs:

$$2\,R\cdot \begin{array}{c} \nearrow RH + R{-}H \\ \searrow R\,R\end{array}$$

Billinge and Gowenlock[67] believe that either path of thermal decomposition of organomercury compounds depends on temperature.

The pyrolytic decomposition of organomercury compounds has no preparative significance. Therefore only particular cases of the decomposition are outlined in the present chapter. As for pyrolysis of most organomercury compounds, only some respective references are presented.

The temperature at which thermal decomposition of the organomercury compound begins, depends on the nature of the compound. At a comparatively lower temperature, the addition products of mercury salts to acetylene are decomposed (these compounds are decomposed not according to the aforementioned schemes (5), (6), and (7), but in compliance with their quasi-complex nature). Di(cis-β-chlorovinyl mercury produces a quantitative yield of acetylene upon heating to 100–105°C for 1 hr:[68]

$$\left(\begin{array}{c} \text{H} \qquad \text{H} \\ \diagdown \quad \diagup \\ \text{C}{=}\text{C} \\ \diagup \quad \diagdown \\ \text{Cl} \end{array} \right)_2 \text{Hg} \xrightarrow{\Delta} 2\ \text{CH}{\equiv}\text{CH} + \text{HgCl}_2$$

The *trans* isomer of (β-ClH=CH)$_2$Hg is not decomposed at its melting point. *trans*-β-Chlorovinylmercury chloride is decomposed at its melting point (124°C), while the *cis* isomer, melting at 79°C, is thermally stable at 123°C and is decomposed into acetylene and HgCl$_2$ at higher temperatures.[69] Among the saturated aliphatic organomercury compounds which are less stable thermally are Alk$_2$Hg compounds having more branched carbons at the mercury atom. Many aliphatic, completely substituted, organomercury compounds having a mercury at a secondary and, particularly, at a tertiary carbon atom, are decomposed at lower temperatures. Thus, di-*tert*-butylmercury is decomposed at 40°C.[70]

Bispentafluorophenylmercury, which is distinguished by its thermal stability, does not undergo any changes during heating to 250°C in a sealed flask for 5 hr.[71] Bispentachlorophenylmercury melts at 383°C.[72] Ethaneoxyhexamercarbide heated to 200°C decomposes with a heavy explosion.[73,74] It is probable that other mercarbides also explode upon heating.

Depending on the temperature at which pyrolysis of organomercury compounds is carried out, not only the ratio of the reaction products but the reaction products themselves may be different. Upon pyrolysis of diisopropylmercury for 60 hr at 100°C, the ratio of propylene to propane is 1:4 and the yield of the latter with respect to the decomposed mercury exceeds 100 mole %.[75] During decomposition of diisopropylmercury for 15 hr at 150°C, the yield of propane drops to 60%, while the yield of 2,3-dimethylbutane increases from 14 to 50%.[75]

During pyrolysis of di–*o*–iodophenylmercury at 600°C, *o*–diiodobenzene, Hg$_2$I$_2$, HgI$_2$, diphenylene, and traces of triphenylene are produced. The formation of the latter compounds proves intermediate formation of benzyne during pyrolysis.[76]

During pyrolysis of RHgX salts, particularly with R = alkenyl, esters of the respective RX acids are produced. During thermal decomposition of acetates of alkenylmercury, the quantity of CH_3COOR produced decreases for R as follows:[77]

$$CH_2=CH > CH_2=CH-CH_2 > CH_2=CH-CH_2-CH_2$$

Temperatures of 110–120°C are required for the decomposition of $CH_2=$ CH— and temperatures of 150–200°C for the decomposition of 3-butenyl.[78] Pyrolysis of a xanthate (115–125°C, 25 mm) and a thiocyanate (100–200°C, 10 mm) of the vinylmercury produces vinyl esters of the respective acids (in the last case a mixture of thiocyanate and isothiocyanate).[79]

During thermal decomposition of diferrocenylmercury (265°C) the main reaction product is ferrocene, while diferrocenyl is produced in a small amount only.[80]

Presented herein are references relating to the pyrolysis of organomercury compounds in the gas phase: dimethylmercury,[81–105] diethylmercury,[81,83, 105–108] di-*n*-propylmercury,[81,109–111] di-isopropylmercury,[81,83,111–114] di-per-fluoroiso-propylmercury,[115] di-*n*-butylmercury,[111,116,117] di-*sec*-butylmercury,[116] di-*tert*-butylmercury,[70] divinylmercury,[109] di-*cis*-β-chlorovinylmercury,[68] di-*trans*-β-chlorovinylmercury,[117] di-3-oxy-3-phenylpropylmercury,[118] di-3-oxy-3-methylbutylmercury,[118] halides[108,119] and *p*-vinylbenzoate[120] of phenylmercury, dibenzylmercury,[121,122] diphenylmercury,[81,108,123–125] $(XC_6-H_4)_2Hg$,[126] 2,2'-diphenylenmercury,[127] polyfluoroarylmercury compounds,[61a] *n*-propylmercury chloride,[108] vinylmercury acetate,[77] *cis*-β-chlorovinyl-mercury chloride,[69] *trans*-β-chlorovinylmercury chloride,[69] allylmercury acetate,[77] and 3-butenylmercury acetate.[77]

C. Pyrolysis in Solution

Behavior of the radical produced during thermal decomposition of organomercury compounds in solution may differ from that of the same radicals generated from other substances. For example, during decomposition of dibenzylmercury in anthracene (at 190–195°C) among other reaction products produced is a 9-benzylanthracene,[128] the formation of which has not been found during reactions of benzyl radicals from other sources on anthracene.

On the other hand, thermal decomposition of organomercury compounds in solution, depending on the nature of the solvent, may produce completely or partially the same or other products that have been obtained during their photodecomposition. For example, both during thermal (280–300°C) and photodecomposition of diphenylmercury in toluene, benzene, a mixture of *o*-, *m*-, and *p*–methyldiphenyls and mercury are formed.[129] Thermal and photodecomposition of diphenylmercury in methyl alcohol produced benzene, mercury, and formaldehyde.[130] However, pyrolysis of diphenylmercury in cyclohexane (260–280°C, sealed flask, 150 hr) yields mercury, benzene, cyclohexene, phenylcyclohexane, traces of diphenyl and polymer (—C_6H_4Hg)n, but during photolysis of diphenylmercury in the same solvent along with the first five aforementioned reaction products, dicyclohexyl and not the polymer is produced.[130]

Decomposition of organomercury compounds in hydrogen-containing solvents is discussed in Chapter 7; in halide-containing solvents, refer to Chapter 10.

D. Thermal *cis,trans* Isomerizations of Unsaturated Organomercury Compounds

As stated above the addition products of mercury salts to unsaturated compounds are thermally decomposed producing the original unsaturated compound. However, in some cases isomerization occurs in addition to decomposition. Bis(*cis*-α-stilbenyl)mercury heated to 140–160°C for 4 hr (sealed flask) is converted into the bis(*trans*-α-stilbenyl)mercury. *trans*-Stilbene is produced simultaneously.[63]

cis–trans- and *trans–cis* Isomerizations for propenyl compounds of mercury occur at substantially lower temperatures: for R_2Hg heated in benzene (50°C, 3 hr), and for RHgBr during 5 hr of boiling in acetone.[131]

E. Miscellaneous Observations

The role of O_2 has been discussed.[132,133] The photolysis of diethylmercury bisdiazoacetate in cyclohexene,[134,135] cyclohexane,[135] *cis*- and *trans*-butene indicates the formation of free radical carbon intermediates. The reaction products depend upon the wave lengths used. p-$CH_3OC_6H_4CH_2CH_2HgO_2$-CCH_3 decomposes with the formation of arylethynyl acetate at 30°C.[136]

Intermediate dihydro- and substituted dihydrothiophenes are formed by pyrolysis of halogenated thienylmercury.[137] Reactions of $ClHgC(O)OCH_3(I)$ with olefinic complexes of $PdCl_2$ proceed as follows:[138]

$$[CH_2=CH_2]PdCl_2 \xrightarrow[\text{THF or } C_6H_6]{I} CH_2CH_2COOCH_3$$

Pyrolysis of *exo*-bicyclo[2,2,2]-oct-2-enyl-5-mercury chloride has been discussed.[139]

The pyrolysis of bis-[1-bromoacenaphthylen-2-yl-]-mercury at 240°C proceeds only in the presence of tetraphenylcyclopentadienone with the formation of 10,11,12,13-tetraphenylfluoranthene.[140] The pyrolysis of dibenzyl-[141,142] and diphenyl-[142] mercury is carried out in aromatic solvents. The pyrolysis of diethylmercury bisdiazoacetate in cyclohexene and cyclohexane (formation of cyclohexyl acetate) confirms the formation of a free radical carbon intermediate.[135]

REFERENCES

1. G. A. Razuvaev, S. F. Zhil'tsov, O. N. Drushkov, and G. G. Petukhov, *Dokl. Akad. Nauk SSSR*, **152**, 633 (1963).
2. Yu. A. Alexandrov, O. N. Drushkov, S. F. Zhil'tsov, and G. A. Razuvaev, *Dokl. Akad. Nauk SSSR*, **157**, 1395 (1964).
3. Yu. A. Alexandrov, O. N. Drushkov, S. F. Zhil'tsov, and G. A. Razuvaev, *Zh. Obshch. Khim.*, **35**, 1440 (1965).
4. G. A. Razuvaev, S. F. Zhil'tsov, Yu. A. Alexandrov, and O. N. Drushkov, *Zh. Obshch. Khim.*, **35**, 1152 (1965).
5. G. A. Razuvaev, S. F. Zhil'tsov, and L. F. Kudrjavtsev, *Dokl. Akad. Nauk SSSR*, **135**, 87 (1960).
6. G. A. Razuvaev, G. G. Petukhov, S. F. Zhil'tsov, and L. F. Kudrjavtsev, *Dokl. Akad. Nauk SSSR*, **144**, 810 (1962).
7. G. A. Razuvaev, G. G. Petukhov, S. F. Zhil'tsov, and L. F. Kudrjavtsev, *Dokl. Akad. Nauk SSSR*, **141**, 107 (1961).
8. G. A. Razuvaev, N. S. Vjazankin, and E. V. Mitrofanova, *Zh. Obshch. Khim.*, **34**, 675 (1964).
9. G. A. Razuvaev, E. V. Mitrofanova, and N. S. Vjazankin, *Dokl. Akad. Nauk SSSR*, **144**, 132 (1962).
10. G. A. Razuvaev, S. F. Zhil'tsov, O. N. Drushkov, and G. G. Petukhov, *Zh. Obshch. Khim.*, **36**, 258 (1966).
10a. V. Weinmayr, *J. Am. Chem. Soc.*, **81**, 3590 (1959).
11. J. H. Robson and G. F. Wright, *Can. J. Chem.*, **38**, 1 (1960).
11a. D. A. Shearer and G. F. Wright, *Can. J. Chem.*, **33**, 1002 (1955).
12. J. H. Robson and G. F. Wright, *Can. J. Chem.*, **38**, 21 (1960).
12a. A. N. Nesmeyanov, A. E. Borisov, and V. D. Vil'chevskaya, *Izv. Akad. Nauk SSSR, Otd. Khim. Nauk*, **1949**, 578; *Uchen. Zap. Moskov. Gos. Univer.*, **132**, 33 (1950).
12b. I. P. Beletskaya, V. I. Karpov, and O. A. Reutov, *Izv. Akad. Nauk SSSR, Otd. Khim. Nauk*, **1964**, 1707.
13. D. H. Hey, D. A. Shingleton, and G. H. Williams, *J. Chem. Soc.*, **1963**, 1958.
14. G. A. Razuvaev, G. G. Petukhov, and Yu. A. Kaplin, *Dokl. Akad. Nauk SSSR*, **135**, 342 (1960).
15. G. A. Razuvaev, G. G. Petukhov, Yu. A. Kaplin, and O. N. Drushkov, *Dokl. Akad. Nauk SSSR*, **152**, 1122 (1963).
16. G. E. Corbett and G. H. Williams, *Proc. Chem. Soc.*, **1961**, 240.
17. G. A. Razuvaev and G. G. Petukhov, *Zh. Obshch. Khim.*, **23**, 37 (1953).
18. C. Heitz and J. P. Adloff, *J. Organometal. Chem.*, **2**, 59 (1964).
19. A. F. Trotman-Dickenson and E. W. R. Steacie, *J. Phys. Chem.*, **55**, 908 (1951).
20. M. K. Phibbs and B. de B. Darwent, *Trans. Faraday Soc.*, **45**, 541 (1949).
21. G. M. Harris and E. W. R. Steacie, *J. Chem. Phys.*, **13**, 559 (1945).
22. B. de. B. Darwent and M. K. Phibbs, *J. Chem., Phys.*, **22**, 110 (1954).
23. R. E. Rebbert and E. W. R. Steacie, *Can. J. Chem.*, **31**, 631 (1953).
24. R. E. Rebbert and E. W. R. Steacie, *J. Chem. Phys.*, **21**, 1723 (1953).
25. R. Holroyd and W. A. Noyes, Jr., *J. Am. Chem. Soc.*, **76**, 1583 (1954).
26. K. H. Müller and W. D. Walters, *J. Am. Chem. Soc.*, **76**, 330 (1954).
27. H. E. Gunning and E. W. R. Steacie, *J. Chem. Phys.*, **14**, 534 (1946).
28. D. M. Miller and E. W. R. Steacie, *J. Chem. Phys.*, **19**, 73 (1951).
29. R. Gomer and G. B. Kistiakowski, *J. Chem. Phys.*, **19**, 85 (1951).
30. R. B. Martin and W. A. Noyes, Jr., *J. Am. Chem. Soc.*, **75**, 4183 (1953).

31. R. A. Marcus and E. W. R. Steacie, *Z. Naturforsch.*, **4a**, 332 (1949).
32. R. Gomer and W. A. Noyes, Jr., *J. Am. Chem. Soc.*, **71**, 3390 (1949); **72**, 101 (1950).
33. R. Gomer, *J. Chem. Phys.*, **18**, 998 (1950).
34. R. E. Rebbert and E. W. R. Steacie, *Can. J. Chem.*, **32**, 113 (1954).
35. R. E. Rebbert and E. W. R. Steacie, *Can. J. Chem.*, **32**, 40 (1954).
36. R. I. Gomer, *J. Am. Chem. Soc.*, **72**, 201 (1950).
37. I. W. Linnett and H. W. Thompson, *Trans. Faraday Soc.*, **33**, 501 (1937).
38. H. W. Thompson and I. W. Linnett, *Trans. Faraday Soc.*, **33**, 874 (1937).
39. C. R. Masson and E. W. R. Steacie, *J. Chem. Phys.*, **19**, 183 (1951).
40. M. K. Phibbs and B. de B. Darwent, *Can. J. Res.*, **B28**, 395 (1950).
41. R. A. Marcus, *J. Chem. Phys.*, **20**, 364 (1952).
42. G. B. Kistiakowski and E. K. Roberts, *J. Chem. Phys.*, **21**, 1637 (1953).
43. R. W. Durham and E. W. R. Steacie, *J. Chem. Phys.*, **20**, 582 (1952).
44. M. Szwarc and I. S. Roberts, *Trans. Faraday Soc.*, **46**, 625 (1950).
45. M. G. Evans and M. Szwarc, *Trans. Faraday Soc.*, **45**, 940 (1949).
46. E. W. R. Steacie and B. de B. Darwent, *Discussions Faraday Soc.*, **2**, 80 (1947).
47. A. O. Allen and C. E. Bawn, *Trans. Faraday Soc.*, **34**, 463 (1938).
48. R. E. Rebbert and P. Ausloos, *J. Am. Chem. Soc.*, **85**, 3086 (1963).
49. R. E. Rebbert and P. Ausloos, *J. Am. Chem. Soc.*, **86**, 2068 (1964).
50. K. Y. Ivin and E. W. R. Steacie, *Proc. Roy. Soc.*, *(London)*, **A208**, 25 (1951).
51. J. N. Bradley, H. W. Melville, and J. C. Robb, *Proc. Roy. Soc. (London)*, **A236**, 318 (1956).
52. W. A. G. Graham and A. R. Gatti, *Ultrapuri Semicond. Mater., Proc. Conf., Boston. Mass.*, **1961**, 106; *Chem. Abstr.*, **57**, 5756 (1962).
53. B. A. Thrush, *Proc. Roy. Soc. (London)*, **A243**, 555 (1957).
54. W. I. Moore and H. S. Taylor, *J. Chem. Phys.*, **8**, 396 (1940).
55. L. C. Fischer and G. J. Mains, *J. Phys. Chem.*, **68**, 2522 (1964).
56. J. Banus, H. J. Emeléus, and R. N. Haszeldine, *J. Chem. Soc.*, **1950**, 3041.
57. E. J. Coule and E. W. R. Steacie, *Can. J. Chem.*, **29**, 103 (1951).
58. H. W. Melville and J. C. Robb, *Proc. Roy. Soc. (London)*, **A196**, 479 (1949).
59. W. J. Moore and L. A. Wall, *J. Chem. Phys.*, **17**, 1325 (1949).
60. K. Erhard, *Naturwiss.*, **49**, 417 (1962).
61. G. J. Fonken, *Chem. Ind.*, **1961**, 716.
61a. R. N. Haszeldine, A. R. Parkinson, and J. M. Birchall, USP 3156715; *Chem. Abstr.*, **62**, 16296 (1965).
62. A. N. Nesmeyanov, A. E. Borisov, and A. N. Abramova, *Izv. Akad. Nauk SSSR, Otd. Khim. Nauk*, **1947**, 289.
63. A. N. Nesmeyanov, A. E. Borisov, and N. A. Vol'kenau, *Izv. Akad. Nauk SSSR, Otd. Khim. Nauk.*, **1954**, 992.
64. G. A. Razuvaev, private communication.
65. H. O. Pritchard, *J. Chem. Phys.*, **25**, 267 (1956).
66. M. E. Russell and R. B. Bernstein, *J. Chem. Phys.*, **30**, 607, 613 (1959).
67. H. M. Billinge and B. G. Gowenlock, *Proc. Chem. Soc.*, **1962**, 24.
68. A. N. Nesmeyanov. A. E. Borisov, and A. N. Gus'kova, *Izv. Akad. Nauk SSSR, Otd. Khim. Nauk*, **1945**, 639.
69. A. N. Nesmeyanov and A. E. Borisov, *Izv. Akad. Nauk SSSR, Otd. Khim. Nauk*, **1945**, 146.
70. C. S. Marvel and H. O. Calvery, *J. Am. Chem. Soc*, **45**, 820 (1926).
71. R. D. Chambers, G. E. Coates, J. G. Livingston, and W. K. Musgrawe, *J. Chem. Soc.*, **1962**, 4367.

72. F. E. Paulik, S. I. E. Green and R. E. Dessy, *J. Organometal. Chem.*, **3**, 229 (1965).
73. K. A. Hofmann, *Ber.*, **31**, 1905 (1898).
74. K. A. Hofmann, *Ber.*, **33**, 1328 (1900).
75. G. A. Razuvaev, O. N. Drushkov, S. F. Zhil'tov, and G. G. Petuchov, *Zh. Obshch. Khim.*, **35**, 174 (1965).
76. G. Wittig and H. F. Ebel, *Ann. Chem.*, **650**, 20 (1961).
77. D. J. Foster and E. Tobler, *J. Org. Chem.*, **27**, 834 (1962).
78. D. J. Foster and E. Tobler, *J. Am. Chem. Soc.*, **83**, 851 (1961).
79. E. Tobler and D. J. Foster, *Z. Naturforsch.*, **17B**, 136 (1962).
80. M. D. Rausch, *Inorg. Chem.*, **1**, 414 (1962).
81. H. O. Pritchard, *J. Chem. Phys.*, **25**, 267 (1956).
81a. K. B. Yerrick and M. E. Russell, *J. Phys. Chem.*, **68**, 3752 (1964).
82. M. E. Russell and R. B. Bernstein, *J. Chem. Phys.*, **30**, 607, 613 (1959).
83. G. E. Harris and A. W. Tickner, *J. Chem. Phys.*, **15**, 686 (1947).
84. G. M. Harris and A. W. Tickner, *Can. J. Res.*, **26B**, 343 (1948).
85. B. G. Gowenlock, I. C. Polanyi, and F. E. Warhurst, *Proc. Roy. Soc. (London)*, **A218**, 269 (1953).
86. L. Radich, I. P. Kravchuk, and R. E. Mardaleyshvili, *Dokl. Akad. Nauk SSSR*, **136**, 657 (1961).
87. K. H. Ingold and F. P. Losing, *J. Chem. Phys.*, **21**, 368, 1135 (1953).
88. F. P. Lossing and W. A. Tickner, *J. Chem. Phys.*, **20**, 907 (1952).
89. F. P. Lossing and I. J. Tanaka, *J. Chem. Phys.*, **25**, 1031 (1956).
90. M. Krech and S. Price, *Can. J. Chem.*, **41**, 224 (1963).
91. K. H. Müller and D. Walters, *J. Am. Chem. Soc.*, **73**, 1458 (1951).
92. L. M. Yeddanapalli, R. Srinivasan, and V. J. Paul, *J. Sci. Ind. Research (India)*, **13B**, 232 (1954); *Chem. Abstr.*, **49**, 4387 (1955).
93. J. P. Cunningham and H. S. Taylor, *J. Chem. Phys.*, **6**, 359 (1938).
94. L. H. Long, *Trans. Faraday Soc.*, **51**, 673 (1955).
95. J. E. Longfield and W. D. Walters, *J. Am. Chem. Soc.*, **77**, 6098 (1955).
96. R. Z. Ganesan, *Z. Phys. Chem.*, **31**, 328 (1962).
97. L. M. Yeddanapalli and C. C. Schubert, *J. Chem. Phys.*, **14**, 1 (1949).
98. C. M. Laurie and L. H. Long, *Trans. Faraday Soc.*, **51**, 665 (1955).
99. R. I. Ganesan, *Sci. Ind. Res. (India)*, **20B**, 228 (1961).
100. K. W. Saunders and H. S. Taylor, *J. Chem. Phys.*, **9**, 616 (1941).
101. S. I. W. Price and A. F. Trotman-Dickenson, *Trans. Faraday Soc.*, **53**, 939 (1957).
102. J. Cettanach and L. H. Long, *Trans. Faraday Soc.*, **56**, 1286 (1960).
103. L. H. Long, *J. Chem. Soc.*, **1956**, 3410.
104. R. J. Srinivasan, *J. Chem. Phys.*, **28**, 895 (1958).
105. H. S. Taylor and W. H. Jones, *J. Am. Chem. Soc.*, **52**, 1111 (1930).
106. W. H. Pasfield, *Dissertation. Abstr.*, **15**, 1325 (1955).
107. D. A. Armstrong and C. A. Winkler, *Can. J. Chem.*, **34**, 885 (1951).
108. H. V. Carter, E. I. Chappell, and E. J. Warhurst, *J. Chem. Soc.*, **1956**, 106.
109. A. F. Trotman-Dickenson and G. J. Q. Verbeke, *J. Chem. Soc.*, **1961**, 2580.
110. H. T. Y. Chilton and B. Q. Gowenlock, *Trans. Faraday Soc.*, **50**, 824 (1954).
111. B. H. M. Billinge and B. Q. Gowenlock, *J. Chem. Soc.*, **1962**, 3252.
112. B. H. M. Billinge and B. Q. Gowenlock, *Trans. Faraday Soc.*, **59**, 690 (1963).
113. H. T. Y. Chilton and B. G. Gowenlock, *Nature*, **172**, 73 (1953); *J. Chem. Soc.*, **1953**, 3232.
114. H. T. Y. Chilton and B. G. Gowenlock, *Trans. Faraday Soc.*, **49**, 1451 (1953).

115. P. E. Aldrich, E. G. Howard, W. J. Linn, N. J. Middleton and W. H. Sharkey, *J. Org. Chem.*, **28**, 184 (1963).
116. H. F. Y. Chilton and B. G. Gowenlock, *J. Chem. Soc.*, **1954**, 3174.
117. A. N. Nesmeyanov and R. Kh. Freidlina, *Izv. Akad. Nauk SSSR, Otd. Khim. Nauk*, **1945**, 150.
118. R. Ja. Levina, V. N. Kostin, and V. A. Tartakovski, *Zh. Obshch. Khim.*, **27**, 2049 (1957).
119. M. Cowperthwaite and E. J. Warhurst, *J. Chem. Soc.*, **1958**, 2429.
120. M. M. Koton and T. M. Kiseleva, *Izv. Akad. Nauk SSSR, Otd. Khim. Nauk*, **1961**, 1783.
121. W. Pope and C. S. Gibson, *J. Chem. Soc.*, **101**, 735 (1912).
122. P. Wolff, *Ber.*, **46**, 64 (1913).
123. E. Dreher and R. Otto, *Ann. Chem.*, **154**, 93 (1870).
124. G. A. Razuvaev and M. M. Koton, *Zh. Obshch. Khim.*, **5**, 361 (1935).
125. M. T. Jaquiss and M. Szwarc, *Nature*, **170**, 317 (1952).
126. M. M. Koton and V. F. Martynova, *Zh. Obshch. Khim.*, **24**, 2177 (1954).
127. G. J. Fanken, *Chem. Ind.*, **1961**, 716.
128. K. C. Bass and P. Nababsing, *Chem. Ind.*, **1965**, 307.
129. G. A. Razuvaev, G. G. Petukhov, L. F. Kudrjavtsev, and M. A. Shubenko, *Zh. Obshch. Khim.*, **33**, 2764 (1963).
130. K. H. Müller and W. D. Walters, *J. Am. Chem. Soc.*, **73**, 1458 (1951).
131. A. N. Nesmeyanov, A. E. Borisov, and N. V. Novikova, *Izv. Akad. Nauk SSSR, Otd. Khim. Nauk*, **1959**, 1216.
132. F. R. Jensen and D. Heyman, *J. Am. Chem. Soc.*, **88**, 3428 (1966).
133. B. F. Hegarty, W. Kitching, and P. R. Wells, *J. Am. Chem. Soc.*, **89**, 4816 (1966).
134. O. P. Strausz, T. Do Minh, and J. Font, *J. Am. Chem. Soc.*, **90**, 1930 (1968).
135. T. Do Minh, H. E. Gunning, and O. P. Strausz, *J. Am. Chem. Soc.*, **89**, 6785 (1967).
136. G. Spengler and T. Trommer, *Brennstoff-Chemie*, **48**, 19 (1967).
137. M. D. Rausch, T. R. Criswell, and A. R. Ignatowitz, *J. Organometal. Chem.*, **13**, 419 (1968).
138. T. Saegusa, T. Tsuda, and K. Nishijima, *Tetr. Lett.*, **43**, 4255 (1967).
139. D. S. Matteson and M. L. Talbot, *J. Am. Chem. Soc.*, **89**, 1123 (1967).
140. K. Rasheed, *Tetrahedron*, **22**, 2957 (1966).
141. K. C. Bass, *J. Organometal. Chem.*, **4**, 1 (1966).
142. K. C. Bass and P. Nababsing, *J. Chem. Soc.*, (*C*), 1184 (1966).

Chapter 17

Reactions with Alkalis

As a rule, alkalis do not cause cleavage of the C—Hg bond in organo-mercury compounds. The examples of such cleavages, given below, are exceptions to the rule. For instance, in some Ar_2Hg where Ar = phenol[1] or substituted phenol,[2] alkalis cause cleavage of one C—Hg bond and the formation of the phenolate; for example, the saponification of $(o\text{-}CH_3COOC_6H_4)_2$-Hg with caustic soda results in $o\text{-}NaOC_6H_4HgOH$.

Among the aliphatic compounds of mercury, there are some compounds which are probably unstable to alkali. In such compounds, mercury is connected to a carbon with another functional group or halogen. Some poly-mercurated compounds also belong to this group. For instance, the ethyl ester of di-(acetoxymercury)chloroacetic acid obtained as the result of addition of mercury acetate to chloroketene acetal is decomposed by a 20% aqueous solution of KOH with the formation of a white sediment which separates metallic mercury.[3] The formation of a salt of glycolic acid from the reaction of alkali upon a product obtained from the reaction of monochloroacetic acid and mercury oxide is described in Refs. 4 and 5. Quantitative deposition of mercury occurs upon boiling mercury bis-(chloromercurychloroacetic ethyl ester) with even a weak alkali solution.[6]

Alkali separates a part of mercury from trimercurated acetaldehyde.[7]

Aqueous alkali, added to the product obtained from the addition of mercury to carbon monoxide in alcoholic medium, gives metallic mercury and bicarbonate.[8] Similarly, a 2% solution of NaOH reacts with bromomethyl-mercury hydroxide upon slight heating to deposit mercury:[9]

$$BrCH_2HgOH \xrightarrow[\Delta]{OH^-} Hg + CH_2O + HBr$$

The effect of diluted alkali upon a product of simultaneous nuclear mercuration and oxymercuration of the double bond of cumaric acid and substituted cumaric acid results in the separation of HgX and OR from the double bond. HgX radicals in aromatic nucleus remain unchanged, for instance:[10]

$$CH_3CO_2Hg \quad \text{(ring)} \quad HgO_2CCH_3$$
$$CH(OR)CH(HgO_2CCH_3)COOH \xrightarrow{\text{NaOH}}$$
$$CH_3CO_2Hg \quad \text{(ring)} \quad HgO_2CCH_3 \quad CH=CHCOOH$$

422

Other products of the addition of mercury salt to double bonds are usually resistant to alkali.

REFERENCES

1. F. C. Whitmore and E. Middleton, *J. Am. Chem. Soc.*, **43**, 621 (1921).
2. I. Stieglitz, M. S. Kharasch, and M. E. Hanke, *J. Am. Chem. Soc.*, **43**, 1185 (1921).
3. I. F. Lutsenko and V. L. Foss, *Dokl. Akad. Nauk SSSR*, **98**, 407 (1954).
4. K. A. Hofmann, *Ber.*, **32**, 870 (1899).
5. P. R. Wells and W. Kitching, *Australian Chem. J.*, **16**, 1123 (1963).
6. A. N. Nesmeyanov and G. S. Pofkh, *Zh. Obshch. Khim.*, **4**, 958 (1934); *Ber.*, **67**, 971 (1934).
7. K. A. Hofmann, *Ber.*, **31**, 2212 (1898).
8. W. Schoeller and W. Schrauth, *Ber.*, **46**, 2871 (1913).
9. A. N. Nesmeyanov, R. Kh. Freydlina, and F. A. Tokareva, *Zh. Obshch. Khim.*, **7**, 262 (1937); *Ber.*, **69**, 2019 (1937).
10. P. S. Rao and T. R. Seshadri, *Proc. Indian Acad. Sci., Sect. A*, **4**, 630 (1936); *Zentrbl.*, **1937**, II 3457.

Subject Index

425